Waste as a Resource

Series Editor

Cristina Trois, Civil Engineering, Stellenbosch University,
Stellenbosch, South Africa

The Waste as a Resource series explores state-of-the-art knowledge and innovation in waste management strategies focused on approaches and technologies that ensure maximum value recovery from waste through materials recycling and energy production. The series aims to navigate the transition toward a circular economy by presenting techniques and strategies for waste minimization, effective waste diversion from landfills, and, most importantly, exploring the possible pathways for treating waste as a resource. The series has a global scope, focusing on developing countries and emerging economies. This will assist in filling the gap in current publications that need to adequately explore valorization technologies that are appropriate for and adaptable in low-income countries or informal contexts.

Cristina Trois • Ramesh C. Ray
Editors

Plantation Crop Wastes: Valorization for Economic Sustainability

Springer

Editors
Cristina Trois
Centre for Renewable and Sustainable Energy Studies
Stellenbosch University
Stellenbosch, South Africa

Ramesh C. Ray
Center for Industrial Biotechnology Research
Siksha 'O' Anusandhan (SOA), deemed to be University
Bhubaneswar, Odisha, India

ISSN 2731-8222 ISSN 2731-8230 (electronic)
Waste as a Resource
ISBN 978-3-031-96804-4 ISBN 978-3-031-96805-1 (eBook)
https://doi.org/10.1007/978-3-031-96805-1

This Springer imprint is published by the registered company Springer Nature Switzerland AG
The registered company address is: Gewerbestrasse 11, 6330 Cham, Switzerland

Preface

Plantation crop residues are a significant waste product of modern horticulture. These residues are usually left to decompose, leading to environmental degradation, pollution of groundwater, and health hazards. However, with the growing demand for sustainable agriculture practices, there is a need to find innovative ways to utilize these residues. This book aims to provide a comprehensive exploration of the potential of plantation crop waste valorization, showcasing innovative technologies and applications for transforming waste materials into valuable resources in a circular bioeconomy. Addressing various aspects of the plantation crop waste-to-wealth paradigm, it will serve as a valuable resource for researchers, policymakers, and industry professionals seeking sustainable solutions for plantation crop residue management and the transition from linear to a circular economy. There are 20 chapters in this book divided into sections: Introduction, Edible crops (i.e., mango, apple, date palm, and palmyra palm), Beverage crops (i.e., tea, coffee, and cocoa), Masticatory and nut crops (betel vine, areca nut, and cashew nut), Oil yielding crops (i.e., olive, oil palm, and coconut palm), Industrial crops (i.e., bamboo, rubber, and sugar cane), Economics and policy issues, and Life Cycle Analysis. The individual chapters in this book are written by experts worldwide.

The editors are thankful to the authors for timely submission of manuscripts, to the reviewers for their expert review in strengthening the contents of the manuscript, and to Springer International for publishing the book.

Stellenbosch, South Africa — Cristina Trois
Bhubaneswar, Odisha, India — Ramesh C. Ray

Preface

Contents

Part VII [illegible] Crops

About the Editors

Cristina Trois is the Director of the Centre for Renewable and Sustainable Energy Studies at Stellenbosch University, in South Africa. She also heads the ARUA CoE in Energy and the ARUA-Guild Cluster of Research Excellence in Renewable Energy. Full Professor in Environmental Engineering (since 2015), with a degree in Environmental Engineering (summa cum laude) and a PhD in Geo and Environmental Engineering. She was the former Dean of Engineering (2012-2018) and former DSI/NRF SARCHI Chair in Waste and Climate Change (2018-2023) at the University of KwaZulu-Natal (RSA).

Her main fields of expertise are environmental and civil-engineering, waste management and climate change, bioenergy, waste to energy, biogas and bio-hydrogen from organic waste, systems modelling and decision-support tools, renewable energy from waste and GHGs control from zero waste in Africa and developing countries, and circular economy.

Prof Trois is the author of over 150 peer-reviewed publications and Series Editor for Springer (Nature) for the Series "Waste as a Resource" and co-editor of the Springer book *Waste Management in Developing Countries*. Prof. Trois has over 25 years of working experience in Environmental Sanitary Engineering and has been the principal investigator as well as the project coordinator of many multimillion feasibility and R&D waste management projects with municipalities in South Africa and Africa. She is the developer of the WROSE™ Model /Decision-support Tool, and advisor for climate services, waste and climate change, waste to energy projects, CDM projects to the South African Government, and intergovernmental organizations.

Ramesh C. Ray is working as an Adjunct Professor, at the Center for Industrial Biotechnology Research, Siksha 'O' Anusandhan (SOA) Deemed to be University, in India. He was a former Principal Scientist (Microbiology) and Head of the ICAR-Central Tuber Crops Research Institute (Regional Centre), Bhubaneswar, India. He has 40 years of research experience in agriculture, food, and industrial microbiology, published 145 research and review papers in national and international journals, 75 book chapters, edited/co-edited 26 books by Springer, CRC, and Elsevier, and co-authored three books. He was a Visiting Professor/ Scientist at several universities and research institutes in India and abroad, and a Consultant to the International Potato Center, Peru on sweet potatoes, stationed in India. He is a distinguished fellow of the prestigious National Academy of Agricultural Sciences, New Delhi, India, and the National Academy of Biological Science, Chennai, India. Prof. Ray has recently been listed in the most coveted list of "World Ranking of Top 2% Scientists" published by Stanford University, USA (2021, 2022, 2023, 2024).

Part I
Introduction

Chapter 1
Plantation Crops: Residue Utilization for Industrial Biotechnology and Economic Sustainability: An Overview

Omojola Awogbemi, Ramesh C. Ray, and Ayodeji Usman Akerele

Abbreviations

AD	Anaerobic digestion
CRISPR	Clustered Regularly Interspaced Short Palindromic Repeats
MT	Million tons
POME	Palm oil mill effluent
TALENs	Transcription Activator-Like Effector Nucleases
TILLING	Targeting Induced Local Lesions in Genomes
ZFNs	Zinc Finger Nucleases

1 Introduction

In recent decades, the cultivation of plantation crops has increased significantly. This is due to the increase in demand for food and raw materials occasioned by the rising population and industrial development worldwide. Expectedly, higher cultivation and harvesting of plantation crops have increased waste generation across all facets of utilization and processing. Inappropriate disposal and management of crop residues production without proper treatment negatively impacts the environment and human health [7]. Proper management and utilization of crop residue can be

---

O. Awogbemi (✉)
Department of Mechanical Engineering, Ekiti State University, Ado Ekiti, Nigeria

R. C. Ray
Center for Industrial Biotechnology Research, Siksha 'O' Anusandhan (SOA), deemed to be University, Bhubaneswar, India

A. U. Akerele
Institute of Energy, Ceramic and Polymer Technology, Faculty of Materials and Chemical Engineering, University of Miskolc, Miskolc, Hungary

C. Trois, R. C. Ray (eds.), *Plantation Crop Wastes: Valorization for Economic Sustainability*, Waste as a Resource,
https://doi.org/10.1007/978-3-031-96805-1_1

achieved through biotechnological techniques and ensure environmental protection and economic profitability [8–11]. Residues from plantation crops are rich in lignocellulose, sugars, and other valuable compounds, making them ideal feedstocks for industrial biotechnological applications such as biofuels, bioplastics, enzymes, and biochemicals. The effective valorization of plantation crop residues not only addresses waste management challenges but also offers pathways to enhance economic sustainability for growers and industries alike [9, 38].

Researchers across diverse fields and jurisdictions, including Andrews et al. [3] and Hebbar et al. [22] in their separate studies, investigated the various innovative processing and conversion techniques to derive useful products from plantation crops. Resource recovery, environmental sustainability, economic viability, and fostering of sustainable cities are achievable from appropriate management of plantation crop residues [44, 48]. Honorato-Salazar and Sadhukhan [23] and Ghodke et al. [21] reported the economic importance and environmental sustainability of bioprocessing conversion of agricultural and forest residues for bioenergy and bioproducts. By leveraging the plantation crop residues effectively, industries can address global challenges related to resource efficiency, energy security, and environmental sustainability, paving the way for a greener and more resilient future [27, 40].

Despite this avalanche of studies, it is debatable whether investigations into the various avenues for conversion and utilization of plantation crop residues have been adequately explored. The current chapter gives an overview of the utilization potentials of plantation crop residues with particular reference to industrial biotechnology and economic sustainability, highlighting their role in advancing circular economy principles and contributing to global sustainable development goals. The major pretreatment techniques and plantation crop residue utilization avenues are briefly discussed. Benefits, challenges, and recommendations for effective conversion and utilization of plantation crop residues are equally highlighted before the conclusion section. The combination of the utilization avenues, and industrial biotechnological approach with the economic sustainability of plantation crop residues makes the current intervention novel. The outcome of the chapter is expected to stimulate further studies into the opportunities in plantation crop residue recycling and utilization.

2 Overview of Plantation Crops Residues

2.1 Description

Plantation crops are, by definition, perennial crops cultivated on a large scale for commercial purposes. They are usually cultivated and produced in the tropical or subtropical regions all year round, mainly for industrial processing or exports. The cultivation of plantation crops is usually monoculture involving the growing of a single crop on a large scale and often requires land, machinery, finance, chemicals, and labor. Examples of plantation crops include tea, coffee, cocoa, coconut, areca nut, oil palm, cashew nut, and rubber [33, 40]. Due to its monoculture nature, the scale of production, and the use of chemicals and machinery, the system of agriculture presents social, economic, environmental, labor, and biodiversity implications. Economically, the large-scale cultivation of plantation crops for industrial processing and exports serves as a tremendous source of revenue to farmers and foreign exchange earnings to the country. Large plantations provide both skilled and unskilled employment opportunities for many people, ranging from laborers, drivers, supervisors, accountants, to managers, etc. The cultivation of plantation crops ensures efficient land use through the adoption of innovative agricultural techniques and practices. During the harvesting and processing of plantation crops, huge amounts of wastes are generated which pose sanitary and health hazards [28].

2.2 *Classification and Global Production*

Generally, plantation crops are classified based on their use and characteristics as beverage crops, oil-bearing crops, forest crops, fruit crops, and sugar crops. Beverage crops are cultivated specifically for the production of drinks. They contain starch and sugar, which are usually used in the production of beverages. Beverages are a major part of the food industry, used in the production of alcoholic and nonalcoholic drinks, chocolate confectionery, and major industrial raw materials. Common examples of beverage crops include tea, coffee, cocoa, grapes, and barley. [45]. The global production of tea in 2022 is 6.7 million tonnes (MT) with India, China, and Sri Lanka as top producers. [17].

Oil-bearing crops are plants cultivated primarily to extract oils from their seeds or fruits. These oils are used for cooking, heating, lubrication, fuels, and other industrial applications [37]. Oil palm, coconut, cashew nut, etc., are the major examples of oil-bearing crops. Indonesia and Malaysia are the leading producers of oil palm while a substantial percentage of global coconut is cultivated in the Philippines, Indonesia, and India. Forest crops are plants that are cultivated or harvested from forested areas. These crops can include various products such as timber, fruits, nuts, resins, and medicinal plants. Examples include timber trees, wood pulp, rubber, and bamboo. Figure 1.1 shows the global production of some plantation crops in 2022.

2.3 *Pretreatment of Plantation Crop Residues*

Pretreatment techniques are the series of processing applied to plantation crop residues before the actual conversion. The components in most lignocellulosic biomass, including plantation residues, form a matrix through covalent bonds between the compounds. During biomass pretreatment, the complex structure of the plantation

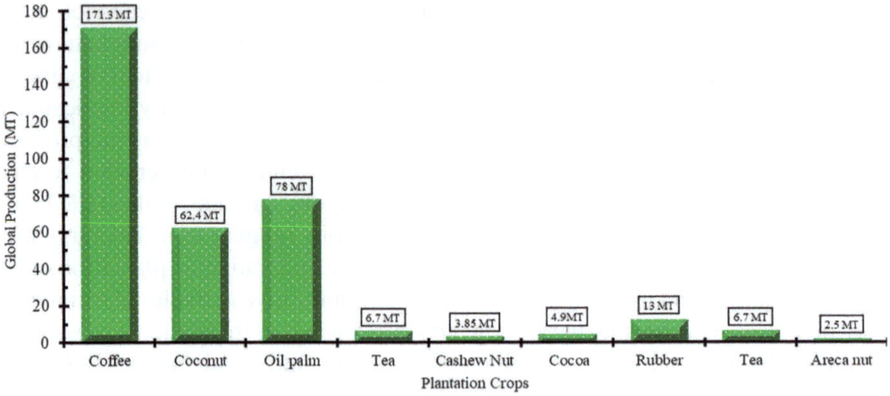

Fig. 1.1 Global production of some plantation crops in 2022. (Data collated from [17, 18])

crop residues is broken down by forces and through chemical or biological methods that remove the lignin, disrupt the hemicellulose, and make the cellulose more accessible for conversion. During the process, there is lignin hydrolysis and a significant increase in the surface area of the biomass, and a breaking down of the recalcitrant structure. The main aim of the pretreatment is to increase the degradation and digestibility of the residues. After pretreating the residues, the residues become degraded and easier to convert, reducing conversion time, costs, and energy consumption. Pretreatment also helps to preserve the sugar in the biomass, reduce the production of inhibitors, and ultimately reduce the conversion efficiency [5].

Various techniques have been used to pretreat biomass, depending on its nature, moisture content, target product, etc.

- **Physical pretreatment** techniques involve sorting and removing impurities to prevent contamination while cutting, grinding, and milling, which reduces fiber particle size and increases the surface area. Other commonly used physical pretreatment techniques include extrusion, microwave, and ultrasound pretreatment. One main feature of physical pretreatment is that the process does not disrupt the biomass and does not alter the chemical composition and properties of the residues.
- In **thermal pretreatment**, the residues are heated to break the covalent bond between the structure of the compounds, degrade the recalcitrant lignin and hemicellulose, and increase access to the cellulose. Heating also enhances the breaking of the hydrogen bonds in crystalline complexes, expands the biomass, and increases the accessible surface area. Wet lignocellulosic biomass is also heated to reduce the moisture content and prepare it for the actual conversion [5].
- **Chemical pretreatment** is the use of chemicals to break down the complex and recalcitrant structure of the biomass by degrading the lignin and the hemicellulose to enhance the processing of the cellulose to fermentable sugars. Commonly used chemicals for pretreatment include acids (H_2SO_4, HCl, etc.), alkalis (NaOH, KOH, etc.), organic solvents (ethanol, acetone, etc.), ionic liquids (pyridinium, etc.), and deep eutectic solvents (urea, acetic acid). The choice of a suitable

chemical for the pretreatment process depends on the crop residue type, cost, anticipated product and its use, and environmental impact. Apart from the environmental impact, the chemical pretreatment process can be costly and constitute additional production costs [24].

- **Biological pretreatment** is the application of microorganisms to break down the structure of plantation crop residues to ensure easier and faster conversion into simpler substances. Various microorganisms, including bacteria, fungi, insects, and worms, or their enzymes, with the unique ability to degrade or drastically reduce lignin and hemicellulose content and enhance the glucose yield, without impacting the cellulose composition, have been used for biological pretreatment. The biological pretreatment method is cost-effective, environmentally benign, requires no chemicals, generates no inhibitors, consumes less energy, and ensures high product yield. However, the process is slow and time-consuming [8]. However, two or more of these basic pretreatment methods are often combined to ensure cost reduction, lower energy consumption, reduce inhibitor formation, and increase process efficiency. The use of physicochemical, thermochemical, biochemical, and other novel pretreatment techniques has gained visibility in recent times owing to their capacity for multipurpose application, lower costs, and efficacy of the process [8–11].

3 Plantation Crops Utilization, Industrial Biotechnological, and Economic Sustainability

3.1 Major Plantation Crops Utilization

Plantation crop residues are the by-products generated from the cultivation, harvesting, and processing of plantation crops. These residues include a variety of materials depending on the crops. Some of the residues commonly generated from plantation crops include leaves, shells, husks, pulps, and pods. The conversion of plantation crop residues to useful products contributes to waste reduction, ensures environmental sustainability, enhances resource recovery, promotes sustainable agricultural practices, and provides additional resources for industrial applications [6, 34]. This section highlights the conversion of major plantation crops, their residues, and the products generated from them. Figure 1.2 summarizes major plantation crop residues and their utilization avenues.

3.1.1 Palm Residues

Palm residues such as palm fronds, empty fruit bunches, waste cooking oil, mesocarp fiber, palm kernel shells, and palm oil mill effluent (POME) are converted to biofuels, biochar, activated carbon, chemicals, nanocellulose, and other bioproducts. Wastewater generated from the washing, extraction, and processing of palm oil is also a low-cost feedstock for the production of useful products. After appropriate

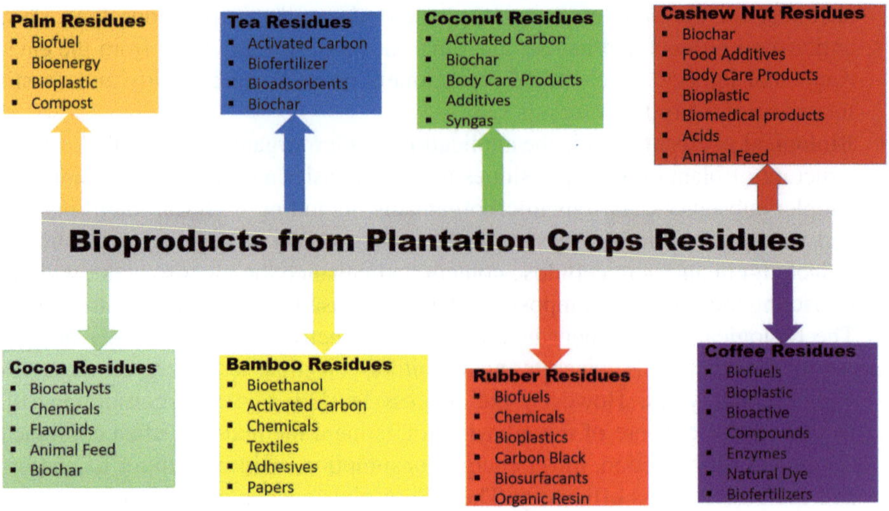

Fig. 1.2 Bioproducts generated from major plantation crop residues

pretreatment procedures, chemical, thermal, biological, biochemical, and thermochemical techniques are deployed to extract useful products from the residues. Products from palm residues are used for clean energy generation, heat production, food additives, and biopharmaceutical products and contribute to resource recovery, circular economy, and environmental sustainability [1, 35].

3.1.2 Tea Residue

With the application of appropriate thermal and bioconversion technologies, residues from tea harvesting, processing, and leftovers after tea preparation are converted to biochar, bio-oil, gases, and other products. Residues such as tea leaves, pruning from tea tree branches, unwanted summer and fall teas, buds, and other byproducts from tea processing are used converted to biochar and activated carbon production, energy storage, chemicals, pharmaceuticals, biofertilizers, bioadsorbents, bioplastics, composite membranes, and for mulching and vermicomposting [49].

3.1.3 Coffee Residues

Coffee residues such as spent coffee grounds, coffee pulp, and coffee huskcontain cellulose, hemicellulose, lignin, lipids, phenolic compounds, and proteins. This unique composition makes it a low-cost feedstock for producing bioethanol, biodiesel, biochar, coffee oil, biopolymers, biocomposites, phenolic compounds, and other bio-based products. Coffee residues are also used for carbon sequestration, soil amendments, and maintenance of biodiversity and natural ecosystems [2].

3.1.4 Cocoa Residues

Cocoa pod husk, cocoa bean shells, wastewater, and other residues from cocoa harvesting and processing can be pretreated and subjected to biochemical, thermochemical, and other procedures to generate useful products. Biochar, biomethane, biofertilizer, chemicals, catalysts, compost, and other bio-based products are generated from the deployment of appropriate conversion processes to cocoa residues [15].

3.1.5 Rubber Residues

Rubber tree bark, latex, and other residues from rubber processing are used as feedstock for the production of biofuels, biochar, adhesives, chemicals, and compost. Rubber sawdust and bark wood were subjected to thermochemical processes such as torrefaction for the production of bioenergy for power generation and an eco-friendly alternative to fossil fuels [25]. Rubber residues are used as feedstock for fuel oil, naphtha-like oil, fuel additives, carbon nanomaterials, plastics, thermoplastic elastomers, asphalt, road surface cement, and other useful products [21].

3.1.6 Bamboo Residues

Waste from bamboo includes bamboo leaves, branches, roots, culm, shoot apex, stems, shoot shells, and sawdust. Most of these residues are high in cellulose, hemicellulose, and lignin and are used as wood-based panels, bamboo-plastic composites, bamboo charcoal, bamboo vinegar, bamboo extracts, and paper products. After appropriate pretreatment processes, bamboo residues are used as feedstock for bioenergy, organic fertilizer, adsorbents, and other bio-based materials [30].

3.1.7 Coconut Residues

During coconut harvesting and processing, residues such as coconut husk, coconut shell, and coconut pith and residues from oil production are used as feedstocks for biofuels, chemicals, and other bio-based products. For example, coconut shells are used for biochar, activated carbon, and bioadsorbents production, while bio-oil, bioethanol, and biodiesel are generated from coconut oil. Other residues from coconut processing are used as low-cost raw materials for the production of chemicals, acids, pharmaceuticals, sunscreen formulations, cosmetics, and body care products [46].

3.1.8 Cashew Residues

Residues from the processing of cashews have been converted into biofuels, animal feed, industrial chemicals, and other bioproducts. Cashew residues such as cashew nut shell liquid, cashew apple, cashew apple bagasse, cashew pulp, and cashew nut testa are feedstocks for biodiesel, green diesel, biochar, dyes, pesticides, pesticides, adhesives, varnishes, biopolymers, animal feed, antioxidants, biodegradable plastics, and bioactive chemicals [12].

3.2 Industrial Biotechnological Conversion of Plantain Crops Residue

Biotechnological conversion majorly entails relying on microorganisms or enzymes to transform carbohydrate substrates derived from sugar- or starch-rich crops and other biomass materials into useful products. On the other hand, industrial biotechnological conversion is the process of microbial conversion of feedstocks to industrial products such as fuels, chemicals, bioproducts, and other similar commodities [43]. The process involves applying modern techniques to create products, prevent pollution, and ensure sustainable utilization of waste materials using living systems such as microorganisms or enzymes. Microorganisms are cultivated in suitable conditions using carbon, nitrogen, salts, vitamins, and other trace elements as raw materials. The microorganisms or enzymes react with and convert carbohydrate substrates from materials rich in starch or sugar to synthesize chemicals, fuels, and other related products. The biotechnological conversion of crop residues in most cases involves biomass pretreatment, hydrolysis of the pretreated biomass, and the actual conversion of reducing sugars into fuels, acids, biochemical, and other products. Biofuels such as bioethanol, biomethane, biodiesel, and biohydogen, acids like citric acid, lactic acid, and succinic acid, and other useful bioproducts are the main products of biotechnological conversion of sugar-rich plantation crops residue [32].

Traditional methods of disposing of plantation crop residues such as open burning and landfilling exacerbate pest infestation, offensive odor, soil degradation, surface and underground water contamination, and air quality deterioration. The biotechnological conversion of plantation crop residue through techniques including fermentation [19], composting [26], anaerobic digestion (AD) [42], gasification [16], enzymatic hydrolysis [13], etc. have been significantly explored. Fermentation is a versatile and commonly used biotechnological conversion technique for the conversion of residues such as sugarcane bagasse, leaves, pulps, and husks into bioethanol, biomethane, biohydrogen, organic acids, chemicals, and other bioproducts [19]. During composting, plantation crop residues such as leaves, pulp, wood chips, and sawdust are converted by aerobic decomposition by microorganisms into nutrient-rich compost. When appropriately applied, compost improves soil fertility, reduces soil erosion, and contributes to increased agricultural productivity and food security [26]. The decomposition of plantation crop residues by AD generated a methane-rich gas called biogas from palm kernel shells, and coconut shells. The biogas is further upgraded to biomethane for power generation and other applications [42]. Gasification of organic materials such as coconut shells, palm kernel shells, and wood chips generates biosyngas, which are used for biofuels and electricity generation [16]. The deployment of enzymatic hydrolysis to convert the complex carbohydrates in crop residues such as husks and bamboo culms to simple sugars, and ferment to biofuels, organic acids, and other chemicals. Plantation crop residues such as rubber tree back, oil palm fronts, woods, sawdusts, are pyrolyzed by heating the materials in the absence of oxygen to produce biosyngas, bio-oil, and biochar [13]. Table 1.1 shows some of the conversion techniques for plantation crop residues and the products generated.

In recent times, certain integrated biotechnological techniques have been developed to convert agricultural residues into bioplastics, biosurfactants, and other biocommodities which are biodegradable and environmentally friendly alternatives to conventional products [47]. Soares et al. [41] and Ashokkumar et al. [4] reported the deployment of advanced biotechnological techniques for the conversion of agricultural waste, including plantation crop residues, into biofuels, bioplastics, biosurfactants, organic fertilizers, organic acids, and other valuable resources. The use of innovative approaches such as Clustered Regularly Interspaced Short Palindromic Repeats (CRISPR), Transcription Activator-Like Effector Nucleases (TALENs), Zinc Finger Nucleases (ZFNs), RNA Interference (RNAi), and Targeting Induced Local Lesions IN Genomes (TILLING) have been used as sustainable pathways from turning waste biomass into valuable resources, promoting environmental sustainability and circular bioeconomy [41].

CRISPR involves the deployment of novel gene-editing tools to modify the DNA of organisms to enhance the efficiency of microorganisms in breaking down crop residues into biofuels, organic acids, chemicals, and organic fertilizers [20].

Table 1.1 Conversion techniques for some plantation crop residues and the products generated

Conversion Techniques	Plantation crop residues	Products generated
AD, gasification, pyrolysis	Empty fruit bunches, palm fronds, palm kernel shells	Biofuels, biogas, biochar, organic fertilizers
Enzymatic hydrolysis, pyrolysis, composting	Husks, shells, leaves	Coir, activated carbon, biochar, compost
Composting, pyrolysis	Pruned leaves, stems	Compost, biochar, organic fertilizers
Fermentation, AD, composting	Coffee pulp, husks, leaves	Bioethanol, biogas, compost, animal feed
Pyrolysis, composting	Leaves, bark	Biochar, compost, organic fertilizers
Fermentation, pyrolysis, enzymatic hydrolysis	Bamboo leaves, branches, shoots, sawdust	Bioethanol, biogas, biochar, activated carbon, compost, bioplastics

AD anaerobic digestion

TALENs are engineered proteins targeted as specific DNA sequences and are used to modify the genetic makeup of plants and microorganisms to improve their ability to convert agricultural waste into biofuels and other valuable bioproducts. ZFNs are synthetic proteins that are capable of creating double-strand breaks in DNA at specific locations to allow for the effective mortification of plantation crop residues and product formation. RNAi technique involves suppressing specific genes to amend the expression of traits in plants and microorganisms to boost the degradation of lignocellulosic biomass and advance the formation of biofuels and other bioproducts [14]. The TILLING conversion method involves inducing mutations in specific genes to create desirable traits in plants. This contributes to the development of crop varieties that are easily converted and generate fewer waste [39]. The advent of these advanced industrial biotechnological approaches is revolutionizing the management and conversion of plantation crop residues and lignocellulosic biomass generally.

3.3 Economic Sustainability of Plantation Crops Residue Conversion and Utilization

Economic sustainability involves practices that support long-term economic growth without negatively impacting social, environmental, and cultural aspects during plantation crop residue conversion and utilization. The biotechnological conversion and utilization of plantation crop residues play a significant role in achieving economic sustainability [50]. However, investigations have shown that many farmers overlook the financial benefits of conversion and utilization of crop residues and resort to open burning. Apart from the threat to environmental sustainability and human health, burning of agricultural residues is prohibited in many countries. While it has been noticed that regulations against open burning are less effective, eco-friendly, and sustainable crop residue management is labor-intensive and costly. The financial implications of residue collection, transportation, sorting, and conversion have been put at between USD 78 and USD 125 per hectare in some parts of India [31]. This cost has also been exacerbated by some harvesting techniques which result in the scattering of residues and impede effective residue collection and sorting. The introduction of innovative technologies for plantation crop harvesting will significantly reduce the labor and cost requirement for crop residue management.

With public awareness of the economic viability of sustainable crop residue management, government, and research institutions have increased investments in novel technologies to douse the interest in traditional open burning of crop residues. Alongside labor costs and availability of appropriate personnel, the deployment of equipment and technologies for low-cost residue management is determined by their farm-household characteristics, public perception and awareness, and how the techniques affect the management practices [29].

4 Benefits, Challenges, and Recommendations

4.1 Benefits

The conversion of the huge residues generated from the cultivation, harvesting, processing, and conversion of plantation crops offers some benefits. This section highlights some environmental, economic, agricultural, energy generation, waste reduction, value addition, and circular addition benefits [4, 8]. These benefits collectively contribute to resource efficiency, making agricultural practices more sustainable and environmentally friendly.

4.1.1 Environmental Sustainability

Converting plantation crop residues into valuable products reduces waste and minimizes environmental pollution. Unlike direct burning of the residues which emit toxic gasses, impact air quality, and contaminate water sources, the deployment of biotechnological techniques to convert plantation crop residues helps in mitigating greenhouse gas emissions, prevents soil degradation, and enhances human health.

4.1.2 Renewable Energy Generation

The global energy scenario is heavily dominated by fossil-based and nonrenewable energy sources with their attendant impacts. The usage of plantation crop residues and other agricultural waste provides low-cost and readily available feedstock for renewable energy resources. Residues from plantation crops can be converted into biofuels, biogas, and biochar, providing renewable energy sources, reducing reliance on fossil fuels, and ensuring energy security.

4.1.3 Soil Fertility and Health Improvement

Composting, AD, and fermentation of crop residues generate biofertilizers. Converting crop residues to organic fertilizer is low-cost, environmentally friendly, and sustainable. The use of crop residues as compost or biofertilizers enhances soil fertility and structure, contributes to food security, and promotes sustainable agriculture. Using residues to produce compost and biofertilizers improves soil health and fertility. This can lead to higher crop yields and reduced expenditure on chemical fertilizers, contributing to long-term agricultural sustainability. Healthy soils are essential for sustainable agriculture and food security.

4.1.4 Economic Benefits and Market Diversification

The high cost of disposal and management of crop residues and processing of agricultural waste impacts the economic viability of the farmers and communities. However, the collection and conversion of crop residues into bioenergy, organic acids, chemicals, and other useful products creates additional revenue streams for farmers and industries and contributes to the local economy. The conversion of residues to organic manure helps farmers save the money used for the procurement of fertilizers to improve crop yield. The conversion of crop residues opens up new markets for bioproducts such as biofuels, bioplastics, and biofertilizers. This diversification helps in stabilizing income for farmers and industries, making them less vulnerable to market fluctuations in primary crop markets. It also encourages innovation and the development of new products and technologies.

4.1.5 Job Creation and Cost Saving

The establishment of biotechnological conversion facilities and the associated supply chain activities create employment opportunities. This includes jobs in the collection, transportation, processing, and distribution of the converted products. The development of these facilities can stimulate local economies and provide livelihoods for many people. Utilizing crop residues for energy production can reduce reliance on external energy sources, leading to significant cost savings. For instance, biogas produced from AD of crop residues can be used for cooking, heating, and electricity generation, reducing the need for fossil fuels. This not only conserves nonrenewable resources but also promotes sustainable energy practices.

4.1.6 Resource Efficiency and Optimization

By converting residues into valuable products, we ensure that resources are used efficiently. This reduces the need for new raw materials and minimizes waste, contributing to a circular economy. Efficient resource use also helps in reducing production costs and improving profitability. For example, using residues to produce compost and biofertilizers recycles nutrients back into the soil, enhancing soil fertility and reducing the need for chemical fertilizers. Advanced biotechnological processes optimize the use of residues, ensuring that every part of the crop is utilized effectively. This maximizes the overall efficiency of resource use in agriculture.

4.1.7 Nutrient Recycling and Circular Economy

Using residues to produce compost and biofertilizers recycles nutrients back into the soil. This enhances soil fertility and reduces the need for chemical fertilizers, which can be resource-intensive to produce. By turning waste into valuable products, residue conversion promotes a circular economy. This approach ensures that resources are used efficiently and sustainably, reducing the need for new raw materials and minimizing environmental impact.

4.1.8 Waste Reduction and Value Addition

By converting agricultural residues into valuable products, we reduce the amount of waste that would otherwise be discarded. This minimizes the environmental impact and makes better use of available resources. Residue conversion adds value to what would otherwise be considered waste. This creates additional revenue streams for farmers and industries, making the agricultural sector more economically efficient.

4.1.9 Revenue Generation

Converting plantation crop residues into valuable products creates additional revenue streams for farmers and industries. For example, residues like sugarcane bagasse can be converted into bioethanol, which can be sold as a renewable fuel. Similarly, palm kernel shells can be used to produce biochar, which has various industrial applications. This diversification helps stabilize income for farmers and industries, making them less vulnerable to market fluctuations in primary crop markets.

4.2 Challenges

Notwithstanding the benefits of plantation crop residue conversion and utilization, obvious drawbacks arising from the process cannot be ignored. This section highlights some challenges [36].

4.2.1 Economic Viability

The high cost of labor, infrastructure, and initial investment for setting up biotechnological conversion facilities can be challenging for small-scale farmers and industries. This is capable of discouraging small and medium-scale enterprises from venturing into waste conversion endeavors. The lack of markets for plantation residues also hinders the economic viability of the sector.

4.2.2 Technological Complexities

Collection, handling, sorting, and conversion of some of the plantation crop residues require some level of infrastructural and technological complexities. Apart from the need to upgrade some of the traditional conversion techniques, the introduction of advanced biotechnological technologies requires specialized knowledge, personnel, and equipment, which may not be readily available in all regions.

4.2.3 Regulatory Obstacles

There are not sufficient and appropriate legal and regulatory frameworks to guide waste conversion and utilization. Getting ethical and administrative approvals for waste conversion processes can be tedious and costly. Strict legal, ethical, regulatory, and policy frameworks can hinder the adoption of new technologies and processes for residue conversion.

4.2.4 Public Acceptance

There may be resistance from the public and stakeholders due to a lack of awareness and understanding of the benefits of residue conversion. Some entrenched sociocultural and religious biases are against waste handling, processing, and recycling, thereby reducing public acceptance of waste management. This tends to scare aware potential investors and laborers from participating in waste-related businesses.

4.2.5 Logistical Issues

Collecting, transporting, and storing large quantities of residues can be logistically challenging and costly. Some of the residues can be toxic or make the operators susceptible to injuries or harm, especially during prolonged exposure. There is a need to provide personal protective equipment and other logistics to protect the operators. Managing emerging labor issues, safety, compensation, and other logistical issues can be challenging and problematic.

4.3 Recommendations

The identified challenges to effective plantation crop residue conversion and utilization are not insurmountable. By implementing the following recommendations, farmers and other stakeholders can maximize the benefits of plantation crop residue conversion and contribute to a more sustainable and circular economy.

4.3.1 Policy and Regulatory Support

Governments should provide incentives and subsidies to encourage the adoption of biotechnological conversion processes and support research and development in this field. There is a need for full implementation of existing laws and enactment of appropriate legal, ethical, and regulatory policies to support waste conversion and utilization. Such policies and guidelines should provide monitoring frameworks and simplify administrative approval processes.

4.3.2 Public Awareness

There are lot of misinformation, disinformation, and outright falsehoods about waste conversion, recycling, and management that are discouraging participation and investment in the sector. There is a need for adequate public enlightenment and education for farmers, potential investors, community leaders, and other stakeholders on the benefits of waste conversion and utilization. Educating farmers, industry players, and the general public about the benefits of residue conversion can help in gaining acceptance and support.

4.3.3 Technological Advancements

The traditional techniques for waste conversion are becoming inadequate and problematic due to low conversion efficiency, high-cost consumption, and serious ecological impacts. The application of innovative biotechnological technologies requires state-of-the-art infrastructure and technological inputs. Investing in research and development and innovative technologies to improve the efficiency and cost-effectiveness of conversion technologies is urgently crucial.

4.3.4 Collaboration and Partnerships

Encouraging collaboration and partnerships between farmers, industries, researchers, policymakers, and international organizations can lead to the successful implementation of residue conversion projects. Conferences, workshops, postgraduate studies, and other capacity building initiatives must be prioritized. The government must encourage and fund collaborative research in waste conversion and utilization and ensure a town-gown handshake to ensure the implementation of research outcomes.

4.3.5 Infrastructure Development

Building infrastructure for the collection, transportation, and processing of residues can help in overcoming logistical challenges. Stakeholders must leverage experienced and developed nations for capacity building, infrastructural development, knowledge sharing, and collaboration and partnerships.

4.3.6 Sustainable Practices

Emphasizing sustainable agricultural practices and promoting the use of renewable energy sources will contribute to the long-term economic sustainability of the agricultural sector. Integrating residue conversion into existing agricultural systems will enhance resource efficiency and reduce environmental impact.

5 Conclusion and Future Research

Increased population, demand for food, and more active industrial sector have led to higher agricultural activities, worldwide. The increment in plantation crop cultivation and production has led to a higher generation of residues from harvesting, consumption, and processing. Consequently, the global generation of leaves, husks, peels, shells, bagasse, and other plantation crop residues has increased significantly. The deployment of conversion technologies to transform residues into biofuel, chemicals, organic acids, and other bioproducts is cost-effective, environmentally benign, and supports the circular economy. Biotechnological conversion technologies involve the deployment of appropriate microorganisms and enzymes to break down the carbohydrates in the residues into simple sugars and other useful products. The industrial biotechnological conversion of plantation crop residues offers a sustainable solution to managing agricultural waste while generating valuable products. The conversion processes, such as fermentation, anaerobic digestion, pyrolysis, and enzymatic hydrolysis, have proven effective in transforming residues into biofuels, biogas, biochar, compost, and other valuable products. The economic benefits of residue conversion include cost savings, job creation, market diversification, and enhanced soil fertility. However, challenges such as high initial investment, technological complexities, regulatory obstacles, and public acceptance need to be addressed to fully realize the potential of residue conversion. The development of integrated biorefineries that can process multiple types of crop residues and produce a variety of bioproducts, maximizing resource efficiency and economic viability is also desirable. Multidisciplinary research into life cycle assessments, techno-economic analysis, sustainable conversion practices, and policy analysis to ensure process improvement, economic viability, energy security, circular economy, and environmental sustainability.

Conflict of Interest The authors declare no conflict of interest.

References

1. Abogunrin-Olafisoye OB, Adeyi O, Adeyi AJ, Oke EO (2024) Sustainable utilization of oil palm residues and waste in Nigeria: practices, prospects, and environmental considerations. Waste Manag Bullet 2(1):214–228. https://doi.org/10.1016/j.wmb.2024.01.011
2. Ahmed H, Abolore RS, Jaiswal S, Jaiswal AK (2024) Toward circular economy: potentials of spent coffee grounds in bioproducts and chemical production. Biomass 4(2):286–312. https://doi.org/10.3390/biomass4020014
3. Andrews EM, Kassama S, Smith EE, Brown PH, Khalsa SDS (2021) A review of potassium-rich crop residues used as organic matter amendments in tree crop agroecosystems. Agriculture 11(7):580. https://doi.org/10.3390/agriculture11070580
4. Ashok Kumar V, Flora G, Venkatkarthick R, SenthilKannan K, Kuppam C, Mary Stephy G, Ngamcharussrivichai C (2022) Advanced technologies on the sustainable approaches for conversion of organic waste to valuable bioproducts: emerging circular bioeconomy perspective. Fuel 324:124313. https://doi.org/10.1016/j.fuel.2022.124313
5. Awogbemi O, Kallon DVV (2022) Pretreatment techniques for agricultural waste. Case Stud Chem Environ Eng 6:100229. https://doi.org/10.1016/j.cscee.2022.100229
6. Awogbemi O, Kallon DVV (2024) Conversion of hazardous waste cooking oil into non-fuel value added products. Int J Amb Energy 45(1):2345253. https://doi.org/10.1080/01430750.2024.2345253
7. Awogbemi O, Kallon DVV, Aigbodion VS (2021) Trends in the development and utilization of agricultural wastes as heterogeneous catalyst for biodiesel production. J Energy Inst 98:244–258. https://doi.org/10.1016/j.joei.2021.06.017
8. Awogbemi O, Kallon DVV, Kumar KS (2024a) Contributions of artificial intelligence and digitization in achieving clean and affordable energy. Intell Syst Appl 22:200389. https://doi.org/10.1016/j.iswa.2024.200389
9. Awogbemi O, Ojo AA, Adeleye SA (2024b) Advanced thermochemical conversion approaches for green hydrogen production from crop residues. J Renew Mater 12(1):1–28. https://doi.org/10.32604/jrm.2023.045822
10. Awogbemi O, Von Kallon DV, Ray RC (2024c) Conversion of agriculture residues for bioenergy production. In: Arora J, Joshi A, Ray RC (eds) Transforming agriculture residues for sustainable development: from waste to wealth. Springer, Cham, pp 273–294
11. Awogbemi O, Von Kallon DV, Ray RC (2024d) Transformation of cassava wastes into useful products. In: Arora J, Joshi A, Ray RC (eds) Transforming agriculture residues for sustainable development: from waste to wealth. Springer, Cham, pp 353–370
12. Barroso TLCT, Ferreira VC, Castro LEN, da Rosa RG, de Souza LGA, Goldbeck R, Forster-Carneiro T (2024) Bioenergy and bioproducts from cashew apple bagasse: a scientometric overview and strategies for sustainable utilization. Biofuels Bioprod Biorefin 1-12:540. https://doi.org/10.1002/bbb.2718
13. Baruah J, Bardhan P, Mukherjee AK, Deka RC, Mandal M, Kalita E (2022) Integrated pretreatment of banana agrowastes: structural characterization and enhancement of enzymatic hydrolysis of cellulose obtained from banana peduncle. Int J Biol Macromol 201:298–307. https://doi.org/10.1016/j.ijbiomac.2021.12.179
14. Bhuyan SJ, Kumar M, Ramrao Devde P, Rai AC, Mishra AK, Singh PK, Siddique KH (2023) Progress in gene editing tools, implications and success in plants: a review. Front Genome Edit 5:1272678. https://doi.org/10.3389/fgeed.2023.1272678

15. Cazier EA, Pham T-N, Cossus L, Abla M, Ilc T, Lawrence P (2024) Exploring industrial lignocellulosic waste: sources, types, and potential as high-value molecules. Waste Manag 188:11–38. https://doi.org/10.1016/j.wasman.2024.07.029
16. Chaves M, Torres C, Tenorio C, Moya R, Arias-Aguilar D (2024) Syngas characterization and electric performance evaluation of gasification process using forest plantation biomass. Waste Biomass Valoriz 15(3):1291–1308. https://doi.org/10.1007/s12649-023-02231-3
17. FAO (2022) OECD-FAO Agricultural Outlook 2022–2031. https://openknowledge.fao.org/handle/20.500.14283/cc0308en
18. FAOSTAT (2021) Agricultural production statistics 2000–2022. https://openknowledge.fao.org/server/api/core/bitstreams/fba4ef43-422c-4d73-886e
19. Faria DJ, Carvalho APA, Conte-Junior CA (2024) Fermentation of biomass and residues from brazilian agriculture for 2G bioethanol production. ACS Omega 9(39):40298–40314. https://doi.org/10.1021/acsomega.4c06579
20. Garg D, Samota MK, Kontis N, Patel N, Bala S, Rosado AS (2023) Revolutionizing biofuel generation: unleashing the power of CRISPR-Cas mediated gene editing of extremophiles. Microbiol Res 274:127443. https://doi.org/10.1016/j.micres.2023.127443
21. Ghodke PK, Anthony C, Sharma AK (2024) Bioprocessing of agricultural and forest waste. In: Patel AK, Sharma AK (eds) Sustainable production innovations: bioremediation and other biotechnologies. Scrivener Publishing LLC, pp 367–393
22. Hebbar K, Pandiselvam R, Beegum P, Ramesh S, Manikantan M, Mathew A (2022) Seventy-five years of research in processing and product development in plantation crops-coconut, arecanut and cocoa. Int J Innov Horti 11(1):103–119. https://doi.org/10.5958/2582-2527.2022.00011.2
23. Honorato-Salazar JA, Sadhukhan J (2020) Annual biomass variation of agriculture crops and forestry residues, and seasonality of crop residues for energy production in Mexico. Food Bioprod Process 119:1–19. https://doi.org/10.1016/j.fbp.2019.10.005
24. Joshi M, Manjare S (2024) Chemical approaches for the biomass valorisation: a comprehensive review of pretreatment strategies. Environ Sci Pollut Res 31(36):48928–48954. https://doi.org/10.1007/s11356-024-34473-6
25. Kasawapat J, Khamwichit A, Dechapanya W (2024) Waste-to-energy conversion of rubberwood residues for enhanced biomass fuels: process optimization and eco-efficiency evaluation. Energies 17(21):5444. https://doi.org/10.3390/en17215444
26. Kavvadias V, Le Guyader E, El Mazlouzi M, Gommeaux M, Boumaraf B, Moussa M, Guimeur K (2024) Using date palm residues to improve soil properties: the case of compost and biochar. Soil Syst 8(3):69. https://doi.org/10.3390/soilsystems8030069
27. Kiehbadroudinezhad M, Hosseinzadeh-Bandbafha H, Rosen MA, Kumar Gupta V, Peng W, Tabatabaei M, Aghbashlo M (2023) The role of energy security and resilience in the sustainability of green microgrids: paving the way to sustainable and clean production. Sustain Energy Technol Assess 60:103485. https://doi.org/10.1016/j.seta.2023.103485
28. Knápek J, Starý O, Vávrová K, Bemš J, Weger J, Horák M (2024) Modelling the impact of business risk on the competitiveness of purpose-grown biomass of annual and perennial crops. Renew Sust Energ Rev 202:114707. https://doi.org/10.1016/j.rser.2024.114707
29. Krishna VV, Mkondiwa M (2023) Economics of crop residue management. Ann Rev Resour Econ 15(1):19–39. https://doi.org/10.1146/annurev-resource-101422-090019
30. Liu H, Li Q (2025) Preparation of green concrete from bamboo biochar (BB) and concrete slurry waste (CSW): preparation method and performance evaluation. Constr Build Mater 462:139964. https://doi.org/10.1016/j.conbuildmat.2025.139964
31. Lopes AA, Viriyavipart A, Tasneem D (2020) The role of social influence in crop residue management: evidence from northern India. Ecol Econ 169:106563. https://doi.org/10.1016/j.ecolecon.2019.106563
32. Michailidou F (2023) The scent of change: sustainable fragrances through industrial biotechnology. Chembiochem 24(19):e202300309. https://doi.org/10.1002/cbic.202300309

33. Nadaraja D, Lu C, Islam MM (2021) The sustainability assessment of plantation agriculture—a systematic review of sustainability indicators. Sustain Prod Consumpt 26:892–910. https://doi.org/10.1016/j.spc.2020.12.042
34. Ojo AA, Awogbemi O, Oni T (2024) Physicochemical assessment of vegetable oil used to fry some Nigerian food items. Sci Technol Energy Trans 79:87. https://doi.org/10.2516/stet/2024086
35. Parthasarathy P, Alherbawi M, Shahbaz M, Mackey HR, McKay G, Al-Ansari T (2024) Conversion of oil palm waste into value-added products through pyrolysis: a sensitivity and techno-economic investigation. Biomass Convers Biorefinery 14(8):9667–9687. https://doi.org/10.1007/s13399-022-03144-2
36. Sarkar S, Skalicky M, Hossain A, Brestic M, Saha S, Garai S, Brahmachari K (2020) Management of crop residues for improving input use efficiency and agricultural sustainability. Sustain For 12(23):9808. https://doi.org/10.3390/su12239808
37. Shahin SM, Jaleel A, Alyafei MAM (2021) The essential oil-bearing plants in The United Arab Emirates (UAE): an overview. Molecules 26(21):6486. https://doi.org/10.3390/molecules26216486
38. Sikiru S, Abioye KJ, Adedayo HB, Adebukola SY, Soleimani H, Anar M (2024) Technology projection in biofuel production using agricultural waste materials as a source of energy sustainability: a comprehensive review. Renew Sust Energ Rev 200:114535. https://doi.org/10.1016/j.rser.2024.114535
39. Singh D, Chaudhary P, Taunk J, Singh CK, Chinnusamy V, Sevanthi AM, Pal M (2024) Targeting induced local lesions in genomes (TILLING): advances and opportunities for fast tracking crop breeding. Crit Rev Biotechnol 44(5):817–836. https://doi.org/10.1080/07388551.2023.2231630
40. Smalley R (2013) Plantations, contract farming and commercial farming areas in Africa: a comparative review. www.future-agricultures.org
41. Soares C, Dorosh O, Fernandes F, Grosso C, Moreira M, Delerue-Matos C (2024) Biotechnological approaches for agriculture waste transformation. In: Arora J, Joshi A, Ray RC (eds) Transforming agriculture residues for sustainable development: from waste to wealth. Springer, Cham, pp 85–107
42. Suhartini S, Nurika I, Paul R, Melville L (2021) Estimation of biogas production and the emission savings from anaerobic digestion of fruit-based agro-industrial waste and agricultural crops residues. Bioenergy Res 14(3):844–859. https://doi.org/10.1007/s12155-020-10209-5
43. Sun M, Gao AX, Liu X, Bai Z, Wang P, Ledesma-Amaro R (2024) Microbial conversion of ethanol to high-value products: progress and challenges. Biotechnol Biofuels Bioprod 17(1):115. https://doi.org/10.1186/s13068-024-02546-w
44. Thanigaivel S, Vickram S, Dey N, Gulothungan G, Subbaiya R, Govarthanan M, Kim W (2022) The urge of algal biomass-based fuels for environmental sustainability against a steady tide of biofuel conflict analysis: is third-generation algal biorefinery a boon? Fuel 317:123494. https://doi.org/10.1016/j.fuel.2022.123494
45. Tireki S (2021) A review on packed non-alcoholic beverages: ingredients, production, trends and future opportunities for functional product development. Trends Food Sci Technol 112:442–454. https://doi.org/10.1016/j.tifs.2021.03.058
46. Vieira F, Santana HE, Jesus M, Santos J, Pires P, Vaz-Velho M, Ruzene DS (2024) Coconut waste: discovering sustainable approaches to advance a circular economy. Sustain For 16(7):3066. https://doi.org/10.3390/su16073066
47. Vučurović D, Bajić B, Trivunović Z, Dodić J, Zeljko M, Jevtić-Mučibabić R, Dodić S (2024) Biotechnological utilization of agro-industrial residues and by-products—sustainable production of biosurfactants. Food Secur 13(5):711. https://doi.org/10.3390/foods13050711
48. Wang X, Zhang Y, Xia C, Alqahtani A, Sharma A, Pugazhendhi A (2023) A review on optimistic biorefinery products: biofuel and bioproducts from algae biomass. Fuel 338:127378. https://doi.org/10.1016/j.fuel.2022.127378

49. Wang Z, Ahmad W, Zhu A, Zhao S, Ouyang Q, Chen Q (2024) Recent advances review in tea waste: high-value applications, processing technology, and value-added products. Sci Total Environ 946:174225. https://doi.org/10.1016/j.scitotenv.2024.174225
50. Zhou Q, Le QV, Yang H, Gu H, Yang Y, Sonne C et al (2022) Sustainable conversion of agricultural biomass into renewable energy products: a discussion. Bioresources 17(2):3489–3508. https://doi.org/10.15376/biores.17.2.Zhou

Part II
Edible Crops

Chapter 2
Date Palm (*Phoenix dactylifera* L.) Fruit Processing Wastes and By-Products: Valorization Using Bioprocess Technology

Zahra Beig-Mohammadi, Sanam Darvish, Behzad Farrokhi, Mohammad Younesi, Anousha Chapnevis, Mobina Mirzaei, Elina Tavakoli, and Kianoush Khosravi-Darani

Abbreviations

ABTS	2,2′-azino-bis(3-ethylbenzothiazoline-6-sulfonic acid
DP(s)	Date palm(s)
DS(s)	Date seed(s)
DPC	Date palm cake
DPB	Date palm bar
DPPH	2,2-diphenyl-1-picrylhydrazyl
DPR	Date palm residue
DPF(s)	Date palm fruit(s)
DPW	Date palm waste
FAO	Food and Agriculture Organization
FRAP	Ferric Reducing Antioxidant Power
MENA	Middle East and North Africa
SCOBY	Symbiotic Culture of Bacteria and Yeast
UDSA	United States Department of Agriculture
DPLF	Date palm leaves and fronds
RGH	Rhodes grass hay
RDF	Rejected date fruits

Z. Beig-Mohammadi (✉) · S. Darvish · B. Farrokhi · M. Younesi · A. Chapnevis
M. Mirzaei · E. Tavakoli
Department of Food Science and Technology, NT. C., Islamic Azad University, Tehran, Iran
e-mail: beigmohammadi@iau.ac.ir

K. Khosravi-Darani
National Nutrition and Food Technology Research Institute, Faculty of Nutrition Sciences and Food Technology, Shahid Beheshti University of Medical Sciences, Tehran, Iran

C. Trois, R. C. Ray (eds.), *Plantation Crop Wastes: Valorization for Economic Sustainability*, Waste as a Resource,
https://doi.org/10.1007/978-3-031-96805-1_2

1 Introduction

Date palms (*Phoenix dactylifera* L.), a member of the Arecaceae family, are critical in the agricultural and economic sustainability of arid and semi-arid regions. With a history of cultivation exceeding 5000 years, particularly in the Middle East and North Africa (MENA), date palms (DPs) have long been a source of food, income, and environmental benefits. Leading producers such as Saudi Arabia, Egypt, Iran, and the United Arab Emirates account for over 75% of worldwide palm date output. Because of their remarkable durability, date palms have maintained their popularity. According to Zaid and Arias-Jimenez [88], these trees are particularly well-suited to desert areas vulnerable to desertification because they can withstand harsh environmental factors like high temperatures, scarce water supplies, and saline soils. This adaptability ensures year-round food production, offering nutritional security in areas. Dates, rich in carbohydrates, vitamins, and minerals, serve as a dietary staple and a significant export commodity across MENA. Beyond the fruit, various date palm by-products, such as leaves, fronds, and seeds, contribute additional economic value. These materials are widely repurposed for construction, handicrafts, and animal feed, further enhancing the utility of date palm cultivation [49]. Cultivating date palms is crucial for food production and ecological stability in desert environments, where they serve as "ecosystem engineers" by creating favorable microclimates that support other crops and protect soil from erosion [28]. Despite the benefits of date palm cultivation, the large volumes of waste generated during production and processing pose significant environmental challenges. With worldwide efforts increasingly focused on sustainable agriculture and the creation of a circular bioeconomy, there is an urgent need for new waste management solutions. Bioprocessing technologies, which utilize microorganisms, enzymes, and fermentation processes, offer an effective pathway to reduce waste while converting it into valuable resources [29]. For instance, anaerobic digestion of date palm residues (DPR), such as fronds and date seeds (DSs), can produce biogas, a renewable energy source [45]. Bioethanol, created by fermenting sugars extracted from DS, is a sustainable alternative to fossil fuels [45]. The fibrous material from date fronds holds promise for producing biodegradable bioplastics, which can reduce reliance on synthetic plastics and mitigate environmental pollution [17]. Furthermore, wastewater produced during the processing of date palms is processed to produce biofertilizers, which enhance agricultural sustainability and nutrient recycling [78]. This chapter explores the valorization of date palm waste through bioprocessing technologies, emphasizing bioenergy production, bioplastics synthesis, and bioremediation. It aims to identify innovative biotechnological approaches for sustainably managing date palm waste (DPW) and converting it into valuable products. By reviewing current strategies, the chapter further examines the challenges of scaling these technologies and evaluates their economic and environmental implications.

2 Geographical Origin, Distribution, Production, and Productivity of DPs

The date palm is one of the oldest cultivated fruit crops. Its origin is believed to be traced back to the arid regions of Mesopotamia and North Africa, with evidence suggesting cultivation as early as 4000 B.C. in Mesopotamia and around 3000–2000 B.C. in Egypt's Nile Valley [88]. The date palm's resilience to drought and nutritional value made it a crucial crop in arid and semi-arid regions. As trade routes expanded during antiquity, date palms were introduced to Southern Europe, South Asia, and North Africa, becoming integral to the economies and diets of these regions [77]. By the 19th and 20th centuries, cultivation spread further to the Americas and Australia, where the species adapted well to new environments [66]. Today, date palms are cultivated in over 30 countries across five continents, with the highest production concentrated in the MENA, South Asia, and parts of the Americas. The MENA region, including Egypt, Saudi Arabia, Iran, and Algeria, leads global production due to its ideal climatic conditions and extensive cultivation practices [9]. South Asia, particularly Pakistan and India, is another major contributor, while in the Americas, California, and Arizona dominate United States production, and Mexico focuses on growing dates primarily for export. Australia, an emerging producer, emphasizes sustainable cultivation and the export of premium-quality dates. The global demand for dates has driven significant growth in production over recent decades. Food and Agriculture Organizations' data indicates that global palm date output rose from roughly 1.8 million tons in 1961 to 2.8 million tons in 1985 to 5.4 million tons in 2001. In more recent years, production has continued to rise at an average annual rate of 4%, attributed to the expansion of cultivated areas. By 2022, global palm date production will reach approximately 9.7 million tons, a notable increase from 8.52 million tons in 2018 (FAO 2023). Egypt leads global production, contributing around 1.7 million tons annually, which accounts for 17.7% of the worldwide output. Saudi Arabia, Iran, and Algeria are the major producers. These countries benefit from advanced irrigation systems, high-yield cultivars, and government-supported agricultural programs [26]. Date palm productivity depends on various factors, including soil type, irrigation practices, and cultivar selection. A mature DP typically produces 40–60 kg of fruit annually, though optimal agricultural practices can enhance yields [77]. Advanced techniques, such as tissue culture propagation, have significantly increased productivity by ensuring uniformity among high-yielding varieties. The economic and environmental benefits of date palm farming are further enhanced by press cakes produced from date processing serve as a valuable feedstock for bio-based products. The sustainability of the date palm industry underscores its importance in combating desertification and supporting agricultural livelihoods globally [26, 66, 77]. Its role in promoting sustainable agriculture and reducing the environmental impact of desertification highlights the crop's broader significance [28].

3 Description, Cultivation, Proximate Composition, Nutritional Values, and Uses of Date Palms(DP)

3.1 Description and Cultivation

DP has separate male and female plants. Under favorable conditions, a mature DP can grow to 20–25 meters and live for over 100 years [82]. The tree has a slender trunk crowned with arching pinnate leaves reaching up to 6 meters (Fig. 2.1). Cultivating DPs relies on several fundamental practices ensuring successful growth and productivity.

Propagation Traditionally, farmers multiply DPs through offshoots or suckers, which ensure genetic uniformity. In recent years, tissue culture techniques have gained popularity for large-scale propagation [66].

Irrigation An adequate water supply is critical, particularly during the first fruit setting phases and flowering. Modern irrigation systems, such as drip irrigation, have been widely adopted to enhance water use efficiency [77].

Pollination As dioecious plants, female DPs require artificial pollination to ensure successful fruit production.

Harvesting Dates are typically harvested manually or mechanically, depending on the production scale. The fruits of the DP are oval or cylindrical and vary in color from bright yellow to reddish-brown, depending on the fruit-setting phases and flowering. In modern irrigation systems, such as drip irrigation, the fruits of the DP are oval or cylindrical and vary in color from bright yellow to reddish-brown, depending on the cultivar and stage of ripening. Each fruit contains a single elongated seed encased in a fleshy pericarp (Fig. 2.1), which serves as a valuable food source [77]. The development of date palm fruit (DPF) occurs in four distinct stages: **Kimri** refers to the juvenile green stage, **Khalal** is the full-sized crunchy stage,

Fig. 2.1 Date palm tree (**a**), Date palm fruit and Date palm seed (**b**)

Fig. 2.2 Sample images of date fruits' four distinct stages of maturity

Rutab is the ripe, soft, and partially brown stage, and **Tamr** is the fully mature, dried stage ideal for storage (Fig. 2.2). DPs are primarily grown in arid and semi-arid environments because of their exceptional tolerance for high temperatures (up to 50 °C) and drought conditions. The ideal soil for DP agriculture is sandy loam, characterized by adequate drainage and high water retention [14].

3.2 Proximate Composition and Nutritional Values

DPFs are renowned for their high carbohydrate content, dietary fiber, vitamins, and minerals, making them an excellent energy reservoir (Table 2.1). They are highly nutritious fruits primarily composed of sugars such as glucose, fructose, and sucrose, rich in dietary fiber, which promotes digestive health and helps regulate blood sugar levels [14]. While dates are not a significant source of protein, they provide essential amino acids that complement other dietary protein sources [82]. Low in fat, dates primarily contain unsaturated fatty acids, which benefit heart health [26]. They are also rich in vitamins such as B-complex vitamins (e.g., B6, niacin, and pantothenic acid) and vitamin C, which contribute to metabolic processes and immunity. Moreover, dates are an excellent source of minerals, particularly potassium and magnesium, which regulate blood pressure and support muscle and nerve function, respectively [9]. DP is a high-energy food with an energy content of 280–350 kcal per 100 g, especially valuable in regions with significant energy demands [84]. Besides these valuable components, dates are known as a good source of bioactive phenolic compounds and antioxidants [4, 68]. Bioactive compounds in dates, including phenolic acids, flavonoids, and carotenoids, exhibit potent antioxidant, anti-inflammatory, and anticancer properties [18, 89]. Their antioxidant compounds neutralize free radicals, reducing the risk of chronic diseases such as cardiovascular disorders and cancer [39]. Polyphenols in dates may inhibit cancer cell growth, suggesting their potential role in cancer prevention [47]. Additionally, date extracts exhibit antimicrobial activity against bacteria and fungi, underscoring their utility in natural remedies [18]. Dates also promote gastrointestinal health by enhancing gut microbiota and supporting overall digestion [39]. The

Table 2.1 Nutritional, medicinal, and functional properties of date palm fruits (DPFs)

Category	Amount	Value (based on 100 g DPF)	References
Nutritional properties	Carbohydrates	70–80%	Al-Farsi et al. [14]
	Dietary fiber	6–10%	Al-Farsi et al. [12]
	Protein	1.5–2.5%	Tang et al. [82]; Al-Farsi et al. [13]
	Fats	<0.5%	Aydeniz-Güneşer [26]
	Vitamins	B-complex, C	Al-Alawi et al. [9]
	Minerals	Potassium, calcium, magnesium	Al-Alawi et al. [9]
	Bioactive compounds	Phenolic acids, flavonoids, carotenoids	Ghafoor et al. [42]; Zihad et al. [89]
	Antioxidants	High	Ghafoor et al. [42]; Eid et al. [39]
	Energy content	280–350 kcal/100 g	USDA [84]
Medicinal properties	Antioxidant effects	Reduces oxidative stress	Zihad et al. [89]
	Anti-inflammatory	Modulates inflammation	Eid et al. [39]; Al-Alawi et al. [9]
	Antimicrobial	Effective against bacteria and fungi	Alkhalidy et al. [18]; Abu-Reidah et al. [4]
	Anticancer	Potential anticancer properties	Habib et al. [47]; Al-Alawi et al. [9]
	Gastrointestinal health	Improves gut microbiota	Eid et al. [39]
Functional properties	Functional foods	Energy bars, snacks, beverages	Al-Farsi et al. [13]; Aydeniz-Güneşer [26]
	Bioproducts	Organic fertilizers, soil amendments	Abid et al. [3]; Benabderrahim et al. [30]
	Biomaterials	Biocomposites, biopolymers	Hamza et al. [48]; Mousa et al. [62]
	Animal feed	Animal feed from palm date meal	Boufennara et al. [32]; Kari et al. [54]
	Biofuels	Bioethanol, biogas, biodiesel	Bharath et al. [31]
	Cosmetics	Skincare products from DS oil	Al-Alawi et al. [9]

DS Date seed, *DPF* Date palm fruits

functional properties of dates extend beyond nutrition and health. Dates are increasingly recognized for their functional applications, including use in energy bars, snacks, and beverages due to their natural sweetness and nutritional benefits [60]. DPW is repurposed for agricultural, industrial, and cosmetic applications. For example, DPW-based organic fertilizers and soil amendments enhance agricultural productivity. Similarly, date palm fibers produce biocomposites and biopolymers, offering sustainable alternatives to synthetic materials [31]. DS is a valuable feed additive, offering an environmentally sustainable alternative to traditional feed ingredients [54]. The potential of DPW extends to energy production, where it serves as a resource for bioethanol, biogas, and biodiesel [31]. Finally, in the cosmetics industry, DS oil and extracts are incorporated into skincare products for their antioxidant and moisturizing properties [21].

3.3 Uses of Date Palm

Dates are significant in agricultural and cultural practices, especially in arid and semi-arid regions. Renowned for its resilience and versatility, this tree offers diverse applications. Its edible components include DPF, DS, and sap—a sweet liquid extracted from the trunk of the DP—which provides essential nutrients and is incorporated into a wide array of food products, such as natural sweeteners, energy bars, and beverages. The nonedible components, such as leaves, trunk, wood, and fibers, are used in traditional handicrafts, construction, biofuel production, and biodegradable industrial materials [77]. This diverse utilization underscores the DP's value as a sustainable resource. An overview of the DP's various parts and their specific food and nonfood industry applications is presented in Table 2.2.

3.3.1 Food Applications

The DP offers a variety of edible components, with the fruit being the most prominent. Known simply as dates, fruits are globally recognized for their nutritional richness and culinary versatility. Rich in natural sugars, dietary fiber, potassium, magnesium, and antioxidants, dates are a staple food in numerous cultures [4]. They are consumed fresh and dried, with the latter offering extended shelf life and concentrated nutrients. Dates are also processed into diverse products, including syrups (dibs), pastes, sugars, natural sweeteners in baking, and functional foods. Fermented beverages like date wine or vinegar further expand their food applications [67]. In addition to the fruit, other parts of the DP found significant uses, particularly DSs, which are a by-product of fruit processing. Roasted and ground seeds serve as a caffeine-free coffee alternative while the extracted oil, rich in oleic acid, is utilized in cosmetics and biodiesel production [11, 51]. Milled DSs are also incorporated into baked goods as a dietary fiber supplement [12]. Young leaves and terminal buds, although less commonly consumed, are edible and used as tender vegetables in salads or cooked dishes. However, their use is limited to sustainability concerns since the removal of the terminal bud invariably results in the death of the tree [88].

Table. 2.2 Food and nonfood applications of date palm

Category	Specific Use	Highlights and Details	Source
Food products	Fresh consumption	A staple in MENA diets and fasting rituals; Consumed at the *Tamr* stage; rich in natural sugars, dietary fiber, and essential nutrients	Afolake et al. [5]; Ibrahim et al. [50]
	Dried dates	Longevity and ease of transport; Shelf-stable product used in snacks and desserts; provides instant energy	
	Date syrup (*Dibis*)	Multifunctional sweetener for diverse industries; sweetener extracted thermally or enzymatically; used in baking and marinades.	Almuzainiet al. [20]
	Date paste	Growing demand for healthy snack ingredients; Used in energy bars, pastries, and traditional confections	Tassoult et al. [83]
	Date sugar	Healthier sugar alternatives; natural sweetener made by drying and grinding dates; retains dietary fiber	Afolake et al. [5]; Ibrahim et al. [50]
	Fermented beverages	Production of Kombucha with enhanced probiotic properties; dates are fermented to produce wine, beer, and liquor	Khosravi et al. [56]; Oluwole et al. [67]
	DS coffee substitute	Utilization of by-products; caffeine-free option; Roasted and ground DSs as a caffeine-free coffee	Jamil et al. [51]; Al-Askari et al. [11]
Nonfood products	Biofuels	Renewable energy potential from agricultural waste conversion of DS oils and DPC into bioethanol and biodiesel	Taghizadeh-Alisaraei et al. [81]; Lattieff [58]; Bahkali et al. [27]
	Activated carbon	Environmentally friendly water purification option; produced from DS shells; used for water filtration	Benabderrahim et al. [30]; Din et al. [37]
	Cosmetics; skincare	Antioxidant-rich for skin care and hair care; DS oil is used in lotions and creams	Aydeniz-Güneşer [26]
	Construction materials	Sustainable materials for construction and crafts; date palm wood and leaves are used in building and crafting	El-Mously and Darwish [40]; El-Mously et al. [41]; Dhakal et al. [35]
	Animal feed	Nutrient-rich feed alternatives; DS and DPBP are used as livestock feed	Boufennara et al. [32]
	Handicrafts	Cultural significance and economic value; Production mats and baskets of date palm leaves and fronds	El-Mously and Darwish [40]; El-Mously et al. [41]

MENA Middle East and North Africa, *DS* date seeds, *DPBP* Date Palm by-product, *DPC* date press cake

3.3.2 Nonfood Applications

The nonedible components of the DP offer substantial utility across diverse industries, from construction to environmental sustainability. The fiber-rich leaves and leaflets have a long history in traditional handicrafts and contemporary industrial applications. Traditionally, they are woven into mats, baskets, fans, and other artisanal items, particularly in date palm cultivation regions. Processed leaves and leaflets are also used as roofing and insulation in traditional architecture, especially in resource-limited desert environments [53]. Extracted fibers are utilized in ropes, bags, and biocomposites, finding applications in industries like automotive, where the demand for sustainable alternatives to synthetic materials is growing. The durable trunk and wood of the date palm have diverse uses. In traditional construction, the trunk is used as a building material due to its strength and rot resistance, particularly in arid regions [26, 74]. It is also used as biofuel in areas with limited access to other energy sources, supporting cooking and heating needs [31]. Date palm fibers, derived from various parts such as leaf sheaths, have vital industrial applications. Composted palm fibers improve soil quality and enhance agricultural productivity. They are also incorporated into biodegradable plastics and composites, providing an environmentally sound alternative to petroleum-based products [48]. Date palm fibers are increasingly used in biofuels, bioplastics, and biochar production, promoting environmental sustainability [62]. Leaves and trunks are also utilized in wastewater treatment and soil improvement, further supporting sustainable industrial practices [53]. Beyond its edible applications, DP sap contributes to sustainable agriculture as a biofertilizer, enhancing soil fertility and supporting crop growth in organic farming systems. DSs, constituting 10–15% of the fruit's weight, are rich in fiber, antioxidants, and oils. Crushed seeds are processed into bioethanol and biodiesel, offering renewable energy sources [31].

4 Processing for Food and Beverages, and Residue and By-Product Generation

4.1 Date Palm Fruit (DPF) Processing

DPF is processed in several forms, such as fresh, dried dates, date syrup, date paste, and date powder. The key processing stages include:

- **Harvesting and sorting:** Dates are harvested at various stages of ripeness, typically categorized as Khalal (unripe), Rutab (semi-ripe), and Tamar (ripe). Sorting is essential for quality control, as the dates are grouped by size, color, and ripeness for processing or marketing [76].

- **Cleaning and washing:** Dates are cleaned to remove contaminants such as dirt, debris, and insect residues. This is done using water, air blowers, or steam. This step is vital for both the safety and quality of the product [76].
- **Pitting and packaging:** Pitted dates, which involve mechanical removal of the seeds, are packaged as whole fresh consumption or further processed into products for commercial sale like date paste and syrup. Some varieties are dried before packaging for extended shelf life [76].
- **Value-added products:** While traditionally consumed in their fresh or dried form, dates have become a key ingredient in the food industry, leading to the development of a diverse range of processed products including sweeteners such as syrup and sugar, chocolates, candies, energy bars, and nutrient-rich products, condiments and pickled dates. Low-quality or unsellable dates, which may not be suitable for direct consumption, are increasingly processed into value-added products such as syrups, pastes, and powders. By utilizing these processed products, manufacturers can offer natural sweeteners and functional ingredients with a range of health-promoting properties, making them suitable for traditional and plant-based food formulations [87]. Date by-products have been incorporated into a wide range of food processing and beverage production, such as soft drinks and baked goods. Techniques such as boiling, blending, grinding, and drying are commonly employed to extend the shelf life of these by-products while enhancing their versatility for various culinary and industrial uses. This process not only mitigates waste but also provides nutritional benefits, allowing for broader applications in the food and beverage industry. The fermentation of dates broadens their application, leading to the creation of alcoholic beverages, vinegar through oxidation, and distilled alcohol. Additionally, DSs are utilized for oil extraction and as an alternative to coffee, highlighting their role in waste valorization and promoting sustainable innovation.
- **Residue and by-product generation:** DS, damaged fruits, or low-quality dates, make up most of solid waste. DS typically represent 10–15% of the DPF weight and are repurposed for oil extraction, animal feed, or as raw nutraceutical material [46]. Liquid waste from washing or juice extraction contains sugars, pulp, and other soluble organic compounds, which require proper treatment to prevent environmental contamination [63]. The DPR can be valorized into compost, biofuels, or used to produce functional ingredients [2].

4.2 Food and Beverage Applications

DPF and its by-products have seen growing application in the food and beverage industries, owing to their health benefits and versatility as natural sweeteners. Dates are rich in essential nutrients, including fiber, potassium, and antioxidants, making them an ideal ingredient in various food products. The shift towards plant-based and clean-label products drives the adoption of dates and date-derived products as alternatives to refined sugars. Date products, especially date syrup, are finding innovative applications in beverages offering a natural alternative to refined sugars. One notable example is their use in Kombucha-like beverages, where date syrup is used as a fermentation substrate [34]. Kombucha is a popular fermented tea drink and nonalcoholic beverage obtained by the fermentation of sweetened black tea with a symbiotic culture of bacteria and yeast to convert sugars into beneficial compounds during fermentation. Date syrup was used by Khosravi et al. [56] as a medium for the preparation of Kombucha beverages. The results proved that the Kombucha starter culture is appropriate for developing a novel fermented date syrup beverage that may offer additional health benefits, and dominate the symbiotic culture of bacteria and yeast such as enhanced probiotic properties. Additionally, date juice creates alcoholic drinks such as date wine, which can be further processed into vinegar through acetic acid fermentation [67]. This illustrates the versatile use of dates in beverage production, highlighting their multifaceted potential.

4.2.1 Functional Foods and Nutraceuticals

Dates are rich in bioactive compounds such as antioxidants, dietary fibers, and essential minerals, making them ideal for functional foods and nutraceuticals. These compounds have been shown to provide various health benefits, such as anti-inflammatory, anti-oxidative, and digestive benefits [77]. As consumer preference shifts toward natural and nutrient-dense food alternatives, dates are emerging as a valuable component in health-focused product formulations.

Research and innovation in extraction, fermentation, and fortification techniques are improving the functional and sensory attributes of dates, making them increasingly relevant in the health-focused food and beverage sectors. Consequently, date products are being increasingly incorporated into functional food formulations, such as energy bars, yogurt, and baked goods, where they contribute to both nutrition and taste. For example, date paste can be used to increase the fiber content in baked goods, while date syrup serves as a natural sweetener in dairy products like yogurt. Additionally, date pomace, a by-product of date processing, has significant potential for the production of functional foods and beverages [21].

4.2.2 Sweetening and Flavoring

Dates are increasingly utilized as natural sweeteners and flavor enhancers, especially in clean-label and plant-based products. Date paste and syrup are popular for replacing refined sugars in some products like nutrition bars, ice creams, and beverages. This aligns with the rising consumer preference for plant-based, sugar-free alternatives, driven by growing concerns over the health risks of excessive sugar intake. The rich, caramel-like sweetness of date syrup adds a unique flavor profile, enhancing the overall consumer experience [87].

4.3 Residue and By-Product Generation

The processing of date palm products generates considerable amounts of solid and liquid waste (Fig. 2.3). Proper management of these residues is essential for reducing the environmental footprint and increasing the economic potential of these by-products [55].

4.3.1 Solid Waste

- **Pit:** DS or Pit—accounts for approximately 10–15% of the fruit's weight—is a major by-product of date fruit processing. They can be repurposed for various uses, including oil extraction, flour production, or as raw material for activated carbon production.
- **Date Press Cake (DPC):** The solid residue left after extracting juice or syrup from dates, with high moisture content and crumbly texture consisting of fibrous pulp and seeds.
- **Date Press Bars (DPB):** Dense and solid bar-shaped residue, more compact with Lower moisture content, is produced after further processing of DPC by drying and compressing the residue.
- **Rejected Date Fruits (RDF):** Low-quality or damaged dates unsuitable for consumption.
- **Date Palm Leaves and Fronds (DPLF):** They are often discarded after pruning or harvesting.

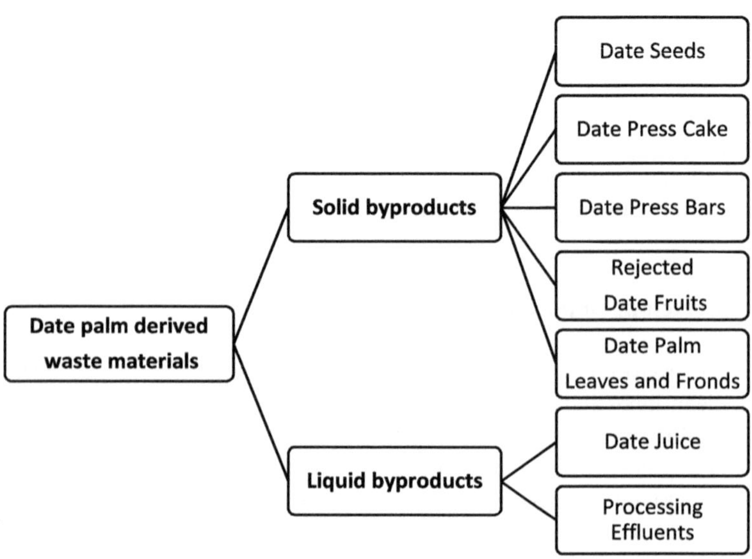

Fig. 2.3 By-products and residual materials generated from date palm fruit

4.3.2 Liquid Waste

- **Date Juice (from Pressing):** The liquid is extracted during the pressing of dates for syrup or juice production.
- **Processing Effluents:** Liquid waste or wastewater generated from cleaning dates or sap processing/syrup production may contain high concentrations of sugars and other organic compounds that require treatment to avoid environmental contamination.

4.4 Valorization of Residues and By-Products

Innovative strategies for utilizing DPRs, which contribute to waste reduction and the enhancement of value derived from these by-products, are outlined in Table 2.3 and Fig. 2.4).

4.4.1 Solid Waste

Date seed (DS) is increasingly recognized for its nutritional and industrial value and can be utilized in multiple ways. The key steps in processing DS include:

Drying and Grinding DSs are dried and ground into a fine powder or flour after harvesting. Ground DS can be processed into gluten-free flour, suitable for baking, or used as a coffee substitute [52]. Carbonized and activated DS are eco-friendly adsorbents in water purification and air filtration, presenting a sustainable alternative to synthetic materials [77].

Oil Extraction DS oil, extracted using cold pressing or solvent extraction methods, is rich in oleic acid and antioxidants. It is valuable in the culinary, cosmetic, and pharmaceutical industries [65]. Oil extraction waste, known as seed cake, can be used as animal feed or fertilizer, contributing to the circular economy [25]. Solvents used in DS oil extraction must be carefully managed to avoid environmental contamination. Proper disposal or recycling of these solvents is crucial [65]. Date press cake is a nutritious feed supplement for livestock due to its high fiber and energy content [28]. Additionally, it can be composted to produce organic fertilizer, improving soil health and fertility. Beyond agriculture, the fibrous material from press cake is also utilized as a feedstock for biogas production or processed into solid biofuels, contributing to renewable energy generation [79]. DPB can be converted into solid biofuels, such as briquettes or pellets, for sustainable energy production [77]. Due to their dense composition, they are also used in biocomposite materials for packaging, construction, and industrial applications, providing an environmentally friendly alternative to conventional synthetic materials [35, 80]. RDF, which does not meet quality standards for human consumption due to over-

Table. 2.3 Valorization of date palm by-products (DPBP) processing

	By-product	Valorization Method	Application	Reference
Solid by-products	**DS**	Oil extraction	Culinary, cosmetics, pharmaceuticals	Niroula et al. [65]
		Grinding into gluten-free flour	Baking industry	Jaouhari et al. [52]
		Roasting as a coffee substitute	Beverage industry	Jaouhari et al. [52]
		Carbonization and activation	Water purification, air filtration	Kavvadias et al. [55]
	DPC	Livestock feed	Animal nutrition	Banadka et al. [28]
		Composting for organic fertilizer	Soil enrichment	Abid and Ammar [2]
		Biofuel production (biogas, solid fuel)	Renewable energy	Bahkali et al. [27]
	DPB	Compaction into briquettes or pellets	Solid biofuel for energy generation	Shi et al. [77]
		Processing into biocomposites	Packaging, construction materials	Dhakal et al. [35]
	RDF	Livestock feed	Alternative animal feed	Banadka et al. [28]
		Composting for organic fertilizer	Soil improvement	Abid and Ammar [2]
		Fermentation into bioethanol	Renewable biofuel	Shi et al. [77]
	DPLF	Weaving into traditional products	Baskets, mats, roofing materials	Tassoult et al. [83]
		Livestock fodder	Animal feed, particularly in feed-scarce regions	Banadka et al. [28]
		Processing into biodegradable materials	Sustainable packaging, construction	Rangaraj et al. [71]
Liquid by-products	**Date juice**	Use as a natural sweetener	Beverages, syrups, desserts	Jaouhari et al. [52]
		Fermentation into alcoholic beverages or vinegar	Beverage industry	Alhomsi and Bayraktar [16]
		Integration into functional foods and dietary supplements	Health and nutrition	Jaouhari et al. [52]
	Wastewater	Treatment and reuse for irrigation	Sustainable agriculture	Subhash et al. [80]
		Extraction of organic matter for liquid fertilizer	Soil and crop improvement	Shi et al. [77]
		Anaerobic digestion for biogas production	Renewable energy generation	Abid and Ammar [2]

DS date seeds, *DPC* date press cake, *DPBP* date palm by-product, *RDF* rejected date fruits, *DPLF* date palm leaves and fronds

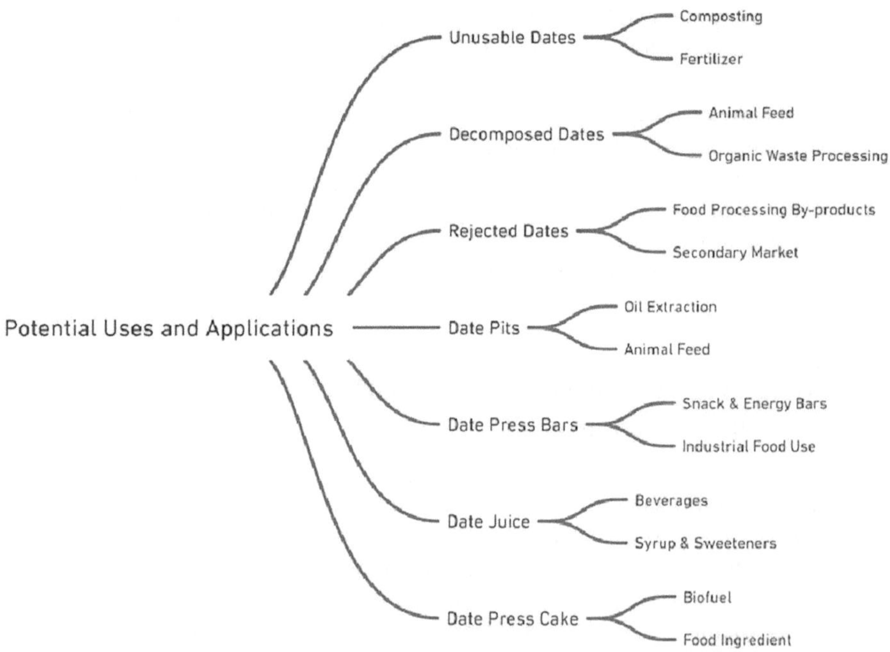

Fig. 2.4 Valorization of by-products materials generated from date palm fruit processing

ripeness, damage, or defects, can be repurposed in various ways. They serve as a nutrient-rich feed for livestock, offering a cost-effective alternative to traditional animal feed [28]. Additionally, these fruits can be composted to produce organic fertilizers, enhancing soil fertility [2]. Moreover, their high sugar content makes them ideal for fermentation into bioethanol, a renewable biofuel that contributes to sustainable energy solutions [77]. Traditionally, the DPLF of the date palm is woven into baskets, mats, and roofing materials [83]. Dried leaves serve as fodder for livestock, particularly in regions facing feed shortages [28]. Moreover, the fibrous material can be processed into biodegradable packaging and construction materials, offering an environmentally responsible alternative to plastic-based products [35].

4.4.2 Liquid Waste

Date juice, rich in natural sugars, vitamins, and minerals, is widely utilized as a natural sweetener in beverages, syrups, and desserts [52]. Additionally, date juice can be fermented to produce alcoholic beverages or vinegar, further enhancing its commercial value [16]. Due to its high antioxidant and nutrient content, it is also incorporated into functional foods and dietary supplements, offering health benefits while reducing waste [52]. The wastewater by-product can be effectively valorized in several ways. Treated wastewater can be reused for irrigation or industrial applications, reducing overall water consumption [80]. Additionally, the organic matter present in wastewater can be extracted and processed into liquid fertilizers, promoting sustainable agricultural practices [77]. Furthermore, the anaerobic digestion of wastewater generates biogas, a renewable energy source that can be used for power generation or heating [2]. Date sap is collected by tapping the trunk of the DP, typically early in the morning to prevent premature fermentation. The sap is consumed fresh or fermented into alcoholic beverages. For its production, the sap is filtered to remove any solid impurities and then pasteurized to extend its shelf life or boiled to concentrate it into a syrup or sugar that can be packaged and sold. The fibrous materials left over from the filtration can be repurposed as compost or animal feed [74].

5 Bio-valorization of Waste and By-Products in Circular Economy

The concept of bio-valorization within a circular economy framework has garnered significant attention as a sustainable strategy for transforming agro-industrial by-products into high-value products [26]. The circular economy concept emphasizes reducing waste, optimizing resource use, and extending the lifecycle of materials through sustainable practices. In the date palm industry, bio-valorization effectively transforms agricultural residues and by-products into valuable products, contributing to both environmental sustainability and economic growth. Table 2.4 demonstrates the potential of DPRs for the production of bioproducts, biomaterials, animal feed, biofuels, biochar, and bio-oil.

5.1 Bioproducts

Processing dates generate substantial quantities of waste—including DS, press cake, and fibrous residues—which are rich in crude fiber, pectin, polyphenols, and bioactive compounds which make it an ideal substrate for microbial fermentation. When properly valorized, these by-products can be converted into high-value products that reduce the environmental burden of waste disposal and promote a circular economy. The extraction of bioactive compounds, such as polyphenols and flavonoids, from DPR has been extensively studied, revealing their strong antioxidant and antimicrobial properties [86, 89]. In addition, the high concentrations of enzymes, organic acids, and microbial polysaccharides present in these by-products have spurred research into their extraction for use in food, cosmetic, and

Table 2.4 Valorization of waste and by-products in a circular economy

Category	Sub-Category	Product/Application	References
Bioproducts	Organic fertilizer	Compost for soil fertility and plant growth	Abid et al. [3]; Benabderrahim et al. [30]
	Soil amendment	Biochar as a soil amendment to improve soil properties	Benabderrahim et al. [30]; Rahmat [69]; Kavvadias et al. [55]
	DS oil	DS oil for cooking, pharmaceuticals, cosmetics, and green fuel production	Habib et al. [46]; Nehdi et al. [64]
Biomolecules	Enzymes	Nanocatalyst derived from DS waste for hydrolytic enzyme production and biohydrogen synthesis.	Bahkali et al. [27]
	Microbial polysaccharides	Polysaccharides like xanthan are produced from date syrup.	Khosravi-Darani et al. [57]
	Essential fatty acids	Arachidonic acid and eicosapentaenoic acid produced from *Mortierella alpina* using date waste.	Asadi et al. [22]
	Low-cost culture media	Use DPBP to cultivate probiotic bacterial *lactobacillus* spp.	Ayad et al. [24]; Di Salvo et al. [36]
Biomaterials	Biosurfactants	Biosurfactants produced from date molasses by *Bacillus subtilis*, and *lactobacillus rhamnosus*	Ghasemi et al. [43]
	Biocomposites	Biocomposites made from DP fibers for sustainable packaging and construction; production of active gelatin films with DPW	Rangaraj et al. [71]
	Bioplastics	Synthesized bioplastics of DP fibers and single-use plastic of DPW biocomposites	Hamza et al. [48]; Mousa et al. [62]; Abed et al. [1]
	Cellulose nanocrystals (CNC)	CNC from DPW and CNC from DS	Raza et al. [73] Raza and Abu-Jdayil [72];
Animal feed	Feed resources	Animal feed made from DPW and DPBP for ruminants; ensiled DPLF for Omani sheep; DS meal as an alternative energy and protein source in poultry	Ghasemi et al. [43]; Boufennara et al. [32]; Mahgoub et al. [59]; Attia et al. [23]

(continued)

Table 2.4 (continued)

Category	Sub-Category	Product/Application	References
Biofuels	Bioethanol	Bioethanol is produced from DPW fermentation.	Taghizadeh-Alisaraei et al. [81]
	Biogas	Biogas is produced from the anaerobic digestion of DPW; Nanocatalyst derived from DS waste for biohydrogen synthesis.	Lattieff [58] Bahkali et al. [27]
	Bio-jet fuel / green diesel	Conversion of DS oil into sustainable aviation fuel and green diesel, contributing to cleaner fuel alternatives.	Bharath et al. [31] Al-Mardeai et al. [19]
	Hydrogen-rich syngas	Syngas (hydrogen-rich) are produced from DSs via pyrolysis.	Rambabu et al. [70]
Biochar		Biochar production from pyrolysis of DPW	Giwa et al. [44]; Usman et al. [85]
Bio-oil		Bio-oil produced from pyrolysis of DPW	Ahmad et al. [6, 7]

CNC cellulose nanocrystals, *DS* date seed, *DP* date palm, *DPBP* date palm by-products, *DPW* date palm waste, *DPLF* date palm leaves and fronds

pharmaceutical applications. These compounds can be incorporated into food preservatives, pharmaceuticals, and cosmetic formulations, thus enhancing the economic viability of DPW processing [38]. For example, Date extract has been explored as a substrate for xanthan gum production using *Xanthomonas campestris* PTCC1473 to produce xanthan gum by optimization of the submerged fermentation process and balancing carbon and nitrogen sources to maximize yield [57]. Date-processing residues have been utilized to produce organic fertilizers and soil amendments, which contribute to enhancing soil structure, improving water retention, and increasing nutrient availability. The conversion of date palm residues into biochar holds considerable promise for sustainable agriculture, offering a compelling solution for both waste management and soil enhancement. Burezq and Davidson [33], in their comprehensive review, illuminate the potential of biochar derived from these residues to not only boost soil fertility, water retention, and microbial activity but also contribute significantly to carbon sequestration. They astutely point out that realizing these potential hinges on technological advancements and supportive policies, which are necessary to overcome current barriers and standardize production protocols in the face of variable biomass quality. Rahmat [69], taking a more focused experimental approach, delves into the chemical intricacies of biochar produced from date seeds under low-temperature pyrolysis. Her work reveals that pyrolysis temperature exerts a significant influence on the nutrient profile of the resulting biochar, particularly affecting the concentrations of potassium and phosphorus. This study underscores the critical role of carefully controlled pyrolysis conditions in tailoring the chemical properties of biochar to meet specific soil amendment needs, especially in nutrient-poor soils. Moving beyond individual applications, Kavvadias et al. [55] explore the synergistic effects of combining compost and biochar derived from DPR. Their research demonstrates that such an integrated approach can dramatically improve soil organic matter, water-holding capacity, and nutrient availability, particularly in the challenging conditions characteristic of arid and semi-arid regions. Their findings showcase the potential of this combined strategy to mitigate soil erosion and enhance moisture retention, offering a vital tool in combating desertification. Collectively, these studies paint a vibrant picture of date palm biochar's potential for sustainable agriculture. However, to fully unlock this potential, future research must prioritize the investigation of the long-term impacts of biochar application across diverse soil types and cropping systems. This research should be coupled with ongoing efforts to refine pyrolysis protocols and, crucially, to foster greater awareness and engagement among agricultural communities, thereby paving the way for the wider adoption of this promising technology.

5.2 Biomaterials

Recent advances in the utilization of date palm biomass underscore its multifaceted potential in the development of sustainable materials. The incorporation of date palm fibers into biodegradable polymers has emerged as an effective strategy for producing biocomposites with enhanced mechanical properties and improved biodegradability. For instance, Saifullah et al. [75] investigated dry-blended biocomposites composed of polycaprolactone and date palm fibers for packaging applications, noting significant improvements in tensile modulus and biodegradability, while Dhakal et al. [35] demonstrated that surface-treated date palm fibers markedly enhance the mechanical properties of polymer composites, thereby broadening their applicability in the automotive and construction sectors. Concurrently, polysaccharides extracted from date residues and DPC are increasingly employed to manufacture biodegradable plastics that offer sustainable alternatives to conventional, petroleum-based polymers in food packaging [38]. Moreover, DPLF rich in cellulose, hemicellulose, and lignin, represents ideal feedstocks for the production of biocomposites, bioplastics, and bioenergy [17]. Collectively, these studies highlight the promise of date palm by-products (DPBP) in advancing eco-friendly materials and technologies.

5.3 Animal Feed

As the demand for alternative and sustainable feed sources rises, particularly in areas facing nutritional shortages during dry seasons, DPBP has emerged as a viable alternative. By-products including DS, DPLF, DPB, and DPB, are emerging as potential feed sources in both ruminant [32] and poultry nutrition [23], particularly in arid regions where traditional feed resources are limited. Studies on their chemical composition show that these by-products are rich in carbohydrates, fats, and essential minerals but are also high in fiber, which can reduce their digestibility. In sheep, diets including ensiled urea-treated DPLF and by-product concentrates were found to offer lower body weight gains and feed conversion efficiency compared to those fed more conventional feeds like Rhodes grass hay and commercial concentrates. However, these by-products remain a valuable option for maintaining livestock during nutritional shortages, especially in desert environments [38], date stone meal has demonstrated favorable effects when included in diets at levels of 5–10%, replacing traditional grains like corn and barley. It has shown improvements in growth performance and feed efficiency, despite its high fiber content [23]. Supplementation with enzymes and oil is often recommended to improve digestibility, date palm by-products such as waste dates have been found to possess higher metabolizable energy and organic matter digestibility than other fibrous feeds like DSs or palm leaves, making them a viable option for ruminants. However, due to their high fiber content, protein supplementation may be required when using these by-products in ruminant diets.

5.4 Biofuels and Bioethanol Production

Another promising avenue for valorizing date processing waste is its conversion into biofuels. Biofuels are a critical resource that can be refined into liquid fuels for applications in both the heating and transportation sectors. The organic-rich nature of date by-products renders them suitable substrates for various bioenergy conversion processes. Several technological approaches—including anaerobic digestion, fermentation, and pyrolysis—have been systematically investigated. In anaerobic digestion, microbial consortia break down organic matter in the absence of oxygen to produce biogas—a mixture of methane and carbon dioxide. Biogas generated from DPC or other residues can be used for electricity generation or as a direct energy source. Fermentation techniques can convert the sugars present in DPFs and wastes into bioethanol. Moreover, the efficiency of fermentation can be enhanced through enzymatic pretreatment, which breaks down complex carbohydrates. Pyrolysis, a thermal decomposition process conducted in an inert atmosphere, converts date waste into bio-oil, char, and syngas. This process is particularly attractive for its ability to generate a range of energy-dense products that can be used as fuel or further processed into chemicals. Furthermore, thermochemical conversion methods, such as gasification and pyrolysis, have been evaluated for processing date palm midribs and seeds, with results indicating promising heating values [38]. Economic feasibility studies indicate that the viability of biofuel production from date waste depends on factors such as feedstock availability, process efficiency, and integration with existing energy infrastructures [35]. Continued advancements in catalyst development, process optimization, and the integration of renewable energy systems are expected to improve the overall efficiency of these technologies. The production of biofuels, such as biogas, bioethanol, and biodiesel, from date residues represents a sustainable energy solution. Biogas is produced through the anaerobic digestion of date palm cake and liquid effluents, serving as a renewable source of cooking fuel and electricity. Bioethanol is generated by fermenting sugars extracted from date pulp, which can be blended with gasoline to reduce carbon emissions. Additionally, oils extracted from DSs are trans esterified into biodiesel, offering a cleaner alternative to conventional diesel fuel. Moreover, bioethanol created by fermenting sugars extracted from DS is recognized as a sustainable alternative to fossil fuels [45]. Overall, DPW is considered a potential source of bioethanol, biogas, and biodiesel.

5.5 Biochar and Bio-Oil

Biochar is a carbon-rich organic material produced through the pyrolysis of biomass in an oxygen-limited environment, widely recognized for its potential to enhance soil fertility, improve moisture retention, and sequester carbon. Date palm commonly cultivated in arid and semi-arid regions, generates substantial agricultural residues that are often discarded or burned, leading to environmental pollution. Converting these residues into biochar presents a sustainable alternative that aligns with global efforts to mitigate climate change and promote soil health [55]. Despite its potential, knowledge about biochar derived from date palm biomass remains limited, necessitating further research and farmer education to optimize its utilization. A concentrated solar energy system for the pyrolysis of DPW was evaluated for its economic and environmental viability compared to conventional electric heating-based pyrolysis. The results indicate that the solar-driven process is more economically feasible, with a shorter payback period and higher financial returns. Additionally, it significantly reduces CO_2 emissions, producing only 38% of the emissions generated by conventional methods. Sensitivity analysis highlights that profitability is more influenced by biochar selling price than feedstock costs. Overall, this approach presents a sustainable alternative for biochar production, lowering both environmental impact and operational expenses [33]. Bio-oil, as defined in the paper, is a liquid product obtained through the pyrolysis of biomass, particularly from the residues of the palm oil industry, such as palm shells, empty fruit bunches, and mesocarp fiber. In this process, the biomass is subjected to high temperatures in an inert atmosphere (e.g., nitrogen), resulting in the thermal decomposition of the material and the formation of bio-oil, bio-char, and non-condensable gases. The bio-oil produced in this study is characterized by its acidic nature and high oxygen content, containing a variety of organic compounds, including acids, aldehydes, ketones, and alcohols, which are the by-products of the breakdown of the biomass. The heating value of the bio-oil, ranging from 10.49 MJ/kg to 14.78 MJ/kg, is relatively low when compared to conventional fuels, indicating that while bio-oil holds promise as a renewable energy source, it may require further processing or upgrading to enhance its energy content.

Despite these challenges, bio-oil presents a potential avenue for the production of renewable fuels, chemicals, and absorbents, although its practical applications may necessitate additional refinement to improve its properties and increase its overall energy efficiency. The system for the pyrolysis of DPW was evaluated for its economic and environmental viability compared to conventional electric heating-based pyrolysis. The results indicate that the solar-driven process is more economically feasible, with a shorter payback period and higher financial returns. Additionally, it significantly reduces CO_2 emissions, producing only 38% of the emissions generated by conventional methods. Sensitivity analysis highlights that biochar selling price influences profitability more than feedstock costs. Overall, this approach presents a sustainable alternative for biochar production, lowering both environmental impact and operational expenses [33].

6 Techno-economic Analyses and Industrial Prospects

Recent techno-economic evaluations underscore the promising potential of valorizing date palm by-products through advanced bioprocessing techniques. Process optimization studies have demonstrated that fermenting date waste with *Saccharomyces cerevisiae* can yield approximately 360 mL of bioethanol per kg of substrate, reducing production costs and supporting renewable energy supply [6]. Furthermore, life cycle assessments comparing conventional and microwave pyrolysis methods have shown that the traditional process generally offers superior financial performance and reduced environmental impacts [2]. Additional studies confirm that integrated bioprocessing—utilizing various residues such as pits fibers, press cake, spikelets, and seeds—can efficiently convert these by-products into high-value commodities like bioethanol, biodiesel, and activated carbon, thereby advancing the principles of a circular bioeconomy [15]. Moreover, recent findings on the utilization of DS oil for biodiesel production highlight the economic viability of complete seed valorization [19]. Collectively, these results provide a robust foundation for scaling up industrial applications that transform DPR into sustainable and economically attractive products.

7 Real-Time Applications

Recent advancements demonstrate that date palm by-products are increasingly translated into market-ready technologies through innovative real-time applications and patenting activities. In water remediation, activated carbon produced from DSs and spikelets has been effectively applied to remove phenolic contaminants and pollutants like methyl orange, with unique developments such as the ZnO@DS AC composite exhibiting enhanced adsorption capacities [15]. Moreover, a green synthesis approach for activated carbon from DS biomass has shown promise not only for environmental remediation but also for potential applications in energy storage [25]. In the energy domain, the utilization of DS oil as a feedstock for biodiesel production is substantiated by its favorable fatty acid profile and oxidative stability [19]. Furthermore, the integration of date palm fibers into polymer matrices has resulted in biocomposites suitable for automotive, packaging, and construction applications [52]. Traditionally, low-quality dates that fail to meet commercial standards are discarded or used as animal feed; however, recent research has demonstrated that these wastes can be repurposed into valuable bioproducts, thereby mitigating environmental impacts while generating economic benefits. For example, Khosravi-Darani et al. [57] optimized a submerged fermentation process using *Xanthomonas campestris* PTCC1473 to produce xanthan gum from date extract, highlighting the importance of balancing carbon and nitrogen sources to maximize yield. In a parallel approach, Asadi et al. [22] employed solid-state fermentation with *Mortierella alpina* CBS 528.72 to simultaneously produce arachidonic acid (ARA) and eicosapentaenoic acid (EPA) from date waste. Recent reviews further indicate that microbial fermentations using date waste can generate a wide range of value-added products, including enzymes, organic acids, and microbial polysaccharides, thereby converting underutilized by-products into economically viable bioproducts while reducing environmental waste [61]. Several studies have extensively examined the antioxidant properties and phenolic compositions of various date palm (*Phoenix dactylifera* L.) cultivars. For instance, Zihad et al. [89] evaluated the Ajwah, Safawy, and Sukkari varieties from Saudi Arabia, employing assays such as 2,2-diphenyl-1-picrylhydrazyl (DPPH) and hydroxyl radical scavenging, alongside total antioxidant capacity assessments. Their analysis, utilizing ultra-performance liquid chromatography coupled with quadrupole time-of-flight mass spectrometry identified 22 bioactive compounds, including phenolics, flavonoids, sterols, and phytoestrogens. The extracts exhibited significant antioxidant activity, with half-maximal inhibitory concentration (IC_{50}) values ranging from 103 to 177 μg/mL for DPPH radical scavenging, 1.1 to 1.55 mg/mL for hydroxyl radicals, and 87 to 192 μg/mL for total antioxidant capacity. Similarly, Alam et al. [10] conducted a comparative analysis of 24 date varieties from the United Arab Emirates, Pakistan, and Tunisia. They quantified total phenolic content using the Folin–Ciocalteu method and assessed antioxidant activity through $ABTS^+$, ferric reducing antioxidant power, and DPPH assays. The total phenolic content varied widely, ranging from 46 to 397 mg gallic acid equivalents (GAE) per 100 g fresh weight. Antioxidant activity measurements also showed considerable variation, with $ABTS^+$ assay results between 0.9 and 4.3 μmol Trolox equivalents (TE) per 100 g, FRAP (Ferric Reducing Antioxidant Power) values from 355 to 2421 μmol TE per 100 g, and DPPH IC_{50} values spanning 0.0 to 1.8 mg/mL. Another study focusing on Iranian date varieties including Kabkab, Rabbi, Zahedi, and Mazafat, by Radfar et al. [68]. They assessed the phenolic content of DS extracts using the Folin–Ciocalteu method and identified compounds via high-performance liquid chromatography. The findings revealed exceptionally high phenolic concentrations, ranging from 1483 to 3377 mg GAE per 100 g dry weight, indicating that date pits are a potent source of bioactive phenolic compounds. Boufennara et al. [32] conducted an in vitro evaluation of the nutritive value of various date palm by-products for ruminant feed, specifically focusing on waste dates, palm leaves, pedicels, and date pits. The study revealed that date palm by-products, particularly waste dates, are highly digestible and provide considerable energy content, making them valuable as a supplement to low-quality roughages. Despite their low protein content, waste dates demonstrated high in vitro ruminal degradability, organic matter digestibility, and volatile fatty acid production, suggesting that they could be used effectively to boost energy levels in ruminant diets. In contrast, date pits were shown to be less digestible, with low ruminal degradability, indicating that they are less suitable as a primary feed resource. Palm leaves and pedicels, although high in fiber, can serve as emergency roughages in low-producing animals, but they require additional protein supplementation to balance the nutritional needs of the animals. The authors concluded that waste dates have the potential to be a viable alternative feed source in regions facing feed shortages during the dry season, particularly for ruminants. Mahgoub et al. [59] investigated the practical application of date palm by-products as feed for Omani sheep, focusing on ensiled date palm fronds and a by-product concentrate formulated from local agricultural products. The study compared the effects of four different feeding strategies: Rhodes grass hay (RGH) combined with a commercial concentrate, RGH with a by-product concentrate, urea-treated palm fronds (UTPF) with a commercial concentrate, and UTPF with by-product concentrate. The results showed that sheep fed the UTPF-based diets had lower feed intake and weight gain than those fed the RGH-based diets. However, feed intake per unit of body weight remained similar across all diet groups, indicating that although UTPF-based diets reduced overall feed consumption, the efficiency of feed utilization was consistent. The average daily weight gain was highest for the commercial concentrate + RGH group (80 g/d) and lowest for the by-product concentrate + UTPF group (10 g/d). The study also found that feeding UTPF and by-product concentrate resulted in lower carcass weight, fat content, and dressing percentage. Despite these reductions, no adverse effects on the health of the animals were reported, and the authors concluded that date palm by-products, particularly in the form of UTPF could be used as a feed resource during periods of nutritional shortage, though they may not be as effective in promoting growth as traditional feeds. Attia et al. [23] provided a comprehensive review of the chemical composition, nutritive value, and potential applications of date palm by-products, specifically date stone meal (DSM), in poultry nutrition. The review highlighted that DSM, while rich in carbohydrates and fat, contains a high level of crude fiber, which limits its use in poultry diets. However, the authors discussed how DSM could be included in poultry diets up to 10% without negatively affecting growth performance. They observed that DSM's inclusion at this level sometimes improved growth performance, possibly due to its bioactive compounds, such as antioxidants and phenolic compounds, which may enhance poultry health. Additionally, the review emphasized that DSM has a comparable protein content to barley and could serve as an alternative energy source in poultry diets. The authors also pointed out that the fiber content of DSM is a critical factor limiting its use in poultry nutrition and suggested that supplementation with oils or fats could help offset its low metabolizable energy. Despite the challenges posed by its high fiber content, the review concluded that DSM could be a valuable feed ingredient for poultry, especially in regions where traditional feed ingredients, such as corn and barley, are scarce or expensive. The authors further recommended future research to develop more accurate formulas for the metabolizable energy of DSM based on its fiber content.

8 Socioeconomic Impacts, Climate Change Mitigation, and Well-being

The valorization of date palm by-products confers extensive socioeconomic and environmental benefits that are pivotal in regions where date palm cultivation is a major agricultural activity. In North Africa and the Middle East, the conversion of agricultural waste into bio-based products not only creates new revenue streams and fosters rural development by generating employment and stimulating small-scale industries but also establishes new value chains that drive economic growth. Moreover, the production of biofuels such as bioethanol and biodiesel from date waste significantly reduces dependency on fossil fuels, thereby lowering greenhouse gas emissions and contributing to climate change mitigation [6]. In parallel, the fabrication of biochar, activated carbon, and biodegradable packaging materials—particularly those reinforced with date palm fibers—enhances soil quality through carbon sequestration and provides cost-effective solutions for water treatment, pollutant removal, and waste management, thereby bolstering energy independence and promoting environmental remediation and public health [8, 77]. Collectively, these integrated initiatives underscore the role of date palm by-product valorization in advancing a circular bioeconomy that supports sustainable resource management and overall societal well-being.

9 Conclusion and Future Perspectives

Recent advancements in the valorization of date palm by-products, particularly in the development of novel bioprocessing techniques, have unlocked a range of diverse applications that offer significant socioeconomic, environmental, and industrial benefits, especially in regions where date palms are a key agricultural resource. These by-products, which include date pits, fibers, and seeds, have been effectively transformed into valuable products like biofuels, biocomposites, activated carbon, and other biomaterials. The integration of these by-products into bioprocessing frameworks not only addresses critical issues related to waste management but also contributes to the promotion of a circular bioeconomy, advancing environmental sustainability and economic development. Technological innovations are increasingly driving the adoption of DPW for applications in multiple sectors. Biofuels derived from DPR present a promising pathway towards sustainable energy independence, particularly in regions heavily reliant on fossil fuel imports. However, challenges remain in scaling up production to meet demand and ensuring cost-competitiveness with traditional fuels. The development of biodegradable packaging materials and biocomposites from date palm by-products offers a compelling solution to the growing problem of plastic waste. Various biocomposite materials have been successfully created from date palm fibers that exhibit comparable strength and durability to conventional plastics, with potential applications in food and nonfood industries. Furthermore, activated carbon derived from DPR has shown considerable promise in water purification and environmental remediation. Microbial fermentation of date palm by-products has yielded valuable biochemicals such as xanthan gum and arachidonic acid, opening up new avenues for their utilization in the food and pharmaceutical industries. Xanthan gum, for example, is a valuable food additive used as a food additive for thickening or stabilizing different products, while arachidonic acid has applications in functional foods or nutritional supplements. The valorization of date palm by-products offers significant socioeconomic benefits, particularly for rural communities in regions like the Middle East and North Africa, where date palm cultivation is central to the local economy. The establishment of biorefineries and processing facilities can create new employment opportunities, and support the growth of small-scale industries. Furthermore, the increased income generated from these activities can contribute to improved livelihoods and reduced poverty. Looking forward, realizing the full potential of date palm by-product valorization requires concerted research and development efforts. Refining fermentation and pyrolysis technologies are crucial for improving the efficiency and cost-effectiveness of biofuel production. Specifically, future research should focus on this area and optimize enzyme activity or explore novel catalysts. Furthermore, comprehensive life cycle assessments are essential to fully evaluate the environmental and economic sustainability of these bioprocessing technologies, considering factors such as water usage, energy consumption, and greenhouse gas emissions. Exploring new applications in health and nutrition, such as the development of functional food ingredients from date palm by-products, represents a promising area for future research. For instance, date palm extracts have shown potential for antioxidant or anti-inflammatory properties in preliminary studies, suggesting their possible use in nutraceuticals or functional foods. However, further clinical trials are needed to confirm these findings. Supportive policy frameworks and increased investment in research and development are crucial for accelerating the industrial adoption of these sustainable technologies. Governments can play a key role by providing incentives for biorefinery development, funding research programs, or establishing standards for biofuel quality. In conclusion, the continued valorization of DPBP holds substantial promise for advancing a sustainable circular bioeconomy, particularly in arid and semi-arid regions where date palm cultivation is prevalent. By fully harnessing the potential of these currently underutilized resources, we can move towards a future characterized by reduced waste, enhanced economic growth, and a diminished impact of climate change. However, realizing this vision requires sustained research efforts, technological innovation, supportive policies, and a commitment to comprehensive sustainability assessments. The transition from pilot-scale projects to widespread industrial adoption will be key to unlocking the full socioeconomic and environmental benefits of date palm by-product valorization.

References

1. Abed RM, Muthukrishnan T, Nassar MM, Al Battashi H, Alzebdeh KI (2024) Degradation of newly developed date palm agro-residues-filled polyethylene biocomposites in the planktonic and benthic zones of a marine environment. Biomass Convers Biorefinery 14(2):1793–1808. https://doi.org/10.1007/s13399-022-02514-0
2. Abid W, Ammar E (2022) Date palm (*Phoenix dactylifera* L.) wastes valorization: a circular economy approach. In: Ramadan MF, Farag MA (eds) Mediterranean fruits bio-wastes. Springer, Cham. https://doi.org/10.1007/978-3-030-84436-3_17
3. Abid W, Magdich S, Mahmoud IB, Medhioub K, Ammar E (2018) Date palm wastes co-composted product: an efficient substrate for tomato (*Solanum lycopersicum* L.) seedling production. Waste Biomass Valori 9:45–55. https://doi.org/10.1007/s12649-016-9767-y
4. Abu-Reidah IM, Gil-Izquierdo Á, Medina S, Ferreres F (2017) Phenolic composition profiling of different edible parts and by-products of date palm (Phoenix dactylifera L.) by using HPLC-DAD-ESI/MSn. Food Res Int 100:494–500. https://doi.org/10.1016/j.foodres.2016.10.018
5. Afolake SB, Victoria TT, Olaitan AZ (2023) Quality assessment of date fruit (as sweetener) and date seed powder (as an alternative to coffee). Int J Agric Res. https://ijojournals.com/index.php/ar/article/view/729
6. Ahmad A, Naqvi SA, Jaskani MJ, Waseem M, Ali E, Khan IA, Faisal Manzoor M, Siddeeg A, Aadil RM (2021) Efficient utilization of date palm waste for the bioethanol production through *Saccharomyces cerevisiae* strain. Food Sci Nutr 9:2066–2074. https://doi.org/10.1002/fsn3.2175
7. Ahmad W, Makkawi Y, Samara F (2024) Co-pyrolysis of date palm waste and *Salicornia Bigelovii*: insights for bioenergy development in arid and semi-arid regions. ACS Omega 9:24082. https://doi.org/10.1021/acsomega.4c02972
8. Al Malki M, Yaser AZ, Hamzah MAAM, Zaini MAA, Latif NA, Hasmoni SH, Zakaria ZA (2023) Date palm biochar and date palm activated carbon as green adsorbent—synthesis and application. Curr Pollut Rep 9(3):374–390. https://doi.org/10.1007/s40726-023-00275-6
9. Al-Alawi RA, Al-Mashiqri JH, Al-Nadabi JSM, Al-Shihi BI, Baqi Y (2017) Date palm tree (*Phoenix dactylifera* L.): natural products and therapeutic options. Frontiers. Plant Sci 8:845. https://doi.org/10.3389/fpls.2017.00845
10. Alam MZ, Alhebsi MS, Ghnimi S, Kamal-Eldin A (2021) Inability of total antioxidant activity assays to accurately assess the phenolic compounds of date palm fruit (Phoenix dactylifera L.). NFS J 22:32–40. https://doi.org/10.1016/j.nfs.2021.01.001
11. Al-Askari GA, Al-Afour MF, AL-Monsef IM, AAl-Sanabani AS, Sinnan AM (2024) Chemical composition study of three varieties of date seeds (Iraqi, Saudi and Yemeni) and their utilization as caffeine-free coffee alternative. J Chem Nutr Biochem 5(1):1–11. https://doi.org/10.48185/jcnb.v5i1.1079

12. Al-Farsi MA, Lee CY (2008) Nutritional and functional properties of dates: a review. Crit Rev Food Sci Nutr 48(10):877–887. https://doi.org/10.1080/10408390701724264
13. Al-Farsi M, Alasalvar C, Morris A, Baron M, Shahidi F (2005) Comparison of antioxidant activity, anthocyanins, carotenoids, and phenolics of three native fresh and sun-dried date (*Phoenix dactylifera* L.) varieties grown in Oman. J Agric Food Chem 53(19):7592–7599. https://doi.org/10.1021/jf050579y
14. Al-Farsi MA, Alasalvar C, Al-Abid M, Al-Shoaily K et al (2007) Compositional and functional characteristics of dates, syrups, and their by-products. Food Chem 104(3):943–947. https://doi.org/10.1016/j.foodchem.2006.12.051
15. Al-Hazeef MS, Aidi A, Hecini L, Osman AI, Hasan GG, Althamthami M et al (2024) Valorizing date palm spikelets into activated carbon-derived composite for methyl orange adsorption: advancing circular bioeconomy in wastewater treatment—a comprehensive study on its equilibrium, kinetics, thermodynamics, and mechanisms. Environ Sci Pollut Res 31(38):50493–50512. https://doi.org/10.1007/s11356-024-34581-3
16. Alhomsi A, Bayraktar MK (2025) Prebiotic properties of dates and their impact on health. Turkish J Agric-Food Sci Technol 13(1):229–236. https://doi.org/10.24925/turjaf.v13i1.229-236.7148
17. Ali Z, Khan A, Hussain M (2020) Bioconversion of agricultural residues into biofertilizers: current status and future prospects. Bioresour Technol 302:122836. https://doi.org/10.1016/j.biortech.2020.122836
18. Alkhalidy H, Al-Nabulsi AA, Al-Taher M, Osaili T, Olaimat AN, Liu D (2023) Date (*Phoenix dactylifera* L.) seed oil is an agro-industrial waste with biopreservative effects and antimicrobial activity. Sci Rep 13(1):17142. https://doi.org/10.1038/s41598-023-44251-y
19. Al-Mardeai S, Aldhaheri M, Al Hashmi A, Qassem M, Al-Zuhair S (2023) Complete utilization of date seeds for biofuel production. Cleaner Eng Technol 17:100698. https://doi.org/10.1016/j.clet.2023.100698
20. Almuzaini H, Faqih E, Alhumaid N, Shabib Z, Abd El-Hady ES, Gadallah MG (2024) Advantages of khalas date products (*Phoenix dactylifera* L.) as a natural substitute for sugar and sweeteners in food processing. Alex Sci Exch J 45(4):661–674. https://doi.org/10.21608/asejaiqjsae.2024.391939
21. Alu'datt MH, Rababah T, Tranchant CC, Al-u'datt DA, Gammoh S, Alrosan M et al (2025) Date palm (*Phoenix dactylifera*) bioactive constituents and their applications as natural multifunctional ingredients in health-promoting foods and nutraceuticals: A comprehensive review. Compr Rev Food Sci Food Saf 24(1):e70084. https://doi.org/10.1111/1541-4337.70084
22. Asadi SZ, Khosravi-Darani K, Nikoopour H, Bakhoda H (2018) Production of arachidonic acid and eicosapentaenoic acid by Mortierella alpina CBS 528.72 on date waste. Food Technol Biotechnol 56(2):197. https://doi.org/10.17113/ftb.56.02.18.5379
23. Attia AI, Reda FM, Patra AK, Elnesr SS, Attia YA, Alagawany M (2021) Date (*Phoenix dactylifera* L.) by-products: chemical composition, nutritive value, and applications in poultry nutrition, an updating review. Animals 11(4):1133. https://doi.org/10.3390/ani11041133
24. Ayad AA, El-Rab DG, Shahbazi A, Worku M, Schimmel K, Ejimakor G et al (2016) Using date palm (*Phoenix dactylifera* L.) by-products to cultivate *Lactobacillus reuteri* spp. https://doi.org/10.5539/jfr.v5n5p77
25. Ayaz M, Shah SS, Younas M, Safder U, Khan I, Aziz MA et al (2025) Green synthesis of activated carbon from biomass waste of date palm seeds: A sustainable solution for energy storage and environmental impact. J Energy Storage 110:115291. https://doi.org/10.1016/j.est.2025.115291
26. Aydeniz-Güneşer B (2022) Valorization of date palm (*Phoenix dactylifera*) wastes and by-products. In: Mediterranean fruits bio-wastes: chemistry, functionality, and technological applications. Springer, Cham, pp 391–402. https://doi.org/10.1007/978-3-030-84436-3_16
27. Bahkali AH, Syed A, Elgorban AM, Abdel-Wahab MA, Srivastava N, Gupta VK (2023) Date seed waste derived nanocatalyst and its application in production of hydrolytic enzyme, fermentative sugars and biohydrogen. Bioresour Technol 390:129837. https://doi.org/10.1016/j.biortech.2023.129837

28. Banadka A, Srinivas K, Bhavsar R, Sanjay S, Shaikh A, Murali N et al (2025) Genetic diversity of date palm (*Phoenix dactylifera* L.) and sustainable utilization. In: Genetic diversity of fruits and nuts. CRC Press, pp 187–226. https://doi.org/10.1201/9781003492207
29. Beig-Mohammadi Z, Hamidi-Esfahani Z, Sahari MA, Khosravi-Darani K (2018) Purification and characterization of extracellular phospholipase A1 from Trichoderma atroviride sp. ZB-ZH292. Biocatal Agric Biotechnol 13:176–181. https://doi.org/10.1016/j.bcab.2017.11.009
30. Benabderrahim MA, Elfalleh W, Belayadi H, Haddad M (2018) Effect of date palm waste compost on forage alfalfa growth, yield, seed yield, and minerals uptake. Int J Recycl Org Waste Agric 7:1–9. https://doi.org/10.1007/s40093-017-0182-6
31. Bharath G, Hai A, Rambabu K, Banat F, Jayaraman R, Taher H et al (2020) Systematic production and characterization of pyrolysis-oil from date tree wastes for bio-fuel applications. Biomass Bioenergy 135:105523. https://doi.org/10.1016/j.biombioe.2020.105523
32. Boufennara S, Bouazza L, de Vega A, Amanzougarene Z, López S, Fondevila M (2016) In vitro assessment of nutritive value of date palm by-products as feed for ruminants. Emir J Food Agric 28(10):695–703. https://doi.org/10.9755/ejfa.2016-01-104
33. Burezq HA, Davidson MK (2023) Biochar from date palm (*Phoenix dactylifera* L.) residues—a critical review. Arab J Geosci 16(2):101. https://doi.org/10.1007/s12517-022-11123-0
34. Cantadori E, Brugnoli M, Centola M, Uffredi E, Colonello A, Gullo M (2022) Date fruits as raw material for vinegar and non-alcoholic fermented beverages. Food Secur 11(13):1972. https://doi.org/10.3390/foods11131972
35. Dhakal R, Shrestha P, Kumar S (2024) Mechanical property enhancement of polymer composites using surface-treated date palm fibers. Constr Automot Mater 22(1):89–99. https://doi.org/10.1002/mame.202400081
36. Di Salvo E, Vadalà R, De Maria L, Genovese C, Cicero N (2024) Supplementation of date palm (*Phoenix dactylifera* L.) residue for growth and lactic acid production of probiotic bacterial *Lactobacillus* spp. Nat Prod Res:1–7. https://doi.org/10.1080/14786419.2024.2365440
37. Din H, Kiran M, Haq F, Osman AI, Khan IA, Aziz T et al (2024) Synergizing date palm seeds-derived oxidized activated carbon: sustainable innovation for enhanced water retention, efficient wastewater treatment, and synthetic dye removal. Chem Eng Res Des 204:212–227. https://doi.org/10.1016/j.cherd.2024.02.040
38. Djaafri M, Salem F, Kalloum S, Desideri U, Bartocci P, Khelafi M et al (2024) A route for bioenergy in the sahara region: date palm waste valorization through updraft gasification. Energies 17:2520. https://doi.org/10.3390/en17112520
39. Eid N, Enani S, Walton G, Corona G, Costabile A, Gibson G et al (2014) The impact of date palm fruits and their component polyphenols, on gut microbial ecology, bacterial metabolites and colon cancer cell proliferation. J Nutr Sci 3:e46. https://doi.org/10.1017/jns.2014.16
40. El-Mously H, Darwish EA (2020) Date palm byproducts: history of utilization and technical heritage. In: Midani M, Saba N, Alothman OY (eds) Date palm fiber composites. Springer, Singapore, pp 3–71. https://doi.org/10.1007/978-981-15-9339-0_1
41. EL-Mously H, Midani M, Darwish EA (2023) Date palm byproducts in construction, insulation and building materials. In: Date palm byproducts: a Springboard for circular bio-economy, pp 179–220. https://doi.org/10.1007/978-981-99-0475-4_7
42. Ghafoor K, Sarker MZI, Al-Juhaimi FY, Babiker EE, Alkaltham MS, Almubarak AK (2022) Extraction and evaluation of bioactive compounds from date (*Phoenix dactylifera*) seed using supercritical and subcritical CO_2 techniques. Food Secur 11:1806. https://doi.org/10.3390/foods11121806
43. Ghasemi A, Moosavi-Nasab M, Behzadnia A, Rezaei M (2018) Enhanced biosurfactant production with low-quality date syrup by Lactobacillus rhamnosus using a fed-batch fermentation. Food Sci Biotechnol 27(4):1137–1144. https://doi.org/10.1007/s10068-018-0366-5
44. Giwa A, Yusuf A, Ajumobi O, Dzidzienyo P (2019) Pyrolysis of date palm waste to biochar using concentrated solar thermal energy: economic and sustainability implications. Waste Manag 93:14–22. https://doi.org/10.1016/j.wasman.2019.05.022

45. Gonçalves C, Pereira JA, Silva JA (2018) Bioethanol production from date palm residues: a review. Renew Sust Energ Rev 92:1–9. https://doi.org/10.1016/j.rser.2018.03.056
46. Habib HM, Kamal H, Ibrahim WH, Al Dhaheri AS (2013) Carotenoids, fat-soluble vitamins, and fatty acid profiles of 18 varieties of date seed oil. Ind Crop Prod 42:567–572. https://doi.org/10.1016/j.indcrop.2012.06.039
47. Habib HM, El-Fakharany EM, Souka UD, Elsebaee FM, El-Ziney MG, Ibrahim WH (2022) Polyphenol-rich date palm fruit seed (*Phoenix dactylifera* L.) extract inhibits labile iron, enzyme, and cancer cell activities, and DNA and protein damage. Nutrients 14(17):3536. https://doi.org/10.3390/nu14173536
48. Hamza H, Bettaieb I, Benltoufa A, Ghorbal A, Ahmed HB, Elfalleh W (2023) Bioplastic reinforced with parthenocarpic date palm fibers: characterization and microorganisms involved in degradation. Food Biosci 51:102352. https://doi.org/10.1016/j.fbio.2023.102352
49. Hassan MA, Khan S, Liu J (2019) Valorization of agricultural waste for energy production: the case of date palm residues. Biotechnol Adv 37(1):1–15. https://doi.org/10.1016/j.biotechadv.2018.12.007
50. Ibrahim SF, Dalek NEM, Raffie QFM, Ain MF (2020) Quantification of physicochemical and microstructure properties of dark chocolate incorporated with palm sugar and dates as an alternative sweetener. Mater Today: Proceed 31:366–371. https://doi.org/10.1016/j.matpr.2020.06.235
51. Jamil NM, Al-Obaidi JR, Saleh NM, Jambari NN (2022) Comparative nutritional and toxicity analyses of beverages from date seed and barley powders as caffeine-free coffee alternatives. Int Food Res J 29(4):786–795. https://doi.org/10.47836/ifrj.29.4.06
52. Jaouhari Y, Disca V, Ferreira-Santos P, Alvaredo-López-Vizcaíno A, Travaglia F, Bordiga M, Locatelli M (2024) Valorization of date fruit (*Phoenix dactylifera* L.) as a potential functional food and ingredient: characterization of fiber, oligosaccharides, and antioxidant polyphenols. Molecules 29(19):4606. https://doi.org/10.3390/molecules29194606
53. Jonoobi M, Shafie M, Shirmohammadli Y, Ashori A, Hosseinabadi HZ, Mekonnen T (2019) A review on date palm tree: properties, characterization and its potential applications. J Renew Mater 7(11):1055–1075. https://doi.org/10.32604/jrm.2019.08188
54. Kari ZA, Goh KW, Edinur HA, Mat K et al (2022) Palm date meal as a non-traditional ingredient for feeding aquatic animals: a review. Aquac Report 25:101233. https://doi.org/10.1016/j.aqrep.2022.101233
55. Kavvadias V, Le Guyader E, El Mazlouzi M, Gommeaux M, Boumaraf B, Moussa M et al (2024) Using date palm residues to improve soil properties: the case of compost and biochar. Soil Syst 8(3):69. https://doi.org/10.3390/soilsystems8030069
56. Khosravi S, Safari M, Emam-Djomeh Z, Golmakani MT (2019) Development of fermented date syrup using Kombucha starter culture. J Food Process Preserv 43(2):e13872. https://doi.org/10.1111/jfpp.13872
57. Khosravi-Darani K, Reyhani FS, Nejad BN, Farhadi GBN (2011) Bench scale production of xanthan from date extract by Xanthomonas campestris in submerged fermentation using central composite design. Afr J Biotechnol 10(62):13520–13527. https://doi.org/10.5897/AJB11.018
58. Lattieff FA (2016) A study of biogas production from date palm fruit wastes. J Clean Prod 139:1191–1195. https://doi.org/10.1016/j.jclepro.2016.08.139
59. Mahgoub O, Kadim IT, Al-Busaidi MH, Annamalai K, Al-Saqri NM (2007) Effects of feeding ensiled date palm fronds and a by-product concentrate on performance and meat quality of Omani sheep. Anim Feed Sci Technol 135(3–4):210–221. https://doi.org/10.1016/j.anifeedsci.2006.07.011
60. Mahomoodally MF, Khadaroo SK, Hosenally M, Zengin G, Rebezov M, Ali Shariati M et al (2023) Nutritional, medicinal and functional properties of different parts of the date palm and its fruit (*Phoenix dactylifera* L.)—a systematic review. Crit Rev Food Sci Nutr 64(22):7748–7803. https://doi.org/10.1080/10408398.2023.2191285
61. Mohammadi ZB, Khoshtinat K, Ghasemi S, Ahmadi Z (2022) Nanotechnology and food grade enzymes. In: Novel food grade enzymes: applications in food processing and preservation industries. Springer, Singapore, pp 455–487. https://doi.org/10.1007/978-981-19-1288-7_17

62. Mousa N, Galiwango E, Haris S, Al-Marzouqi AH, Abu-Jdayil B, Caires YL (2022) A new green composite based on plasticized polylactic acid mixed with date palm waste for single-use plastics applications. Polymers 14(3):574. https://doi.org/10.3390/polym14030574
63. Musa KH, Elnour AA (2024) Advances and future perspectives in biotechnological and bioconversion of dates byproducts. J Agric Food Res 101145:101145. https://doi.org/10.1016/j.jafr.2024.101145
64. Nehdi IA, Sbihi HM, Tan CP, Rashid U, Al-Resayes SI (2018) Chemical composition of date palm (*Phoenix dactylifera* L.) seed oil from six Saudi Arabian cultivars. J Food Sci 83:624–630. https://doi.org/10.1111/1750-3841.14033
65. Niroula A, Alkhyeli A, Abdulsamad F, Rabbani A, Stathopoulos C, Nazir A (2025) Alkali-assisted extraction of date seeds for stabilizing bioactive-rich oil-in-water emulsions. ACS Food Sci Technol 5:640. https://doi.org/10.1021/acsfoodscitech.4c00799
66. Oladzad S, Fallah N, Mahboubi A, Afsham N, Taherzadeh MJ (2021) Date fruit processing waste and approaches to its valorization: a review. Bioresour Technol 340:125625. https://doi.org/10.1016/j.biortech.2021.125625
67. Oluwole O, Kosoko S, Familola O, Ibironke O, Cheikyoussef A, Raheem D et al (2023) Fermented traditional wine from palm trees: microbial, nutritional attributes and health impacts. Front Food Sci Technol 3:1225762. https://doi.org/10.3389/frfst.2023.1225762
68. Radfar R, Farhoodi M, Ghasemi I, Khaneghah AM, Shahraz F, Hosseini H (2019) Assessment of phenolic contents and antioxidant and antibacterial activities of extracts from four varieties of Iranian date palm (Phoenix dactylifera L.) seeds. Appl Food Biotechnol 6(3):173–184. https://doi.org/10.22037/afb.v6i3.23379
69. Rahmat A (2021) Chemical properties of biochar from date palm seed (*Phoenix dactylifera* L.) under low-temperature pyrolysis as soil amendment candidate. Appl Res Sci Technol 1(2):116–120. https://doi.org/10.33292/areste.v1i2.13
70. Rambabu K, Bharath G, Sivarajasekar N, Velu S, Sudha PN, Wongsakulphasatch S, Banat F (2023) Sustainable production of bio-jet fuel and green gasoline from date palm seed oil via hydroprocessing over tantalum phosphate. Fuel 331:125688. https://doi.org/10.1016/j.fuel.2022.125688
71. Rangaraj VM, Rambabu K, Banat F, Mittal V (2021) Effect of date fruit waste extract as an antioxidant additive on the properties of active gelatin films. Food Chem 355:129631. https://doi.org/10.1016/j.foodchem.2021.129631
72. Raza M, Abu-Jdayil B (2025) Extraction of cellulose nanocrystals from date seeds using transition metal complex-assisted hydrochloric acid hydrolysis. Int J Biol Macromol 139477:139477. https://doi.org/10.1016/j.ijbiomac.2025.139477
73. Raza M, Abu-Jdayil B, Banat F, Al-Marzouqi AH (2022) Isolation and characterization of cellulose nanocrystals from date palm waste. ACS Omega 7(29):25366–25379. https://doi.org/10.1021/acsomega.2c02333
74. Saadaoui N, Rouilly A, Fares K, Rigal L (2013) Characterization of date palm lignocellulosic by-products and self-bonded composite materials obtained thereof. Mater Des 50:302–308. https://doi.org/10.1016/j.matdes.2013.03.011
75. Saifullah A, Chacko NG, Dhakal HN, Khan SH, Sarker F, Zhang Z (2024) Valorisation of agricultural residue bio-mass date palm fibre in dry-blended polycaprolactone (PCL) bio-composites for sustainable packaging applications. Waste Biomass Valorization 15(10):5805–5817. https://doi.org/10.1007/s12649-024-02550-z
76. Sarraf M, Jemni M, Kahramanoğlu I, Artés F, Shahkoomahally S, Namsi A et al (2021) Commercial techniques for preserving date palm (*Phoenix dactylifera*) fruit quality and safety: a review. Saudi J Biolog Sci 28(8):4408–4420. https://doi.org/10.1016/j.sjbs.2021.04.035
77. Shi J, Olivares L, Chen C (2023) Advances in date palm processing and waste valorization. J Agric Food Chem 71(4):1745–1758. https://doi.org/10.1021/jafc.3c00455
78. Singh B, Gu S (2021) Recent advances in the biotechnological valorization of date palm by-products. Front Biotechnol 8:302–315. https://doi.org/10.3389/fbioe.2021.678501
79. Soomro AH, Marri A, Shaikh N (2023) Date palm (Phoenix dactylifera): a review of economic potential, industrial valorization, nutritional and health significance. In: Ismail T, Akhtar S, Lazarte CE (eds) Neglected plant foods of South Asia. Springer, Cham. https://doi.org/10.1007/978-3-031-37077-9_13

80. Subhash AJ, Bamigbade GB, Ayyash M (2024) Current insights into date by-product valorization for sustainable food industries and technology. Sustain Food Technol 2(2):331–361. https://doi.org/10.1039/D3FB00224A
81. Taghizadeh-Alisaraei A, Motevali A, Ghobadian B (2019) Ethanol production from date wastes: adapted technologies, challenges, and global potential. Renew Energy 143:1094–1110. https://doi.org/10.1016/j.renene.2019.05.048
82. Tang ZX, Shi LE, Aleid SM (2013) Date fruit: chemical composition, nutritional and medicinal values, products. J Sci Food Agric 94(6):1051–1060. https://doi.org/10.1002/jsfa.6154
83. Tassoult M, Kati DE, Bachir-Bey M, Benouadah A, Rodriguez-Gutiérrez G (2021) Valorization of date palm biodiversity: physicochemical composition, phenolic profile, antioxidant activity, and sensory evaluation of date pastes. J Food Meas Charact 15:2601–2612. https://doi.org/10.1007/s11694-021-00844-7
84. USDA (2021) National Nutrient Database for standard reference. United States Department of Agriculture
85. Usman AR, Abduljabbar A, Vithanage M, Ok YS, Ahmad M, Ahmad M et al (2015) Biochar production from date palm waste: charring temperature-induced changes in composition and surface chemistry. J Anal Appl Pyrolysis 115:392–400. https://doi.org/10.1016/j.jaap.2015.08.016
86. Yazdi APG, Mohammadi ZB, Khoshtinat K, Rousta LK, Jafari SM (2021) Application of spray dried encapsulated bioactives in food products. In: Spray drying encapsulation of bioactive materials. CRC Press, pp 377–421. https://doi.org/10.1201/9780429355462
87. Younas, A., Naqvi, S. A., Khan, M. R., Shabbir, M. A., Jatoi, M. A., Anwar, Fet al. (2020). Functional food and nutra-pharmaceutical perspectives of date (*Phoenix dactylifera* L.) fruit. J Food Biochem, 44(9), e13332. doi:https://doi.org/10.1111/jfbc.13332
88. Zaid A, Arias-Jimenez EDJ (1999) Date palm cultivation (No 156). Food and Agriculture Organization (FAO), Rome
89. Zihad SMNK, Uddin SJ, Sifat N, Lovely F, Rouf R, Shilpi JA, Sheikh BY, Göransson U (2021) Antioxidant properties and phenolic profiling by UPLC-QTOF-MS of Ajwah, Safawy and Sukkari cultivars of date palm. Biochem Biophys Rep 25:100909. https://doi.org/10.1016/j.bbrep.2021.100909

Chapter 3
Turning Trash into Treasure: Unlocking the Value of Palmyra Palm Waste

Sweety Angela Kuldeep, Showmiya Segaruban, Nafija Anjum Abonty, Jefwin Paul, and Paulraj Mosae Selvakumar

1 Introduction to Palmyra Palm

The Palmyra palm, or the Asian palm, has been hailed as a celestial tree for centuries. Tracing its nativity to tropical Africa, it is now cultivated in Bangladesh, Cambodia, China South-Central, India, Jawa, Laos, Lesser Sunda Islands, Malaya, Myanmar, Socotra, Sri Lanka, Sulawesi, Thailand, and Vietnam. Palmyraculture is defined as the plantation and utilization of Palmyra palm for self-reliance and sustainable development in India [1].

Known by various names such as the Asian palmyra palm, toddy palm, sugar palm, or Cambodian palm, it is referred to as *Kerigi* (Soqotra), *Mak tan kok* (Lao), *Panna-maram* (Tamil, Tamil Nadu/Sri Lanka), *Taan*, or *Than* (Thai). Derived from

S. A. Kuldeep · N. A. Abonty
Science and Math Program, Asian University for Women, Chattogram, Bangladesh

S. Segaruban
Environmental Sciences Program and Green Bangle Movement, Asian University for Women, Chattogram, Bangladesh

J. Paul
Panaiyaanmai (Palmyraculture)—The Centre for Self-Reliance and Sustainable Development, Munnetram Green Industry, Kadayam, Tamil Nadu, India

Punitha Arulappar Hr. Sec. School, Avudayanoor, Tamil Nadu, India

P. M. Selvakumar (✉)
Environmental Sciences Program and Green Bangle Movement, Asian University for Women, Chattogram, Bangladesh

Panaiyaanmai (Palmyraculture)—The Centre for Self-Reliance and Sustainable Development, Munnetram Green Industry, Kadayam, Tamil Nadu, India
e-mail: p.selvakumar@auw.edu.bd

C. Trois, R. C. Ray (eds.), *Plantation Crop Wastes: Valorization for Economic Sustainability*, Waste as a Resource,
https://doi.org/10.1007/978-3-031-96805-1_3

the Portuguese Palmeira, the internationally recognized name "Palmyrah" (or Palmyra) has historical significance ([2]; Palmweb). The name "Palmera," attributed to the Portuguese, originates from the resemblance of the tree's leaves to the human hand [3]. The Greek words "Borassus" (fruit with leather covering) and "flabellifer" (fan-bearer) further highlight its etymology. Known in various Indian languages as *Tal*, *Talgachh*, and *Tarkajhar* in Hindi and Bengali; *Lulu* or *Tadi* in Telugu; and *Karimpana* in Malayalam, it is also referred to as the fan palm, Brab tree, toddy palm, and *tala* palm [4]. Table 3.1 presents the detailed taxonomy of the Palmyra palm, emphasizing its scientific classification [5, 6], and Table 3.2 outlines the various names of the Palmyra palm, showcasing its cultural and linguistic diversity [2, 6].

This perennial plant, which can grow to a height of 30 m, has a lifespan of up to 100 years. It begins to bear produce after 15 years in areas rich in water sources and after 25 years in arid regions [4]. It thrives in diverse agroclimatic conditions, from seashores to plains, valleys, and hills at altitudes up to 762 m above sea level. Palmyra was declared the state tree of Tamil Nadu in 1978 due to its wide distribution, significant contribution to rural and cottage industries, and profound impact on rural livelihoods [4].

Table 3.1 Taxonomy of Palmyra palm [5, 6]

Classification level	Details
Kingdom	*Plantae*
Subkingdom	*Tracheobionta*
Superdivision	*Spermatophyta*
Division	*Magnoliophyta*
Class	*Liliopsida*
Subclass	*Arecidae*
Order	*Arecales*
Family	*Arecaceae*
Genus	*Borassus*
Species	*Borassus flabellifer L.*

Table 3.2 Various names of Palmyra palm [2, 6]

Language/region	Name
Hindi (India)	*Taad*
English	*Toddy palm, Palmyra palm, fan palm*
Tamil (India)	*Talam, Panna-maram*
Telugu (India)	*Tatichettu, Lulu, Tadi*
Malayalam (India)	*Karimpana*
Bengali (India)	*Taala, Talgachh, Tarkajhar*
Kannada (India)	*Olegari*
Sanskrit (India)	*Taalah*
Lao	*Mak tan kok*
Thai	*Taan, Than*
Soqotra	*Kerigi*

1.1 Botanical and Cultivation Features

The palmyra palm originates from Africa and is an allotetraploid species. The first frond emerges approximately 5 months after planting, and fan-shaped leaves develop in the second year. Flowering, essential for sweet sap extraction, begins when the tree reaches a height of 12–18 m, typically around 13–15 years [7].

The cultivation of Palmyra is low-cost and labor-efficient, requiring minimal maintenance once the tree matures, typically between 15 and 30 years. Palmyra is uniquely suited to tribal and rural areas due to its ability to grow in wastelands and its low maintenance needs. The tree thrives in sandy, red, black, and river alluvium soils and is well-suited to semi-arid climates with annual rainfall below 750 mm. The ideal planting season is October to November [8, 9].

1.2 Cultural, Mythological, and Historical Significance

In many tribal communities, it is revered as a "tree of life," and its fruits and sap are central to their diet and economy. The Oriya astronomical text *Panjika* describes its sacred status, forbidding its fruit on the seventh lunar day [10]. In Tamil Nadu, Palmyra has a deep cultural resonance. It is compared to the mythical *Kalpaka Vrisha* and is celebrated as a celestial tree. In Tamil literature, the poem *Thalavilasam* catalogs 801 uses of the tree, while ancient scriptures and texts were historically engraved on Palmyra leaves, signifying its role in preserving knowledge. Ancient poems extol their divine origins, narrating how the gods created the Palmyra trees to fulfill humanity's needs for food, medicine, and wealth [9, 10].

The Tamil literary work *Naladiyar* contains verses that emphasize the significance of the Palmyra tree, stating that compared to a male Palmyra (Fig. 3.1a), the female tree (Fig. 3.1b) is far more beneficial. One such verse reads:

நடுவூருள் வேதிகை சுற்றுக்கோட் புக்க
படுபனை யன்னர் பலர்நச்ச வாழ்வார்;
குடிகொழுத்தக் கண்ணுங் கொடுதத்துண்ணா மாக்கள்
இடுகாட்டுள் ஏற்றைப் பனை. -நாலடியார் 96

This verse metaphorically compares the female Palmyra growing in the middle of a village, providing shade and nourishment to charitable and prosperous individuals, while the male Palmyra is likened to the unproductive, lifeless tree in a barren cemetery [9].

The domestication of Palmyra dates back to the Mesolithic period, with mentions in epic Indian texts like the *Ramayana* and *Mahabharata* and ancient agricultural scriptures like *Krishi Geeta*. Ancient texts, including the *Srimad Bhagavata Purana* and *Ramcharitmanas,* reference the Palmyra palm [3]. The Palmyra is one of the five sacred trees of Hindu paradise and is linked to Hindu and Buddhist traditions. Palmyra trees were also integral to royal symbolism in South India, where rulers' crowns were made from their fanned leaves [10].

Fig. 3.1 (**a**) Male and (**b**) female Palmyra trees

1.3 Economic and Practical Importance

The global palm sugar market was valued at US$ 1684.2 million in 2017 and is projected to reach US$ 2205.8 million by 2025, with a compound annual growth rate of 3.4% during this period [4]. The Palmyra palm is renowned as the most productive of the four sugar-producing palms—*Borassus flabellifer* (Palmyra), *Phoenix sylvestris* (Date palm), *Caryota urens* (Sago palm), and *Cocos nucifera* (Coconut). It produces a wide range of edible and non-edible products and plays a pivotal role in the socio-economic well-being of rural communities by providing food, shelter, and employment [8, 11].

Mahatma Gandhi described the Palmyra palm as a "remedy against poverty," underscoring its value as a "whole palm," where every part serves a purpose. In many regions, it serves as a primary source of income for impoverished populations [11, 12].

Symonds [9] highlights a Tamil proverb that states how Palmyra lives for a thousand years and is even utilized for a thousand more years after it dies. Whether it is alive or dead, the Palmyra appears to be of nearly equal use to the natives of those nations where it is found in any abundance. It provides them with a variety of tools to help them pursue their vocations, including food, wine, fuel, shelter, furniture, medicine, and stationery. In this chapter, we will dive into how the Palmyra is an everlasting tree, highlighting its significance and various ways it helps the environment and human sustenance through life and even after death.

2 Edible Uses of Palmyra Palm

Palmyra has the potential to contribute various benefits to the environment, ecosystem, and human community. From root to shoot, every part of a Palmyra tree has edible and non-edible usage, and they are environmentally friendly and sustainable. These products of Palmyra are significant to many communities all around the Asian region, specifically religious, traditional, economic, and livelihood. A deep view of the tree indicates how widely Palmyra is used for edible purposes. In the Asian continent, people use various parts of the Palmyra tree as their essential food, which is rich in nutritional value. Most of its edible parts are used as herbal medicine in many parts of Sri Lanka and Tamil Nadu. Various parts of Palmyra can be used for either raw or processed edible products, for example, haustorium (germinated seeds), tuber (Palmyra sprouts), sap (*neera*), toddy, Palmyra fruit (immature endosperm and mesocarp pulp), etc. In addition to that, products are rich in nutrients, such as vitamins A, B12, C, and E, and minerals, fibers, and antioxidants [13].

2.1 Haustorium (Germinated Seeds)

Palmyra haustorium is the embryo seed of the Palmyra tree, which is in the inner part of the seed, white in color, and has a sweet and sour taste (Fig. 3.2b). Mostly in Sri Lanka and Tamil Nadu, people, particularly children, like to eat Palmyra haustorium. In the article by Kuldeep et al. [1], the authors mention the nutritional value and health benefits of Palmyra haustorium. Haustorium, also known as *Thavun* in Tamil, is rich in nutrition and taste. It is used in the pharmaceutical, commercial, and culinary sectors. For many communities, the plant produces haustorium whose sap is turned into delicious treats and the leftover seed cover has several uses, including cooking and the production of carbon nanotubes and activated charcoal [1, 14]. Before germination, its dry seeds (Fig. 3.3a) can also be put to use in crafts, like coconut shells, as is seen in the example of a pen stand and paper weight made out of carved dried seeds (Fig. 3.3b).

The nutritional value and health benefits of Palmyra haustorium are spectacular, such as fiber, carbohydrates, crude fat, minerals including sodium, potassium, calcium, zinc, magnesium, phosphorus, and copper, as well as antioxidant materials. It also has the potential to scavenge free radicals [15]. There are countless health benefits from Palmyra haustorium, widely available in nature; such as in women and children, Palmyra haustorium mitigates malnutrition problems, and people who face digestive issues of lactose can use Palmyra haustorium as an alternative source since it contains carbohydrates, and food products made of Palmyra haustorium are

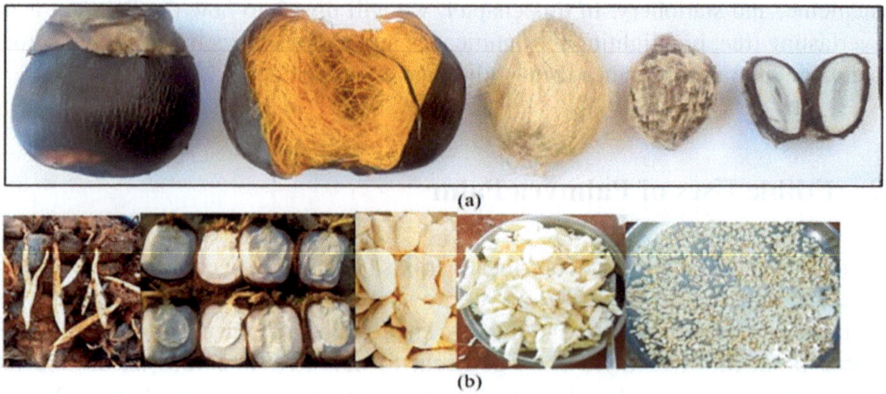

Fig. 3.2 (**a**) Palmyra palm fruit, fiber, fresh seed, and the cross-section view of the seed; (**b**) Germinated Palmyra seed cut open to represent a white, spongy haustorium/seed embryo. Panaiyaanmai (Palmyraculture)—the Centre for Self-reliance and Sustainable Development, Kadayam, Tenkasi, Tamil Nadu, India [1]

Fig. 3.3 (**a**) Dried Palmyra seeds; (**b**) decorative pen stand and paper weight crafted from Palmyra seeds by Crafttary in Kopai, Jaffna, Sri Lanka

one of the dietary foods for obesity, diabetes, hypertension, and heart problems [16]. Palmyra haustorium can also be used as an alternative source of many fast and synthetically processed foods available on the market. The nutrition can be obtained naturally, and at the same time, it is cost-effective so that all people from any society can easily access it.

2.2 *Tuber (Palmyra Sprouts)*

Palmyra tuber shows uniqueness among all the other products obtained from the Palmyra tree. The tuber is known for its nutritional, socio-economic, and health benefits. The seed growth can be natural or planted by people for various purposes, like small-scale businesses with Palmyra tubers during the season [17]. The tuber of the Palmyra palm, called *Panamkizhangu*, is used as food in the north and east parts of Sri Lanka and Tamil Nadu [18]. The mature tuber is strong and thick, can be broken easily, and is a source of carbohydrates and fibers. After peeling off, the tuber will be cooked and eaten by people directly, and the roasted one is mostly used for making flour, which is mixed with wheat flour for baking purposes [17]. People boil the tuber and eat it with coconut and dried tuber called *odiyal.* It is rich in fiber, blood sugar regulation, and natural detox and comprises many vitamins, minerals, and antioxidants [18, 19]. Palmyra tuber *laddu* (sweet balls), Palmyra tuber-*soya laddu* (sweet balls), Palmyra tuber *kesari* (sweet), Palmyra tuber *payasam* (kheer), Palmyra tuber *idli* (steamed buns), Palmyra tuber *upma* (tiffin), Palmyra tuber porridge, and Palmyra tuber *pakora* (crunchy hot and spicy fritters) are made using Palmyra tuber.

2.3 Sap

Another one of the economically significant products is Palmyra sap, which is obtained from the inflorescences of male and female Palmyra trees through the process of tapping. Sap is also known as *neera* or *pathaneer* in the Tamil language. It is famous in the North and East parts of Sri Lanka and Tamil Nadu [20]. People from local areas use Palmyra sap as medicine for various diseases, and it is one kind of entertainment drink among the locals. Palmyra toddy attracts attention because it is drunk equally without bias; thus, it is not the same as many alcoholic drinks that come with so-called social norms of drinking. Moreover, pregnant women drink toddy right after they give birth to reduce the delivery pain. Many food items are made with sap, which is good for our health and can be eaten by all age groups of people. For instance, palm jaggery, palm sugar, tamarind candy, chili, palm honey, and treacle. The nutritional profile of Palmyra sap is considered a healthy drink and contains more nutrients, such as mineral-rich, low-acid, nutrient-dense, low-fat sap that possesses antioxidant properties such as total phenolic content and radical scavenging activity.

2.4 Palmyra Fruit

Young and ripped Palmyra fruit is used to make many food products. Juice is made from young Palmyra fruit, and people eat the jelly-type part from the young Palmyra fruit. Palmyra juice is made from Palmyra pulp. There are many other food items made of Palmyra pulp, such as Palmyra spread, Palmyra toffee, *barfi*, Palmyra *puttu, idly*, *palakaram* (sweet famous in Sri Lanka and Tamil Nadu), Palmyra cake, Palmyra ice cream, and *taaler pitha* (prominent in Bangladesh). There is another significant food that is very common in the northern part of Sri Lanka: *Panattu*—dried jelly from Palmyra fruit pulp, which has a slightly bitter and sweet taste. It used to be eaten with rice [21]. According to another study by Uluwaduge et al. [22], there is potential for Palmyra *pinattu* to be developed into a functional food for the management of diabetes, which would add to the dietary approaches for blood sugar regulation [22]. Furthermore, in Indonesia, people make a dish called *dawet,* a mixture of fruits, which is one of the traditional desserts of Tuban in Indonesia. Equally raw or boiled, mashed, and filtered fruit purple is eaten with corn or porridge in Indonesia [23]. Figures 3.4 and 3.5 show the various food items that can be prepared using the Palmyra fruit pulp.

One research study talks about the nutritional composition, antioxidant properties, and biological activity of granulated sugar derived from the Palmyra palm. Granulated sugar is made of concentrated dried palm syrup from Palmyra sap; the sweetness level of the sap is very high. Similarly, granulated sugar derived from Palmyra palm is commonly used in cooking and baking, such as cookies, cakes, muffins, pancakes, waffles, and beverages [24]. Moreover, pickle is made by pickling small Palmyra fruits in vinegar, and canned palm is made of exportation; semi-ripe, soft-shelled palm seeds are canned in transparent, lightly sweetened water, resembling jelly [4].

Fig. 3.4 Various food items made using Palmyra fruit pulp

Fig. 3.5 (**a**) Palmyra fruit pulp; (**b**) Panagai Paniyaram

3 Non-edible Uses of Palmyra Palm

Various parts of the Palmyra tree are used for different purposes for enhancing human lifestyle and maintaining a sustainable environment. From the root to the shoot, the Palmyra tree can be used as a valuable product. These can be used by all people because it is abundant in nature, cost-effective, and helpful for small-scale businesses to big industries, and each part of the Palmyra tree can be used for various purposes. Also, instead of using many other plastic and synthetic products on the market, the materials from Palmyra will be the perfect alternative for health by reducing the contamination, side effects, and toxicants in nature itself. Thus, the waste parts from the Palmyra tree are also converted into valuable products in different ways. Let's take a deeper look into the various parts of the Palmyra tree with their significance in making the waste materials from valuable products.

3.1 Leaf Stalk (Petiole)

- The leaf stalk yields four types of fiber: the strongest from the edges, the weakest from the stalk's center, the second from the upper side, and the third from the lower side. These fibers are used to make cables, cordage, ropes, yarn, and twine.

The leaf stalk is also strong enough to serve as palings for fences and cross sticks for small house roofs [9].

- Leaf stalks are widely used for fencing and yield three types of fibers for utility items [8]. In the local areas, people use Palmyra stalks to make fences to protect their land and crops from other animals. They line up the stalks in a line and tie them using rods (Fig. 3.6a).
- Stalk gives support to the leaves and protection to the tree from tropical conditions such as cyclones. The leaf part of Palmyra has comprehensive economic uses among industries, while the stalk will be the waste material.
- Because of their strength, flexibility, and lightweight properties, Palmyra leaf stalks have also been used to make handles for hammers (Fig. 3.6b).

Fig. 3.6 (**a**) Palmyra stalk suitable to make fences; (**b**) Palmyra stalk used to make hammer handles

- Also, the edges of the stalk have some sharp thorns, which help the locals to protect their land. When the predator enters the land or crop field, drones never let them inside and attack them, as well as they can be used as sharp weapons.
- On the other side, it is used as a cricket bat in rural places. Since it's very strong and thick, children clean the throne edges and use it as a cricket bat, which is a cost-effective and easier alternative for them to use as photo frames, brooms, small furniture, and small paper holds can be made by the Palmyra stalk.

3.2 Fibers

- Fresh leaf stalks can be hand-crushed or beaten to produce rope and threads. These fibers are also used to make brooms and painting brushes [25]. They can be obtained from various parts of the Palmyra, like leaves, timber, and fruits, and are used for multiple purposes in daily life.
- Obtained fibers are valuable for making hats, mats, and baskets, particularly in Ceylon (Sri Lanka) [9].
- The Palmyra palm produces five types of fibers: a loose fiber from the leaf stalk base, a long fiber from the leaf stalk, fiber from the stem interior, coir from the pericarp, and fibrous material from the leaves. These are used to create basketware, cord, rope, and matting. Leaf-based fiber is a significant export product [4, 8].
- Petiole fibers are split longitudinally for use as ropes in biofencing, thatched house construction, baskets, and winnowing fans. The serrated edges of the stalk are used to make durable handicrafts like jaggery baskets and mesh for cots and chairs [3, 26].
- In the northern part of Sri Lanka, people use Palmyra fibers for body washing and maintaining their hygiene. One of the studies indicates that Palmyra fruit fibers can be an alternative to synthetic fibers in the clothing industry. Also, in place of synthetic fibers, natural fibers can be used to strengthen polymeric materials in the creation of affordable, sustainable, and eco-friendly composites (Md. Raju [27]).
- The flexible fiber obtained from Palmyra leaves as petiole, also called *Eerku* in Tamil, is used to make flexible household objects like a mat or bed, called *a panai* (Palmyra) leaf bed, which is constructed from the flexible fibers. Another product, winnower, called *solavu* in Tamil, is a handmade object used to separate sand and small waste particles from grains and rice, create innovative household pen stands, and many more.
- Another study about Palmyra fibers from sprout (tuber of Palmyra) proves that along with biosilica from red matta rice husk ash, it enhances materials made of epoxy resin. This study suggests that this composite demonstrates notable gains in strength and durability; the resistance of the composite is much better in terms of wear resistance. These improved qualities collectively imply that biosilica and Palmyra sprout fiber can be efficient, environmentally beneficial substitutes for synthetic materials in a range of applications, including home and automobile goods [28].

3.3 Roots

- The Palmyra's roots can reach extraordinary depths, often matching the tree's height, allowing it to access underground moisture during dry months [9].
- Properly seasoned roots from century-old Palmyra trees are valuable for construction. The bole at the root can be repurposed into water buckets or drums [9].
- Young white roots, when roasted or boiled, are consumed as vegetables. Boiled roots can also be ground into flour and mixed with ingredients like fish, salt, and chilies. Old roots are used in traditional remedies for urinary diseases [9].
- The root network stabilizes soil, prevents erosion, and acts as a windbreak to reduce sand dune formation [8].
- Roots are also used to craft baskets [8].

3.4 Web

- A Palmyra traveler's torch, consisting of 15 webs, burns red and is wind-resistant. Webs are excellent fuel and are used to strain the palm's saccharine sap [9].
- Singhalese doctors use the light brown, cotton-like substance found at the base of the fronds to treat superficial wounds [9].

3.5 Fruit

- Palmyra fruits are large and three-cornered in shape, eventually ripening to a nearly globular form. Each fruit is roughly the size of the head of a typical young child. Poor communities consume them raw or roasted [9].
- Tender fruit is used as cattle feed and is believed to increase butter and milk secretion in cows [8, 9].
- Toys and high-end products are made from the fiber extracted from fruits [11].

3.6 Trunk

- A trunk is a black, strong, tremendous part of the Palmyra tree that holds the entire tree and supports it during every environmental condition. The trunk is a massive part of the Palmyra tree, which has tremendous benefits. Mostly, it is used as firewood in many places. Moreover, it used to make a big musical instrument called *parai* (drum) in Tamil, which is very big and strong.
- The female tree's wood is harder and preferred for construction, while male tree wood is used only when mature. The wood is valued for its strength, durability, and resistance to cross-strain. It is used for rafters, beams, pillars, and even small boats [8, 9].

- The wood of older Palmyra trees is harder, blacker, and more resilient; it weighs roughly 65 pounds per cubic foot and lasts for 80 years. The trunk is divided into four sections for rafters and eight sections for reapers. The wood splits easily lengthwise but can withstand significant cross-strain. The dark exterior wood is used in Europe for umbrella handles, walking sticks, paper rulers, fancy boxes, wafer stamps, and more. In Sri Lanka, it has been used for rafters, pillars, and posts, as well as veneering and inlaying in England [9].
- In the sandy areas of Jaffna, hollow Palmyra trunks are used as wells. Historically, seagoing vessels were planked with Palmyra wood, but the iron fastenings were prone to rapid destruction. Boats made with this wood were common on the Godavari until the mid-nineteenth century. Two stems lashed to spars were commonly used for crossing lakes and rivers, with the root end forming the head of the canoe [9].
- Trunks are used extensively for construction across East Java, Indonesia. In Tuban, farmers use trunks to make wood canoes. In Sumenep, farmers sell trunks for crafting [23].
- Palmyra trunks are sliced and used in construction sites to give support to the buildings (Fig. 3.7a, b).
- In addition to making the main body of the common musical instrument drum and *Takudi* (a spindle for spinning thread), the matured trunk was used to craft *Seni* (roof plates) and *Rua* (rafters) for thatched houses, which are known for their durability and are also often adjusted to support the roofs of houses (Fig. 3.7c). Long walking sticks and small handy sticks were also made and were often used by older individuals and morning walkers [3].
- In some parts of Bangladesh and Bengal, boats, or *dongas,* are made of Palmyra trunks. These boats, which have a lifespan of over 10 years, are still utilized in Bengal's low-lying areas (Fig. 3.8) [29].
- The trunk of the Palmyra palm is also crafted into a divine spear associated with Tamil God Muruga and other utility items (Fig. 3.9).

Fig. 3.7 (**a, b**) Palmyra tree trunks adjusted to be used for construction purposes; (**c**) Trunks sliced to support roofs

Fig. 3.8 (**a**, **b**) Palmyra tree trunks used as *donga* (boats) [29]

Fig. 3.9 Divine spear associated with Tamil God Muruga, crafted from Palmyra trunk

- Another fascinating use of the Palmyra waste trunk is hives. People grow honey bees in dead Palmyra trunks. It gives security and shelter to the honey bees for honey production and reproduction. The beekeepers wait until the termites clean the inner pith, which gives enough space to grow the bees inside the dead trunk of the Palmyra tree [30].

*3.7 Sheath (*Pannadai*)*

- The other part, called *pannadai* or *chillattai* (sheath), is a thick, cloth-like structure surrounding the leaf stems of palm trees. It is known as this because the warp and weft are arranged freely. This part of the tree is considered a waste product (Fig. 3.10a).

Fig. 3.10 (a) *Pannadai* (sheath); (b) parts used for firewood

- Among the Tamil community, people use the word *pannadai* (sheath) to indicate a person who is very lazy and dumped because they portray the person as waste material as *pannadai* (sheath).
- This part has various kinds of benefits and is used for various purposes, such as making mosquito nets to protect us from mosquito attacks. It is an effective alternative to synthetic or plastic mosquito nets, which are widely used everywhere. There might be high chance of microplastic erosion from plastic mosquito nets while using them for a long period, so this will be a suitable, sustainable, and eco-friendly alternative for the environment as well as human health.
- It is used as a filter in many rural areas to filter the *pathaneer* (pure sap) and water. Only the liquid part is allowed to pass through the small hole in the sheath while the waste products will be filtered out. At the same time, it is a simple idea of a cheap, temporary, effective filter, and a natural way for filtration.
- In some parts of rural areas, sheath is used for covering the fence to protect the crops from cows, goats, and some other animals. Moreover, there are some other usages from *pannadai,* for example, broom, firewood (Fig. 3.10b), etc. [19].
- Palmyra sheaths are also used to make everyday craft products. Figure 3.11 showcases items crafted from Palmyra sheath fiber by MD Yunus and his family in Bogura, Bangladesh.

3.8 Peduncles

- Another part of the Palmyra tree, which is frequently considered waste material from Palmyra, is the outer part of the ice apple (immature fruit), also known as *nungu* in Tamil (Fig. 3.12). The endosperm of immature Palmyra fruit is used as food, and the mesosperm, or outer part of the fruit, eventually as waste material. They will be fed to animals like cows and goats.

Fig. 3.11 Palmyra sheath fiber-based products made by MD Yunus and family, Bogura, Bangladesh

Fig. 3.12 Palmyra fruit peduncles

- Fruit peduncles of the Palmyra tree are often regarded as agricultural waste, and their inadvertent disposal through open dumping on land can cause various environmental effects. Also, it not only increases value but also lessens the amount of waste that needs to be processed and managed, which lessens the load on the solid waste management community, and numerous industries, including electronics, food packaging, medical, and automotive, use the cellulose fibers that are derived from this source.
- Palmyra fruit wastes can be used as pots to grow small plants in house gardens and make toys for children [31]. In addition, it is also used as a fertilizer when it decomposes and will be a natural fertilizer that enhances the fertility of the soil.

3.9 Leaf

The leaves of the Palmyra palm have diverse applications due to their large size, durability, strength, and suppleness. The long, parallel fibers and fibrovascular bundles make them easy to split, contributing to their versatility [3].

- **Thatching and Roofing**: The leaves are extensively used for thatching houses, cattle sheds, and hut roofs. Tribes from various states use them for these purposes, especially during dry spells [8, 10, 23].
  - Small hut-type shelters called *kodappu* or *kudil* (the small hut) in Tamil, made of Palmyra leaves and trunks, are used for cattle or livestock in village areas to protect them from predators and natural disasters. In some other places, people also use them for temporary or permanent shelter.
- **Handicrafts and Household Items**: The leaves are crafted into mats, baskets, fans, hats, umbrellas, buckets, and sandals. The fiber extracted from leaves is used to make brushes [3, 11]. Figures 3.13, 3.14, and 3.15 show examples of these Palmyra tree parts that are used as various household and creative craft items.
- **Utility and Fancy Articles**: Tender leaves are used for utility items and fancy handicrafts. The leaf butt provides fibers for brushes, while the leaf rachis is used for household articles [8].
- **Rain Gear**: Narrow strips of leaves are stitched to overlap, forming *kongeni*, a rain cover that provides effective shelter. Shepherds use this to sleep comfortably outdoors during intense rain [9].
- **Water Vessels**: A single folded leaf can serve as a well bucket, and other water vessels are crafted from it [9].

Fig. 3.13 Hand fans made using Palmyra tree leaves

Fig. 3.14 Palmyra leaf and fi ber crafts: (**a**) leaf bird, (**b**) leaf-woven hammer, (**c–f**) traditional woven baskets and cultural utility items, (**g**) diversifi ed and innovative value-added craft products

Fig. 3.15 Creative Palmyra leaves-based interactive periodic table games for chemistry and environmental science students of AUW in Palmyraculture research group and Green Bangle Movement

- **Organic Fertilizer and Fuel**: After use as thatching and fencing, leaves are repurposed as organic fertilizer or cooking fuel. Senesced leaves are also used as firewood. Dried leaves and fruit mesocarps are commonly used as fuel in East Java, Indonesia [3, 11, 23].

- **Fishing Implements**: In Indonesia, leaves are used to make fishing implements, which are tied and woven into nets [4, 10].
- **Food Wraps and Weaving**: In East Java, farmers use Palmyra palm leaves for food wraps, including traditional foods like *ketupat*, *lepet*, and *dumbek*. The leaves are also woven into mats and baskets. Mats are often used for wrapping dried tobacco leaves. Baskets serve as crop harvesting containers or jaggery molds. Balinese use raw young leaves for offerings, palm-leaf writing, handicrafts, and other traditional purposes [23]. Green Palmyra leaves are used to eat food in Sri Lanka. The two edges of the leaves fold and tie together, and the inner part is where the food is served.
- **Special Applications in Odisha**: In rural Odisha, leaves are used to make small mats (*chatei*), long mats for beds, baskets (*bhogai*), hand fans, traditional umbrellas, decorative hats, toys, crackers, and traditional invitation cards [3].
- **Writing and Stationery**: The leaves are historically significant as writing material. Dried palm leaves are used for manuscripts, preserved Hindu literature, and administrative purposes. The slips taken from young leaves are dried and smoothed for writing [8, 9].
- **Decorative Items:** During festival time, young Palmyra trees are used for making decorations in houses, which is called *thoranam*.

Table 3.3 highlights the diverse non-edible applications of the Palmyra palm, detailing how various tree parts are utilized for construction, handicrafts, and traditional practices.

Table 3.3 Non-edible applications of Palmyra palm across its different parts

Palmyra palm parts	Application	Description
Leaf stalk (Petiole)	Fiber production	Yields four types of fibers for making cables, ropes, cordage, yarn, and twine; also used for fencing and roof supports
	Protection and crafts	Sharp edges provide protection against animals; used to make fences, cricket bats, photo frames, small furniture, and brooms
	Local innovations	Used as inexpensive cricket bats by children and for making household items like paper holders
Fibers	Utility items	Used for ropes, threads, mats, and baskets; fibers from various parts of the tree have applications in household and industrial use
	Hygiene and clothing	Body-washing fibers; potential for use in clothing industries and eco-friendly composites
	Enhanced durability	Fibers strengthen polymeric materials, improve wear resistance, and create sustainable alternatives to synthetic materials

(continued)

Table 3.3 (continued)

Palmyra palm parts	Application	Description
Roots	Soil stabilization	Prevents erosion, anchors soil, and reduces sand dune formation
	Construction and crafts	Used to make water buckets, drums, and baskets
	Edible uses	Boiled or roasted roots are consumed as vegetables and ground into flour for recipes
Web	Fuel and straining	Used as wind-resistant fuel and for straining palm sap
	Medicinal uses	Cotton-like substance at the fronds' base used for treating wounds
Fruit	Cattle feed	Tender fruits enhance milk secretion in cows
	Crafting	Fiber extracted from fruits is used for toys and high-end products
	Traditional uses	Fruits used raw or roasted in poor communities; outer shell used for plant pots and toys
Trunk	Construction	Timber is used for rafters, beams, pillars, wells, boats, and musical instruments
	Crafting and beekeeping	Long walking sticks, musical instruments, and beehives crafted from dead trunks; trunks provide security for honey production
	Innovative uses	Handles for hammers; hollow trunks used in water wells and small boats
Pannadai (Leaf Sheath)	Mosquito nets	Used as eco-friendly mosquito nets, preventing microplastic erosion
	Filtration	Used as a natural filter for sap and water
	Protection and fuel	Covering for fences to protect crops; also used as broom and firewood
Peduncles	Agricultural waste valorization	Cellulosic fibers from peduncles used in composites for electronics, food packaging, and medical industries
	Planting and fertilizer	Used as pots for small plants; decomposed peduncles enhance soil fertility
Leaves	Thatching and roofing	Widely used for thatching houses and cattle sheds; utilized during dry spells
	Handicrafts	Crafted into mats, baskets, fans, hats, and umbrellas
	Food wraps and weaving	Used for traditional food wraps (e.g., *ketupat*); woven into baskets for jaggery molds
	Writing materials	Historically used for manuscripts and preserved Hindu literature
	Rain gear and decorations	*Kongeni* rain covers and festival decorations

3.9.1 Palmyra Leaves as Manuscripts and Odisha's Legacy

The palm leaf, with its intricate vein structure and supple texture, has long been revered as a canvas for recording manuscripts, symbolizing the intellectual and cultural ethos of Odisha and the broader region. These manuscripts, meticulously etched and preserved, reflect a heritage that is both artistic and scholarly.

Crafting these manuscripts began with young, partially opened palm leaves. The preparation process was intricate: the leaves were sun-dried, immersed in water for weeks, and air-dried to maintain their flexibility and durability. They were then cut to size, the stalks removed, and holes punched for threading cords. Using a stylus, artists and scribes inscribed texts and illustrations, enhanced by carbon black pigment to ensure visibility. To protect the manuscripts from insects and extend their longevity, a unique paste of bean leaves, coconut shell charcoal, til oil, and turmeric were applied. This not only preserved the leaves but also added a smooth finish [32]. The manuscripts bore engravings of kings, deities, flora, and architecture, signifying the societal values. The Oriya script took on a rounded, cursive nature; this was probably because of their natural alignment along the veins of the leaves, thereby lending it greater utility and beauty [32, 33].

In the *Kavya Mimansa*, the poet, Rajasekhara of the tenth century, says:

तस्य सम्पुटिका सफलकरवटिका समुद्गक,
सलेखनोकमवी भाजनानि ताड़िपत्राणि
भूजंत्वचोबा, सलोहकण्टकानि सुसम्मृष्टामितयः
सतत सन्निहिताः ।

English translation: A poet must always have near him a box, a board and chalk sticks, leaves of the *Tadi* tree or bark of the *Bhoja* tree, pens and inkpots, leaves of the *Tala* tree along with iron nails, and well-rubbed plates [33].

Historically, these manuscripts housed significant works such as the *Amarusataka*, *Sarvanga Sundari*, and *Gita Govinda* (Fig. 3.16b, c), showcasing the region's literary and philosophical depth. Beyond literature, palm leaves recorded horoscopes, ceremonial records, and even invitations, underscoring their integral role in societal functions [33].

Odisha's cultural exchanges with South India, dating back to the Kalinga Emperor Kharavela in the first century BC, greatly influenced its manuscript traditions. The skills of artisans and scribes ensured that many of these documents remained in excellent condition. Palm leaves are rich in biochemical compositions, which are important for resilience, allowing them to endure for centuries [32].

Yet, this treasured tradition faces challenges in the modern era. The knowledge and skills of rural artisans and scribes, once vital for this art form, are fading into obscurity. Nonetheless, the surviving manuscripts stand as a testament to Odisha's enduring legacy and a reminder of the importance of preserving this cultural heritage for future generations (Fig. 3.16a).

Fig. 3.16 (**a**) Display of information on Palm Manuscript; (**b, c**) and Gitagobinda on Mala Shape (Odisha State Museum, Bhubaneswar, Odisha)

Fig. 3.17 Traditional narrative palm-leaf engravings from Odisha: (**a**) tree of life story panel, (**b**) deity-centered mythological panel, (**c**) a tree of life composition with community scenes, (**d**) a botanical panel featuring stylized tree forms, and (**e**) mythological storytelling scroll

Figures 3.17 and 3.18 below showcase the beautiful artistic creations using its leaves. These illustrated bookmarks and manuscripts depict deities and fascinating historical and mythological stories. These creations shed light on the efforts to conserve the historical significance of the *tala* leaves.

Fig. 3.18 (**a**, **b**) *Oolai chuvadi* or modern Palm-leaf manuscript; (**c**, **d**) Palm-leaf bookmarks made by local artisan Satya Narayan Barik, Bhubaneswar

4 Value Addition of Palmyra Fruit Pulp Products

The Palmyra palm is a treasurable asset for waste valorization, whose pulp and by-products when processed become functional and commercially viable.

4.1 Palmyra Pulp Preservation

Most Palmyra fruits are wasted due to the absence of proper preservation techniques. Some of the processes adopted for the fruit pulp of Palmyra include preservation methods to be carried out to enhance its shelf life while retaining its nutritional vigor and reducing the chances of wastage [34, 35]. These include:

- Heating: The pulp is thickened by heating for 10–30. Then it's cooled to preserve its texture, thus prolonging its shelf life.

- Drying: A thorough process reduces the pulp to powder, lightweight, and thus limits spoilage.
- Canning: During the season very close to commercial distribution, the pulp is put in syrup or brine, maintaining a fresh color.
- Freezing: This helps with preservation for long periods without a compromise on nutritional value.
- Fermentation: This helps with the production of value products like vinegar, wine, or alcohol.

4.2 Nutritional and Functional Properties of Palmyra Pulp

Palmyra fruit pulp is obtained from ripe Palmyra, containing typical and nutrient-rich constituents. The pulp is a source of carbohydrates, proteins, vitamins, and minerals for numerous food applications while possessing thickening, swelling, and hydration properties that lend it excellent versatility in industrial usage [36–38].

Table 3.4 presents the physicochemical and nutritional properties of Palmyra pulp while highlighting its functional attributes pertinent to its diverse applications for food and in industrial processing.

Table 3.4 Physicochemical and nutritional composition of Palmyra pulp

Parameter	Value
Color	Light orange
pH	5.5–6
Moisture (%)	74.77
Ash (g)	1.2
Fat (g)	0.8
Total carbohydrate (g)	22.5
Reducing sugar (g)	9.5
Protein (g)	1.24
Ascorbic acid (mg)	16
Calcium (mg)	8.76
Energy (kcal)	102.83
Phosphorus (mg)	15
Iron (mg)	1.2
Magnesium (mg)	10.2
Sodium (mg)	20
Potassium (mg)	21.5

4.3 *Value-Added Product from Palmyra Palm Pulp*

The Palmyra palm tree has been hailed as a multipurpose tree for generations. It is closely linked to our cultural, economic, and ecological aspects of life. Table 3.5 outlines the various edible and non-edible value-added products that we get from the Palmyra fruit pulp along with their storage/preservation techniques and shelf lives [34, 35, 39, 40].

Table 3.5 Edible and non-edible value-added products made using the Palmyra fruit pulp

Category	Product	Description	Storage and preservation	Shelf life
Edible products	Palm jaggery	A natural sweetener made from the sap of the palm tree	Store in a cool, dry place	Several months if stored properly
	Palmyra fruit jelly	A sweet jelly made from the pulp, often mixed with other fruits	Refrigeration recommended; store in sterilized jars	2–3 weeks in the fridge
	Palmyra fruit syrup	A concentrated syrup used as a sweetener or beverage base	Store in sealed bottles in a cool, dry place	–
	Palmyra fruit pudding	Dessert made with pulp, sugar, milk, and flavoring agents	Refrigerate; consume within 2–3 days	2–3 days refrigerated
	Palmyra palm candy	Traditional candy made by boiling pulp with sugar	Store in airtight containers in a cool, dry place	Around 6 months
	Palmyra flour	Made from dried pulp, used in baked goods	Store in an airtight container in a cool, dry place	6–12 months
	Palm spread	Sweet spread made with pulp, sugar, milk powder, and cardamom	Store at ambient temperature for up to 9 months; 12 months under refrigeration	9–12 months
	Palm toffee	Chewy confection made with pulp, sugar, and milk products	Store at room temperature	Around 8 months
	Palm burfi	Sweet made with pulp, khoa, butter, and sugar	Best consumed fresh; refrigeration extends shelf life	Short-term refrigerated
	Neera (Toddy)	Fresh sap drink; ferments into alcohol	Best consumed fresh; refrigeration slows fermentation	Short-term refrigerated
	Palm cola	Soft drink made with palm sugar and cola essence	Refrigeration recommended	Few months
	Preserved nungu	Tender endosperm of palm fruit, preserved in syrup	Store in jars; refrigeration extends shelf life	Several months
	Ice cream	Ice cream made with Palmyra pulp as a base	Store in the freezer	Several months

(continued)

Table 3.5 (continued)

Category	Product	Description	Storage and preservation	Shelf life
Non-edible products	Cosmetic oils	Extracts used in skin care formulations	Store in dark bottles away from sunlight	–
	Animal feed	Residual pulp processed into feed	Keep dry and pest-free	Several months
	Palmyra soap	Soap made from palm oil and pulp extract	Store in a cool, dry place	1–2 years
	Pharmaceutical extracts	Traditional medicine uses for digestive issues	Store in sealed containers away from sunlight	Up to 1 year
	Cosmetic creams/lotions	Skincare products with pulp extracts	Store in airtight containers in a cool, dark place.	6 months to 1 year
	Natural dye	Extract used for coloring textiles and cosmetics	Store in sealed containers away from sunlight and moisture	Up to 1 year

5 Palmyra Palm for Biofuels and Sustainable Energy

The Palmyra palm shows promise for a source of biofuels that are capable of solutions for renewable energy. The process begins with the extraction of biodiesel from crushed and cleaned seeds. In this process, hexane solvent is used to remove the oil, which is refined via degumming and neutralization to remove impurities. Subsequently, transesterification of the refined oil occurs, in which it is reacted with a catalyst and alcohol to make biodiesel and glycerol, and the resulting biodiesel is purified by washing and dehydration before being subjected to viscosity, flashpoint, and cetane number tests (Fig. 3.19) [41]. The Palmyra biofuels significantly reduce greenhouse gas emissions and support sustainable energy production, making it an eco-friendly alternative to conventional fossil fuels [41].

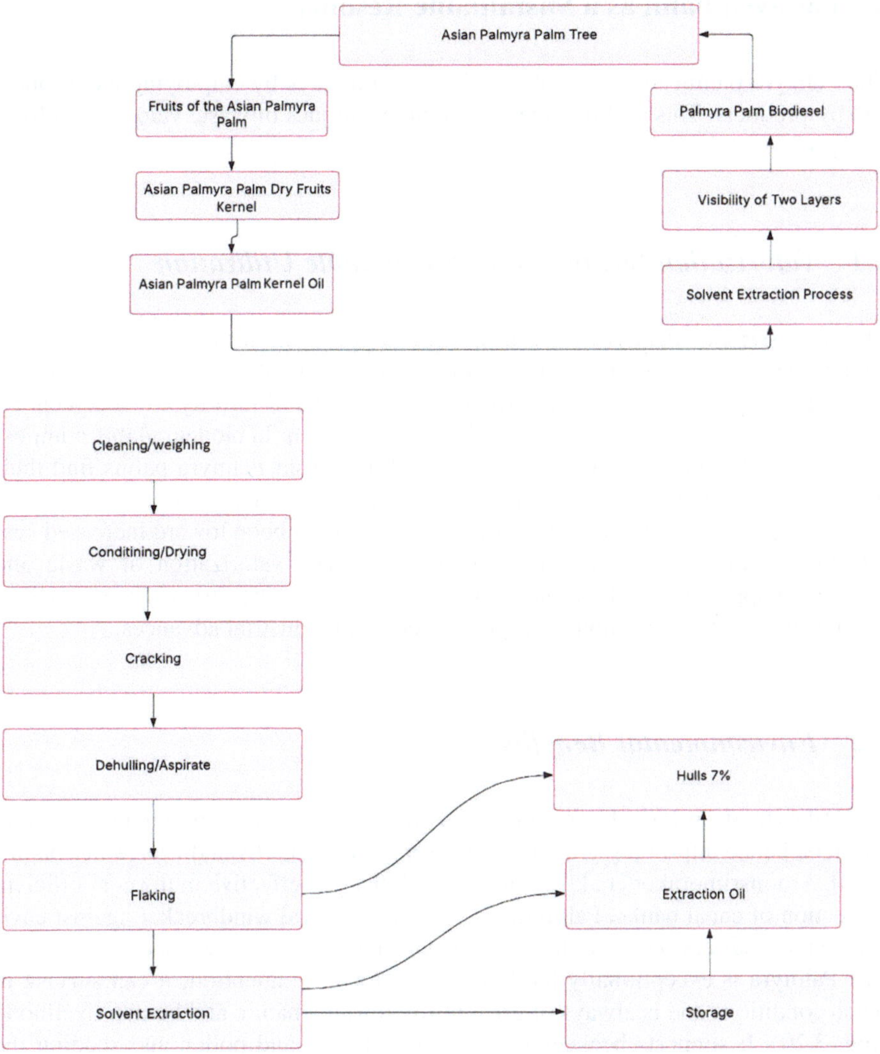

Fig. 3.19 Depiction of the biodiesel extraction process

6 Palmyra Palm as a Sustainable Resource

The Palmyra palm can be flexibly used in several ways by employing its products and by-products. This section notes different techniques of waste valorization from the species.

6.1 *Valorization Strategies for Sustainable Utilization*

The processing of Palmyra's waste part for the production of charcoal briquettes from husks, fruit calyx, and shells provides a method of valorization that assures the use of a low-emission fuel. The carbon retention and soil fertility enhancement by biochar make it promising as an agricultural amendment. In biodegradable composites and eco-friendly construction materials, fibers from Palmyra palms find their use. Circular economy practices enhance resource use efficiency by converting agricultural waste into high-value products as an additional boon toward increased sustainability and economic development [42–46]. The valorization of waste and innovative applications, therefore, is where the Palmyra tree truly drives environmental sustainability, economic empowerment, and industrial advances.

6.2 *Environmental Benefits*

The Palmyra, deep-rooted in nature, can store water very efficiently in underground aquifers, hence improving groundwater reserves when cultivated near water bodies [4, 8]. Aroonsrimorakot et al. [25] mentioned that it is effective in the cost-efficient protection of canal banks. Palmyra also acts as a natural windbreaker against environmental disasters and paves the way for sustainable development and self-reliance. The Palmyra is exceptionally hardy and asks for little attention; it can survive in harsh conditions and is always of great utility towards nature and human livelihood (Fig. 3.20). It supports beekeeping, providing nectar and pollen and shading the hives, facilitating construction and brood rearing [4, 30, 47].

6.3 *Socio-Economic Benefits*

Palmyra-based industries are crucial to regional economies; in Tamil Nadu, Andhra Pradesh, and West Bengal, India, leaves, veins, and roots of Palmyra form the basis of handicrafts that sustain livelihoods in the rural areas [8]. The advent of value addition into Palmyra products has tremendously improved marketability, storage life, and year-round availability [4, 11]. In Phetchaburi, Indonesia, Palmyra sugar cake production evolved from a local subsistence project to a commercially viable business with very high returns on revenue and a benefit-cost ratio of 11.30 [48]. In Odisha, scheduled caste communities such as *Chamar* and *Pana* realize their

Fig. 3.20 Palmyra palm demonstrating resilience and adaptability with (**a**) a bent young palm regrowing under dry conditions and (**b**) an exposed root mass continuing to survive and develop

livelihoods through the harvesting of fruits and the creation of products woven from its fibers [3]. By being combined with modern innovations, traditional knowledge will amplify Palmyra's socio-economic prospects.

7 Innovative Utilization and Sustainable Practices

Along with demonstrating the versatility of Palmyra in valorizing waste, sustainable architecture and craft modernization have also benefited the environment and local communities.

7.1 Palmyra House

A multi-practice architecture by Bijoy Jain and his team from Studio Mumbai, Palmyra House showcases sustainable construction. This building in Nandgaon, near Mumbai, is naturally ventilated but prevents direct sunlight and rain by making use of the properties of locally sourced Palmyra wood. It opens up towards a 25-foot-wide courtyard with a swimming pool in between the two structures and gives a seamless impression to the surroundings. Structural framing is done with Ain wood (referred to as Marutham in Tamil) so as not to aid the spread of fire, while the slatted louvre-like openings made of Palmyra bark improved the air circulation. The roof was made from Palmyra leaves, which are much more long-lasting: four times so than coconut leaves. The house was built with sensitive respect for ecological principles and cost around INR 2 crores, assisting in the preservation of local craftsmanship [49].

7.2 Modernization of Palmyra Wicker Crafts

Creative social enterprises have revived wicker crafts from Palmyra in East Nusa Tenggara, Indonesia, which involves artisans undergoing traditional and modern weaving. Such enterprises better provide access to markets, create financial independence, and facilitate community resilience through cultural preservation and economic empowerment. Palmyra industries continue to raise livelihoods while preserving the traditional knowledge and craftsmanship [50].

8 Palmyra Palm as a Gift to Communities

The Palmyra tree has a significant connection with many communities all around Asian countries; people treat the Palmyra tree as the identity of their community and culture. These people rely on the Palmyra plant for their economy, livelihood, and various other reasons. Commonly, people from Tamil Nadu and the North and East parts of Sri Lanka have deep connotations with the Palmyra tree and Palmyraculture. These people worship the Palmyra tree and get many valuable benefits out of it. The Palmyra palm is revered by several Tamil Nadu communities as *Panjam Pokki* and *Karpaga Tharu*, who believe the tree to be sacred [4]. Moreover, in the article by Hari Babu and Ramasamy [51], the authors talk about the religious and cultural connection between the Palmyra tree and the Tamil community in Tamil Nadu. Among this community, the Palmyra tree is seen as a divine tree. Most of the temples around Tamil Nadu, such as Thiruppanandal and Thirupanaiyoor, carry significant symbols of the Palmyra palm. Many villages and small gods of villages are named after Palmyra trees (*panai*). Similarly, the medicinal value of the Palmyra tree is also crucial to the community. Most of the parts of the Palmyra tree have medicinal benefits, which heal people from various ailments.

8.1 Ethnomedicinal and Traditional Healing Applications

The Palmyra palm has been deeply embedded in traditional medicine and folklore across various cultures. From root to shoot, all of the parts of Palmyra are used for various purposes. Palmyra itself has the potential to treat many diseases. The water found in the pulp cavities of unripe fruit, typically encountered between March and June, is used to prevent hiccups and as a treatment for nausea and vomiting [9].

Palmyra products such as *neera*, palm candy, and roots are widely recognized for their therapeutic uses. *Neera* contains essential minerals and vitamins and acts as a laxative and diuretic, aiding digestion and cooling the body. Palm candy, when added to milk, is a remedy for dry cough and sore throat. The spadix and roots are also traditionally used for various ailments [8].

The ethnomedicinal uses of the Palmyra palm span multiple regions in India. Some tribes mix the juice of young leaves with water as a traditional cure for dysentery [23]. Palmyra palm-leaf secretions also enhance antibacterial and wound healing properties [52]. The palm is used to treat skin infections, syphilis, diarrhea, jaundice, snake bites, earaches, tuberculosis, and more [10].

Traditional uses include its application as an abortifacient and treatment for burns, convulsions, earache, headache, epilepsy, hysteria, gonorrhea, menstrual disorders, rheumatism, scabies, sores, syphilis, and ulcers. Its anti-inflammatory, diuretic, aphrodisiac, and sedative properties are well documented. Fresh sap is a source of vitamin B-complex, while fruits are known to treat fever, general debility, and skin diseases [3, 10].

People in Asian countries use Palmyra palm for healing purposes. Siddha medicine follows the traditional way to treat diseases, very common in Sri Lanka and Tamil Nadu. There are many medicines from the Palmyra tree for different types of diseases such as *Eggu parpam, Karuvanga parpam, Naga parpam, Rasa mezhugu, Kumatti mezhugu, Nandhi mai, Vaan mezhugu,* etc. Additionally, the young blossoms of the palm are crushed, and their drops are left into the eyes of persons affected with chicken pox. The palm hand fan is used in cases of fever: *Vatha, pitha, kabha, kabavatha, pithakaba, pithavatha,* and *vatapitha* conditions. Palm tree blossom is used for bleeding piles, inflorescence rheumatic dyspepsia, tooth diseases, and chronic fever. The water from young Palmyra fruit kernels is said to provide relief from prickly heat, and eating the kernel with its skin helps treat dysentery. Its fruit parts are used for treating eczema, scabies, constipation, and UTIs, while toddy is used for aphrodisiac action, thirst, and purifying arsenic compounds. The root cures syphilis, has refrigerant action, and gives beauty to the skin. Toddy with limestone applied on the pot, a sweet drink is collected called *patha neer*. Drink continuously for 40 days; it cures venereal diseases [53].

Ethnic communities in Bangladesh, India, Indonesia, Sri Lanka, and Nigeria use raw fruits and various parts of the Palmyra palm to treat numerous illnesses. In Dungkek, Indonesia, male inflorescence is also used to prepare traditional medicines for reproductive ailments [23]. Indigenous communities have relied on Palmyra products for their stimulant, diuretic, and antiphlogistic properties. Fruits are considered aphrodisiacs, sedatives, and laxatives, while juice and toddy are known for treating inflammatory reactions, bleeding, and edema. The root paste has been used for menstrual regulation and the treatment of prolonged cycles [3, 10].

8.2 Scientific and Nutritional Benefits of Palmyra Palm

The Palmyra palm contains several bioactive compounds such as gums, albuminoids, steroidal glycosides, fats, and carbohydrates like sucrose, as well as spirostane-type steroids like borassosides and dioscin. These help contribute to its extensive health benefits [4, 54–56].

The root of the Palmyra palm has compounds with anti-inflammatory, anticancer, antioxidant, and antimicrobial properties [10]. The male inflorescence exhibits strong anti-inflammatory properties, while the young roots are known for their diuretic and anthelmintic effects. A decoction made from the roots is used for respiratory conditions [4]. The seed coat extract has notable antimicrobial activity. Its bark's decoction mixed with salt can be served as a mouthwash, while charcoal derived from the bark is used as a dentifrice [4]. Various parts of Palmyra palm are also used for the synthesis of varieties of nanomaterial which exhibit different chemical and biological properties [14, 19].

Its fruit pulp is a low-calorie sugar alternative. The pulp's high fiber content supports digestive health, while its glycemic index makes it suitable for diabetics [57]. Palmyra pulp is also known for its yellowish-orange color in visible light and green fluorescence under UV light. This fluorescence property is used to detect Fe(II) and Cd(II) ions [58]. The sprouts are rich in magnesium, iron, calcium, and protein, promoting bone health, improving cardiovascular function, and enhancing oxygen flow in the body [57]. A decoction of palm sugar, pepper, dry ginger, and coriander seeds is a traditional remedy for cough and bloating. Palm sugar also enhances energy levels and overall health [7]. Table 3.6 shows the discussed medicinal benefits of the Palmyra palm, highlighting its role in traditional remedies and modern health practices.

Table 3.6 Medicinal benefits of Palmyra palm

Specific benefits	Description
Ethnomedicinal applications	Palmyra products (e.g., sap, palm candy, roots) are widely used in traditional medicine for digestive health, skin ailments, and respiratory issues
Treatment for skin ailments	Palm fruit jelly relieves itching, treats prickly heat, and soothes chickenpox symptoms
Digestive and respiratory aid	Sap acts as a laxative and diuretic; root decoctions relieve respiratory and digestive conditions
Anti-inflammatory properties	Fruit pulp, roots, and male inflorescence reduce inflammation and swelling
Antimicrobial properties	Ash from spadix, toddy, and seed coat extracts exhibit antimicrobial and antibacterial activity
Traditional remedies	Used for hiccups, nausea, vomiting, diarrhea, dysentery, jaundice, tuberculosis, menstrual disorders, snake bites, syphilis, burns, convulsions, and more
Scientific validation	Contains bioactive compounds like flavonoids, saponins, and phenolic compounds, supporting antioxidant, anti-inflammatory, and antimicrobial effects

9 Scopes and Challenges

Palmyra trees are valuable resources for waste valorization, earning potential from handicrafts, medicines, and food products. Their merits of adaptation to semi-arid regions become sustainable options for future developments. However, challenges continue to exist in terms of attained traditional knowledge as the skills and expertise offered by older generations gradually fade. Poor infrastructure and a lack of advanced processing technologies further stand as hindrances to Palmyra-based industries. Efforts must also be directed towards integrating traditional knowledge systems with modern innovations to ensure better processing and marketing access. Strengthening value addition and marketing will maximize the economic and ecological contributions of Palmyra while assuring sustainable utilization with the minimum of waste.

10 Conclusion

Palmyra palm is also known as "nature's gift to the community." This tropical giant survives in different arid climates and is a powerhouse in many ways. The Palmyra is noted for being one whose shoots to roots make great utility, totally embracing an integrated way of resource utilization, even after its death. Transforming waste into wealth perfectly resonates with the global push toward circular economy practices and sustainable development. Cropping up in attention now in the age of the circular economy is looking into what use these plants can have to help this, coming together through a blend of old wisdom and modern innovation to address any immediate challenges. The Palmyra palm stands as a sign of nature's ability to function towards sustainable solutions, be it for environmental, social, or economic progress. There is mounting pressure to harness its entire lifetime's potential through innovative thinking and broaden options for a future where our approach to waste valorization and sustainable development is much welcomed.

Acknowledgments We would like to thank the following organizations for their amazing support of our research on Palmyraculture: Panaiyaanmai—The Centre for Self-Reliance and Sustainable Development, Kadayam, Tamil Nadu; the Asian University for Women, Chittagong; the Palmyrah Development Board (PDB), Jaffna; the Bangladesh Sugar Crop Research Institute (BSRI), Pabna; and Green Bangle Movement. PMSK acknowledges Naam Tamilar Katchi and Govt. of Tamil Nadu for their contribution in palmyrah palm tree plantation and conservation related activities. Author suggests that the Tamil Nadu government should lift the ban on Asian Palmyra toddy which is a nutritional, medicinal, and traditional drink for Tamilians and many Asian community people. We also thank all of the Palmyra Warriors, such as climbers, toddy tappers, and artisans, for their self-sufficiency, environmentally conscious way of life, and genuine efforts to preserve and make use of the Palmyra palms in Asia. We also express our gratitude to farmers, activists, scientists, and ecopreneurs for their support in promoting Palmyraculture.

References

1. Kuldeep SA, Vijayakumar V, Segaruban S, Sinnathurai Y, Hossain MS, Begum K, Pau J, Selvakumar PM (2023) An insight into the molecules and materials of the haustorium of the Asian Palmyra palm (Borassus flabellifer) and their nutraceutical values. Food Nutr Chem 1(3):1–10. https://cae.usp-pl.com/index.php/fnc/article/view/115
2. Bayton RP (2007) *Borassus flabellifer L.*, Bayton, Palmweb, 1187(1753). https://www.palmweb.org/cdm_dataportal/taxon/025ac0de-6d00-4243-bacb-16196d41b0a4
3. Panda T, Mishra N, Pradhan BK, Rahimuddin S, Mohanty RB (2021) Palmyra palm (*Borassus flabellifer*, Arecaceae) as a valuable resource for livelihood security in Odisha, India. Adv Res Life Sci 5(1):64–70. https://doi.org/10.2478/arls-2021-0031
4. Krishnaveni TRS, Arunachalam R, Chandrakumar M, Parthasarathi G, Nisha R (2020) Potential review on Palmyra (*Borassus flabellifer L.*). Adv Res 21(9):29–40. https://journalair.com/index.php/AIR/article/view/738
5. Garaga S, Shaik RA, Naga J, S. (2018) Borassus flabellifer fruit versatile pharmaceutical application: an overview. Int J Adv Res Med Pharm Sci (IJARMPS) 3(4):12. http://www.ijarmps.org/wp-content/uploads/2018/04/v3.i4.2.Borassus-Flabellifer-Fruit-Versatile-Pharmaceutical-ApplicationAn-Overview.pdf
6. Priyagummadi V, Rao Battu G, Diyya MSK, Manda K (2016) Review on Palmyra palm (*Borassus flabellifer*). Int J Curr Pharm Res 8(2):975–7066. https://journals.innovareacademics.in/index.php/ijcpr/article/view/12102/4414
7. Murugesan R, Vathsala V (2023) Palmyra palm (*Borassus flabellifer*): a multifaceted plant with diverse uses and cultivation practices. 2(11):440–444. https://www.researchgate.net/publication/376649351_Palmyra_Palm_Borassus_flabellifer_A_Multifaceted_Plant_with_Diverse_Uses_and_Cultivation_Practices
8. Sankaralingam A, Hemalatha G, Ali AM (1999) A treatise on palmyrah. Central Plantation Crops Kasaragod Research Institute, pp 1–38. https://cpcri.icar.gov.in/filemgr/webfs/publication/CP2817.pdf
9. Symonds WA (1908) The Palmyra palm and its uses. pp 237–244. https://books.google.co.in/
10. Thakur NS, Attar SK, Gunaga RP, Chouhan RS, Hegde HT, Bhusara JB (2016) Mythology and ethno-pharmacology of Asian Palmyra palm (*Borassus flabellifer L.*). In: National conference on Palmyra Palm, pp 80–93. https://www.researchgate.net/profile/N-S-Thakur/publication/298788493_Anecdote_about_mythology_and_ethno-pharmacolgy_of_Asian_palmyra_palm_Borassus_flabellifer_L/links/59cce88daca272d761c2aff8/Anecdote-about-mythology-and-ethno-pharmacolgy-of-Asian-palmyra-palm-Borassus-flabellifer-L.pdf
11. Srivastava A, Bishnoi SK, Sarkar PK (2017) Value addition in Palmyra palm (*Borassus flabellifer*): a potential strategy for livelihood security and poverty alleviation. Rashtriya Krishi 12(1) e-ISSN-2321-7987. https://www.researchgate.net/publication/320908186
12. Dalibard C (1999) Overall view on the tradition of tapping palm trees and prospects for animal production. Livest Res Rural Dev 11(5) http://www.lrrd.org/lrrd11/1/dali111.htm
13. Arunachalam K, Saravanan S, Parimelazhagan T (2011) Nutritional analysis and antioxidant activity of palmyrah (*Borassus flabellifer L.*) seed embryo for potential use as a food source. Food Sci Biotechnol 20(1):143–149. https://doi.org/10.1007/s10068-011-0020-y
14. Selvakumar PM, Nuzhat S, Quadrey MM, Monichan S, Samdavid Thanapaul RJR, Muthukumar Nadar MSA (2022) Green nanomaterials: design, synthesis, properties, and industrial applications. In: Handbook of smart materials, technologies, and devices: applications of Industry 4.0. Springer, Cham, pp 1–19. https://doi.org/10.1007/978-3-030-58675-1_98-1
15. Narayanankutty A, Job JT, Kuttithodi AM, Sasidharan A, Benil PB, Ramesh V, Elsadek MF, Rizwana H, El-Din MME (2023) Proximate composition, antioxidant, anti-inflammatory and antidiabetic properties of the haustorium from coconut (*Cocos nucifera L.*) and Palmyra palm (*Borassus flabellifer L.*). J King Saud Univ Sci 35(1):1–4. https://www.sciencedirect.com/science/article/pii/S1018364722005857

16. Prasad ARB, Anandharaj A, Sivanandham V, V CD, Baskaran N (2022) Exploring the nutritional profiling and health benefits of Palmyra haustorium (*Borassus flabellifer*). S Afr J Bot 151:228–237. https://www.sciencedirect.com/science/article/pii/S0254629922000291
17. Vengaiah PC, Prasad KR, Murthy GN, Sumitha S, Jerard BA (2024) Palmyrah (*Borassus flabellifer L.*) tuber in India: present status and scope. Asian J Curr Res 9(2):73–80. https://ikprress.org/index.php/AJOCR/article/view/8596
18. Sandhiyadevi P, Abul Kalam Azad S, Kavin Kumar D, Logu P, Naveen Kumar P (2021) Nutritional analysis of Palmyra tuber. Int J Innov Res Sci Eng Technol 10(3):2205–2212. https://www.ijirset.com/upload/2021/march/127_NUTRITIONAL_NC.pdf
19. Monichan S, Thevamirtha C, Thanapaul RJRS, Selvakumar PM (2021) Palmyraculture: an insight into the nano medicines from Palmyra palm (*Borassus flabellifer L.*). SCIRP 5(11):143–150. https://actascientific.com/ASMS/ASMS-05-1079.php
20. Sivaji M, B A (2021) Antioxidant activity and nutritional properties of freshly tapped Palmyra (*Borassus flabellifer*) sap. Adv Technol 1(2):300–306. https://journals.sjp.ac.lk/index.php/ait/article/view/4880/3838
21. Deepa (2020) Panaattu & Paani Panaattu: Sri Lankan Palmyrah fruit treats. Paticheri. https://www.paticheri.com/2020/12/20/panaattu-paani-panaattu-sri-lankan-palmyrah-fruit-treats/
22. Uluwaduge I, Parera A, Jansz E, Thabrew I (2008) A pilot study on palmyrah pinattu (dried fruit pulp) as an anti-diabetic food component. Int J Biol Chem Sci 1(3):250–254. https://www.ajol.info/index.php/ijbcs/article/view/39713
23. Saidi IA, Efendi N, Azara R, Hudi L (2021) Uses of Palmyra palm plant parts in three regions of East Java, Indonesia. Afr J Food Agric Nutr Dev 21(5):18055–18065. http://ijrphr.com/files/articles/1c09a65311a920739f6b75c3f4b3e1df.pdf
24. Le DHT, Lu W, Li P (2020) Sustainable processes and chemical characterization of natural food additives: Palmyra palm (*Borassus flabellifer Linn.*) granulated sugar. Sustainability 12(7):2650. https://doi.org/10.3390/su12072650
25. Aroonsrimorakot S, Bhaktikul K, Metadilogkul O (2020) Palmyra palm and the protection of landslide along the canal. Naresuan Univ J Sci Technol (NUJST) 29(1):121–131. https://doi.nrct.go.th/admin/doc/doc_654657.pdf
26. Franco FM, Samuel G, Francis T (2020) Mutualism between humans and palms: the curious case of the palmyra palm (*Borassus flabellifer L.*), and its tapper. *Working Paper Series*, pp 8–27. https://www.researchgate.net/profile/F-Merlin-Franco/publication/344347248_Mutualism_between_Humans_and_Palms_The_Curious_Case_of_the_Palmyra_Palm_Borassus_flabellifer_L_and_its_Tapper/links/5f6ab93392851c14bc8e2f4e/Mutualism-between-Humans-and-Palms-The-Curious-Case-of-the-Palmyra-Palm-Borassus-flabellifer-L-and-its-Tapper.pdf
27. Ahmed MR (2014) Palm fruit fiber extraction, spinning, dyeing, and characterization. Textile Today. https://www.textiletoday.com.bd/palm-fruit-fiber-extraction-spinning-dyeing-and-characterization
28. Alshahrani H, Pathinettampadian G, Gujba AK, Vincent Rethnam AP (2022) Effect of Palmyra sprout fiber and biosilica on mechanical, wear, thermal, and hydrophobic behavior of epoxy resin composite. J Ind Text 52:1–20. https://journals.sagepub.com/doi/full/10.1177/15280837221137382
29. Bhattacharyya S (2017) নৌকোঘর “*Noukoghar*” abode of Bengal Boat [Photograph]. Bengal Boat. https://bengalboat.blogspot.com/2017/
30. Mehta A, Sinha SK, Behera L, Pathak J (2016) Palmyra palm and beekeeping. ASPEE College of Horticulture and Forestry, NAU, Navsati (Guiatd), pp 32–36. https://www.researchgate.net/publication/301324852
31. Balasubramani V, Nagarajan KJ, Karthic M, Pandiyarajan R (2024) Extraction of lignocellulosic fiber and cellulose microfibrils from agro waste-Palmyra fruit peduncle: water retting, chlorine-free chemical treatments, physio-chemical, morphological, and thermal characterization. Int J Biol Macromol 259(2) https://www.sciencedirect.com/science/article/abs/pii/S014181302400076X

32. Constantinescu AG, Marinescu AH (2020) Asia& Europe in social sciences: connections, representations, interpretations. Editura Universităţii din Bucureşti-Bucharest University Press. https://www.researchgate.net/publication/343237547_Asia_and_Europe_in_Social_Sciences_From_Orientalism_to_Post-Colonial_Discourse_Introduction_In_book_ASIA_EUROPE_IN_SOCIAL_SCIENCES_Connections_representations_interpretations_Editors_Alexandra_Gabriel
33. Patnaik DP (1989) Palm-leaf etchings of Orissa. https://books.google.co.in/
34. Chaurasiya AK, Chakraborty I, Saha J (2011) Value addition of Palmyra palm and studies on the storage life. J Food Sci Technol 51(4):768–773. https://link.springer.com/article/10.1007/s13197-011-0561-3
35. Currylines (2020) Ripe Palmyra fruit—*Borassus flabellifer*. https://currylines.com/ripe-palmyra-fruit-borassus-flabellifer
36. Behera S (2022) Phytochemical constituents and nutritional potential of Palmyra palm: a review. Rev Contemp Sci Acad Stud 2(12):1–7. https://thercsas.com/wp-content/uploads/2022/12/rcsas2122022003.pdf
37. Vijaya Kumara B VP, Prasad KR MG (2015) Physico-chemical properties of Palmyrah fruit pulp (*Borassus flabellifer*). J Nutr Food Sci 5(5):1–4. https://doi.org/10.4172/2155-9600.1000391
38. Vijaya kumari B, Kiranmayi P, Vengaiah PC (2017) Estimation of vitamins, minerals and amino acids in Palmyra palm (*Borossus flabellifer L.*) fruit pulp. Int Res J Pharm 7(12):70–73. https://doi.org/10.7897/2230-8407.0712150
39. Rayappa MK (2023) Palmyrah palm (*Borassus flabellifer*) non-centrifugal sugar—current production practices as a natural sugar and a promising functional food/additive. J Agric Food Res 14:100829. https://www.sciencedirect.com/science/article/pii/S2666154323003368
40. Thevamirtha C, Balasubramaniyam A, Srithayalan S, Selvakumar PM (2023a) An insight into the antioxidant activity of the facial cream, solid soap and liquid soap made using the carotenoid extract of palmyrah (*Borassus flabellifer*) fruit pulp. Ind Crop Prod 195:116413. https://doi.org/10.1016/j.indcrop.2023.116413
41. Elumalai PV, Sivakandhan C, Parthasarathy M, Mohamed Iqbal S, Arunkumar M (2021) Investigation on the mitigation of environmental harmful emissions by incorporating nanoparticles to biofuel water nano emulsion in low heat rejection engine. Heat Mass Transf 57:1235–1250. https://link.springer.com/article/10.1007/s00231-021-03028-7
42. Arifianti QAMO, Anggarini U, Nafis A, Rosita Sari E, Fidiya Nugrahani E (2020) The effect of activated carbon addition on woody cutting waste briquette combustion quality. E3S Web Conf 190:00029. https://www.e3s-conferences.org/articles/e3sconf/pdf/2020/50/e3sconf_icorer2020_00029.pdf
43. Chumsang C, Upan P (2014) Production of charcoal briquettes from Palmyra palm waste in Kirimat District, Sukhothai Province, Thailand. Appl Environ Res 36(3):29–38. https://doi.org/10.35762/AER.2014.36.3.3
44. Faiad A, Alsmari M, Ahmed MM, Bouazizi ML, Alzahrani B, Alrobei H (2022) Date palm tree waste recycling: treatment and processing for potential engineering applications. Sustainability 14(3):1134. https://doi.org/10.3390/su14031134
45. Hafemann E, Battisti R, Marangoni C, Machado RAF (2019) Valorization of royal palm tree agroindustrial waste by isolating cellulose nanocrystals. Carbohydr Polym 218:188–198. https://doi.org/10.1016/j.carbpol.2019.04.086
46. Wichchuki S, Chantaratepthimakul S, Thongkaew C, Fuengfoo M (2023) Application of low methoxyl pectin powder derived from Palmyra palm pulp wash water to enhance toddy palm cake quality. Agric Nat Resour 57(6):1015–1024. https://doi.org/10.34044/j.anres.2023.57.6.12
47. Rahman AN, Sayanhika S, Parameshwaram PM, Huseynova L, Prakash A, Chakma S, Selvakumar PM (2025) The multifarious ecosystem services of Asian Palmyra palm. In: Global warming: problems and precautions, Part 4, pp 33–46. https://books.akademisyen.net/index.php/akya/catalog/download/3453/10572/80840

48. Nujnetra A (2016) The process of changing traditional knowledge into value for money: Palmyra sugar cake from *Borassus flabellifer* in Phetchaburi. RMUTI J Human Soc Sci 4(1):1–11. https://so05.tci-thaijo.org/index.php/RMUTI_SS/article/view/189596
49. Zachariah SS (2020) Made of local Palmyra wood, this Mumbai house was built by hand & needs no ACs! The Better India. https://thebetterindia.com
50. Gumulya D, Purba JT, Hariandja ES, Pramono R (2022) Modernization of Palmyra wicker crafts as a byproduct of creative social enterprise's innovation capability. Int J Des Nat Ecodyn 17(2):209–219. https://www.iieta.org/journals/ijdne/paper/10.18280/ijdne.170207
51. Hari Babu P, Ramasamy R (2023) Developing a Palmyra palm sacred grove wellness tourism circuit in Tamil Nadu, India involving active stakeholders' participation. https://doi.org/10.21203/rs.3.rs-2547263/v1
52. Mariselvam R, Ignacimuthu S, Ranjitsingh A, Mosae SP (2021) An insight into leaf secretions of Asian Palmyra palm: a wound healing material from nature. Mater Today Proc 47(3):733–738. https://doi.org/10.1016/j.matpr.2020.05.393
53. Ramya S (2018) Review on traditional and phyto-pharmacological aspects of *Borassus flabellifer* (Palmyra tree). Int J Reverse Pharmacol Health Res 1(1):6–11. http://ijrphr.com/files/articles/1c09a65311a920739f6b75c3f4b3e1df.pdf
54. Sayanhika S, Selvakumar PM (2024) An insight into the cosmetic and dermatologic applications of the molecules of Palmyra palm. Cosmetics 11(6):196. https://doi.org/10.3390/cosmetics11060196
55. Selvakumar PM, Thanapaul RJRS (2020) An insight into the polymeric structures in Asian Palmyra palm (*Borassus flabellifer Linn*). Org Polym Mater Res 2(2):16–21. https://doi.org/10.30564/opmr.v2i2.2639
56. Thevamirtha CTA, Kuldeep SA, Dewanjee S, Kazi E, Tharmalingam S, Dhanaraj P, Paulraj MS (2023b) An insight into the structure and functions of flabelliferins and borassosides; nutraceuticals in Asian palm (*Borassus flabellifer*). Food Nutr Chem 1(2):1–8. https://doi.org/10.18686/fnc.v1i2.31
57. Vijayanchali SS (2024) Proximate composition and health benefits of Palmyra (*Borassus flabellifer*): an overview (Chapter 8). In: Nurturing sustainability: exploring agricultural dimensions, pp 111–123. https://www.researchgate.net/publication/382641622
58. Varadaraju C, Paulraj MS, Selvan GT, Vijeindran SS, Mariselvam R (2021) An insight into Asian Palmyra palm fruit pulp: a fluorescent sensor for Fe^{2+} and Cd^{2+} ions. Mater Today Proc 47(3):747–750. https://doi.org/10.1016/j.matpr.2020.06.532

Chapter 4
Palmyra Palm Fruit Processing Wastes and By-products: Valorization Using Bioprocess Technology

Aanjana Shetty, Prasanna Bhalerao, Ashish Dabade, and Sachin K. Sonawane

Abbreviations

CFRP	Carbon Fiber Reinforced Polymer
CMC	Carboxy methyl cellulose

1 Introduction

Beyond the well-known fruit, the Palmyra palm (*Borassus flabellifer* L.) has several highly valued uses, particularly in traditional medicine, industry, and agriculture. Processing wastes and by-product handling are certainly among the largest challenges associated with palmyra palm cultivation. These residues, the major components of which are fruit shells, leaf stalks, and other biomass, continuously threaten the environment if not utilized and result in inefficiency to the economy. But under the prism of bioprocess technology, these wastes become precious resources ready to be valued rather than liabilities. The present chapter deals with how wastes/by-products from palmyra palm fruit processing can be used as a potential tool to change the scenario using bioprocess technology. Anaerobic digestion, enzymatic hydrolysis, and fermentation are some of the biological processes for converting such agricultural wastes into several high-value products. Bioprocesses open the doors to sustainable solutions to satisfy current resource efficiency and

A. Shetty · A. Dabade · S. K. Sonawane (✉)
School of Biotechnology and Bioinformatics, D. Y. Patil deemed to be University, Navi Mumbai, MH, India
e-mail: sachin.sonwane@dypatil.edu

P. Bhalerao
School of Engineering, Ajeenkya D. Y. Patil University, Pune, MH, India

C. Trois, R. C. Ray (eds.), *Plantation Crop Wastes: Valorization for Economic Sustainability*, Waste as a Resource,
https://doi.org/10.1007/978-3-031-96805-1_4

environmental stewardship needs from renewable energy sources, including biogas, to bioactive chemicals in medicines and functional foods.

The palmyra palm is an important cultural and economic plant species, and its historical and geographical origins are in the tropical parts of South and Southeast Asia. The palmyra palm is native to the Indian subcontinent; it is prevalent in India, Sri Lanka, and parts of Southeast Asia, like Thailand, Myanmar, and Cambodia. It is tolerant of a drought cycle and does well in hot, dry climates. This palm thrives in lowland and coastal settings due to its easy adaptability to sandy or saline soils. The palmyra palm is not confined to its native regions. It has been introduced and is now grown in many tropical countries around the globe. These include regions in the Caribbean and Pacific Islands where it becomes an economic livelihood and culture, and parts of Africa where it was introduced through prehistoric trade routes. Its main production source is the fruit of the palmyra palm, picked for its pulp and seeds. Many recipes use this palm tree fruit, and its sap is collected for making traditional sweeteners like jaggery or toddy. Leaves are woven into thatch for roofing and carpets, thus enriching regional economies in the building and craft industries. Fully grown trees can generate high yields of fruit and sap. The environmental conditions and farm practices would, however, govern the actual productivity. Individual palms may yield several hundred fruits per year, sufficient for commercial and subsistence cultivation. The productivity of palmyra palms refers to their resilience and adaptability, indicating this species' importance as a renewable source of material, food, and income in various geographical locations.

This chapter presents the techniques, processes, and applications driving the value-addition of palmyra palm wastes. The current scientific bases of bioprocesses are discussed, pointing out the global successes and debating financial and environmental gains from using such technologies. It also intends to query how existing bioprocess methodologies could profit from a synergy between traditional practices and cutting-edge biotechnological developments. Bioprocess technology makes it a means to reduce waste, enhance resource efficiency, and create economic opportunities for communities dependent on palmyra palm production by converting agricultural leftovers into assets.

2 Various Parts of the Palmyra Palm Tree and Their Associated Applications

The sturdy *Borassus flabellifer* tree is around 30 m (98 feet) high in its growth. The withering leaves remain attached to the tree for several years before they gracefully fall off; the trunk is stout, gray-colored, and ringed with leaf scars. The leaves have heavy black teeth at the petiole margins and are fan-shaped, around 3 m (9.8 feet) long. *B. flabellifer*, like all *Borassus* species, is dioecious, with male and female flowers on separate plants. Male flowers are less than 1 cm long and form a semicircular cluster concealed beneath scale-like bracts within the catkin-like

Fig. 4.1 (**a**) Palmyra Palm tree (**b**) cluster of fruits (**c**) grittier fiber and (**d**) fine fibers. (Adopted from [42]; with permission, License Number 5956990408499)

inflorescence. The female flowers are golf ball-sized and solitary, sitting upon the surface of the inflorescence axis. Figure 4.1 represents the various parts of palmyra palm [42].

2.1 *Fruit*

Clusters of fruits can range in diameter from 10 to 18 cm. Its pericarp is black-colored. Cut off the top end of the fruit to reveal translucent, jelly-like, pale-white, and sweet seed sockets. They have a similar appearance and feel to lychee pits but have a softer flavor. The fruit has two, three, or four sweet jelly seed sockets. There is a thin, brownish-yellow skin covering the jelly portion of the fruit. It is well known that inside their fleshy white bodies is a watery fluid. These seed sockets served as the model for several Bengali candies named Jalbhara, or Sandesh [21].

2.2 *Leaves*

In many tropical places, the leaves of the palmyra palm are vital to many elements of daily life. They form an integral part of traditional construction, as their leaves are highly insulating and resistant to the elements. This is mainly for thatching roofs. Besides serving as building materials, artisans transform Palmyra palm leaves into various items such as fans, mats, baskets, and caps. More usefully, artisans twine the leaves into robust fibers for use in brushes, ropes, and twine, which are suitable for both hard applications and domestic use [21]. These leaves are integral to rituals and traditions in various cultures, holding significant religious and ritual values. Folk medicine utilizes its medicinal properties, leading to a multifaceted role in practical and cultural aspects.

2.3 Sap

The palmyra palm tree is a valuable resource with numerous uses in almost all tropical areas where it grows. The sap, harvested from the trunk or the inflorescence, significantly contributes to local economies and culinary traditions. People drink fresh palmyra palm sap, known as toddy when tapped fresh, as a sweet, refreshing, nonalcoholic drink. Besides this, the sap undergoes natural fermentation to turn into a slightly alcoholic palm wine, deeply associated with local traditions. In addition to drinks, local cuisines cook the sap of palmyra palms to obtain highly valued and unique palm sugar. In many tropical regions, the sap of the Palmyra palm tree holds significant value for various applications. Extracted from the trunk or inflorescence, the sap of this tree forms the backbone of regional economies and most of the cultural food habits [15]. People consume the fresh palmyra palm sap, whether tapped fresh, neeru, or toddy, as a cool, nonalcoholic drink. More interestingly, the self-fermenting sap eventually becomes palm wine, a mildly alcoholic drink richly ingrained in the tapestry of local culture. Beyond functional drinks, locals boil down the sap from palmyra palms to prepare palm sugar, enhancing local dishes' flavor.

3 By-products Preparation

Palmyra palm growing areas provide valuable by-products for daily life and the local economy. Among the by-products is palm sugar, which is acquired by boiling sap up to its crystallization or caking and is used as a natural sweetener in product development. Furthermore, the sap undergoes fermentation to produce toddy, a locally popular and intricately crafted cultural beverage with a low alcoholic content. The sturdy leaves of the Palmyra palm are woven into fans, hats, mats, and baskets, showcasing traditional craftsmanship and providing helpful household items. Building and furniture-making uses firm and durable hardwood, which the palm trunk produces. In addition, manufacturing brushes and ropes from the trunk's fibers contributes to domestic and agricultural purposes. Feeding animals with palm fronds and other parts contributes to efforts to enhance the sustainability of farming. In traditional treatments, various concoctions are prepared using portions of palmyra palms for therapeutic purposes. These different by-products indicate the versatility and importance of palmyra in supporting regional economies, cultural traditions, and ecological equilibrium [11]. Furthermore, extracting raw plant material from the palmyra tree involves systematic processes for separating and purifying cellulose. First, the cellulose-rich parts of the tree, such as the trunk and leaves, are collected, thoroughly washed to remove dirt or other contaminants, and then dried to lower its moisture content. After drying, the trunks and leaves are pulverized into a fine powder to enhance their surface area, facilitating the attachment of subsequent chemical treatments. After soaking in water, these powdered materials are filtered to remove the solid waste containing cellulose and all soluble impurities such as sugars and proteins.

The solid waste, usually under an alkali treatment using sodium hydroxide (NaOH), disintegrates and dissolves noncellulosic materials like hemicelluloses and lignin. After treating the residual, it is rinsed with water to neutralize the alkali and remove any remaining chemicals. Finally, the filtered cellulose is dried in an oven at a known temperature to halt deterioration and obtain a high-purity cellulose product. The cellulose is then chemically modified to create carboxymethyl cellulose (CMC), which has numerous industrial applications.

3.1 Carboxymethyl Cellulose (CMC) from Palmyra Palm

Due to its wide applicability, the food industry commonly uses CMC as an emulsifier, thickener, and stabilizer. It improves several food products' uniformity, texture, and quality. Such products include bakery items, ice cream, sauces, salad dressings, and beverages. In ice cream, CMC prevents the formation of ice crystals and provides a refined texture. CMC is water-soluble and has a gel-like property [22, 23]. The USA Food and Drug Administration considers CMC safe (Generally Regarded as Safe) when applied according to acceptable manufacturing practices.

3.1.1 Carboxymethylation

After dispersing dried cellulose in an organic solvent such as ethanol or isopropanol, sodium hydroxide (NaOH) at 20–40% of the cellulose weight is carefully added to encourage swelling. Chloroacetic acid, at a concentration of 0.5–1.5 times the weight of the swollen cellulose, aids the substitution reaction by adding carboxymethyl groups. The mixture is constantly agitated for several hours at high temperatures (50–70 °C) to ensure a successful reaction. Finally, acetic acid is added to neutralize the mixture, to stop the reaction, and stabilize the final product. After the process, the mixture is filtered to separate the solid CMC from the filtrate. The solid CMC is cleaned thoroughly using ethanol and water to ensure the finished product is free of impurities and by-products. The pure CMC is obtained and dried in a low-temperature oven for a final refined product.

3.2 Antioxidant Extraction from Pericarp

Extraction of antioxidants from the pericarp of a palmyra palm is a complex process and involves a systematic process of separation and purification of bioactive compounds with antioxidant properties. While harvesting the palmyra palm fruit for the first time, the outer layer of the fruit known as pericarps is separated. The pericarp is cleaned to remove dust or other pollutants and then dried. Drying can be done either in a low-temperature drying oven or in the shade to protect the heat-sensitive antioxidants in the pericarp. The dried pericarps are ground into fine powder using a grinder or blender. This increases the surface area and makes it easier to extract the antioxidants.

Subsequently, after grinding the pericarp powder, a solvent such as ethanol or methanol, or their aqueous mixture, is added to conduct the solvent extraction. The mixture is constantly churned and allowed to macerate for an appropriate time, usually several hours or overnight. This would help the dissolution of all antioxidants present in the pericarp in the solvent. After the maceration period, the mixture is filtered to remove solid residue and the liquid extract that contains antioxidants. This filtrate is then concentrated by a rotary evaporator, whereby the low temperature and low pressure preserve the antioxidant chemicals by evaporating the solvent.

Further drying techniques such as freeze-drying or vacuum-drying can be used for the antenna of a dry antioxidant extract. Further chromatographic techniques, such as column chromatography, can isolate monomolecular antioxidants if more specialized applications or higher purity are required [35]. Because such excellent antioxidants have acceptable health benefits and act against oxidative stressors, these high-quality antioxidants are extracted from the pericarp of the palmyra palm for use in a wide array of food preservation, nutraceutical, and cosmetics applications.

3.3 Fiber

The palmyra palm fiber is prepared through a multistep process, segregating and refining the fibrous material from different plant components. The first step involves harvesting eligible parts of the Palmyra palm, like its leaves or trunk. These sections were chosen due to the significant content of cellulose, the chief constituent of natural fibers. Plants that are collected are cleaned thoroughly to remove dirt, dust, and other contaminants. Oven drying or sun drying follows to reduce the moisture content and enhance handling for further processing. Fiber from this palmyra palm is prepared through an all-inclusive process that seeks to extract and purify the fibrous material from the parts of this plant [45]. First, parts of the palmyra palm, either its leaves or trunk, are appropriately harvested. The reason for choosing these parts is because they contain high contents of cellulose, which is the chief constituent of any natural fiber. Afterward, the plant material thus harvested is cleaned thoroughly of dirt, dust, and other impurities. The material is sun-dried or dried in a drying oven to reduce its moisture content, making it easily handled for further processing. The material is subjected to mechanical processing, which pulverizes or grinds the materials into smaller fragments. This mechanical action helps the fibers and nonfibrous components to separate from each other. The ground material is left soaked in water for some time to soften and eventually separate the fibers from it. Further, the soaked material undergoes a series of retting processes, wherein pectins and hemicelluloses, which bond the fibers together, get broken down by treating them with water or an averaged alkali solution. It can be done in various ways: enzymatic, chemical, or aqueous—these methods wholly rely on the final qualities and properties one wants in the fiber product. After retting, it is cleaned with clean water to remove any contaminants and remains of chemicals that can still exist. After cleaning, the material is combed or carded to align the strands and break those short and weak, leaving strong and long fibers. These aligned fibers undergo additional refinement processes, such as bleaching and drying, to enhance their purity and whiteness. Depending on their intended use, the refined fibers undergo either mat treatment or yarn spinning [41]. People value the resultant palmyra palm fiber for its strength, resilience, and adaptability, making it suitable for various applications in producing textiles, rope, and various biodegradable items. This is a careful preparation process that ensures that the produced fibers are of very high quality and meet standards for various commercial and industrial uses.

3.4 Palmyra Palm Leaf Stalk/Carbon Fiber-Based Composites

Palmyra palm is an abundant supply of renewable materials that may be valorized and utilized innovatively to obtain high-value by-products and advanced composite materials. The palmyra palm leaf stalks for extraction of natural fibers are first processed through the retting process. Retting refers to the procedure whereby nonfibrous materials are destroyed using chemicals or microbes to make the decortication easier, which will occur later. In decortication, mechanically separated fibers undergo cleaning and drying to produce long, robust fibers suitable for various uses. Such palmyra palm fibers are ideal construction material, handicrafts, textiles, and paper, and their propension is for being tough and strong. They are used in handicrafts, which are made into mats, baskets, and ropes, and in textiles to be spun into yarns and further into fabrics. Applied in the paper industry, palmyra palm fibers make high-grade paper of excellent tear strength. They further show their potential in reinforcing concretes and other building materials by increasing their flexibility and tensile strength.

Beyond fiber extraction, the stalks of palmyra palms are pyrolyzed to obtain biochar, a carbon-rich substance utilized in water filtration systems to substitute and raise soil fertility. Biochar produced from palmyra palms helps raise output in agriculture by improving the water and nutrient retention ability of the soil. In addition, because it is porous, it filters out pollutants from water to make it clean for drinking in rural areas and resource-constrained locations. Carbon fibers are mixed with a polymer matrix, like epoxy resin, to generate the carbon fiber reinforced polymer (CFRP). This strong and light composite material mixture is created by filling the fibers with resin through compression molding or resin infusion, which is then cured by the application of heat and pressure. A few examples of users of CFRP manufactured from palmyra palm fibers include aerospace, automobile, and sporting goods industries. They require materials with very high performance. A variety of noteworthy advantages for sports applications include improved performance of sports equipment, improved structural integrity in auto parts, and, for aeronautical applications, fuel economy [20].

3.5 *Torrefaction of Shell and Red Gram Stalk*

A specially designed thermal treatment process in the torrefaction of shell and red gram stalks from a palmyra palm tree to increase the biomass's amenability to energy generation and other applications. This process considers all such elements, from biomass collection to the properties of the torrefied biomass and its various applications, to emphasize its significance for the sustainable use of natural resources [26, 38]. The first component of the torrefied biomass production process is collecting and preparing the raw feedstock. The raw biomass of the palm fruit shells and fruit stalk is a good feedstock that can be effectively incorporated into torrefaction. These residues are usually wasted or used irrationally and are assumed to be agricultural waste. They can be utilized, however, to increase their value, and they have a significant energy capacity.

To ensure that the torrefaction process is sufficiently competent and efficient, biomass preparation is required. Since high moisture content may bring about bad energy conversion by hindering the torrefaction process, drying raises the energy density of the biomass due to less water that must be heated and evaporated during torrefaction. Biomass can be chipped or ground for minor product use, thus increasing the surface of the particles; this will enable the heat to be distributed uniformly during the torrefaction process. Torrefaction converts biomass into a more stable and energy-dense material. It mainly involves heating biomass at moderate temperatures, in most cases, ranging between 200 °C to 300 °C, under the absence of oxygen, to modify the biomass structure—both physically and chemically. Torrefaction is carried out under controlled conditions to avoid combustion. No risks of burning biomass occur due to the treatment of an inert atmosphere; oxygen is not a part of the environment. The temperature range is carefully chosen to gain maximum desirable properties from the torrefied biomass, like energy density and grindability. The time required for torrefaction varies with the biomass type, particle size, and final product specifics. Torrefaction processes are either within a few hours or 30 min.

Volatile organic compounds and moisture are removed from the biomass during torrefaction with the aid of heat. The process therefore reduces the moisture contents in the torrefied biomass and increases its energy density [12]. There are structural changes in the lignocellulosic biomass structure. The decomposition of the lignin, hemicellulose, and cellulose components produces carbonaceous chemicals [38]. This increases the carbon content of the biomass and its combustibility. Organic components included in the biomass undergo degradation as well as transformation while going through chemical processes in the torrefaction process. However, results attributed to the formation of carbon-rich molecules have explained the stability and slightly high energy content in torrefied biomass. Several properties of torrefied biomass are more desirable than those of raw biomass and thus become suitable for application in power generation, among other aspects. These features include the following:

- **Energy Density:** Torrefied biomass has higher energy density than raw biomass because some fractions of moisture and volatile components are eliminated. The result yields a higher calorific value, which is most suited for applications related to energy production [12, 13].
- **Grindability:** Through torrefaction, biomass's grindability increases, facilitating easier handling and processing. This is crucial for biomass briquettes and pellets used in gasification and combustion systems [6, 34].
- **Hydrophobicity**: The torrefaction process significantly reduces the hygroscopicity of biomass compared to raw biomass; it absorbs moisture much more slowly. That way, the risk of biomass degradation in storage and transportation is reduced, and storage stability is improved [1, 14].
- **Combustion Properties**: Torrefied biomass is burned cleaner and more efficiently than raw biomass. It has also reached better environmental sustainability in combustion applications, as it has decreased the amount of particulate matter, volatile organic compounds, and other contaminants [18].

3.6 Kernel Hydrogel

A semisolid texture is developed by mixing a solid matrix as a continuous phase in an aqueous solvent as a dispersed phase, also known as gel, a colloidal system [5]. The structure of these gels is mainly provided by the biopolymers that comprise them. The two most utilized biopolymers are proteins and polysaccharides. These biopolymers give rise to a range of semisolid food items. Networks of different shapes, such as blocks, particles, or fibers, make up these biopolymer-based gels, which behave differently under mechanical stress [33]. The kernel hydrogel from the palmyra palm tree needs a quite laborious preparation procedure as it takes advantage of the kernel's peculiar qualities, serving many purposes. The hard-shelled palmyra palm fruit is important in tropical countries; its kernels are used to prepare hydrogels as they contain starches, fibers, and other bioactive compounds. In the first step, the ripe fruits of the palmyra palm tree are collected; in the second step, the kernels are obtained after the removal of their shells. Meanwhile, the fruits of the palmyra palm are collected upon full maturity. The kernels are extracted mechanically or manually while ensuring they remain intact and clean [40].

3.6.1 Preparation of Hydrogel from Palmyra Palm Kernel

Kernels are initially subjected to lipid extraction using organic solvents or supercritical fluid extraction techniques. The technique extracts the lipids from the kernel, leaving a lipid-deficient kernel residue in its wake. Kernel constituent alterations can be done using enzymatic hydrolysis or treatment, which helps to degrade the complex polysaccharides to yield simpler compounds in the kernels' residue, including oligosaccharides and peptides. Solubility and bioavailability of the kernel components are enhanced in this state. Modified kernel components undergo a gelation process, conventionally done by either physical temperature- or pH-induced gelation or through the use of cross-linking agents. The latter is added to establish a stable hydrogel network, although other cross-linking agents include calcium ions, and chemical cross-linkers include glutaraldehyde [40]. Palmyra palm kernel hydrogel exhibits remarkable swelling behavior and 62% encapsulation effectiveness when loading the anticancer medication 5-fluorouracil (5-FU), according to Rao et al. [40]. It also has a denser structure and a rough surface. Additionally, it suggested using palmyra palm kernel hydrogel in combination with the medication 5-fluorouracil (5-FU) to treat colon cancer.

3.7 Biogas Production

Biogas is produced from the biomass of palmyra palm trees by using their organic components to provide a renewable source of energy. Palmyra palm is cultivated mainly for fruit, but the amount of resulting biomass is large in terms of fruit shells, stalks of leaves, and other agricultural wastes. After processing the fruits, these residues, otherwise regarded as waste material, are rich in organic matter that can be utilized efficiently to produce biogas through anaerobic digestion [31].

The pretreatment of biomass is done to ensure maximal effectiveness in anaerobic digestion. Therefore, Small pieces are cut or shredded from the collected biomass materials, like leaf stems and fruit shells. This provides a large surface area for the biomass, thereby encouraging more microbial digestion and consequently an increased rate of biogas production. Various types of biomass residuals can be mixed to produce a balanced feedstock. This ensures that the anaerobic digester receives various organic compounds and nutrients, supporting healthy microbial activity and biogas production [48].

Anaerobic digestion is a biological process by which organic matter is degraded by microbes without oxygen. This degradation yields biogas. This process normally takes place within a completely sealed anaerobic digester, under exactly controlled conditions, to enhance the optimum activity of microbes and methane production. The digester is sealed to provide anaerobic conditions necessary for fermentation by microbes. A consortium of anaerobic bacteria within the digester begins breaking down the complex organic molecules of the biomass. This temperature-dependent process, by specific microbial species working in the digester and ambient conditions, produces CH_4 and CO_2 gases as by-products of biochemical reactions. The microbial activity should be under the optimum conditions so that the biogas formation will occur maximally. Temperatures are usually kept around 25 °C and 55 °C in the digester, contingent upon the type of bacteria involved and the reaction rates to be achieved. pH is monitored and adjusted regularly to achieve a neutral to slightly acidic pH between 6.5 and 7.5, conducive to microbial digestion. Solids retention time: The actual amount of time biomass spends in the digester, which varies with feedstock characteristics, the design of the digester, and the projected biogas production rate. Digestion, i.e., the natural breakdown of organic matter material, with the help of microorganisms, can vary between several days and weeks [2, 4].

The major fraction of the biogas production from palmyra palm biomass is methane and carbon dioxide, where in normal cases it contains 50 to 70% methane content in the gas mixture [27]. Other contaminants may be present in traces in biogas, and traces of H_2S may also be contained, which needs purification before this gas can be used in some applications. The produced biogas in the anaerobic digester is collected with great care so as not to leak gas and ensure safety. In all cases, the gas is typically either stored in a gas holder or used directly for energy applications. In summary, the process of methane-rich biogas generation from palmyra palm tree biomass is an integrated process starting from the collection and preparation of biomass to the anaerobic digestion process. It is an alternative renewable energy source for heating and electricity generation, cooking, building heating, and vehicle fuel. In addition, it supports economic growth, environmental sustainability, and energy security. Future developments in the infrastructure and technology of biogas boost its effectiveness and increase its use in many industries, thus helping the world turn to a low-carbon economy with a sustainable energy source [24].

3.8 Activated Carbon from Peel and Kernel Shell

Processing of this palmyra palm peel into activated carbon produces an exceedingly useful and versatile material from an otherwise agricultural waste product. Prepared activated carbon from palmyra palm peel is applied in air filtration, water purification, and many more industrial processes because of the large surface area and adsorption properties. Palmyra palms are primarily cultivated for their fruit; in this context, fruit peels, after processing, are disposed of as agricultural waste. The farms or processing plants where the fruits are harvested and cleaned are where these peels are sourced. Following a good cleaning to remove any dirt or impurities, the peels are dried to bring down their moisture content. Drying can be done in the sun or by industrial driers to dry the peels without putrefaction [36].

The dried palmyra palm peels are subjected to carbonization, heating the organic material without oxygen to produce carbon. The dried peels are heated in a furnace or kiln up to 400 °C to 600 °C, depending on the apparatus and the required qualities. Due to the high temperature in the absence of oxygen, volatile chemicals are driven out of the organic matter and converted into material full of carbon [46]. The material obtained, charcoal or biochar, is allowed to cool slowly to reach room temperature after carbonization. This results in the stabilization of the carbon structure and makes it ready for activation. The carbonized palmyra palm peel that allows for higher adsorption capacity can be made into activated carbon forming a porous structure using chemical or physical activation methods.

Steam activation In this process, the carbonized peel is activated by steam at a temperature ranging from 800 °C to 1000 °C under an inert gas environment, such as nitrogen. This method of activation tends to increase the surface of the material and the number of adsorption sites through the development of a pore network on the surface of the carbon [10, 17].

Gas activation It relies on steam, but other gases such as carbon dioxide or air can be used depending on the type of pore structure and surface properties expected of the activated carbon [10, 17].

Chemical activation Chemical agents, including phosphoric acid or potassium hydroxide, are impregnated on the carbonized peel before heating. A change in the surface chemistry of the activated carbon and better pore formation occurs with chemical activation [49].

After activation, the excess ash or other impurities are removed by flushing the activated carbon with water. This washing improves the quality and degree of purity of the activated carbon in the final product [8]. The activated carbon is dried to remove excess moisture following the washing process. The activated carbon is dried enough for proper packaging and storage through air drying or industrial drying processes. It involves some steps in converting the palmyra palm peel into activated carbon starting from collection, drying, carbonization, activation, washing, and finally drying. The processes of converting agricultural waste into a useful substance were high in adsorption capacity. Hence, it is used in many industrial operations, environmental remediation, air filtration, and water treatment [36]. The production of activated carbon from the palmyra palm peel offers numerous technical solutions to the issues of resource management, pollution reduction, and economic and environmental sustainability using locally available and renewable biomass resources. It enhances economic development and sustainability. Improved efficiency optimization and expansion of the application of activated carbon in many spheres benefit global efforts toward a clean, sustainable future based on current and future research and innovation in activated carbon production [36]. The palm kernel shell created activated carbons utilizing physical activation techniques using various precursors. Palm kernel powder was first cleaned of contaminants before being thoroughly cleaned with distilled water. A grinding machine ground the main raw components into a consistent powder. After being sieved, the resulting powder was used to make activated carbon. A leaching operation was carried out to purify the primary powder and reduce the amount of ash in the almond shell by a de-ash procedure [44].

Water vapor was utilized for activation using the physical activation approach. This involved filling the flask with distilled water, connecting the argon gas hose from the flask's input, and setting the flask on the heater; after setting the temperature so that the distilled water began to steam gradually, argon gas and steam were simultaneously released into the furnace from the top of Erlen using special hoses. The gas and steam had to be linked until the temperature reached T = 600 °C. After

3 h at this temperature, the heater was switched off, allowing only gas to enter the furnace and stay there for 3 h before the carbon was removed [44].

4 Applications

The by-products derived from various tree parts are highly important in domestic, industrial, and agricultural applications. The fruit produces nutrient-rich pulp used in food processing, making products such as candies and jams, while the seeds can be used to mill starch-rich flour utilized in baking and brewing. Its leaves are important for handicrafts and thatching due to their flexibility and resilience. The extracted sap from the flower can also be used in making palm sugar and alcoholic drinks. Its hard trunk shell is turned into fuel and utensils, while its fibrous character makes it useful for construction work. Its processing rejects can be organic manure and fodder, therefore additionally proving the adaptability of palm to diversified environmental conditions. These are some aspects of why the Palmyra palm plays a significant role in helping rural livelihood and sustainable agriculture.

4.1 Active Packaging for Food Preservation

Active packaging is defined as using active substances or components in packaging materials to produce safe food products and improve quality and safety conditions while prolonging the food's shelf life [3]. Antimicrobial, antioxidant, and moisture-absorbing characteristics inherent in unique ingredients derived from the palmyra palm make it relevant for active packaging [9]. For example, the husk of palmyra palm fruit contains many phenolic chemicals, including tannins, which are natural antioxidants. Isolation of these chemicals and their incorporation into packaging materials will result in the scavenging of free radicals and the termination of the oxidation process that causes the degradation of food [25]. Additionally, the antibacterial traits of palmyra palm extracts could enhance microbiological stability in packaged goods since they prevent the growth of harmful bacteria and fungi [32]. The seeds of the palmyra palm plant can also be used to make biodegradable film-forming material for active-based packaging applications. These starch films can then further be added to antioxidants or antimicrobials of other palm components to enhance their use in maintaining food safety and quality [28]. Abundance in the growing regions and hitherto unused by-products make palmyra palm active packaging quite consistent with sustainable ways of use. The use of palmyra palm in making active packaging to enhance general food safety and reduce food loss, other than just the shelf-life improvement of the food products from its inherent bioactive ingredients, is well-documented [29]. Further research and development can bring innovation and environmentally friendly food preservation technology to this area. Packaging materials with extracts from the husk of the palmyra palm fruit, rich in phenolic compounds such as tannins, are antioxidants. These antioxidants delay lipid oxidation and extend the shelf life of products that are prone to rancidity, such as fatty meals and oils.

Organic antibacterial agents present in fruit extracts and the bioactive substances from the leaves of a palmyra palm used in packaging—the growth of bacteria and fungi can be restricted, thereby reducing spoilage and enhancing food safety [39]. Biodegradable films and food packaging coatings are developed from the starch of palmyra palm fruit, coming from the seeds of the palm tree [16]. The films do not only act as a barrier to moisture and oxygen, just like traditional plastic packaging, but after being disposed of, they naturally degrade and thus minimize environmental impact. Moisture-controlled packaging materials use fiber and palmyra palm extracts to maintain maximum freshness for moisture-sensitive food items like snacks and dried fruits. In areas rich with palmyra palms, the locals have used palms' leaves to wrap edibles for thousands of years. These natural wraps can be great alternatives to their modern-day counterparts, offering a more sustainable method of packaging [43].

4.2 Palmyra Palm Pericarp Syrup

Palmyra palm pericarp syrup is a special and versatile ingredient made from the outer husk of the palmyra palm fruit, with many culinary and nutritional uses. Its application as syrup from the pericarp of palmyra palms is generally applied as a natural sweetening agent. It has unique flavor tones with a mild sweetness and a rich background of caramels. It is increasingly used within juices, smoothies, and beverages to add a sweet depth to the drink. Adding more in-depth flavor, the syrup creates a greater sweetness of juices, smoothies, and beverages. The cooking and baking process brings this mystery sweetness or flavoring in sauces, salads, and defers, perfectly marrying the flavors of both savory and sweet dishes [19].

This syrup made of the pericarp of palmyra palm is rich in natural sugars, such as fructose, glucose, sucrose, and dietary fiber, which offers energy to the body and ensures good digestive health. Besides, the syrup also contains phytochemicals such as antioxidants and phenolic compounds, which can positively affect health. The pericarp syrup reflects cultural importance in the areas of origin of the palmyra palm, parts of the two subcontinents of South and Southeast Asia. It forms an integral part of regional culinary tradition and traditional medicine, reflected in its use in traditional foods, beverages, and remedies. Extracting syrup from the pericarp of the palmyra palm supports sustainability by reducing wastes to the least minimal level and, at the same time, maximizing palm fruit for use [19]. Therefore, it encourages food production in a sustainable manner and environmental responsibility. Overall, palmyra palm pericarp syrup's inherent taste, nutritional, and sociocultural value establish it as a premium ingredient for traditional and modern culinary applications, exhibiting its versatility and tremendous potential as a novel, sustainable food product.

4.3 *Textile Applications*

The Palmyra palm plays cultural importance, along with its intellectual values, within the textile field. The fibers are derived from the hard trunk and leaf sheaths of the palm tree and have highly noticeable mechanical properties in terms of high tensile strength and good durability. These fibers need a lot of processing, cleaning, and spinning to be made into yarns for modern textile inventions and traditional handicrafts. In countries and regions where such fibers are abundant, especially in South Asia and parts of Africa, local craftsmanship involves weaving these fibers into crafts, mats, and baskets with great dexterity. This way, they keep up old traditions and cultural heritage. Apart from being an eco-friendly source of natural dyes, the palmyra palm is also a work of art since its colors are sophisticated and lively, coming from its fruit and leaves. This eco-friendly artificial dye alternative is in sync with worldwide trends of promoting sustainable textile processes [30]. The next feature of biodegradability concerning the palmyra palm fibers is that it reflects the circular economy concept that emphasizes how effectively resources are used, after which they are returned to the environment. Current development keeps being made with these palm fibers for high-performance textiles used in construction, agriculture, and even aerospace after further research and innovation. It makes the textiles more sustainable and contributes to regional economic growth and cultural resilience.

4.4 *Medicinal Applications*

The Palmyra palm has a rich set of medical applications almost entirely based on traditional wisdom and has recently been explored through modern research. Roots, Fruits, and even the sap of this palm tree have been used for medicinal purposes. The juice extracted from the spathe of the palmyra palm is converted to palm sugar and jaggery, which are unrefined sweeteners and contain minerals, including potassium, calcium, and iron. It is highly valued in traditional medicine as a natural way to boost the body's energy levels and supply it with the required nutrients. Considerable quantities of flavonoids, tannins, and phenolic compounds have been identified from the fruit and pericarp of the Palmyra palm. Flavonoids and phenolics are two main classes of phytochemicals with antioxidant activity, known to help defend against the impact of free radicals on the body. The palmyra palm is a good source of dietary fiber extracted from the trunk and leaf sheaths. The fiber eases bowel movement and gives way to normal defecation, which aids the body's digestive process. Traditionally, it is used to maintain proper digestion and as a cure for constipation. Extracts from this palmyra palm's leaves, fruits, and other parts have shown resistance against many bacteria and fungi. These antibiotic properties have helped in traditional medicine in the treatment of minor infections and for healing wounds. Infusion and preparations using the derived parts of the various sections of the palmyra palm tree are used to treat skin conditions, colds, and coughs in cultures where the palm is native [35].

The pharmacological potential of palmyra palm, in fields like the treatment of diabetes, cardiovascular health, and skincare, is under research by contemporary researchers. The nutraceutical and bioactive potentials in palm oil, including its nutritional benefits, provide a platform for newer, safer, and more effective pharmacological activities that will be explored in confirming its traditional use and forging a new therapeutic application. Palmyra palm is a rich botanical resource with much medicinal potential. The current scientific works on its bioactive compounds and health impacts reach a synergy that can complement its use in traditional medicine. Apart from protecting the traditional ways of life, embracing and exploring these medicinal qualities would lead to the development of functional foods and natural therapies that guarantee overall wellness [35].

4.5 *Treatment of Industrial Wastes*

Under sustainable practices and environmental care, industrial waste generated through the process of Palmyra palm has many interesting applications in waste treatments and resource recovery. One of the important applications is in using palm waste, including remaining fruit fibers and shells, in anaerobic digestion to produce biogas [48]. In a controlled environment, organic matter in these wastes can be metabolized by microorganisms to yield methane-rich biogas. The use of such biogas shall potentially mitigate greenhouse gases and dependency on fossil fuels for sustainable energy in both heat and power production. Another valuable by-product of palmyra palm is the composting of waste. This nutrient-rich compost prepared from palm trash has the potential to support healthy plant growth and higher agricultural yields using high fertility and good soil texture. Further, it is an organic product and replaces the synthetic edaphic materials or fertilizers that deteriorate soil and contaminate water bodies. Activated carbon is also produced from the carbonaceous shell of palmyra palm fruits. Now, it is being utilized in removing impurities, pollutants, and odor from exhaust gas and other industrial wastes because of its vast surface area and consequent high adsorbing capacity. Biochar can be obtained through pyrolysis of palmyra palm material at high temperatures without oxygen [36]. With increasing soil fertility and water-holding capacity, biochar is also in a more stable carbon form. It sequesters atmospheric carbon dioxide and stores it in the soil, reducing negative climate change effects.

The palmyra palm offers good potential for phytoremediation, in which plants are used to clean soil and water. Its viability for the process is excellent since it grows in saline soils and complex environments, and its biomass could hold and absorb high molecular weight contaminants. Areas to which the palm tree can offer potential include accumulating and storing heavy metals and other pollutants from contaminated zones and the extensive root system to stabilize the soil and control erosion. These applications highlight a circular economy strategy and deal with

Table 4.1 Valaorized product developed from the palmyra palm wastes

Palmyra palm waste	Valorized product	Processing method	Applications	Reference
Sprouts peels, fruit husk	Carboxymethyl cellulose (CMC)	Carboxymethylation	Biodegradable packaging	Raja Rajeshwari et al. [37]; Kaewprachu et al. [22, 23]
Palmyra fruit peduncles	Lignocellulosic fibers and cellulose microfibrils	Extraction	Engineering structural components, medical devices, and food packaging	Balasubramani et al. [7]
Fruit shell	Low-cost microporous activated carbon adsorbents	Pyrolysis	For CO_2 capture	Swapna et al. [47]
Pericarp	Syrup	Cooking	Functional drink	Hutabarat and Nugroho [19]

waste management challenges. The main industries will minimize disposal to landfills, conserve resources, and further explore business opportunities in waste treatment and resource recovery sectors by maximizing and adding further value to the derived waste streams of Palmyra palm. The development of valorized products and their application is summarized in Table 4.1.

5 Future as Sustainable Technology

The Palmyra palm processing industry generates an enormous amount of trash and by-products. Under well-regulated conditions, these can be converted into valuable resources by bioprocess technology. This practice not only hacks a solution for a significant, looming waste management issue but also keeps the ecological footprint at a minimum while maximizing resource efficiency, all contributing to sustainable development. The significant jet technique in the bioprocess technique when it comes to making use of palmyra palm trash to generate biogas is anaerobic digestion. Biogas produced in this way primarily consists of carbon dioxide and methane. Apart from this, fruit shells and fibers can be used in anaerobic digestion due to the inherent compounds. This biogas can be used as a biofuel or converted into renewable energy for heating, production of electricity, and other uses related to heat and power generation, hence mitigating greenhouse gas emissions and dependency on fossil fuels. Additionally, bioprocessing could be used in the production of bioethanol and biobutanol as alternative biofuels for the transportation sector, from sugars originating from Palmyra palm residues. Generally, this process is performed by fermentation of the sugar derived from the sugar-rich palmyra palm residues, utilizing either microbes or enzymes. These renewable fuels will replace the conventional gasoline and diesel used in transportation. They make the environment better for energy security and lessening greenhouse gas emissions. The by-products from palmyra palms can be used in the recovery of significant bioactive substances through bioprocess technology. These bioactive find applications in the food, cosmetic, and pharmaceutical industries, including antioxidants and other phytochemicals. Bioprocessing, therefore, imparts added value to palm-left-over materials by recovering and purifying these bioactive materials, which also provide sustainable alternatives to synthetic chemicals and additives.

Because of the immense surface area and adsorption ability, activated carbon, resulting from the pyrolysis of carbon-rich fruit shells of Palmyra palms, is applied massively in water and air purification. The activated carbon prepared from palm wastes may efficiently remove impurities, pollutants, and odors from industrial wastewater and exhaust gasses, giving a long-term solution to environmental pollution problems. Biotechnological methods can help regain nutrients such as nitrogen, phosphate, and potassium from palmyra palm trash. The thus-recovered nutrients could be converted into organic fertilizers and soil amendments to enhance fertility and productivity in agriculture. This way, bioprocessing reduces our dependence on artificial means for nutrient inputs while trying to practice sustainable agriculture by recycling nutrients back into the soil. The circular economy principles can be directly applied through palmyra palm processing industry bioprocess technology. It will underpin social cohesion, secure economic expansion, and ensure environmental sustainability in palm-growing regions.

6 Conclusion

This chapter explored the numerous potentials of wastewater from processing and by-products of Palmyra palm fruit using bioprocess technology. The huge diversity of by-products involved in palmyra palm processing in fibers, shells, residues, juices, etc., all have appreciable potential for valorization. These wastes, which look minor, turn out to be a huge reservoir of resources with innovative bioprocessing techniques. Available methods for this bioconversion include fermentation, enzymatic treatments, and microbial processing, which have envisioned a true microbial potential because of the conversion of the by-products to valuable products such as biofuels, bioplastics, and bioactive molecules. They further present additional significant advantages: the generation of new revenues and reduction in economic losses by reducing investments and operational costs in managing the waste. The adoption of these bioprocess technologies under existing frameworks for palm fruit processing might lead to more sustainable and efficient practices. The path further guarantees stronger follow-up of broader environmental goals for waste reduction, lower carbon footprints, and increased circular economy principles. As research and development further this platform, the search for new applications and the optimization of existing methods must create a continuum. The future development shall be directed toward scaling bioprocess technologies in a way that becomes economically viable, adaptable to existing processing setup, and all other possible permutations and combinations to harness the maximum potential of palmyra palm fruit by-products to ensure more sustainable and economically viable palm fruit industries. In conclusion, the valorization of the processing wastes of palmyra palm fruit by bioprocess technology helps address critical issues in waste management and achieve a more sustainable future. The acceptance of these innovative solutions will play the first and most important foremost role in the improvement of environmental stewardship and economic development related to palm fruit processing.

References

1. Acharjee TC, Coronella CJ, Vasquez VR (2011) Effect of thermal pre-treatment on equilibrium moisture content of lignocellulosic biomass. Bioresour Technol 102:4849–4854. https://doi.org/10.1016/j.biortech.2011.01.018
2. Adou KE, Kouakou AR, Ehouman AD, Tyagi RD, Drogui P, Adouby K (2022) Coupling anaerobic digestion process and electrocoagulation using iron and aluminium electrodes for slaughterhouse wastewater treatment. Sci African 16:e01238. https://doi.org/10.1016/j.sciaf.2022.e01238
3. Ahmed MW, Haque MA, Mohibbullah M, Khan MSI, Islam MA, Mondal MHT, Ahmmed R (2022) A review on active packaging for quality and safety of foods: current trends, applications, prospects and challenges. Food Packag Shelf Life 33:100913
4. Alengebawy A, Ran Y, Osman AI et al (2024) Anaerobic digestion of agricultural waste for biogas production and sustainable bioenergy recovery: a review. Environ Chem Lett 22:2641–2668. https://doi.org/10.1007/s10311-024-01789-1
5. Ali A, Ahmed S (2018) Recent advances in edible polymer-based hydrogels as a sustainable alternative to conventional polymers. J Agric Food Chem 66:6940–6967. https://doi.org/10.1021/acs.jafc.8b01052
6. Arias B, Pevida C, Fermoso J, Plaza MG, Rubiera F, Pis JJ (2008) Infuence of torrefaction on the grindability and reactivity of woody biomass. Fuel Process Technol 89:169–175. https://doi.org/10.1016/j.fuproc.2007.09.002
7. Balasubramani V, Nagarajan KJ, Karthic M, Pandiyarajan R (2024) Extraction of lignocellulosic fiber and cellulose microfibrils from agro waste-Palmyra fruit peduncle: water retting, chlorine-free chemical treatments, physio-chemical, morphological, and thermal characterization. Int J Biol Macromol 259:129273
8. Bandosz TJ (2006) Activated carbon surfaces in environmental remediation. Academic Press, New York
9. Behera S, Nayak B (2022) Phytochemical constituents and nutritional potential of Palmyra palm: a review. Rev Contemp Sci Acad Stud 2(12). https://doi.org/10.55454/rcsas.2.12.2022.003
10. Bergna D, Hu T, Prokkola H, Romar H, Lassi U (2020) Effect of some process parameters on the main properties of activated carbon produced from peat in a lab-scale process. Waste Biomass Valori 11:2837–2848. https://doi.org/10.1007/s12649-019-00584-2
11. Chaurasiya AK, Chakraborty I, Saha J (2014) Value addition of Palmyra palm and studies on the storage life. J Food Sci Technol 51:768–773
12. Chen W-H, Zhuang Y-Q, Liu S-H, Juang T-T, Tsai C-M (2016) Product characteristics from the torrefaction of oil palm fber pellets in inert and oxidative atmospheres. Bioresour Technol 199:367–374. https://doi.org/10.1016/j.biortech.2015.08.066
13. Ciolkosz D, Wallace R (2011) A review of torrefaction for bioenergy feedstock production. Biofuels Bioprod Biorefin 5:317–329. https://doi.org/10.1002/bbb.275

14. Colin B, Dirion JL, Arlabosse P, Salvador S (2017) Quantification of the torrefaction effects on the grindability and the hygroscopicity of wood chips. Fuel 197:232–239. https://doi.org/10.1016/j.fuel.2017.02.028
15. Davis TA, Johnson DV (1987) Current utilization and further development of the Palmyra palm (*Borassus flabellifer* L., Arecaceae) in Tamil Nadu state, India. Econ Bot 41(2):247–266
16. da Costa RDS, Flôres SH, Brandelli A, Vargas CG, Ritter AC, da Cruz Rodrigues AM, da Silva LHM (2023) Development and properties of biodegradable film from peach palm (Bactris gasipaes). Food Res Int 173:113172
17. Fu J, Zhang J, Jin C, Wang Z, Wang T, Cheng X, Ma C (2020) Effects of temperature, oxygen and steam on pore structure characteristics of coconut husk activated carbon powders prepared by one-step rapid pyrolysis activation process. Bioresour Technol 1(310):123413. https://doi.org/10.1016/j.biortech.2020.123413
18. Gil, M.V., García, R., Pevida, C., and Rubiera, F. (2015) Grindability and combustion behavior of coal and torrefied biomass blends. Bioresource technology 191:205–212. Https:// doi. Org/ 10. 1016/j. biortech. 2015. 04. 117
19. Hutabarat DJC, Nugroho YDS (2024) Effect heating time and sugar on physicochemical and antioxidant activity of Palmyra palm pericarp syrup. In: IOP conference series: earth and environmental science, vol 1324, No. 1. IOP Publishing, p 012120
20. Islam MT (2023) Reinforced composites by palmyra palm leaf stalk fiber: a basic review on parameters, method, and performance. Trends Textile Eng Fashion Technol 9:1049–1051. https://doi.org/10.31031/TTEFT.2023.09.000708
21. Jana H, Jana S (2017) Palmyra palm: Importance in Indian agriculture. Rashtriya Krishi 12(2):35–40
22. Kaewprachu P, Jaisan C, Klunklin W, Phongthai S, Rawdkuen S, Tongdeesoontorn W (2022a) Mechanical and physicochemical properties of composite biopolymer films based on carboxymethyl cellulose from young Palmyra palm fruit husk and rice flour. Polymers 14(9):1872
23. Kaewprachu P, Jaisan C, Rawdkuen S, Tongdeesoontorn W, Klunklin W (2022b) Carboxymethyl cellulose from young Palmyra palm fruit husk: synthesis, characterization, and film properties. Food Hydrocoll 124:107277
24. Karidio Daouda Idrissa OK, Tsuanyo D, Kouakou AR, Yao Kouassi B (2023) The effect of pH variation in biogas production: impact of cocoa pods and empty Palmyra palm bunches. Waste Biomass Valori 14(7):2267–2274
25. Kuai L, Liu F, Chiou BS, Avena-Bustillos RJ, McHugh TH, Zhong F (2021) Controlled release of antioxidants from active food packaging: A review. Food Hydrocoll 120:106992
26. Kumar A, Jain S (2023) Torrefaction of biomass for energy applications: A comprehensive review. Renew Sust Energ Rev 125:109771. https://doi.org/10.1016/j.rser.2020.109771
27. Kwietniewska E, Tys J (2014) Process characteristics, inhibition factors and methane yields of anaerobic digestion process, with particular focus on microalgal biomass fermentation. Renew Sustain Energy Rev 34:491–500. https://doi.org/10.1016/j.rser.2014.03.041
28. Martins MP, Dagostin JLA, Franco TS, de Muñiz GIB, Masson ML (2020) Application of cellulose nanofibrils isolated from an agroindustrial residue of peach palm in cassava starch films. Food Biophys 15:323–334
29. Mukkadan MJ, Preetha R, Sreejit V (2024) Development of active packaging film reinforced with nano cellulose extracted from Palmyra palm (*Borassus flabellifer*) residues for prolonging the shelf life of chicken meat. Int J Food Sci Technol 59(3):2033–2041
30. Muniyandi M, Govindaraj P (2021) Potential removal of methylene blue dye from synthetic textile effluent using activated carbon derived from Palmyra (palm) shell. Mater Today Proc 47:299–311
31. Murugesan R, Vathsala V (2023) Palmyra palm (*Borassus flabellifer*): a multifaceted plant with diverse uses and cultivation practices. Agric Mag 2:440–444
32. Natarajan M, Elizabeth AA (2024) Borassus flabellifer seed coat: an in vitro exploration of its antibacterial and antifungal properties with preliminary phytochemical screening. Indian J Pharm Sci 86(4)

33. Nazir A, Asghar A, Maan AA (2017) Food gels: gelling process and new applications. In: Ahmed J (ed) Advances in food rheology and its applications. Woodhead Publishing, Sawston, pp 335–353
34. Phanphanich M, Mani S (2011) Impact of torrefaction on the grindability and fuel characteristics of forest biomass. Bioresour Technol 102:1246–1253. https://doi.org/10.1016/j.biortech.2010.08.028
35. Prasad AB, Vignesh S, Elumalai A, Anandharaj A, Chidanand D, Baskaran N (2023) Nutritional and pharmacological properties of Palmyra palm. Food Humanity 1:817–825. https://doi.org/10.1016/j.foohum.2023.07.030
36. Prastuti OP, Kurniati Y, Septiani EL, Pratiwi NF, Ardiansyah MP, Ismadji S et al (2023) Activated carbon from Palmyra palm peel as an alternative adsorbent for removing heavy metal ions Fe (III) and Cr (VI) in industrial waste. Int J Adv Sci Eng Inform Technol 13(1):13.124. https://doi.org/10.18517/ijaseit.13.1.16869
37. Raja Rajeshwari E, Sathanya PS, Vignesh S, Chandrasekar V, Baskaran N (2025) Valorization of *Borassus flabellifer* sprout peel: synthesis and characterization of carboxymethyl cellulose for biodegradable packaging. Biomass Convers Biorefinery 1-24. https://doi.org/10.1007/s13399-025-06500-0
38. Ramakrishna V, Singh AK, Bayen GK, Masto RE, Meena RK, Singh PK et al (2021) Torrefaction of agro-wastes (Palmyra palm shell and redgram stalk): characterization of the physicochemical properties and mechanical strength of binderless pellets. Biomass Convers Biorefinery 1-19:6953. https://doi.org/10.1007/s13399-021-01720-6
39. Renuka K, Devi VR (2018) Subramanian, S.P. Phytochemical screening and evaluation of in vitro antioxidant potential of immature Palmyra palm (*Borassus flabellifer* Linn.) fruits. Int J Pharm Pharm Sci 10:77
40. Rao KM, Rao KSVK, Palem RR, Uthappa UT, Ha CS, Han SS (2023) pH sensitive drug delivery behavior of Palmyra palm kernel hydrogel of chemotherapeutic agent. Gels 9(1):38. https://doi.org/10.3390/gels9010038
41. Rasnatun Ferdous MT, Halder ER, Jianju L (2021) Introduction of Palmyra palm fiber as an alternative natural fiber in textile industry. North Am Acad Res 4(12):48–57
42. Reddy KO, Maheswari CU, Shukla M, Song JI, Rajulu AV (2013) Tensile and structural characterization of alkali treated Borassus fruit fine fibers. Compos Part B 44(1):433–438
43. Saidi IA, Efendi N, Azara R, Hudi L (2021) Uses of Palmyra palmplant parts in three regions of East Java, Indonesia. Afr J Food Agric Nutr Dev 21(5):18055–18065
44. Shahcheragh SK, Bagheri Mohagheghi MM, Shirpay A (2023) Effect of physical and chemical activation methods on the structure, optical absorbance, band gap and urbach energy of porous activated carbon. SN Appl Sci 5:313. https://doi.org/10.1007/s42452-023-05559-6
45. Srinivasababu N, Kumar JS, Reddy KVK (2014) Manufacturing and characterization of long Palmyra palm/*Borassus flabellifer* petiole fibre-reinforced polyester composites. Procedia Technology 14:252–259
46. Sultana M, Rownok MH, Sabrin M, Rahaman MH, Alam SMN (2022) A review on experimental chemically modified activated carbon to enhance dye and heavy metals adsorption. Clean Eng Technol 6:100382. https://doi.org/10.1016/j.clet.2021.100382
47. Swapna S, Parusha Ramudu M, Vishnu Vardhan Reddy P, Ravi Kumar G, Hima Bindu G, Prathap C, Mallesh D (2024) Synthesis of low-cost microporous activated carbon adsorbents for CO_2 capture from Palmyra palm fruit shell waste biomass. J Chin Chem Soc 71(12):1435–1447
48. Tepsour M, Usmanbaha N, Rattanaya T, Jariyaboon R, O-Thong S, Prasertsan P, Kongjan P (2019) Biogas production from oil palm empty fruit bunches and palm oil decanter cake using solid-state anaerobic co-digestion. Energies 12(22):4368. https://doi.org/10.3390/en12224368
49. Zhou X, Xu D, Yan Z, Zhang Z, Wang X (2022) Production of new fertilizers by combining distiller's grains waste and wet-process phosphoric acid: synthesis, characterization, mechanisms and application. J Clean Prod 20(367):133081. https://doi.org/10.1016/j.jclepro.2022.133081

Chapter 5
Avocado, Jackfruit, and Apple: General Aspects, Waste Production, and Biotechnological Applications

Luciana Francisco Fleuri, Milene Stefani Pereira-Vasques, Murillo Cesar Holtz Dorta, Gabriela Contesa Gomes, and Giovana Rafaela Stelzer Monar

Abbreviations

ABPM	Associação Brasileira de Produtores de Maçã (Brazilian Association of Apple Producers)
BNDES	Banco Nacional de Desenvolvimento Econômico e Social (National Bank for Economic and Social Development)
EMBRAPA	Empresa Brasileira de Pesquisa Agropecuária (Brazilian Agricultural Research Corporation)
FAO	Food and Agriculture Organization
IBGE	Instituto Brasileiro de Geografia e Estatística (Brazilian Institute of Geography and Statistics)
MAPA	Ministério da Agricultura e Pecuária (Ministry of Agriculture and Livestock)

L. F. Fleuri (✉) · M. S. Pereira-Vasques · G. C. Gomes · G. R. S. Monar
São Paulo State University (UNESP), Institute of Biosciences,
Department of Chemistry and Biochemistry-UNESP, São Paulo, Brazil
e-mail: luciana.fleuri@unesp.br; milene.pereira@unesp.br;
gabriela.contesa@unesp.br; g.monar@unesp.br

M. C. H. Dorta
São Paulo State University (UNESP), Botucatu Medical School, Nursing Departament, São Paulo, Brazil
e-mail: murillo.holtz@unesp.br

C. Trois, R. C. Ray (eds.), *Plantation Crop Wastes: Valorization for Economic Sustainability*, Waste as a Resource,
https://doi.org/10.1007/978-3-031-96805-1_5

NOAA	National Oceanic and Atmospheric Administration
IPEA	Instituto de Pesquisa Econômica Aplicada (Institute for Applied Economic Research)
SESI	Serviço Social da Indústria (Social Service of Industry)
USP	Universidade de São Paulo (University of São Paulo)

1 Introduction

A significant amount of food is wasted from harvest to final processing, with most of this waste ending up in landfills or being incinerated [15, 19, 43]. In Brazil, food waste reaches 49 million tons per year, with fruit waste accounting for nearly half of this amount [31, 95]. The complete utilization of fruits can occur at the industrial level through the establishment of biorefineries [75] and can also be integrated into daily household practices, where waste is generated from the manual processing of fruits consumed in their natural or minimally processed.

To explore alternative uses for this waste, it is essential to understand its potential [17]. This understanding allows us to predict possible applications on various scales and in different contexts, particularly in biotechnological applications that emphasize the use of the entire fruit within a framework of responsible and sustainable consumption [19].

Avocado, jackfruit, and apple trees play a role in the formation of successional agroforestry systems. This concept, proposed by Ernst Götsch in 2013, aims to bridge agricultural production chains (such as soybeans) with natural ecosystems (such as fruit trees) [50].

In addition to the projected increase in the production of fruits like apples, with an estimated growth of 21.3% over the next 10 years (from 2022/23 to 2032/33) [81], the production of other valuable products, such as wood and meat from jackfruit [50] and oils from avocados and olives [5], has garnered interest from diverse audiences.

Scientific research reflects this growing interest in these fruits, particularly their waste. The number of scientific publications on apple, avocado, and jackfruit waste has increased significantly over the past few decades, with 7 and 204 publications on apple waste, 0 and 67 on avocado waste, and 0 and 34 on jackfruit waste in 2000 and 2022, respectively (Fig. 5.1).

Based on the above, this review will describe these fruits, the waste generated from both industrial and domestic processing, and their biotechnological contexts, with a particular focus on the extraction of bioactive molecules and compounds.

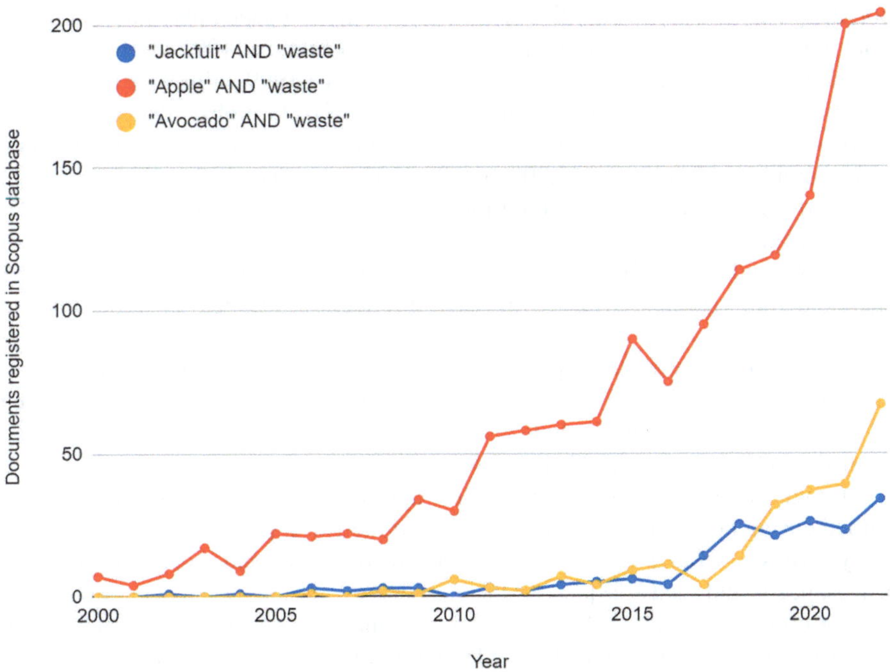

Fig. 5.1 Number of scientific publications on avocado, jackfruit, and apple waste from 2000 to 2022, as recorded in the Scopus Database

2 Geographical Origin, Distribution, Production, Productivity, and Description

2.1 Avocado

The avocado, a fruit of the avocado tree (*Persea americana* L.), belongs to the genus *Persea* within the Lauraceae family, which includes approximately 150 species. Avocados thrive in tropical and subtropical climates, demonstrating good tolerance to cold and an early maturation process [37, 124]. As a climacteric fruit, the avocado begins its ripening process on the tree and continues after harvesting due to ethylene production during its final growth stage. At room temperature, the fruit typically ripens within 5–8 days, reaching maturity when it yields to gentle pressure [98].

Native to the Americas—specifically Central America, Mexico, Guatemala, and the Antilles—avocado consumption in these regions dates back approximately 9000 years [54]. The Hass variety, which is of growing interest today, originated in California (USA) from a hybridization of Mexican and Guatemalan varieties. It is popularly known as avocado [93]. The texture of Hass avocado pulp, attributed to its lipid content, correlates with its high consumer acceptance [109]. Global avocado production has increased by approximately 52% since 1999, with 90% of this production consisting of the Hass variety [148]. In 2021, global avocado exports reached 2.5 million tons, with Mexico, Peru, and Chile being the top exporters, contributing 1.396 million, 542,000, and 98,000 tons, respectively [42].

In Brazil, despite favorable conditions for avocado production, commercial cultivation began only in 1925 [83]. Today, production is concentrated in the states of São Paulo and Minas Gerais, reaching approximately 4000 tons in 2020 [135]. Including all varieties, Brazil's avocado production reached 338,238 tons in 2022 [56].

Commercially, avocado pulp receives the most attention, whether for fresh consumption or for processing into oils and other food products [114]. Products derived from avocado processing include frozen and dehydrated slices, sauces, purees, canned goods, and juices [107, 136, 146].

Despite the extensive use of avocados, approximately 40% of the fruit is discarded during both domestic consumption and industrial processing. The three primary wastes generated throughout the avocado production and processing chain are leaves, peels, and seeds [28].

The utilization of avocado seeds is significant, considering the volume of fruit produced and harvested, and the seed's contribution of 13–18% of the total fruit weight. According to a review by Tesfaye et al. [136], global avocado production is estimated at around 4 million tons annually. Given the seed's proportion of the total fruit weight, this translates to 520,000 to 720,000 tons of seeds being discarded annually, as they are seldom used in most countries. These seeds often end up in landfills, posing potential contamination and ecological issues [39, 136].

Global avocado productivity averages around 9.5 tons/ha, with higher values recorded in major producing countries such as Mexico, Colombia, Peru, and Indonesia, where production ranges from 9.5 to 13.8 tons/ha. The Dominican Republic stands out with a productivity of 30 tons/ha [54].

In Brazil, the 2017 Agricultural Census by the IBGE reported 3.459 million avocado trees across 4427 establishments. In 2022, the average avocado yield in Brazil was recorded at 17.4 tons/ha [55].

Finally, avocado trees typically take about 5 years to bear their first fruit when grown from seeds. Once fruiting begins, each tree can produce an average of 50 fruits per cycle, with production increasing as the tree matures. By the time the tree reaches 10 years of age, it can yield around 1000 fruits per year [12].

2.2 *Jackfruit*

Jackfruit (*Artocarpus heterophyllus* L.), originating from Asian countries such as Thailand, India, and Indonesia, is a fruit characteristic of tropical regions. It is widely cultivated in other countries across Asia, as well as in northern Australia, Africa, and Latin America, particularly in Brazil. In Brazil, jackfruit is grown both in the Amazon and along the tropical coast, from the state of Pará to Rio de Janeiro [9, 34].

According to the 2017 Agricultural Census by the IBGE, Brazil had approximately 298,000 jackfruit trees, producing more than four million fruits, with the state of Bahia leading in production. Globally, FAO projections indicated that between 2015 and 2017, jackfruit production would reach 3.8 million tons [9, 57]. By 2026, the global market for jackfruit is expected to exceed 359 million dollars, with an annual growth rate of 3.3% between 2021 and 2026, and fresh consumption continues to be the most popular among consumers [59].

Jackfruit is known for its large size, with trees capable of producing between 10 and 200 fruits annually, depending on the tree's age. This corresponds to an average productivity of 25.71 tons per hectare. A fully developed jackfruit can weigh between 10 and 25 kg, though some fruits have been known to weigh up to 45 kg. However, only about 20% to 30% of the fruit is typically used for food. Jackfruit has gained prominence in the food market as a plant-based alternative to animal products, with "jackfruit meat" becoming increasingly popular [34, 121]. According to EMBRAPA [40], the word *meat*, as a food, has its definition varied according to what is wanted to be expressed. It can be synonymous with muscle or all edible animal tissues sold in a butcher shop, whether in nature or processed. On the other hand, the term *jackfruit meat* is popularly and commercially used, since the product coming from unripe jackfruit replaces products of animal origin due to its texture, consistency, and appearance characteristics, which resemble shredded meat [52].

Jackfruit meat is derived from the fruit's pulp, particularly the still greenish-yellow bulbs. In addition to the meat substitute, these bulbs are used to produce sweets, cookies, jellies, and juices. The bulbs encase the jackfruit seeds and are attached to the central axis of the fruit, all covered by a thick peel [69].

Jackfruit is composed of several parts and is classified as a pseudo fruit due to its structure around a central axis formed by the fusion of multiple flowers. The peel is hard, fibrous, and rich in calcium and pectin, with thorns that flatten as the fruit matures. The true fruits, resulting from fertilized flowers, are represented by the bulbs, which contain the edible pulp and seeds. The pulp is fleshy, sweet, and yellow, its color resulting from the conversion of chlorophyll, anthocyanins, and carotenoids during ripening. The seeds are covered by a membrane that turns brown as the fruit matures, becoming soft within the bulb if harvesting is delayed. The jackfruit's bagasse, which intertwines the bulbs, is composed of unfertilized flowers that develop into string-like structures. The central axis, or core, is the axis of the mature inflorescence, a rigid structure with a latex-producing vascular system that helps anchor the bulbs [70, 108, 128].

Due to its high volume of nonconsumable parts, jackfruit consumption generates substantial agricultural waste, including peel, bagasse, seeds, and the central axis. These wastes can become environmental contaminants if not properly managed, as bacteria degrade the biomass, potentially polluting water bodies, producing unpleasant odors, and emitting greenhouse gases. However, jackfruit waste has shown significant potential in research, including the production of seed flour and (bio)chemical transformations of the peel, which can be used to obtain bioactive compounds, formulate functional ingredients, and even produce biopolymeric films [32–34].

One notable jackfruit residue is the central axis, also known as the core or nucleus, where the bulbs are anchored. This part of the fruit is highly resistant, making it generally unacceptable for consumption and lacking commercial value. The size of the central axis varies among fruits but is proportional to the number and size of the bulbs, as it supports their weight [87].

2.3 Apple

The apple tree belongs to the Rosaceae family, subfamily Pomoideae, genus *Malus*, and species *Malus domestica* L.. It is one of the most cultivated fruits in the world, originating from Asia and Europe and thriving in temperate climates. The tree requires minimal cold exposure, allowing it to develop in mild winters [46]. It is a woody fruit species that is deciduous and suited for temperate climates [78].

In terms of morphology, the apple tree has a trunk with brown, smooth bark and a rounded canopy structure. The fruit of the apple tree (Fig. 5.2) is a pseudo fruit known as a pome, where a large receptacle covers the ovaries, with the endocarp being leathery or stony and containing the seeds, which are seldom consumed [64]. The fruit is globular and varies in size and shape according to the cultivar. The part we consume develops from the floral receptacle, comprising the skin and pulp of the fruit. Generally, the pulp is white or cream-colored, crispy, firm, and juicy, with a high content of sugars and water and moderate acidity. These characteristics can differ from cultivar to cultivar and are also influenced by climate and ethylene [110].

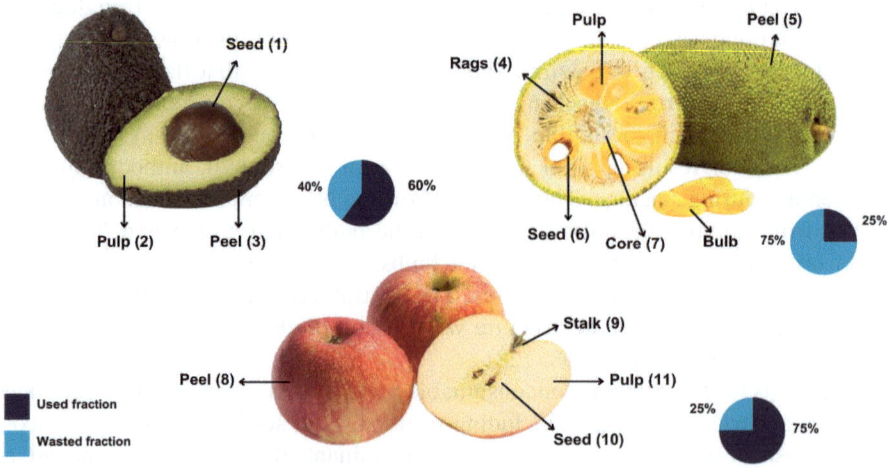

Fig. 5.2 Description of the botanical parts[a] of avocado, jackfruit, and apple, including the percentage of each fraction utilized and discarded during the industrial processing of each fruit[a]. Each part/residue has been enumerated to correspond with Table 5.2. The image was created based on the works of Jolly [64], Luchi [78], Salazar-López et al. [114] and Cruz-Casillas et al. [34]

The peel can be green, yellowish, red, or pink, featuring an internal hydrophilic layer made up of cellulose, hemicellulose, and pectin, as well as a hydrophobic membrane primarily composed of cutin [53]. In some varieties, the peel displays a uniform color, while others exhibit stripes. The peduncle varies in size and thickness and is inserted into the peduncular cavity [38]. Each fruit can contain up to ten relatively small seeds that start light in color and later turn dark; they are typically avoided due to their bitter taste and the presence of amygdalin, a substance that can release cyanide [53].

There are about 7500 registered species and varieties, with the most well-known and cultivated being Anna, Brasil, Catarina, Condessa, Granny, Eva, Gala, Golden Delicious, and Fuji [22]. In Brazil, the Gala and Fuji cultivars stand out, accounting for 90% of the planted area [119].

According to the Family Budget Survey by IBGE, this fruit ranks fourth in consumption in Brazil, following only bananas, oranges, and watermelons. Studies show that apples contain significant levels of vitamins, minerals, fiber, proteins, and phytochemicals [23], which can promote various health benefits.

Initially, apple cultivation in Brazil aimed to supply the domestic market for fresh consumption. However, market conditions began to favor not only cultivation but also pectin extraction, the development of by-products such as concentrated juice, and export opportunities [14, 142].

In 2010, according to FAO data, 69,569,612 tons of apples were produced on 4,728,333 hectares. The world's largest apple producer was China, with 33,265,186 tons, accounting for almost 50% of total production. Following China were the United States, with 4,212,330 tons; Turkey, with 2,600,000 tons; Italy, with 2,204,970 tons; and India, with 2,163,400 tons [24].

Brazil produces approximately 1.1 million tons of apples annually and directly employs around 50,000 people. In terms of productivity, the crop reached 39,461 kg/ha in 2021 [88]. The cold and humid climate, particularly in mountainous areas above 800 m, provides ideal conditions for apple cultivation. Santa Catarina concentrates 51% of Brazil's apple plantations and was expected to harvest more than 550,000 tons during the 2020/21 harvest, followed by Rio Grande do Sul (44%) and Paraná (5%). Over the last three decades, apple production has grown by more than 6000%, as reported by Brazilian Association of Apple Producers [1].

The primary form of apple consumption is fresh; however, apples are also utilized for various processing methods, including making jellies, preserves, sweets, sauces, flours, beverages, and vinegar. In the industrial sector, juice is the main product produced and is highly valued in the market. Additionally, apple wine can be consumed directly or used in cider production, along with other distillates. Consequently, the fruit has a wide range of uses, including table applications, industrial processing, and/or fermentation.

Approximately 25% of the apple is considered waste due to strict market standards regarding the fruit's physical appearance [104]. About 70% of the fruits rejected during selection and classification processes are deemed suitable for industrial use and are referred to as "industrial apples." Even in these conditions, they maintain their quality parameters [1, 92, 149]. According to Sebrae, the average yield for apple juice extraction in industries is 65% juice and 35% pomace [120].

When juice is obtained through conventional or industrial processes, it generates a residue that consists of peels, peduncles, leaves, seeds, and pulp, collectively known as apple pomace. This residue corresponds to approximately 15% to 30% of the fruit's weight before processing [44, 143]. The composition of the residue is typically 94.5% peel and pulp, 4.4% seeds, and 1.1% cores [118].

3 Proximate Composition, Nutritional Values and Applications

3.1 *Avocado*

The nutritional composition of avocados in an average portion (considered as 100 g raw) contains 160 kcal and is a good source of fiber and monounsaturated fats, providing 6.7 g and 9.8 g, respectively. The contents of protein, carbohydrates, and total lipids in the fruit are 2.0 g, 8.53 g, and 14.6 g, respectively. Notably, the sodium-to-potassium ratio is significant, with values of 7 mg and 485 mg, as well as the presence of vitamin E (2.07 mg) and vitamin K (21 μg). Avocados are particularly rich in B-complex vitamins, specifically thiamine (B1), riboflavin (B2), niacin (B3), pyridoxine (B6), and folate (B9) [54].

The peels, leaves, and seeds of avocados are abundant in macro and micronutrients and serve as important sources of bioactive compounds. The avocado peel features an average thickness and a leathery texture, characterized by a rough and rocky surface. During ripening, its color transitions from green to black [124]. Studies have identified around 70 distinct phenolic compounds in the peel, primarily belonging to the (epi)catechin and flavonoid derivatives families, particularly quercetin derivatives. These phenolic compounds exhibit beneficial properties, including antifungal, antibacterial, and antioxidant actions, which support applications in food and human health [86, 107].

In the seeds, researchers have found lipids (including linoleic acid, oleic acid, and palmitic acid), proteins, carbohydrates (including dietary fibers) [36], antinutritional compounds (such as tannins and oxalates), and antioxidants [39]. It is important to note that the seed has high energetic and mineral value, characteristics likely crucial for the germination and survival of the fruit [124].

Regarding the presence of enzymes in avocado waste, Kupnik et al. [73] investigated the activity of cellulase, lipase, peroxidase, polyphenol oxidase, protease, transglutaminase, and superoxide dismutase in the seeds of the Hass variety. The results indicated that enzyme extraction may vary depending on the method and solvent used [73]. Therefore, it is essential to consider the methods that provide the best yield and efficiency, depending on the intended application.

Since the avocado tree has evergreen leaves, identifying the bioactive compounds present in this part of the plant is also important [152]. Recent studies have identified molecules, such as flavonoids and other phenolic compounds, in avocado leaves, that exhibit antioxidant activities beneficial to the body [12, 28].

In a different context, domestic processing of the avocado primarily involves consuming the pulp or using it in various recipes. However, there are limited applications for the seed and peel, which are often discarded as organic waste. Therefore, it is essential to understand the botanical, physiological, sensory, and toxicity characteristics of these inedible parts to ensure the comprehensive utilization of the fruit. This understanding is particularly important for exploring biotechnological applications of the generated waste, in alignment with the principles of responsible food consumption and environmental sustainability.

3.2 *Jackfruit*

Jackfruit is recognized for its nutritional benefits, serving as a source of volatile compounds, phenolic compounds, carotenoids, and various antioxidants. In its ripe state, water is the primary component, followed by carbohydrates, which include monosaccharides, disaccharides, and polysaccharides. Additionally, jackfruit contains vitamins such as vitamin C, vitamin A, and B-complex vitamins (including niacin, riboflavin, and thiamine), though the levels of these vitamins tend to decrease during the final stages of ripening. The mineral content of the fruit, which includes potassium, zinc, calcium, phosphorus, and iron, follows a similar trend of reduction [66, 140].

It is noteworthy that jackfruit pulp also presents free amino acids, including tryptophan, lysine, and glutamine. The protein content in the edible portion is approximately 2% [140]. Although phytochemicals are abundant in jackfruit compared to other fruits, they contribute to its bioactivity, offering benefits such as combatting free radicals and anti-diabetic properties [4, 66].

The central axis of jackfruit has been studied to explore its characteristics and compounds. In research conducted by Adan et al. [3], the residue was analyzed to quantify phytochemical compounds using solvent extraction methods. The study found that while the central axis had lower amounts of phenolic compounds and flavonoids compared to the fibers and peel, significant values were still observed. Furthermore, the antioxidant and antimicrobial of the central axis were noted, along with its mineral content, primarily potassium, calcium, zinc, and magnesium [3]. Other components, such as pectin, have also been successfully extracted and characterized [7].

The jackfruit seed, which is uniquely anchored to the central axis and covered by the edible portion of the fruit, is still underutilized. In popular consumption, it is typically cooked or heated, resulting in a low content of fiber, protein, and minerals such as potassium, calcium, and sodium. Depending on the handling technique, the attributes of jackfruit seeds may allow them to serve as a substitute for cocoa powder in terms of aroma and flavor [133]. Despite remaining gaps in research, jackfruit seed flour has been proven to be a source of proteins, fibers, starch, lipids, and aromatic compounds, indicating the presence of flavonoids [111]. Additionally, jackfruit seeds exhibit biological activities, including antimicrobial action related to the presence of sulfur and its derivatives [144].

The peel, on the other hand, is the outermost layer of the jackfruit, encompassing the entire fruit and featuring protruding points. Among all jackfruit by-products, it is the only one that cannot be consumed by humans, even after processing. This limitation directs its potential use towards applications such as fertilizer or incorporation into animal feed, as it contains high levels of carbohydrates, proteins, and fibers [25].

In addition to these nutrients, the peel is known to contain various bioactive molecules, including steroids, tannins, alkaloids, flavonoids, and other phenolic compounds. The abundance of polyphenols in the peel can largely be attributed to environmental stresses faced by the plant, such as sunlight, UV exposure, and climatic variations [26]. Another significant component found in jackfruit peel is pectin, which is associated with antioxidant, antitumor, and anti-inflammatory properties, making it widely utilized in the food and pharmaceutical industries [151].

Overall, most current studies in the literature focus primarily on the edible part of the jackfruit, specifically its bulb, investigating its potential as a source of biomolecules and its role in combating glycemic disorders and obesity [90, 154].

From the observations made, it is evident that jackfruit waste has a substantial potential for environmental contamination, alongside negative social and economic impacts. Therefore, one effective strategy for mitigating such waste would be to explore the reuse or utilization of these parts, as they have already been studied and demonstrate high added value (Table 5.1) and significant potential for the development of new products [25, 101].

In terms of home consumption, there is still limited data available, which may be attributed to the low acceptance of the fruit among the general population. However, when discussed in the literature, the edible parts typically mentioned are the seeds (after cooking or roasting) and the pulp [80]. In contrast, the bagasse and central axis are generally discarded and not commonly consumed, although some alternatives, such as flour production, have been noted. The peel, on the other hand, is often considered by the population for use as animal feed, thereby promoting a more complete utilization of the fruit [21].

It is frequently observed on websites dedicated to healthy and sustainable eating—such as Sustentarea (USP) and Alimente-se Bem (SESI)—that jackfruit is portrayed as a versatile yet underutilized fruit. Some reasons cited for its low consumption include its perceived perishability and lack of sensory appeal to consumers. Nevertheless, these resources provide suggestions for maximizing the use of the fruit, such as cooking the seeds or core to enhance palatability, roasting or grinding the seeds for inclusion in salads and flours, and even utilizing the peel as fertilizer to prevent the disposal of typically inedible parts [122, 134].

Table 5.1 Uses/applications of parts and waste of avocado, jackfruit and apple

Fruit	Parts/ Waste[a]	Uses/Applications
Avocado	1. Seed	Natural pigments [146], starch, biogas [136], and source of enzymes [73].
	2. Pulp	Oil extraction and fresh consumption [114].
	3. Peel	Phytochemicals [107].
Jackfruit	4. Rags	Biosurfactant [102], pectin [30], and animal feeding [8].
	5. Peel	Biofuel [117], adsorbent materials [61], pectin [151], nitrocellulose [106], biochar [138], fertilizer, and animal feeding [26].
	6. Seed	Bio-oil [117], fibers, and starch [111].
	7. Core	Phytochemicals [3], pectin [7], and animal feeding [21].
Apple	8. Peel	Nanopapers [141], fermentation process [32, 33, 45], fibers and phytochemicals [23], animal feeding [46, 155], and pectin [145].
	9. Stalk	Fermentation process [32, 33, 45] and animal feeding [46, 155].
	10. Seed	Fermentation process [32, 33, 45] and animal feeding [46, 155].
	11. Pulp	Starch, protein, fibers and phytochemicals [23, 32, 33, 45], animal feeding [46, 155], and pectin [145].

[a]The numbers of "Parts/Waste" correspond to the botanical parts assigned in Fig. 5.2 for each fruit

3.3 *Apple*

Apples are a popular and highly nutritious fruit, composed of approximately 85–88% water, 13–15% carbohydrates, 2–4% fiber, and 10–13% natural sugars (primarily fructose, glucose, and sucrose). They contain less than 1% protein and a negligible amount of fat (less than 0.5%). In 100 g of apple, there are approximately 52 calories, 14 g of carbohydrates, including 10 g of sugars and 2.4 g of dietary fiber, as well as 0.3 g of protein and 0.2 g of total fat. Apples also provide 1 mg of sodium, 107 mg of potassium, 4.6 mg of vitamin C (8% of the recommended Daily Value), 54 IU of vitamin A, 6 mg of calcium, and 0.12 mg of iron [23]. Among their benefits, apples are rich in fiber, which aids digestion and promotes a feeling of satiety, and they contain antioxidants such as polyphenols that protect the body against oxidative stress. The composition of apples can vary slightly depending on the variety and growing conditions [71].

Apple pomace, one of the main waste products generated in large quantities, consists of sugars (such as fructose), fibers, phenolic compounds, lipids, proteins, vitamins, and minerals [145]. Apple pulp—which is also a component of the pomace—is rich in polyphenols (including catechin, epicatechin, chlorogenic acid, procyanidins B1 and B2, and phloridzin), antioxidants, vitamins C and E, and pectin [23]. The peel itself has high concentrations of antioxidants and fiber and contains cutin [71].

One of the suggested applications for apple pomace is in the medical field due to its high concentration of phenolic compounds and antioxidant activity, which support research into its effects against cancer cells and its potential preventative role in cardiovascular diseases [65]. Additionally, apple pectin can be considered for applications in the textile industry as raw materials for sustainable creations, as well as for producing biopolymers due to the solubilization of protopectin [77]. Another use for apple pomace is the production of flour, which can serve as a fiber-rich ingredient with higher nutritional value in various foods [11], such as incorporating it into biscuits [116].

However, there are few applications for apple seeds, which are typically discarded due to the presence of amygdalin. Therefore, understanding the botanical, sensory, and toxicity parameters can facilitate a more comprehensive use of apples and their by-products.

Other uses/applications of parts and waste of avocado, jackfruit, and apple can be found in Table 5.1.

4 Bio-valorization of Waste and By-products in the Circular Economy

Given the descriptions and as shown in Table 5.1, there is a clear potential for utilizing avocado, jackfruit, and apple wastes as sources of various molecules and biological activities, thus minimizing waste and preventing improper disposal. In this context, efforts are being made to enhance techniques for harnessing the biochemical potential of these wastes, whether through the extraction of bioactive compounds or their reuse as additives in food [126].

The circular economy emerges as an alternative for waste management in the medium and long term, aiming to prevent, reuse, or recover natural resources and reintroduce waste into the production line as raw materials for creating new products with added value. This can be achieved through biotechnology or sustainable technologies for component extraction, thereby strengthening already established agro-industrial production chains [97].

The principles of the circular economy can be implemented through biorefineries, which seek to exploit biomass resources for the coproduction of bioenergy alongside products derived from reused raw materials. This model generates high-value products that enhance both energy efficiency and economic performance [79].

Fruit wastes have already been tested within this framework; however, challenges remain concerning scalability. The processes involved in extracting bioactive compounds generate liquid and solid wastes that can be subjected to anaerobic digestion to produce biogas, fermented to yield bioethanol, or combusted to generate heat and combined energy [79].

One potential use for avocado wastes, although not extensively covered in the literature, is in solid-state fermentation. For instance, avocado seeds have been utilized as a fermentation substrate with *Aspergillus niger* GH1 in an optimized study aimed at releasing antioxidant compounds. This fermentation process resulted in an increase in antioxidant capacity by up to 77%, alongside a significant rise in cellulase and xylanase activities [153]. Additionally, avocado wastes (specifically peels and seeds) demonstrate potential for use as fuel when processed using fluidized bed technology in conical reactors. In experiments, these wastes exhibited key parameters such as high combustion efficiency, effective heat transfer, and low emissions of harmful gases [115].

Recent reviews have also highlighted the potential of avocado wastes for producing toxin adsorbents designed to remove dyes, metal ions, antibiotics, and pharmaceutical residues from wastewater. The goal of utilizing these fruit wastes is to develop ecological adsorbents that are cost-effective and can aid vulnerable socioeconomic populations in accessing treated clean water. The authors advocate for more research on using these by-products as adsorbents against actual industrial effluents, gaseous pollutants, anions, and radionuclides, as well as studies focusing on competitive adsorption techniques and cost estimates [6]. To enhance the valorization of avocado wastes within a biorefinery context, Rodríguez-Martínez et al. [112] mention that the peels and seeds, generated during the production of oils, guacamole, and sauces, can be repurposed for biofuels, pharmaceuticals, cosmetics, chemical compounds, fertilizers, nutrients, and bioactive compounds like oligo/polysaccharides and phenolic compounds.

Turning to jackfruit, applying solid-state fermentation to its fruit wastes is increasingly becoming a focus of scientific research. In most cases, powdered seeds are used as a fermentation substrate, either independently or in combination with other substrates. Various fungi, such as *Fusarium culmorum* [85], *Monascus purpureus* [18], *Lactobacillus* spp. [89], and *Aspergillus niger* [84] have been shown to produce enzymes [84, 85], pigments [18], and amino acids [10].

So far, a methodology has been developed to enrich protein from jackfruit peel using solid-state fermentation with *Saccharomyces cerevisiae* as the microorganism, aiming to formulate a nutritional supplement for animals [131]. This highlights an area of opportunity that is still ripe for exploration.

In addition to this fermentation, there are alternative methods for transforming jackfruit wastes through physicochemical techniques, such as producing biochar from seeds [67] or peels [58]. Both approaches aim to utilize biochar for treating heavy metals in effluent waters. Beyond the already tested wastes, further research into other components for biochar production would be beneficial, as well as exploring additional applications in the energy sector, soil enrichment for cultivation, or as a catalyst for industrial reactions. In the energy sector, jackfruit peels and seeds have been investigated as sources for the microbial production of biofuels such as bio-oil, methane, and bioethanol [117].

Other technological transformations of jackfruit wastes have been described by Jayarajan et al. [61], who converted fruit peel into a nanoporous adsorbent material capable of removing Rhodamine dye from industrial wastewater. Additionally, Rajasekharan et al. [106] produced completely biodegradable nanocellulose from jackfruit peel for incorporation into bioplastics.

Moreover, jackfruit wastes can also serve as a source of fertilizers, enhancing the availability of micro and macronutrients. This can be achieved through composting processes or by creating fertilizers formulated with other compounds. Silpa and Vijayalakshmi [127] utilized jackfruit and cocoa wastes for plant growth, resulting in increased root numbers, leaf development, fresh mass, and other indicators of growth optimization compared to untreated plants.

Apple wastes, due to their high content of fermentable compounds [45], can be used as substrates for liquid fermentation to produce ethanol, and as substrates in solid-state fermentation to obtain aromas, natural gas, citric acid, pectins, enzymes, and mushrooms [92]. A recent study employed apple pomace in conjunction with *Aspergillus niger* during solid-state fermentation to produce pectinase [32, 33].

Furthermore, apple wastes can be utilized in other processes to extract fibers and charcoal [92] and as coatings associated with apple yeast to reduce post-harvest deterioration of fruits caused by *Penicillium expansum* [123]. Apples and their pomace are also often used as organic fertilizers, particularly for plants growing in acidic soils, and serve as a supplement for animal feed [46]. However, fermentation of these wastes can lead to alcohol poisoning and negatively affect livestock health [155].

Oliveira [96] has also utilized apple waste to produce biochar, highlighting that implementing this product contributes to the diversification of wastes used as raw materials for various intended purposes.

Table 5.2 complements Table 5.1 by providing information that connects the use of fruit parts within the context of the Circular Economy, highlighting the extraction of molecules and biological activities, as well as potential applications.

Table 5.2 Use of avocado, jackfruit, and apple parts with highlighting the extraction of molecules and biological activities and potential applications

Fruit	Part	Molecules and biological activities	Applications	Reference
Avocado var. Hass	Seed	Phenolic compounds, proanthocyanidins, enzyme proteins (cellulases, lipase, peroxidase, polyphenol oxidase, protease, transglutaminase, and superoxide dismutase), antioxidant, and antimicrobial activities	Recovery of bioactive compounds for industrial application	Kupnik et al. [73]
		Starch	Production of biopolymers for textile, medical, electronics, packaging, and agricultural purposes	Tesfaye et al. [136]
	Seed and peel	Proteins, carbohydrates, fiber, lipids, ash, phenolic compounds, and antioxidant activity	Obtaining antioxidants for food to replace synthetic ones	Daiuto et al. [35]
Avocado var. Hass e Fuerte	Seed and peel	Phenolic compounds, antioxidant, and anti-inflammatory activities	Reduction of pro-inflammatory molecules in in vitro studies	Tremocoldi et al. [139]
Jackfruit	Peel	Pectin, anti-inflammatory activity	Production of nanocomposites for bone healing	Govindaraj et al. [49]
		Polysaccharides, antioxidant, and hypoglycemic activities	Potential as a functional food supplement	Wang and Jiang [147]
	Core or nucleus	Carbohydrates, lipids, amino acids, antioxidant, and cell protection activities	Potential as a functional food supplement	Li et al. [76]
	Seed coating	Pectin, proteins, and phenolic compounds	Potential in the food, pharmaceutical, and cosmetics industries	Kumar et al. [72]
Apple	Peel	Amylase	Production of the alpha-amylase enzyme for application in the food and biofuels sector	Singh et al. [129]
		Phenolic compounds and antioxidant activity	Increased antioxidant power through fermentation	Gulsunoglu et al. [51]
	Pomace	Soluble and insoluble fibers	Production of sustainable and nutritionally enriched food products	Qin et al. [103]
		Phenolic compounds and antioxidant activity	Production of nanopapers coated with polyhydroxyalkanoate	Urbina et al. [141]

5 Techno-economic Analyses and Industrial Prospects

Techno-economic analyses of fruits have been conducted in various contexts and regions around the world. Solarte-Toro et al. [130] examined how the scenario, processing scale, and product portfolio influence the techno-economic aspects of small and bigger avocado biorefineries in Colombia. The authors identified smaller biorefineries, and opportunities for regional avocado production focusing on oil, feed, and electricity production. They also described that the best operations for bigger biorefineries focus especially on the production of levulinic acid and furfural from fruit waste.

In Rwanda, Ntabakirabose et al. [94] analyzed the profitability of avocados across three districts, highlighting that agriculture is predominantly a subsistence sector in the country. Their findings revealed that for trees between 3 and 10 years old, the best profitability was for varieties Fuerte, Hass, and traditional, in this sequence. For trees aged 11 years, the Ettinger variety emerged as the most profitable, followed by Fuerte and Hass. Notably, avocado cultivation proved to be more profitable per cultivated area compared to crops like beans, corn, wheat, and potatoes. The authors noted too for avocado production that various factors related to the plant—including variety, age, and perishability—as well as soil characteristics (type and fertility), climate conditions, and management practices, such as inputs, technology, good agricultural practices, and market dynamics, must all be considered for successful production.

A techno-economic analysis of jackfruit is discussed by Aravind et al. [13] within the framework of an integrated coffee-jackfruit agroforestry system. This study aims to achieve negative emission levels by intercropping these two crops in coffee farms. Integration can be achieved through the production of biofuel and biochar from jackfruit by-products, as well as utilizing jackfruit timber. The primary focus is on commercializing the resulting biomass in the form of timber and the production of biochar and biofuel. These processes have the potential to sequester approximately 2, 3.5, and up to 7 tons of CO_2-eq/ha in a single year from the jackfruit crop alone, depending on effective cultivation and management practices. Moreover, this approach helps mitigate anthropogenic CO_2 emissions when considering the production of biofuel [13].

Moreover, prospective applications for jackfruit by-product utilization are frequently identified within the food industry, primarily due to their nutritional characteristics. While not yet widely applied, several examples of jackfruit waste utilization exist. These include pectin, which is used for gelification and stability; essential oils and starch recovered from the seeds; and flavonoids present in various parts of the fruit, which can be incorporated into functional food formulations. Furthermore, the low glycemic index of jackfruit makes it a suitable ingredient for food products designed for individuals with diabetes [16].

Given the versatility of these wastes, Chakraborty et al. [29] analyzed them to explore their transformation into biorefinery products applicable across various industrial sectors, depending on stakeholder interests. The conversion of jackfruit wastes into biomaterials, such as biofilms and polymers, nutraceutical components, functional foods, pharmaceutical precursors (including pectin for bone healing), and biofuels (including biogas, biochar, and syngas) has already been demonstrated at the laboratory scale. This advancement paves the way for integrating waste streams into industrial processes, thereby promoting a more ecological approach to production [26, 29].

Fadavi, Keyhani and Mohtasebi [41] conducted research on apple tree production in the province of West Azarbaijan, Iran, during the 2008–2009 biennium. They reported that packaging and irrigation accounted for the highest energy expenditures, at 57% and 16%, respectively, resulting in total indirect energy consumption that is primarily nonrenewable and therefore inefficient from an environmental perspective. Labor (34%) and packaging (30%) comprised the largest expenses, contributing to a significant use of inputs that produced counterproductive repercussions on both the environment and human health.

Pandey et al. [99] evaluated 20 years of apple production and marketing data in the Darchula district of Nepal. Their findings indicated that fruit cultivation is economically viable due to the modernization of agricultural practices, including improved irrigation techniques and enhanced transport systems aimed at expanding commercialization. However, the authors emphasized the importance of considering other factors, such as market trends and consumer preferences, to further increase the profitability and sustainability of the fruit production chain.

6 Real-Time Applications

Araújo et al. [12] reported a total of 18 patents related to avocados, with ten of these focusing on the food industry. These patents highlight the use of avocado seeds as ingredients for teas or beverages and as components for the preparation of culture media. The authors also mention patents for products such as avocado paste, avocado wine, and fruit juice within the food sector. In the healthcare field, avocado-derived lipids have been registered for applications in leukemia treatment, topical mosquito control products with sunscreen, and a permanent hair repair process, among others. Bangar et al. [20] emphasize the presence of flavonoids in avocado seed extract and catechin for creating personal care products. Additionally, four patents have been registered in the cosmetics area, including an avocado facial cleanser.

The utilization of jackfruit waste for real-time technologies remains an unaddressed gap in the existing literature. Conversely, it has been demonstrated that fresh jackfruit can be converted into carbon dots using hydrothermal technology, exhibiting acetone sensing activity. The study's objectives were successfully achieved, and photoluminescence was observed, thus endowing the material with sensing characteristics in both vapor and aqueous environments [137].

Several patents related to jackfruit technologies focus on its cultivation or post-harvest manipulation. In terms of jackfruit waste, the seeds have garnered the most patent attention. They have been employed as a component in buccal tablets, which serve as a dietary fiber-rich and low-calorie snack [62]. Similarly, there exists a patent for a food-related product based on the wet fermentation of jackfruit seeds, aimed at creating a nutritional feed with desirable digestive and absorptive characteristics. Additionally, a fermentation process has been applied in a patent targeting the production of probiotic components for tofu preparation [63]. In Brazil, a patent describes a process for transforming roasted jackfruit seed flour into a chocolate powder substitute, citing its sensory attributes as justification for this substitution [27].

Another innovative approach involves converting jackfruit peel into hydrochar via a hydrothermal carbonization process. The resulting carbon material was subsequently subjected to chemical treatment to optimize its surface properties. The efficacy of this process was evaluated for the removal of dyes from industrial wastewater systems, with significant success observed in the elimination of azo dyes, including methylene blue [68].

As for the apple, patents are scarcer. In the case of the 2016 patent by Rupasinghe and Robertson [113], the authors describe the production of apple peel extracts rich in flavonoids for the prevention and treatment of neurodegenerative diseases and others resulting from oxidative stress and/or inflammation.

7 Socioeconomic Impacts, Climate Change Mitigation, and Well-Being

Decreasing atmospheric carbon is crucial for mitigating climate change, which is exacerbated by rising CO_2 levels that have increased from 280 μg/ml in 1850 to over 420 μg/ml in 2024, contributing to global warming and biodiversity loss [48, 91, 125]. High carbon emissions, primarily from fossil fuel consumption and deforestation, lead to severe environmental issues such as soil erosion, drought, and flooding [48].

Fruit orchards play a significant role in carbon sequestration, capturing atmospheric CO_2 through photosynthesis and storing it in biomass and soil [105, 125]. By enhancing carbon storage in trees and soils, we can lower atmospheric CO_2 levels, which is essential for sustainable development and environmental protection [48, 105]. Furthermore, the cultivation of fruit trees serves multiple purposes, as it aids in carbon capture, supports biodiversity, and improves soil health, making it a vital strategy in combating climate change [48, 125] (Fig. 5.3).

Monitoring the production chain of fruit trees through sustainable practices involves assessing various factors, including carbon sequestration, soil health, and management techniques. Fruit trees, such as avocado [74], jackfruit [100], and apple [150], have demonstrated significant carbon-sink capabilities, with studies showing that orchards can sequester substantial amounts of carbon that are comparable to young forest systems [105, 125]. Effective management practices,

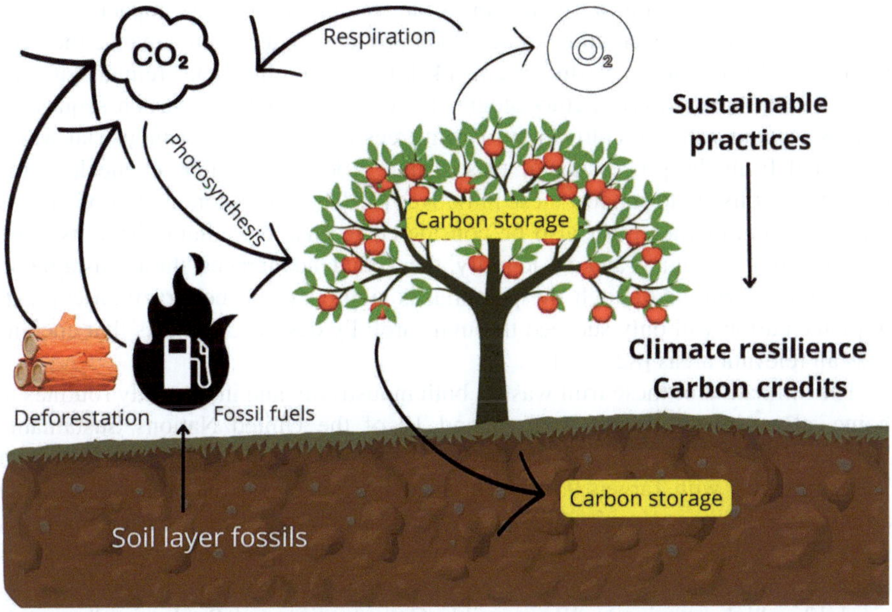

Fig. 5.3 Role of fruit orchards in carbon sequestration, climate resilience, and carbon credits

including intercropping and organic amendments, can enhance carbon pools in both biomass and soil, thereby promoting sustainability [48]. Additionally, the dynamics of photosynthesis in fruit trees can be optimized through elevated CO_2 levels, leading to increased biomass and carbon capture [150]. However, it is crucial to consider the carbon emissions associated with orchard management practices, as these can offset the benefits of carbon sequestration [125]. By integrating these monitoring strategies, stakeholders can ensure that fruit production aligns with sustainable agricultural practices, contributing to ecological balance and climate resilience [48].

The reuse of production chain wastes from fruit cultivation can significantly aid in reducing CO_2 emissions through various mechanisms. By incorporating organic waste, such as fruit peels and pulp, into soil management practices, the carbon stored in these materials can enhance soil organic matter, thereby increasing its carbon sequestration potential [125]. Additionally, the application of composted fruit waste can improve soil health, promoting microbial activity that further aids in carbon capture [48]. Moreover, fruit trees themselves, such as mangoes and avocados, have been shown to sequester substantial amounts of carbon in their biomass, with estimates indicating significant aboveground biomass in orchards [74]. This dual approach—utilizing waste for soil enhancement while simultaneously leveraging the carbon storage capacity of fruit trees—creates a sustainable cycle that mitigates CO_2 emissions and supports ecosystem health and agricultural productivity [105, 150].

In Brazil, the production of carbon credits from fruit trees and the management of fruit waste presents a path toward sustainability and the viability of a circular economy. Utilizing fruit waste not only reduces environmental impact but also offers the potential for creating value-added products, which can enhance the economic benefits of carbon credit systems [47]. However, challenges remain, such as the need for standardized methodologies to accurately quantify carbon sequestration associated with the cultivation of fruit trees and the reuse of industrial waste generated from the processing of these fruits. Additionally, market mechanisms must be established to support these initiatives [2]. The socioeconomic implications for local communities must also be considered to ensure equitable benefits from carbon credit schemes and to identify the social advantages these sustainable changes can bring. Although the potential is significant in the Brazilian context, implementation will only succeed through carefully designed projects that encompass all relevant areas [82, 132].

The utilization of these fruit wastes, both industrially and in our daily routines at home, aligns directly with Goals 9 and 12 of the United Nations Sustainable Development Goals for 2030. Goal 9 addresses "Industry, Innovation, and Infrastructure," with Targets 9.4 and 9.5 focusing on sustainable processes and the adoption of clean and environmentally sound technologies. In the context of this work, these goals align with the utilization of fruit wastes obtained in various ways to extract important bioactive molecules and compounds for diverse applications. Goal 12, which pertains to "Responsible Consumption and Production," encompasses sustainable food management, efficient use of natural resources, the reduction of food waste, and the proper handling of generated waste. In line with this goal is the principle of responsible eating, which promotes the potential use of whole foods (including what is currently considered waste) [60].

8 Conclusion and Future Perspectives

In conclusion, avocado, jackfruit, and apple wastes represent valuable sources of biomolecules and bioactive compounds, as well as raw materials for biotechnological processes. However, it is important to acknowledge both the general characteristics and the specific attributes unique to each type of waste. By correlating the botanical parts of avocado, jackfruit, and apple with the percentage of each fraction utilized and discarded during industrial processing, we can more clearly identify the biomolecules and bioactive compounds that can be derived from these fruits. This correlation depends on both the botanical and physiological aspects of the fruits and their components. Furthermore, the scale of waste production allows for the envisioning of applications that may require either smaller or larger volumes, as well as the potential for expanding or limiting the range of utilization possibilities. The avocado seed currently stands out as the primary source of enzymes, antioxidant compounds, and biomolecules, making it a key candidate for obtaining these valuable substances. Among fruit wastes, jackfruit rind is the most frequently manipulated and processed, while the fruit seed is notably used for its high sugar content, often converted into flour. Among apple parts, the peel exhibits the highest nutritional relevance, being rich in bioactive compounds, vitamins, and fibers. This suggests significant potential for biotechnological applications, including enzyme production and uses in both animal and human nutrition.

Beyond avocado seeds and jackfruit and apple peels, other residues from these fruits warrant further exploration in scientific research to achieve comprehensive utilization. This exploration can lead to the discovery of new biomolecules and bioactive compounds, ultimately resulting in products that serve various applications across different industries. Additionally, particularly for avocado and apple wastes, which have been underexplored in fruit and vegetable processing, there is potential for use as substrates in fermentation processes aimed at generating biomolecules, bioactive compounds, and enhancing protein content. Although research has been conducted using jackfruit peel as a substrate in solid-state fermentation, satisfactory results have not yet been reported. Moreover, the reuse of avocado, jackfruit, and apple wastes can be investigated for energy applications, especially given the pressing need to transition current energy sources to more sustainable options. Another critical area to address is the gap that exists between research on waste utilization and its actual application in industry.

These wastes can be viewed as raw materials for new products, thereby linking various supply chains and reinforcing the principles of a circular economy within current food production and usage practices. On a household level, composting remains a common destination for organic waste; however, raising awareness about the bioactive compounds present and the associated nutritional benefits of these wastes can open new perspectives for their use. Therefore, it is essential to strengthen research focused on avocado, jackfruit, and apple wastes, directing efforts toward effective biotechnological transformations. This approach will enhance their productive chains and associated ramifications while emphasizing the need to raise awareness of their biological and social value, which is intricately linked to environmental sustainability.

References

1. [ABPM] Associação Brasileira de Produtores de Maçã (n.d.) Dados estatísticos sobre a cultura da macieira. https://www.abpm.org.br
2. Abrahão, K.A., Blanger, A.T., Boehl, J.O., Figueiredo, A., Nunes, R. da S. and Domingos, J.P. (2023). Carbon credit market in the mobility sector: opportunities in the Brazilian context. E-tech 16: (3) 1–25. https://doi.org/10.18624/etech.v16i3.1279
3. Adan AA, Ojwang RA, Muge EK, Mwanza BK, Nyaboga EN (2020) Phytochemical composition and essential mineral profile, antioxidant and antimicrobial potential of unutilized parts of jackfruit. Food Res 4(4):1125–1134. https://doi.org/10.26656/fr.2017.4(4).326
4. Afotey B, Yuorkuu E, Akinie SY, Eshun FF, Sufyan MA (2024) Determination of health and nutritional benefits of jackfruits (*Artocarpus heterophyllus*). Journal of the Ghana Institution of Engineering 24(1):33–40. https://doi.org/10.56049/jghie.v24i1.149
5. Aguiar J, Alberto M, Pandolfi C, Torquato L (2020) Análise de mercado do óleo de abacate. Revista Interface Tecnológica 17(1):352–362. https://doi.org/10.31510/infa.v17i1.759

6. Ahmad T, Danish M (2022) A review of avocado waste-derived adsorbents: characterizations, adsorption characteristics, and surface mechanism. Chemosphere 296:134036. https://doi.org/10.1016/j.chemosphere.2022.134036
7. Ahmmed R, Inam AKMS, Alim MA, Sobhan MM, Haque MA (2017) Extraction and characterization of pectin from jackfruit (*Artocarpus heterophyllus* Lam) waste. J Pharm Biol Sci 32:57–66
8. Akter F, Haque M (2020) Jackfruit waste: A promising source of food and feed. Ann Bangladesh Agric 23(1):91–102. https://doi.org/10.3329/aba.v23i1.51477
9. Alves JLF, Da Silva JCG, Mumbach GD, Domenico MD, Da Silva Filho VF, De Sena RF, Machado RAF, Marangoni C (2020) Insights into the bioenergy potential of jackfruit wastes considering their physicochemical properties, bioenergy indicators, combustion behaviors, and emission characteristics. Renew Energy 155:1328–1338. https://doi.org/10.1016/j.renene.2020.04.025
10. Anusree M, Nampoothiri KM, Pandey A (2015) Sugarcane bagasse as an inert support for L-lysine production in solid-state fermentation. Indian J Biotechnol 14(2):193–199
11. Aprikian O, Levrat-Verny MA, Besson C, Busserolles J, Rémésy C, Demigné C (2001) Apple favorably affects parameters of cholesterol metabolism and anti-oxidative protection in cholesterol-fed rats. Food Chem 75(4):445–452. https://doi.org/10.1016/S0308-8146(01)00235-7
12. Araújo RG, Rodriguez-Jasso RM, Ruiz HA, Pintado MME, Aguilar CN (2018) Avocado by-products: nutritional and functional properties. Trends Food Sci Technol 80:51–60. https://doi.org/10.1016/j.tifs.2018.07.027
13. Aravind PV, Champatan V, Gopi G, Vijay V, Smit C, Pande S, van den Broeke LJP, John TD, Illathukandy B, Sukesh A, Shreedhar S, Nandakishor TM, Purushothaman SJ, Posada J, Lindeboom REF, Nampoothiri KUK (2022) Negative emissions at negative cost-an opportunity for a scalable niche. Front Energy Res 10:806435. https://doi.org/10.3389/fenrg.2022.806435
14. Arthey D, Ashurst PR (1997) Procesado de frutas. Acribia 1:247–256
15. Arun KB, Madhavan A, Sindhu R, Binod P, Pandey A, R, R. and Sirohi, R. (2020) Remodeling agro-industrial and food wastes into value-added bioactives and biopolymers. Ind Crop Prod 154:112621. https://doi.org/10.1016/j.indcrop.2020.112621
16. Aswin GA, Bhasin A, Mazumdar A (2022) Utilization of jackfruit by-products and application in food industry. Pharma Innov J 11(7):2993–2999
17. Babbar N, Oberoi HS, Sandhu SK (2015) Therapeutic and nutraceutical potential of bioactive compounds extracted from fruit residues. Crit Rev Food Sci Nutr 55(3):319–337. https://doi.org/10.1080/10408398.2011.653734
18. Babitha S, Soccol CR, Pandey A (2007) Solid-state fermentation for the production of *Monascus* pigments from jackfruit seed. Bioresour Technol 98(8):1554–1560. https://doi.org/10.1016/j.biortech.2006.06.005
19. Bancal V, Ray RC (2022) Overview of food loss and waste in fruits and vegetables: from issues to resources. In: Ray RC (ed.) Fruits and vegetable wastes: valorization to bioproducts and platform chemicals. Springer, Singapore, pp 3–32
20. Bangar SP, Dunno K, Dhull SB, Siroha AK, Changan S, Maqsood S, Rusu AV (2022) Avocado seed discoveries: chemical composition, biological properties, and industrial food applications. Food Chem X 16:100507. https://doi.org/10.1016/j.fochx.2022.100507
21. Basso AM, Moura MDFVD (2017) Jaca: um estudo de sua química. Editora IFRN 102:il. color. http://memoria.ifrn.edu.br/handle/1044/1515
22. Bleicher, J., 2002. História da macieira—A cultura da macieira
23. Bondonno NP, Bondonno CP, Ward NC, Hodgson JM, Croft KD (2017) The cardiovascular health benefits of apples: whole fruit vs. isolated compounds. Trends Food Sci Technol 69:243–256. https://doi.org/10.1016/j.tifs.2017.04.012

24. [BNDS] Banco Nacional de Desenvolvimento Econômico e Social (2010) Informativo Técnico SEAGRI, n. 2, nov. 2010. https://web.bndes.gov.br/bib/jspui/bitstream/1408/2534/1/InformativoSEAGRI_n.2%2C%20nov.2010.pdf
25. Brahma R, Ray S (2022a) In-depth analysis on potential applications of jackfruit peel waste: A systematic approach. Food Chem Adv 1:100119. https://doi.org/10.1016/j.focha.2022.100119
26. Brahma R, Ray S (2022b) A comprehensive review on the recent advances in the valorization of jackfruit waste for the development of value-added products. J Food Technol Res 9:120–134. https://doi.org/10.18488/jftr.v9i2.3124
27. Brazaca, S. G. C., Souza, M. C. and Gutierrez, E. M. R. (2015). Process for obtaining chocolate substitute, use of jackfruit seed as a chocolate substitute and chocolate flavored food compositions. BR Patent # 102013019333A2
28. Castro-López C, Bautista-Hernández I, González-Hernández MD, Martínez-Ávila GCG, Rojas R, Gutiérrez-Díez A, Medina-Herrera N, Aguirre-Arzola VE (2019) Polyphenolic profile and antioxidant activity of leaf purified hydroalcoholic extracts from seven Mexican *Persea americana* cultivars. Molecules 24:1. https://doi.org/10.3390/molecules24010173
29. Chakraborty A, Panda J, Mishra AK, Sivakumar S, Srinivasan LV, Nongbet A, Mohanty SS, Rustagi S, Nath PC, Wagh MS, Mohanta YK (2024) Jackfruit waste for the production of sustainable biorefinery products: a step towards bio-economic advancement. Biomass Convers Biorefinery 15:9801. https://doi.org/10.1007/s13399-024-05954-y
30. Chan SW, Han CE, Tan CP, Khor YP, Lou Z (2023) Ultrasound-assisted extraction of pectin from jackfruit (*Artocarpus heterophyllus*) rags: optimization, characterization, and application in model food gel. Curr Res Nutr Food Sci J 11(3):991–1007. https://doi.org/10.12944/CRNFSJ.11.3.07
31. Cohim EB, Leão AS, Silva H d A, Santos GS (2021) Water loss associated with food loss and waste in Brazil. Revista Brasileira de Ciências Ambientais 56(2):305–317. https://doi.org/10.5327/z21769478885
32. Costa LA, Diógenes ICN, Oliveira MDA, Ribeiro SF, Furtado RF, Bastos MDSR, Silva MAS, Benevides SD (2021a) Smart film of jackfruit seed starch as a potential indicator of fish freshness. Food Sci Technol 41(2):489–496. https://doi.org/10.1590/fst.06420
33. Costa LFF, Fracasso DS, dos Santos MG, Maffessoni D (2021b) Evaluation of pectinase production potential from apple hand residue by solid state fermentation. Brazil J Dev 7(3):27797–27811. https://doi.org/10.34117/bjdv7n3-472
34. Cruz-Casillas FC, García-Cayuela T, Rodriguez-Martinez V (2021) Application of conventional and non-conventional extraction methods to obtain functional ingredients from jackfruit (*Artocarpus heterophyllus* Lam.) tissues and by-products. Appl Sci 11(16):7303. https://doi.org/10.3390/app11167303
35. Daiuto ÉR, Tremocoldi MA, Matias De Alencar S, Vieites RL, Minarelli PH (2014) Chemical composition and antioxidant activity of the pulp, peel, and by products of avocado 'hass. Rev Bras Frutic 36(2):417–424. https://doi.org/10.1590/0100-2945-102/13
36. Dávila JA, Rosenberg M, Castro E, Cardona CA (2017) A model biorefinery for avocado (*Persea americana* mill.) processing. Bioresour Technol 243:17–29. https://doi.org/10.1016/j.biortech.2017.06.063
37. Del Castillo-Llamosas A, del Río PG, Pérez-Pérez A, Yáñez R, Garrote G, Gullón B (2021) Recent advances to recover value-added compounds from avocado by-products following a biorefinery approach. Curr Opin Green Sustain Chem 28:100433. https://doi.org/10.1016/j.cogsc.2020.100433
38. Ebert A, Kreuz EL, Zaffari GR, Petri JL (1988) Raleio de frutos em macieira no Alto Vale do Rio do Peixe em Santa Catarina. EMPASC 13
39. Ejiofor NC, Ezeagu IE, Ayoola MB, Umera EA (2018) Determination of the chemical composition of avocado (*Persea Americana*) seed. Adv Food Technol Nutr—Open J SE SE:S51–S55. https://doi.org/10.17140/aftnsoj-se-2-107

40. [EMBRAPA] Empresa Brasileira de Pesquisa Agropecuária (1999) Qualidade da carne bovina, in: Curso Conhecendo a Carne de Você Consome. Campo Grande, p 25
41. Fadavi R, Keyhani A, Mohtasebi SS (2011) An analysis of energy use, input costs and relation between energy inputs and yield of apple orchard. Res Agric Eng 57(3):88–96
42. [FAO] Food and Agriculture Organization (2022) World Food and Agriculture—Statistical Yearbook 2022. https://doi.org/10.4060/cc2211en
43. [FAO] Food and Agriculture Organization (2019) Major tropical fruits, statistical compendium. FAO, Rome
44. Fávaro VR, Rech Â (2022) Utilização de resíduos agroindustriais na alimentação de ruminantes. Agropecuária Catarinense 35(2):14–16. https://doi.org/10.52945/rac.v35i2.1134
45. Fertonani CR, Scabio AH, Schemin HC, Borba MB, Carneiro E, Nogueira A, Wosiacki G (2006) Influence of acid concentration on extraction and quality of apple pomace pectin. Semina: Ciências Agrárias 27(4):599–612. https://doi.org/10.5433/1679-0359.2006v27n4p599
46. Freire CJ d S, Camelatto D, Flores Cantillano RF, Kovaleski A, Fortes JF (1993) Cultura da maca. EMBRAPA-SPI
47. Gama-Rodrigues EF, Gama-Rodrigues AC, Vicente LC, Alvarenga LCBR, Müller MW, Partelli FL, Gonçalves JM, Freitas LS, Alvaristo DM, Cruz IB, Souza ING, Faitanin MA (2022) Perspectives on carbon footprint of agricultural land-use in Brazil. Carbon Footprints 1:1–6. https://doi.org/10.20517/cf.2022.01
48. Gelaye Y, Getahun S (2024) A review of the carbon sequestration potential of fruit trees and their implications for climate change mitigation: the case of Ethiopia. Cogent Food Agric 10(1):2294544. https://doi.org/10.1080/23311932.2023.2294544
49. Govindaraj D, Rajan M, Hatamleh AA, Munusamy MA (2018) From waste to high-value product: jackfruit peel derived pectin/apatite bionanocomposites for bone healing applications. Int J Biol Macromol 106:293–301. https://doi.org/10.1016/J.IJBIOMAC.2017.08.017
50. Guimarães LA d OP, de Mendonça GC (2017) Conceitos e Princípios Práticos da Agrofloresta Sucessional Biodiversa. In: Seagro: Plantando hoje as riquezas do futuro, pp 108–123
51. Gulsunoglu Z, Purves R, Karbancioglu-Guler F, Kilic-Akyilmaz M (2020) Enhancement of phenolic antioxidants in industrial apple waste by fermentation with *Aspergillus* spp. Biocatal Agric Biotechnol 25:101562. https://doi.org/10.1016/j.bcab.2020.101562
52. Gupta A, Marquess AR, Pandey AK, Bishayee A (2023) Jackfruit (*Artocarpus heterophyllus* Lam.) in health and disease: A critical review. Crit Rev Food Sci Nutr 63(23):6344–6378. https://doi.org/10.1080/10408398.2022.2031094
53. Heredia-Guerrero JA, Heredia A, García-Segura R, Benítez JJ (2009) Synthesis and characterization of a plant cutin mimetic polymer. Polymer 50(24):5633–5637. https://doi.org/10.1016/j.polymer.2009.10.018
54. Hurtado-Fernández E, Fernández-Gutiérrez A, Carrasco-Pancorbo A (2018) Avocado fruit—*Persea americana*. ExoticFruits 37–48. https://doi.org/10.1016/b978-0-12-803138-4.00001-0
55. [IBGE] Instituto Brasileiro de Geografia e Estatística (2024) Produção de Abacate. https://www.ibge.gov.br/explica/producao-agropecuaria/abacate/br
56. [IBGE] Instituto Brasileiro de Geografia e Estatística (2022) Produção de Abacate. https://www.ibge.gov.br/explica/producaoagropecuaria/abacate/br
57. [IBGE] Instituto Brasileiro de Geografia e Estatística (2017) Produção de jaca. https://www.ibge.gov.br/explica/producao-agropecuaria/jaca/br
58. Ibrahim HB, Abid MK, Zain HHM (2020) Heavy metal ions adsorption from aqueous solution by jackfruit peel as activated biochar low-cost adsorbent. J Adv Res Fluid Mech Therm Sci 67(2):154–165
59. IndustryArc (2023) Jackfruit Market: forecast (2023–2028). https://www.industryarc.com/Research/Jackfruit-Market-Research-507377
60. Instituto de Pesquisa Econômica Aplicada [IPEA] (2022) Objetivos de Desenvolvimento Sustentável. https://www.ipea.gov.br/ods/

61. Jayarajan M, Arunachala R, Annadurai G (2011) Agricultural wastes of jackfruit peel nano-porous adsorbent for removal of rhodamine dye. Asian J Appl Sci 4(3):263–270. https://doi.org/10.3923/ajaps.2011.263.270
62. Jiang S (2020) Jackfruit seed buccal tablet and preparation method thereof. CN Patent #110839877A
63. Jinan C, Di H, Yu X, Yu Y (2022) Method for preparing jackfruit probiotic lactone tofu. WO Patent # 2022205855A1
64. Jolly AB (1993) Botânica: introdução à taxonomia vegetal. In A baixada. Editora Nacional
65. Kassim, M., Achoui, M., Mustafa, M. R., Mohd, M. A., & Yusoff, K. M. (2010). Ellagic acid, phenolic acids, and flavonoids in Malaysian honey extracts demonstrate in vitro anti-inflammatory activity. Nutrition Research (New York, N.Y.), 30(9), 650-659. https://doi.org/10.1016/j.nutres.2010.08.008. PMid:20934607.
66. Kaur J, Singh Z, Shah HMS, Mazhar MS, Hasan MU, Woodward A (2023) Insights into phytonutrient profile and postharvest quality management of jackfruit: A review. Crit Rev Food Sci Nutr 64(19):6756–6782. https://doi.org/10.1080/10408398.2023.2174947
67. Khadem M, Husni Ibrahim A, Mokashi I, Hasan Fahmi A, NoemanTaqui S, Mohanavel V, Hossain N, Baba Koki I, Elfasakhany A, Dhaif-Allah MAH, Soudagar MEM, Syed AA (2023) Removal of heavy metals from wastewater using low-cost biochar prepared from jackfruit seed waste. Biomass Convers Biorefin 13:14447–14456. https://doi.org/10.1007/s13399-022-02748-y
68. Khan, M. A., Alqadami, A. A., Siddiqui, M. R. and Alothman, Z. A. (2020). Synthesis of hydrochar from jackfruit. U.S. Patent # 10557098B1
69. Khan MA, Hossain MM, Qadeer Z, Tanweer S, Ahmad B, Waseem M (2023) Jackfruit (*Artocarpus heterophyllus*): an overview of nutritional and functional food properties. In: Ismail T, Akhtar S, Lazarte CE (eds) Neglected plant foods of South Asia. Springer International Publishing, Cham, pp 411–451
70. Kishore K, Ding P, Sahu A (2023) Jackfruit: botany, production and uses. In: Flowering, fruit set and fruit development, pp 195–209. https://doi.org/10.1079/9781800622319.0009
71. Kolattukudy PE (2002) Cutin from plants. Biopolymers Online. https://doi.org/10.1002/3527600035.bpol3a01
72. Kumar M, Potkule J, Tomar M, Punia S, Singh S, Patil S, Singh S, Ilakiya T, Kaur C, Kennedy JF (2021) Jackfruit seed slimy sheath, a novel source of pectin: studies on antioxidant activity, functional group, and structural morphology. Carbohydr Polym Tech Appl 2:100054. https://doi.org/10.1016/j.carpta.2021.100054
73. Kupnik K, Primožič M, Kokol V, Knez Ž, Leitgeb M (2023) Enzymatic, antioxidant, and antimicrobial activities of bioactive compounds from avocado (*Persea americana* L.) seeds. Plan Theory 12:5. https://doi.org/10.3390/plants12051201
74. Kuyah S, Muthuri C, Wakaba D, Cyamweshi AR, Kiprotich P, Mukuralinda A (2024) Allometric equations and carbon sequestration potential of mango (*Mangifera indica*) and avocado (*Persea americana*) in Kenya. Trees Forest People 15:100467. https://doi.org/10.1016/j.tfp.2023.100467
75. Leong YK, Chang JS (2022) Valorization of fruit wastes for circular bioeconomy: current advances, challenges, and opportunities. Bioresour Technol 359:127459. https://doi.org/10.1016/j.biortech.2022.127459
76. Li Z, Lan Y, Miao J, Chen X, Chen B, Liu G, Wu X, Zhu X, Cao Y (2021) Phytochemicals, antioxidant capacity and cytoprotective effects of jackfruit (*Artocarpus heterophyllus* Lam.) axis extract on HepG2 cells. Food. Bioscience 41:100933. https://doi.org/10.1016/J.FBIO.2021.100933
77. Liu H, Kumar V, Jia L, Sarsaiya S, Kumar D, Juneja A, Zhang Z, Sindhu R, Binod P, Bhatia SK, Awasthi MK (2021) Biopolymer poly-hydroxyalkanoates (PHA) production from apple industrial waste residues: A review. Chemosphere 284:131427. https://doi.org/10.1016/j.chemosphere.2021.131427

78. Luchi VL (2002) Botânica e fisiologia- A cultura da macieira. Empresa de Pesquisa Agropecuária e Extensão, pp 59–102
79. Manhongo TT, Chimphango AFA, Thornley P, Röder M (2022) Current status and opportunities for fruit processing waste biorefineries. Renew Sust Energ Rev 155:14–18. https://doi.org/10.1016/j.rser.2021.111823
80. Manianga A, Ekuban CAA, Lukambagire I, Bose C, Sabarinath S (2024) Examining underutilized jackfruit (*Artocarpus heterophyllus*) seeds as a potential source of human subsistence in the Byse community, India. Front Sustain Food Syst 94(5):395–400. https://doi.org/10.3389/fsufs.2024.1377076
81. [MAPA] Ministério da Agricultura e Pecuária (2023) Brasil 2022/23 a 2032/33. Projeções de Longo Prazo 108p
82. Marcondes LP, Parisotto IRS, Faria SGS, Zucco FD, Farah OE (2013) Obtaining carbon credits through energy cogeneration projects from sugarcane bagasse in a sugar and alcohol agroindustry in the northwest region of the state of São Paulo. RMS—Revista Metropolitana de Sustentabilidade 3(3):116–136
83. Massafera G, Braga M, Oliveira J (2010) Composição de ácidos graxos do óleo do mesocarpo e da semente de cultivares de abacate (*Persea americana* Mill.) da região de Ribeirão Preto, SP. Braz J Food Nutrition 2(21):325
84. Mathew JJ, Vazhacharickal PJ, Sajeshkumar N, John N (2016) Comparative study of the activity of amylase produced by *Aspergillus niger* through Solid State Fermentation (SSF) using various starchy materials. Indian J Plant Sci 5(3):79–90
85. Meghavarnam AK, Janakiraman S (2017) Solid state fermentation: an effective fermentation strategy for the production of L-asparaginase by *Fusarium culmorum* (ASP-87). Biocwatal Agric Biotechnol 11:124–130. https://doi.org/10.1016/j.bcab.2017.06.001
86. Melgar B, Dias MI, Ciric A, Sokovic M, Garcia-Castello EM, Rodriguez-Lopez AD, Barros L, Ferreira ICRF (2018) Bioactive characterization of *Persea americana* Mill. by-products: a rich source of inherent antioxidants. Ind Crop Prod 111:212–218. https://doi.org/10.1016/j.indcrop.2017.10.024
87. Mijin S, Ding P (2020) Growth development and structural changes of Malaysian jackfruit cv. TekamYellowsyncarp. Scientia Horticulturae 272:109594. https://doi.org/10.1016/j.scienta.2020.109594
88. Monteiro HSA (2014) Agricultura regenerativa. Campos e negócios Hortifrúti
89. Munishamanna KB, Ajey G, Veena R, Palanimuthu V (2020) Evaluation of different strains of yeast and lactic acid bacteria for nutritional improvement of jackfruit waste under solid state fermentation. Mysore J Agric Sci 54(2):44–50
90. Natta S, Pal K, Kumar Alam B, Mondal D, Kumar Dutta S, Sahana N, Mandal S, Bhowmick N, Sankar Das S, Mondal P, Kumar Pandit G, Kumar Paul P, Choudhury A (2023) In-depth evaluation of nutritive, chemical constituents, and anti-glycemic properties of jackfruit (*Artocarpus heterophyllus* Lam) clonal accessions with flake colour diversity from eastern sub-Himalayan plains of India. Food Chem 407:135098. https://doi.org/10.1016/j.foodchem.2022.135098
91. NOAA, National Oceanic & Atmospheric Administration (2024) Trends in Atmospheric Carbon Dioxide (CO2). Available at: https://gml.noaa.gov/ccgg/trends/gl_trend.html. Acessed 13 Aug 2024
92. Nogueira A, Paganini C, Silva NCC, Wosiacki G (2005) Aproveitamento de bagaço de maçã para a produção de álcool e obtenção de fibras alimentares 29:179–185
93. Nogueira-de-Almeida CA, Da Veiga Ued F, De Almeida CCJN, Almeida ACF, Del Ciampo LA, Ferraz IS, Oliveira D, da Silva LF, Zambom CR, De Oliveira AF (2018) Nutritional profile and benefits of avocado oil (*Persea americana*): an integrative review. Braz J Food Technol 21. https://doi.org/10.1590/1981-6723.21417
94. Ntabakirabose G, Tuyisenge JC, Musoni P, Ndaruhutse F, Majuga JCN, Izamuhaye JC, Hakizimana F, Joshua OO (2021) An economic analysis of the factors influencing avocado production, profitability and sustainability in Rwanda: a case study of Burera, Gicumbi, and Musanze districts. Int J Finance Banking Res 7(6):164–173. https://doi.org/10.11648/j.ijfbr.20210706.14

95. Okino-Delgado CH, Prado DZ, Fleuri LF (2018) Brazilian fruit processing wastes as a source of lipase and other biotechnological products: a review. An Acad Bras Cienc 90:2927–2943. https://doi.org/10.1590/0001-3765201820180095
96. Oliveira A (2021) Estudo da produção e caracterização de Biochar oriundo da pirólise do resíduo de Maçã e suas aplicações. Universidade Federal de Uberlândia. https://doi.org/10.14393/ufu.di.2021.524
97. Osorio LLDR, Flórez-López E, Grande-Tovar CD (2021) The potential of selected agri-food loss and waste to contribute to a circular economy: applications in the food, cosmetic and pharmaceutical industries. Molecules 26(2):515. https://doi.org/10.3390/molecules26020515
98. Ozdemir F, Topuz A (2004) Changes in dry matter, oil content and fatty acids composition of avocado during harvesting time and post-harvesting ripening period. Food Chem 86(1):79–83. https://doi.org/10.1016/j.foodchem.2003.08.012
99. Pandey KR, Neupane MP, Joshi YR, Paudel A, Khatri S (2023) Financial analysis of apple (*Malus domestica*) production in Darchula district of Nepal. Cogent Food Agric 9(2):2272490. https://doi.org/10.1080/23311932.2023.2272490
100. Pakhom MNJ, Rahman MS, Baril MS, Dhar M (2020) Carbon sequestration potentiality of different cropland agroforestry systems in Dinajpur district. J Sci Technol 18:43–52
101. Pathak N, Singh S, Singh P, Singh PK, Singh R, Bala S, Thirumalesh BV, Gaur R, Tripathi M (2022) Valorization of jackfruit waste into value added products and their potential applications. Front Nutr 9(4):1061098. https://doi.org/10.3389/fnut.2022.1061098
102. Patowary R, Patowary K, Kalita MC, Deka S, Lam SS, Sarma H (2022) Green production of noncytotoxic rhamnolipids from jackfruit waste: process and prospects. Biomass Convers Biorefinery 4375–4388:4375. https://doi.org/10.1007/s13399-022-02427-y
103. Qin S, Shekher Giri B, Kumar Patel A, Sar T, Liu H, Chen H, Juneja A, Kumar D, Zhang Z, Kumar Awasthi M, Taherzadeh MJ (2021) Resource recovery and biorefinery potential of apple orchard waste in the circular bioeconomy. Bioresour Technol 321:124496. https://doi.org/10.1016/j.biortech.2020.124496
104. Radenkovs V, Kviesis J, Juhnevica-Radenkova K, Valdovska A, Püssa T, Klavins M, Drudze I (2018) Valorization of wild apple (*Malus* spp.) by-products as a source of essential fatty acids, tocopherols and phytosterols with antimicrobial activity. Plan Theory 7:4. https://doi.org/10.3390/plants7040090
105. Rajan R, Mirza AA, Pandey K (2022) Fruit based system- A viable alternative for carbon sequestration. Ecol Environ Conserv 28(1):263–265. https://doi.org/10.53550/eec.2022.v28i01.036
106. Rajasekharan R, Bahuleyan AK, Madhavan A, Philip E, Sindhu R, Binod P, Kumar Awasthi M, Pandey A (2023) Neem extract–blended nanocellulose derived from jackfruit peel for antibacterial packagings. Environ Sci Pollut Res 30(4):8977–8986. https://doi.org/10.1007/s11356-022-20382-z
107. Ramos-Aguilar AL, Ornelas-Paz J, Tapia-Vargas LM, Gardea-Béjar AA, Yahia EM, Ornelas-Paz J d J, Ruiz-Cruz S, Rios-Velasco C, Escalante-Minakata P (2021) Effect of cultivar on the content of selected phytochemicals in avocado peels. Food Res Int 140:110024. https://doi.org/10.1016/j.foodres.2020.110024
108. Ranasinghe RASN, Maduwanthi SDT, Marapana RAUJ (2019) Nutritional and health benefits of jackfruit (*Artocarpus heterophyllus* Lam.): a review. Int J Food Sci 2019:1–12. https://doi.org/10.1155/2019/4327183
109. Razeto B, Romero F, Araya E (2004) Influencia de algunas propiedades organolépticas en la aceptabilidad del fruto de palto (*Persea americana* Mill.). Agricultura Técnica 64(1):89–94. https://doi.org/10.4067/s0365-28072004000100011
110. Reyes FMS (1994) Manejo de cosecha y postcosecha de manzanas 'Braeburn' 15:95–101
111. Rodríguez IST, Alcántara Quintana LE, Algara Suarez P, Ruiz Cabrera MA, Lagunes AG (2021) Physicochemical properties, antioxidant capacity, prebiotic activity and anticancer potential in human cells of jackfruit (*Artocarpus heterophyllus*) seed flour. Molecules 26(16):4854. https://doi.org/10.3390/molecules26164854

112. Rodríguez-Martínez B, Romaní A, Eibes G, Garrote G, Gullón B, del Río PG (2022) Potential and prospects for utilization of avocado by-products in integrated biorefineries. Bioresour Technol 364:128034. https://doi.org/10.1016/j.biortech.2022.128034
113. Rupasinghe HPV, Robertson GS (2016) Phenolic compositions from apple skin and uses thereof. U. S. Patent # 9,511,107 B2
114. Salazar-López NJ, Domínguez-Avila JA, Yahia EM, Belmonte-Herrera BH, Wall-Medrano A, Montalvo-González E, González-Aguilar GA (2020) Avocado fruit and by-products as potential sources of bioactive compounds. Food Res Int 138:109774. https://doi.org/10.1016/j.foodres.2020.109774
115. San José MJ, Alvarez S, López R (2023) Conical spouted bed combustor to obtain clean energy from avocado waste. Fuel Process Technol 239:107543. https://doi.org/10.1016/j.fuproc.2022.107543
116. Santos VS, Machado AR, Araújo PF, Rodrigues RS (2008) Avaliação sensorial de biscoitos elaborados com resíduo de polpa de amora preta (*Rubus* spp.). In: XVII Congresso de Iniciação Científico Da UFPEL
117. Sarangi PK, Srivastava RK, Singh AK, Sahoo UK, Prus P, Dziekański P (2023) The utilization of jackfruit (*Artocarpus heterophyllus* L.) waste towards sustainable energy and biochemicals: the attainment of zero-waste technologies. Sustain For 15(16):12520. https://doi.org/10.3390/su151612520
118. Sato MF, Vieira RG, Zardo DM, Falcão LD, Nogueira A, Wosiacki G (2010) Apple pomace from eleven cultivars: an approach to identify sources of bioactive compounds. Acta Scientiarum— Agronomy 32(1):29–35. https://doi.org/10.4025/actasciagron.v32i1.3176
119. Schmidt S, Silva AL (2000) Maçã: renova-se a exigência pela qualidade. Informativo SBF, Jaboticabal. Editora Acribia 3
120. Sebrae (n.d.) Cultivo e o mercado da maçã. https://www.sebrae.com.br/sites/PortalSebrae/artigos/o-cultivo-e-o-mercado-da-maca,ea7a9e665b182410VgnVCM100000b272010aRCRD
121. Segundo JF d L, Jerônimo HMÂ, Viera VB, Nascimento CMS d A (2021) Desenvolvimento de hambúrguer vegano adicionado da farinha de couve folha: avaliação físico-química, microbiológica e sensorial. Res Society Dev 10(1):e42510111761. https://doi.org/10.33448/rsd-v10i1.11761
122. [SESI] Serviço Social da Indústria (2023) Alimente-se bem. Sementes de jaca: benefícios à saúde, receitas e curiosidades. https://alimentesebem.sesisp.org.br/arquivos/noticia/semente-de-jaca-beneficios-a-saude-receitas-e-curiosidades
123. Settier-Ramírez L, López-Carballo G, Hernández-Muñoz P, Fontana-Tachon A, Strub C, Schorr-Galindo S (2022) Apple-based coatings incorporated with wild apple isolated yeast to reduce *Penicillium expansum* postharvest decay of apples. Postharvest Biol Technol 185:111805. https://doi.org/10.1016/j.postharvbio.2021.111805
124. Shaffer, B., Wolstenholme, B.N. and Whiley, A.W. (2013). The avocado: botany, production and uses
125. Sharma S, Rana VS, Prasad H, Lakra J, Sharma U (2021) Appraisal of carbon capture, storage, and utilization through fruit crops. Front Environ Sci 9:700768. https://doi.org/10.3389/fenvs.2021.700768
126. Shirahigue LD, Ceccato-Antonini SR (2020) Agro-industrial wastes as sources of bioactive compounds for food and fermentation industries. Ciência Rural 50(4):e20190857. https://doi.org/10.1590/0103-8478cr20190857
127. Silpa M, Vijayalakshmi A (2023) Impact of cocoa shell and jack fruit peel waste biocompost influence on the growth and yield attributes of *Vigna unguiculata* subsp. *sesquipedalis* (L.). Agric Sci Dig 43(2):150–156. https://doi.org/10.18805/ag.D-5488
128. Singh J, Chauhan PS, Kavita A, Bhatnagar P, Meena RR (2015) Jackfruit: a food of promise. HortFlora Research Spectrum 4:277–281

129. Singh R, Langyan S, Sangwan S, Gaur P, Khan FN, Yadava P, Rohatgi B, Shrivastava M, Khandelwal A, Darjee S, Sahu PK (2022) Optimization and production of alpha-amylase using *Bacillus subtilis* from apple peel: comparison with alternate feedstock. Food Biosci 49:101978. https://doi.org/10.1016/j.fbio.2022.101978
130. Solarte-Toro JC, Ortiz-Sanchez M, Restrepo-Serna DL, Piñeres PP, Cordero AP, Alzate CAC (2021) Influence of products portfolio and process contextualization on the economic performance of small- and large-scale avocado biorefineries. Bioresour Technol 342:126060. https://doi.org/10.1016/j.biortech.2021.126060
131. Sousa APM, Macedo ADB, Silva AP d F, Costa JD, Dantas DL, Apolinário M d O, Santana RAC, Campos ARN (2020) Enriquecimento proteico dos resíduos da jaca por fermentação semissólida. Braz Appl Sci Rev 4(3):987–997. https://doi.org/10.34115/basrv4n3-019
132. de Souza CR, Baldoni AB, Tonini H, Maia VA, dos Santos RM, Luvison M, dos Santos JP (2022) Ecological patterns and conservation opportunities with carbon credits in Brazil nut groves: a study-case in the Southeast Amazon. Cerne 29
133. Spada FP, Balagiannis DP, Purgatto E, do Alencar SM, Canniatti-Brazaca SG, Parker JK (2021) Characterisation of the chocolate aroma in roast jackfruit seeds. Food Chem 354:129537. https://doi.org/10.1016/j.foodchem.2021.129537
134. Sustentarea (2018) Panc: jaca verde. <https://www.fsp.usp.br/sustentarea/2018/08/27/panc-jaca-verde/>
135. Taguchi V (2021) Maior produtor de avocado do Brasil espera manter exportações, mesmo com produção menor. https://revistagloborural.globo.com/Noticias/Agricultura/Hortifruti/noticia/2021/03
136. Tesfaye T, Ayele M, Gibril M, Ferede E, Limeneh DY, Kong F (2022) Beneficiation of avocado processing industry by-product: a review on future prospect. Curr Res Green Sustain Chem 5:100253. https://doi.org/10.1016/j.crgsc.2021.100253
137. Thongsai N, Panichakorn J, Sumana K, Insik I, Peerasak P (2019) Utilization of carbon dots from jackfruit for real-time sensing of acetone vapor and understanding the electronic and interfacial interactions using density functional theory. Appl Surf Sci 487:1233–1244. https://doi.org/10.1016/j.apsusc.2019.04.269
138. Ton-That L, Huynh TNL, Duong BN, Nguyen D, Nguyen N, Pham VH, Ho TH, Dinh VP (2023) Kinetic studies of the removal of methylene blue from aqueous solution by biochar derived from jackfruit peel. Environ Monit Assess 195(11):1266. https://doi.org/10.1007/s10661-023-11867-6
139. Tremocoldi MA, Rosalen PL, Franchin M, Massarioli AP, Denny C, Daiuto ÉR, Paschoal JAR, Melo PS, De Alencar SM (2018) Exploration of avocado by-products as natural sources of bioactive compounds. PLoS One 13(2):1–12. https://doi.org/10.1371/journal.pone.0192577
140. Umennadi, R.O., Igwilo, I.O, Umennadi, P.U., Nweze, F. C., Lebechukwu, B.C., Asogwa, K. K. and Ezeobi, P.U. (2023). Comparative studies on the nutrient composition and antioxidant activities of the leaf, pulp, and seed extracts of jack-fruit tree (*Artocarpus heterophyllus*). Biosci J 11: 270–283. https://doi.org/10.54117/the_bioscientist.v11i3.158
141. Urbina L, Eceiza A, Gabilondo N, Corcuera MÁ, Retegi A (2019) Valorization of apple waste for active packaging: multicomponent polyhydroxyalkanoate coated nanopapers with improved hydrophobicity and antioxidant capacity. Food Packag Shelf Life 21:100356. https://doi.org/10.1016/j.fpsl.2019.100356
142. Varnam AH, Sutherland JP (1997) Bebidas: Tecnología, Química y Microbiología
143. Vendruscolo F, Ribeiro CS, Espósito E, Ninow JL (2009) Tratamento biológico do bagaço de maçã e adição em dietas para alevinos. Revista Brasileira de Engenharia Agrícola e Ambiental 13(4):487–493. https://doi.org/10.1590/s1415-43662009000400018
144. Waghmare R, Memon N, Gat Y, Gandhi S, Kumar V, Panghal A (2019) Jackfruit seed: an accompaniment to functional foods. Braz J Food Technol 22:e2018207. https://doi.org/10.1590/1981-6723.20718

145. Wang H, Feng H, Luo Y (2007) Control of browning and microbial growth on fresh-cut apples by sequential treatment of sanitizers and calcium ascorbate. J Food Sci 72:1. https://doi.org/10.1111/j.1750-3841.2006.00210.x
146. Wang W, Bostic TR, Gu L (2010) Antioxidant capacities, procyanidins and pigments in avocados of different strains and cultivars. Food Chem 122:1193–1198. https://doi.org/10.1016/j.foodchem.2010.03.114
147. Wang Y, Jiang Z (2022) Rapid large-scale preparation of polysaccharides from jackfruit peel waste by high-speed countercurrent chromatography and their antioxidant and hypoglycemic activities. J Sep Sci 45:771–779. https://doi.org/10.1002/jssc.202100636
148. Wong M, Eyres L, Ravetti L (2014) Modern aqueous oil extraction-centrifugation systems for olive and avocado oils. Green vegetable oil processing: revised first edition. AOCS Press. https://doi.org/10.1016/B978-0-9888565-3-0.50005-4
149. Wosiacki G, Pholman BC, Nogueira A (2004) Apple varieties growing in subtropical areas: the situation in Paraná—Brazil. Ciências e Tecnologia de Alimentos 24:347–352. https://doi.org/10.1590/s0101-20612004000300007
150. Wu T, Wang Y, Yu C, Chiarawipa R, Zhang X, Han Z, Wu L (2012) Carbon sequestration by fruit trees—Chinese apple orchards as an example. PLoS One 7(6):e38883. https://doi.org/10.1371/journal.pone.0038883
151. Xu SY, Liu JP, Huang X, Du LP, Shi FL, Dong R, Huang XT, Zheng K, Liu Y, Cheong KL (2018) Ultrasonic-microwave assisted extraction, characterization and biological activity of pectin from jackfruit peel. LWT 90:577–582. https://doi.org/10.1016/j.lwt.2018.01.007
152. Yasir M, Das S, Kharya M (2010) The phytochemical and pharmacological profile of *Persea americana* Mill. Pharmacogn Rev 4(7):77–84. https://doi.org/10.4103/0973-7847.65332
153. Yepes-Betancur DP, Márquez-Cardozo CJ, Cadena-Chamorro EM, Martinez-Saldarriaga J, Torres-León C, Ascacio-Valdes A, Aguilar CN (2021) Solid-state fermentation—assisted extraction of bioactive compounds from hass avocado seeds. Food Bioprod Process 126:155–163. https://doi.org/10.1016/j.fbp.2020.10.012
154. Zeng S, Cao J, Wei C, Chen Y, Liu Q, Li C, Zhang Y, Zhu K, Wu G, Tan L (2023) Polysaccharides from *Artocarpus heterophyllus* Lam. (jackfruit) pulp alleviate obesity by modulating gut microbiota in high fat diet-induced rats. Food Hydrocoll 139:108521. https://doi.org/10.1016/j.foodhyd.2023.108521
155. Zhang F, Wang T, Wang X, Lü X (2021) Apple pomace as a potential valuable resource for full-components utilization: a review. J Clean Prod 329:129676. https://doi.org/10.1016/j.jclepro.2021.129676

Chapter 6
Biotechnological Potentialities and Valorization of Mango Peel and Stone Waste: Their Potential Applications

V. Swetha, G. C. Jeevitha, and Aanchal Mittal

Abbreviation

AES	Alkali extraction method
CE	Catechin equivalent
CP	Cold press
DPPH	2,2-diphenyl-1-picrylhydrazyl
GAE	Gallic acid equivalent
HCl	Hydrochloric acid
kHz	Kilohertz
LE	Leucoanthocyanidin equivalents
QE	Quercetin equivalents
TMA	Total monomeric anthocyanins
WES	Water extraction method
WSWS	Water-ultrasonic extraction method

1 Introduction

The mango (*Mangifera indica* L.) has been grown for around 4000 years and is considered the "king" of fruits. Much like its popularity, its production and consumption have surged [1]. The leading mango-producing nation in the world, India, contributes 54.2% of the global mango crop [2]. Mango is a tropical fruit of

V. Swetha · G. C. Jeevitha (✉) · A. Mittal
School of BioSciences and Technology (SBST), Vellore Institute of Technology, Vellore, India
e-mail: jeevitha.gc@vit.ac.in

C. Trois, R. C. Ray (eds.), *Plantation Crop Wastes: Valorization for Economic Sustainability*, Waste as a Resource,
https://doi.org/10.1007/978-3-031-96805-1_6

global importance, with a yearly production of about 38 million tons [3]. The waste generated during mango processing is about 600,000 tons annually [4]. Mango by-products have been shown in numerous studies to be an excellent source of fiber [5, 6], pectin [7], and antioxidants [8, 9]. Applications of unripe mangoes include chutneys, pickles, and powdered forms, which are acidic and savory dishes. On the other hand, ripe mangoes yield pulp, juice, and canned or frozen fruit. Ripe mangoes are also processed into fruit bars, powders, milk, juices, purees, and breakfast cereals [10]. Mango processing results in kernel and peel as by-products which constitute approximately 17–22% of the whole fruit [11]. After being processed industrially, the majority of these by-products are now disposed of as waste. The likelihood of creating added value is diminished by a lack of knowledge about the characteristics and composition of by-products [12]. By creating higher-value applications for its by-products, tropical fruit processing could become economically sustainable [13]. The significance of mango seeds as a source of ingredients with added value is quite promising. It has a high biological value, such as being rich in protein, devoid of trans fatty acids, and is a good source of natural antioxidants [12]. Mango polyphenolic components have been shown to exhibit potent antioxidant properties [14] and demonstrate bioactivity in lung, colon, liver, breast, leukemia, prostate, cervix, and other carcinoma cell line models [15, 16]. The bioactive potential of the mango seed kernel offers the processing industry a unique opportunity to establish green valorization methods that will produce extracts rich in polyphenols with additional value and health benefits. Therefore, methods that emphasize the effective selection of extraction solvents and the use of novel green extraction techniques can aid in achieving the objectives of the green extraction of natural products [17].

2 Geographical Origin, Distribution, Production, and Productivity

2.1 General Description

Seventy species of *Mangifera* are found largely in tropical Asia [18]. Mango trees are evergreen, deeply rooted plants with several branches that eventually grow into massive trees. Mangoes grow well in tropical and subtropical regions from sea level to 1200 m above the equator (between 25°N and 25°S), but they cannot be cultivated commercially above 600 m [19]. About 125 cm of rain, primarily during the monsoon, is ideal. The ideal temperature of 24–27 °C paired with high humidity is suitable for the cultivation of mangoes. Since mangoes are perennial plants, they can be cultivated successfully in a variety of soil types, from lateritic to alluvial. Mangoes grow well in loamy, alluvial, and sandy loam soils [18].

2.2 Geographical Origin

The mango's origins can be traced back to 4000 BC in the Himalayas, including Eastern India, Burma, and the Andaman Islands. The genus *Mangifera* has approximately 49 species, with 41 recognized and 8 uncertain. They are found across Malaysia, from India and Ceylon to the Philippines and New Guinea, and in the Indian Ocean, including the Sunda archipelago and Sulu Archipelago [20].

2.3 Composition of Waste

2.3.1 Different Types of Mango Waste and Their Composition

Mangoes are classified into three distinct parts: the mesocarp, epicarp, and endocarp which are the pulp, peel, and seed kernel respectively (Fig. 6.1). Peels and kernels from processed mangoes are organic wastes that possess nutritional value. 50–60% of the weight of the entire fruit is made up of mango pulp. A waste by-product is formed during processing, which renders about (25–40%) of the pulp that was originally generated [21]. 15% to 20% of the peel is wasted commercially, which pollutes landfills [22]. Protocatechuic acids, mangiferin, and β-carotene pigment are among the peel's functional components that are well-known for their anti-inflammatory, anti-microbial, anti-diabetic, and anti-carcinogenic attributes [22]. Regardless of the cultivar, the mango seed kernel makes about 35–55% of the fresh fruit's weight [23]. The kernel (68%), shell (29%), and testa (3%) are the three distinct components of the mango seed [24, 25].

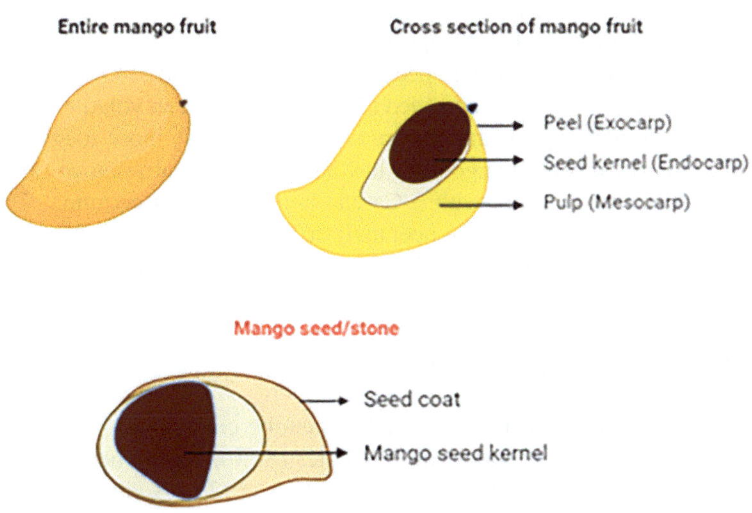

Fig. 6.1 Schematic representation of mango parts

2.4 *Components of Mango Waste*

2.4.1 Nutritional Profile of Mango Peel

Mango peels contain several nutritious substances, such as dietary fiber, carotenoids, polyphenols, and enzymes [8, 9]. The powdered mango peel comprises 86.4% carbohydrates, 2.6% fat, 8.9% crude fiber, and 3.8% protein. It has an abundant amount of β-carotene (5600 μg/g), calcium (4.15 mg/100 g), phosphorous (0.45 mg/100 g), iron (4.62 mg/100 g), and zinc (2.85 mg/100 g) [1]. The principal phenolic substances observed in the mango peel extract were mangiferin pentoside, ellagic acid, quercetin, and syringic acid [5]. The overall, soluble and insoluble dietary fiber in mango peel powder was between 40.6–72.5, 12.8–23, and 27.8–49.5%, respectively. The dietary fiber from mango peels contains bonded phenolic acids including protocatechuic acid, ferulic acid, gallic acid, and syringic acid. The bound flavonoids including kaempferol, rutin, and quercetin are also found in mango peel dietary fiber [6]. Mango peels can be utilized to make functional foods since they are high in phenolics and dietary fiber. Furthermore, they have abundant benzophenone derivatives, flavonol O [26], and xanthone C-glycosides [27]. Mango peels also contain both lipid and water-soluble pigments such as carotenoids and anthocyanin which impart yellow orange and red color to the peel, respectively. Green and ripe mango peels that were hot air-dried had carotenoid concentrations ranging from 7.69 to 16.06 mg/100 g [28]. Mango peels from two distinct cultivars, Janardhan Pasand and Arka Anmol, have a total carotenoid concentration ranging from 0.74 to 31.18 μg/g [29]. Mango peels include dietary fiber in the form of cellulose, hemicellulose, lignin, pectin, and gums [30]. Several cultivars, including Kent, Tommy Atkins, Palmer, Ataulfo, and Nam Dokmai, have varying total dietary fiber contents: 36.0, 36.3, 28.7, 54.53, and 42.4% respectively [31]. The powder obtained from mango peel is rich in fiber that may be utilized to make bakery and other processed items, which can improve their nutritional value and shelf life [27]. The monoterpene content of mango peels was 80.15 μg/g, substantially greater than that of mango pulp (6.37 μg/g) and seed (3.02 μg/g) [27]. The volatile aroma compounds found in mango peel samples are more intense than those found in mango pulp [32]. Comparatively, mango peel has a higher concentration of odor-active compounds than mango pulp, with 3-Carene being the most common. The additional components that give mango peel their distinct smells include 1-octen-3-one, nonanal, ß-damascenone, limonene, 2-nonenal, and decanal [27].

2.4.2 Nutritional Composition of Mango Seed

Mango seeds are abundant in antioxidants, phenolic compounds, fiber, and carotenoids [27]. Mango seed contains carbohydrates (81%), crude fiber (22%), protein (7%), lipids (5.4%), hemicellulose (14%), cellulose (2.9%), and lignin (2%) as shown in Fig. 6.2 [12]. The fiber concentration is higher in the testa and endocarp,

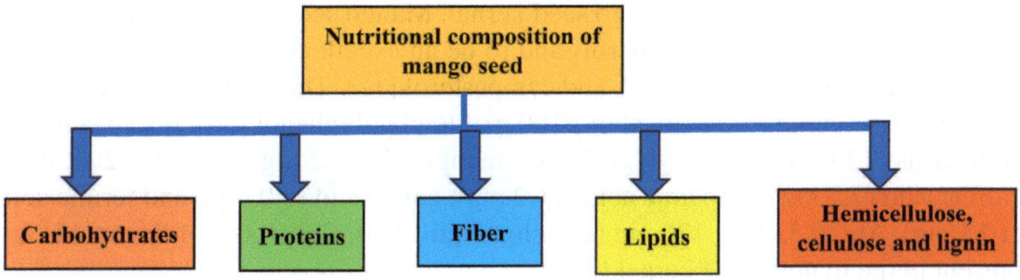

Fig. 6.2 Nutritional composition of mango seed

Table 6.1 Nutritional composition of mango seed

S. No.	Compound	Examplcs
1	Minerals	Potassium, calcium, phosphorous, Sulphur, iron, silicate, and chloride
2	Fatty acids	Octadecanoic acid and 9-(z)-octadecenoic acid
3	Macronutrients	Carbohydrates, crude fiber, protein, lipids, hemicellulose, cellulose, and lignin
4	Fiber	Cellulose

which cover the kernel. The endocarp of mango seeds contains a considerable amount of nutritious fiber, especially cellulose (22%), which could improve their application in the food and biotechnology industries. The minerals present in mango seeds include potassium, calcium, and phosphorous. In addition, it also constitutes other minerals such as sulphur, iron, silicate, and chloride (< 3%). Octadecanoic acid (24.22–32.80%) and 9-(z)-octadecenoic acid (46.37–58.59%) are the primary fatty acids prevalent in mango seeds, which additionally have the characteristic qualities of vegetable butter [33]. The nutritional composition of mango seed is given in Table 6.1.

2.4.3 Nutritional Composition of Mango Seed Kernel

Mango stone is decorticated to extract the mango kernel [25]. Kernels are composed of 77% carbohydrate,11% fat, 2% ash, 6% protein, and 2% crude fiber. Mostly depending on the cultivar, 6–13% protein content is present in the mango seed kernel. The proportion of key amino acids in mango seed kernel protein includes tyrosine (2.7%), isoleucine (4.4%), valine (5.8%), leucine (6.9%), phenylalanine (3.4%), threonine (3.4%), lysine (4.3%), and methionine (1.2%) [22]. Antioxidants like vitamin E and C, and other vital vitamins A, B1–3, B5,6, B9, and B12, K are also present [22]. As reported by Odunsi [24], the kernel has the mineral composition as follows, calcium-4.5 g/kg, phosphorus-2.3 g/kg, potassium-0.6 g/kg, magnesium-9.8 g/kg, sodium-1.5 g/kg, copper-40.2 μg/g, iron-53 μg/g, zinc-26.3 μg/g, and manganese 61.1-μg/g. The compounds that promote health such as triacylglycerols and fatty acids [34], xanthones, mangiferin [14], gallotanins [35], phenolic acids and flavonoids are found in mango seed kernel. Mango seed kernels have also been demonstrated to contain tocopherols and to be an excellent source of phytosterols, such as stigmasterol, ß-sitosterol, and campesterol [11]. The mango seed kernel has 112 mg GAE (gallic acid equivalent)/100 g of total polyphenol [36]. It also contains coumarins (12.6 mg), tannins (20.7 mg), mangiferin (4.2 mg), vanillin (20.2 mg), caffeic acid (7.7 mg), cinnamic acid (11.2 mg), gallic acid (6.0 mg), and ferulic acid (10.4 mg). About 3325 ± 120 mg (catechin equivalent, CE)/100 g of flavonoids are present in the mango seed kernel [37].

3 Valorization Techniques

The various valorization techniques of mango waste are discussed in the following sections.

3.1 Various Techniques for Valorizing Mango Waste

The fruit processing firms can generate more revenue while minimizing pollution by utilizing the mango by-products [27]. Compounds isolated from mango by-products can make various functional foods with additional value. The adaptation of various types of valorization techniques can resolve the waste disposal issue [38].

3.1.1 Bio-Ethanol Production

The bioethanol extracted from mango peel can be a promising alternative to gasoline since it is both environmentally and economically sustainable [39]. The ethanol yield of 9.68% was achieved by using the mango peel as a substrate for fermentation; however, production and productivity were considerably lower than mango pulp [40]. The application of thermotolerant and alcohol-tolerant yeasts at 40 °C for 120 hr. resulted in the production of 13 g/L of ethanol concentration [41].

3.1.2 Wine Production

The immobilized biocatalyst system with yeast and mango peel with strong operational stability was used to ferment the mango wine, which resulted in (1.53–3.29) ethanol productivity [42]. The developed system was found to be more appropriate for low-temperature winemaking with superior quality.

3.1.3 Enzyme Production

Mango peel was used as a substrate for the synthesis of carboxy methyl cellulase (7.8 U/mg) using *Paenibacillus polymyxa* as it has an abundant amount of lingo cellulosic fiber [43–45]. Mango peels are also a rich source of pectin; hence these peels are used as an ideal substrate for the synthesis of pectinases [43, 44]. Polygalacturonase and pectin lyase are produced by *Aspergillus foetidus* utilizing mango peel as a substrate.

3.2 Mango Peels—Valorization for Functional Compounds

3.2.1 Pectin

The mango peels had a pectin content between 10–15% [38]. Pectin can be extracted from mango peel by using various techniques such as conventional hydrochloric acid, microwave-assisted HCl, ultrasound-assisted HCl, sulfuric acid, citric acid, ultrasound-assisted citric acid, acetic acid, precipitation, subcritical water, ultrasound-assisted deep eutectic solvent-based extraction methods [27]. Pectin (22.67 and 21.94%) was extracted from unripe and ripe mango peels using conventional acid (HCl) at different pH ranges (1.5–3) and temperatures (70–90 °C) [46, 47]. The pectin yields from the traditional acid extraction and microwave-assisted acid extraction were 8.8% and 8.6–10.5%, respectively [48]. The pectin yield from the ultrasound-assisted extraction was 13–30% at 37–80 kHz ultrasound frequencies [49]. At 100 °C for one hour at pH 2, sulfuric acid-assisted extraction yielded 21.7% of the pectin yield [50]. Higher pectin yields (24.1–31.7%) were observed in all Ghep, Cat Chu, and Hoa Loc mango cultivars at the ripening stage [51]. The anthocyanin-extracted residue yielded a higher amount of pectin (120–230 mg/g) than the conventional acid extraction (163 mg/g) [52]. The yields of fractional alcoholic precipitation (47.6%), isopropanol-based direct precipitation (47.7%), and hot acid extract-induced selective pectin precipitation (50.8%) were identical [53]. Subcritical water-based extraction resulted in 18.34% of pectin [54]. Pectin extracted using deep eutectic solvent was determined to be 30–38.72% [55]. The methods used for the extraction of compounds from mango peels are shown in Table 6.2.

3.2.2 Phenolic Compounds

The traditional methods for extracting phenolic compounds are Soxhlet, maceration, and hydro distillation. Hydro distillation requires more processing duration and demands excessive energy. These traditional methods have some drawbacks, like reduced extraction yield and efficiency. Mangiferin, gallic acid, and catechins

Table 6.2 Methods used for extraction of compounds from mango peel

S. No.	Compound	Method of extraction
1	Pectin	Conventional hydrochloric acid, microwave-assisted HCl, ultrasound-assisted HCl, sulfuric acid, citric acid, ultrasound-assisted citric acid, acetic acid, precipitation, subcritical water, ultrasound-assisted deep eutectic solvent
2	Flavonoids	Gravimetric method, ultrasound, enzymatic, alkali, ethanol, water, high hydrostatic pressure.
3	Pigments	Ethanol/ acetic acid, hexane and acetone, acidic methanol, supercritical CO_2 extraction, ultrasound, and supercritical fluid extraction.
4	Dietary fiber	Ultrasound-assisted enzymatic extraction, alkaline, ethanol, ultrasound, enzymatic, and hot water extraction
5	Phenolics	Ultrasound-assisted water, water, methanol, ethanol, acetone, organic solvents, ultrasound-assisted methanol, subcritical water extraction and soxhlet, ultrasound-assisted methanol and acetic acid, microwave-assisted DES(deep eutectic solvent) and ethanol, ultrasound-assisted DES and ethanol, pulsed electric field, and membrane filtration.
6	Aroma	Simultaneous distillation-extraction

are the three main phenolic compounds found in mangoes [56]. Phenolic extraction from mango peels has several methods which include organic solvents, water, deep eutectic solvents, ultrasound and microwave-assisted extraction, membrane filtration, pulse electric field, subcritical water extraction, and Soxhlet.

Increased antioxidant capacity (8–28 g TE/100 g) was obtained using organic solvents such as methanol, acetone, and ethanol [57]. Compared to Alphonso, Rumani, Malgua, and Banisha mango peels, Sindhura mango peels exhibited greater total phenolic content (87 mg GAE/g) [58]. Ultrasound-assisted water treatments for 15, 30, and 45 min resulted in phenolic content of 972, 939, and 885 mg/100 g [59]. The phenolic content of the ethanol-based extraction with ultrasound assistance was recovered to be about 67.6 mg GAE/g of extract [60]. The ethanolic extract from mango peels that was microwave-assisted had a greater antioxidant capacity than the one obtained using conventional solvent extraction [61]. The deep eutectic solvent (DES)-based extraction resulted in three times more phenolic content extraction compared with ethanol [62]. Free radical scavenging activity was 82.64% in the DES-based extraction and 80.13% in the ethanol-based extraction [63]. Raspuri ripe mango peel has greater levels of phenolic content (29.5 mg GAE/g) [6]. Membrane-ultrafiltration and nanofiltration yielded approximately 81.6 μmol TE/100 mL of antioxidant capacity and 62.5 mg GAE/g of phenolic content, respectively [64]. After subcritical water and Soxhlet extraction, the phenolic contents were 30.62 and 21.76 mg GAE/g, respectively [65]. The food industry could potentially use the subcritical water method since it is reported to be a convenient and effective technique for saving time that provides a high yield without any harmful solvent residues [38].

3.2.3 Total Flavonoid Content

The total flavonoid contents of raw Raspuri, ripe Raspuri, raw Badami, and ripe Badami cultivars were 0.101, 0.332, 0.273, and 0.392 mg catechin equivalents/g [6]. The microwave exposure of freeze-dried peels with 50% ethanol and 75 °C resulted in 0.7 g of flavonoids/100 g [66]. Two distinct cultivars, Langar and DM Chosa, yielded 90.89 mg catechin equivalent/g and 92.55 mg catechin equivalents/g, respectively, from the methanol-based extraction of flavonoids [67].

3.2.4 Total Monomeric Anthocyanins

At 75 °C, microwave-assisted extraction of freeze-dried peel produced a yield of 3.3 g cy-3-glu/100 g of total monomeric anthocyanin [66].

3.2.5 Proanthocyanidins

A yield of 228 g LE (leucoanthocyanidin equivalents)/100 g was obtained by microwave-assisted extraction at pH 3 with 5% ethanol for 120 min and a solid-to-solvent ratio of 1:10 [57].

3.2.6 Dietary Fiber

There are different techniques for fiber extraction from mango peel such as the gravimetric method, hydrostatic pressure, ultrasound, enzymatic, alkali, ethanol, and water-based extraction [27]. Using the gravimetric method, the dietary fiber content of the mango peels from Raspuri and Badami ranged from 40.6–72.5% [68]. Treatment with high hydrostatic pressure increased the amount of soluble dietary fiber from 37.4% to 45.7% [69]. Ultrasound-assisted enzymatic extraction produced a greater yield of dietary fiber (79%) than that of ultrasound (60%), hot water (26%) extraction methods, alkaline (59%), enzymatic (55%), and ethanol (44%) [70].

The functional compounds of mango peels is represented in Fig. 6.3.

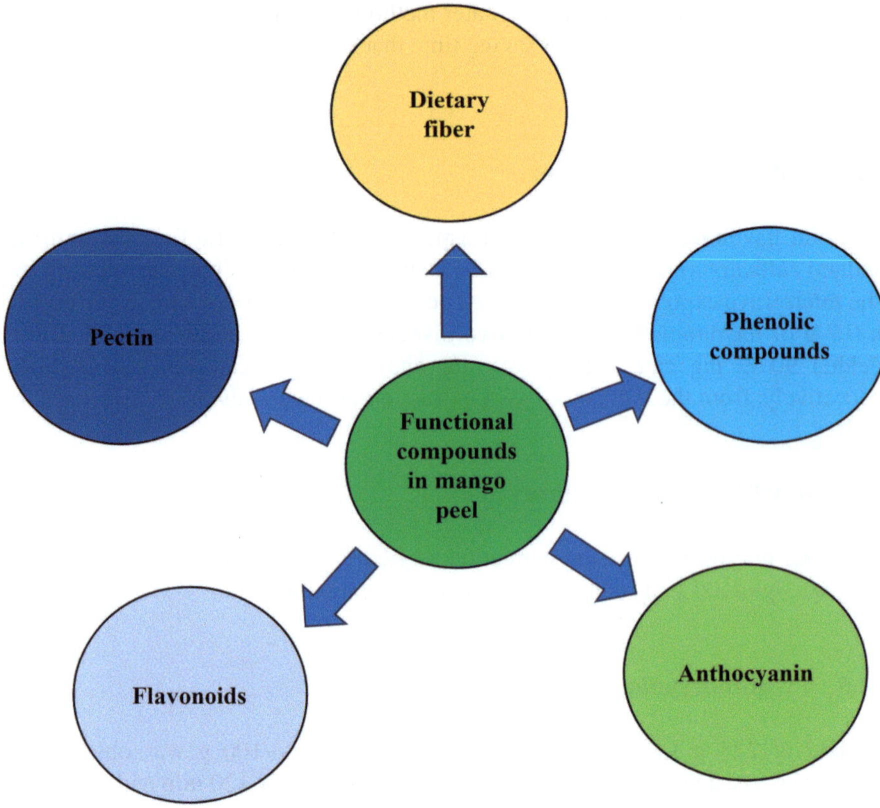

Fig. 6.3 Functional compounds in mango peel

3.3 Valorization of Mango Seed Kernel

3.3.1 Oil Extraction

Oil from mango kernel seeds was extracted using supercritical carbon dioxide and conventional solvents (acetone, hexane, ethanol, and petroleum ether) [71]. The oil yield from mango seed kernel employing traditional solvents like ethanol, acetone, hexane, and petroleum ether are 19.86, 18.7, 9.85, and 8.75%, respectively. Cold press and Soxhlet extraction techniques are used to obtain mango kernel oil [72]. Mango kernel oil yields were greater with solvent extraction (10.5%) than with cold press (4–5%). The hexane-assisted extraction of mango kernel resulted in a maximum oil yield of 15.20% after 90 min of extraction [25]. The Soxhlet extraction apparatus and n-hexane solvent were used to extract the oil from the mango seed kernel. Significant variation was observed in oil yields among the mango varieties, with Dasehri yielding the highest (9.88%) and Sindhri the lowest (6.33%) [73].

3.3.2 Fatty Acid Extraction

The fatty acid profile obtained by hot water, ultrasound, and subcritical water extraction were comparable, and they included high concentrations of saturated fatty acids such as palmitic, stearic, and myristic acids [74]. The acetone-based extracted oil had lower unsaturated fatty acid content. The oil extracted using supercritical carbon dioxide exhibited a higher percentage of saturated fatty acid (47.01%). Oleic acid and stearic acid are the two main unsaturated and saturated fatty acids, respectively. The hexane, petroleum ether, acetone, ethanol, and supercritical carbon dioxide extracted oil contains unsaturated (49.67, 48.85, 43.71, 44.34, and 37.588%, respectively) and saturated fatty acids (48.44, 48.18, 51, 31.87, and 47.41%, respectively) [71].

3.3.3 Proanthocyanidins

Microwave-assisted extraction using 5% acetone and a solid-to-solvent ratio of 1:10 at pH 8 produced a yield of 1.2 g LE/100 g [75].

3.3.4 Phenolics and Flavonoid Extraction

Mango seed phenolic components were extracted using ethyl acetate, alkaline hydrolysis, maceration, and acid/alkaline hydrolysis. Compared to the other processes, the macerated extract had a higher concentration of phenolic components (flavonoids - 86.59 mg quercetin equivalents (QE)/g and phenols - 484.42 mg GAE/g) [76]. In a microwave-assisted extraction at 75 °C, the total flavonoid content of the freeze-dried seed kernel with 50% acetone was almost 1.3 g CE/100 g [66]. The results of a flavonoid extraction using methanol ranged from 45.56 to 48.43 mg CE/g [67].

3.3.5 Starch

The starch from the seeds of mangoes was extracted using centrifugation, alkaline, sedimentation, and distillation methods. The alkali approach yielded the maximum amount of starch (91.72%), compared to starch separated by centrifugation (74.56%), sedimentation (74.82%), and distillation (81.87%) [77]. The ethanol-based precipitation method gave a starch yield of about 13.93% [78]. Water, ultrasound-assisted water, and the alkali extraction method of starch from the Chokanan mango seed kernel resulted in a starch yield of 15.51–28.59% [79]. Mieles-Gómez et al. [80] reported that ultrasound-assisted extraction resulted in the highest starch yield of 54%. Fresh mango seed kernels (47.37 ± 2.56%) resulted in a lower starch yield compared to the dried mango seed kernel (64.41 ± 5.51%) [81].

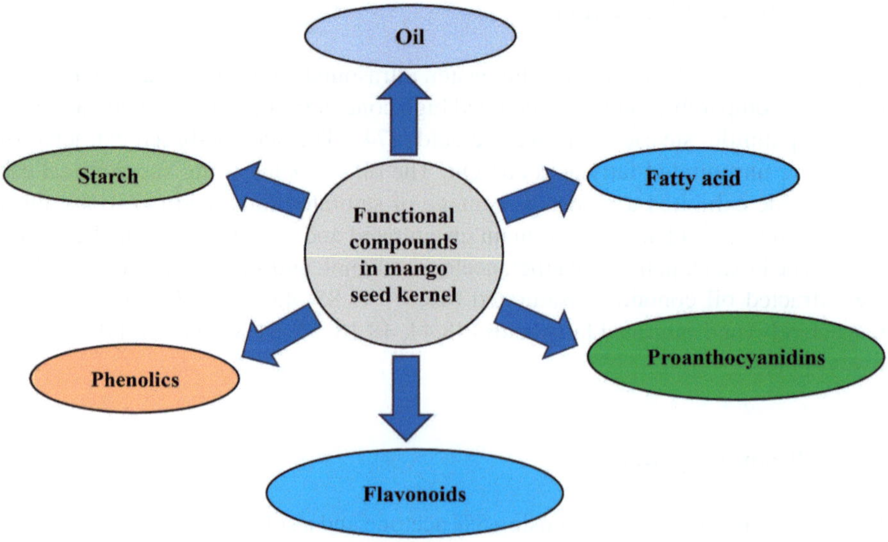

Fig. 6.4 Functional compounds in mango seed kernel

The functional compounds present in mango seed kernel are provided in Fig. 6.4.

4 Food Applications

4.1 Application of Mango Seed

Mango seeds have been studied by Kittiphoom [82] as a source of starch, inherent antioxidants, cocoa butter, and antimicrobial compounds. The seed was subjected to additional processing to become flour, which may be used as a component in bread [83]. The fat that was isolated from the mango seed kernel by immersing it in supercritical carbon dioxide together with other vegetable fats was identified as a high-quality analog fat for cocoa butter [84]. Muchiri et al. [85] revealed that the oil from mango seeds had similarities with vegetable butter oil. In addition, the seeds contain a significant amount of carbohydrates. [86] stated that the seed of mango has also been utilized for poultry feed. Mango seeds have been discovered to be useful as biosorbents for removing the frequently used food coloring dye malachite green [87]. Processing waste from mango kernels can be used to extract starch, which can be preserved for 2 months, and substitute up to 50% in soup mixes having maize starch without significantly altering the soup's sensory qualities [88].

4.2 Application of Mango Peel

By-products from the mango processing sector are used to develop sustainable and functional food items. Herbal tea is prepared with powdered mango peel. Products made with the mango peel herbal tea were visually appealing and scored highly in both the sensory and storage phases. The mango peel herbal tea has the bonus of gradually becoming more flavorful. By lowering fruit deterioration rates, preventing fruit waste in the market, and effectively utilizing sparse, highly nutritious fruits in the form of refreshments all while maintaining the micronutrients and phytopharmaceutical attributes of fruits for several years this method can help entrepreneurs deal with their problems [89]. Mango peel, a by-product of the mango industry, can be used to make macaroni with improved nutritional value Ajila et al. [5].

The nutritional profile of macaroni can be enhanced without compromising its cooking, textural, or sensory qualities by adding mango peel powder. Mango peels can be utilized to make functional foods since they are high in dietary fiber and bound phenolics [6]. The study by García-Magaña et al. [4] highlight the characteristics of dietary fiber, functional properties, and phenolic content present in industrial residues, indicating a possible application in food items, primarily as by-products of the Ataulfo Mexican variety. By-products from mangoes include vitamin C, pectin, carotene, and phenolic compounds. These bioactive substances may have anti-inflammatory, immunomodulatory, antiviral, anti-cancer, antiallergenic, antibacterial, and analgesic properties [90] which offers them a plentiful supply of polyphenols that may help prevent conditions like diabetes and cancer.

4.3 Enzymes

The enzymes extracted from the mango by-products such as cellulase enzyme and carboxymethylcellulose could be used by the food sector as a thickening or viscosity modifier, as well as stabilize emulsions in various products such as ice cream, laxative, diet pills [91]. Pectinase enzymes extracted may be employed in techniques that break down plant components, particularly in the production of wine [91].

4.4 Phenolic Compounds

Phenolic compounds can enhance the qualities of packaging materials and prolong food shelf life, which makes them valuable in active food packaging [27]. Several investigations on the addition of phenolic compounds to a range of perishable foods have demonstrated that these compounds can be utilized to preserve food and may be a viable substitute for artificial preservatives [92]. The phenolic extract from mango peels can improve the color of food products. The model solution comprising strawberries and red radishes became a more vibrant color when phenolic chemicals isolated from mango peel were added, due to the intermolecular co-pigmentation [93].

4.5 Aroma Compounds

Dried mango peel powder adds flavor to yogurt and offers it an appealing aroma [94]. The volatile oil from mango peels can be used to make nutritious drinks with instant mango flavor [95]. The essential oil derived from the peels of Zebdeya, Hindi, and Cobaneya mangoes has shown cytotoxicity towards the cell lines that represent liver, colon, and breast cancer [27].

4.6 Fiber

Mango by-products could be utilized in baked goods like noodles, sponge cakes, bread, and biscuits as they are abundant in carotenoids, polyphenols, and vitamins with various health-promoting qualities [39]. Mango peel flour was used to partially substitute wheat flour in varying amounts (5–30%) to make the sponge cake. In comparison to the control, the prepared cake had more dietary fiber and had lower fat, calories, and glycemic index. The quality of cake made with 10% mango peel flour had superior sensory attributes [96]. Better sensory qualities are found in jellies made from 1% pectin that has been extracted using acid [97]. Chinese sausage that has 5% pectin added as a fat substitute has a textural profile and sensory characteristics that are comparable to control [97]. Films suitable for use in medicine delivery systems and food packaging systems are made with pectin [46, 47]. The other applications include gene therapy, tissue engineering, and lowering blood cholesterol levels. Mango by-products have a significant hydration capacity and adequate fiber content, which renders them suitable for use in the production of foods that are rich in fiber [91]. The synthesis of lactic acid from mango by-products has the potential for a major benefit since the method is financially viable due to the low cost of the raw material [91].

5 Challenges and Opportunities Associated with Valorization

Currently, at least 50 enterprises recover beneficial compounds from food waste as a result of the technology's adoption by businesses all around the world throughout time. They convert them into food components such as natural preservatives or functional compounds to meet the shelf-life demands of processed food products. These compounds do not alter the texture or flavor of the actual product [98]. Implementation of full-scale mass production and pilot plants, and establishment of standard application methodologies can overcome the various issues associated with commercialization. For example, food wastes are frequently available in great amounts due to their seasonality, but they are also susceptible to microbial deterioration [99]. However, trash can be handled by cold storage and the addition of preservatives if the collection procedure is properly managed. Waste by-products can be changed by adding a pretreatment step, but they still pose a significant risk because of their vast diversity. Additionally, the problem of demanding a longer time for heating and mixing, handling more materials, a higher level of inspection, and air integration comes when batch processes are switched to continuous processes. As higher concentrations of food elements recovered industrially are required, this raises the cost of production. Process costs consequently rise as a result of using industrially recovered compounds in food formulations at larger quantities than laboratory-recovered ones [100].

Consequently, resolving these problems is essential to assure process sustainability, revenue generation, and continual availability of manufactured products in the market [98]. In an increasingly competitive market, the developed items ought to endure up to the high expectations of the consumers. In contrast to the demands of satisfying customer expectations and reducing environmental effects, developers must also make sure that the finished product and manufacturing process adhere to certain standards, such as offering clean-label components without compromising flavor or texture [99]. Mango peel has the potential to be separated from these active components and used as a component in processed food items such as bars, breakfast cereals, baked goods, and pasta products [1]. Traditional process technology for agro-food waste valorization needs further research and development as it would pave opportunities for industrialization and scale-ups [99].

6 Conclusion

A viable way to make profitable use of leftover mango peel waste is through biotechnological techniques. The mango processing firm can benefit financially by using technological developments to transform, this environmentally hazardous waste into lucrative products. Mango peels, seeds, and kernels are rich in important nutrients and bioactive chemicals that can improve food products and benefit consumers' health. Through the application of diverse valorization methodologies, these marginalized by-products can be converted into valuable ingredients for innovative and environmentally conscious culinary uses. To maximize the potential of mango waste valorization, issues related to large-scale production and marketing must be resolved. Sustained research and development endeavors are strongly recommended to enhance extraction procedures, investigate innovative uses, and guarantee the financial and ecological durability of this promising sector.

References

1. Baddi J, Durgannavar NA, Chandru R (2015) Mango peel: a potential source of natural bioactive phyto-nutrients in functional food. J Dairying Foods Home Sci 34(1):75–77. https://doi.org/10.5958/0976-0563.2015.00016
2. Tharanathan RN, Yashoda HM, Prabha TN (2006) Mango (*Mangifera indica* L.), "the king of fruits"—an overview. Food Rev Intl 22(2):95–123. https://doi.org/10.1080/87559120600574493
3. Mitra SK (2016) Mango production in the world–present situation and future prospect. Acta Hortic 1111:287–296. https://doi.org/10.17660/ActaHortic.2016.1111.41
4. de Lourdes García-Magaña M, García HS, Bello-Pérez LA, Sáyago-Ayerdi SG, de Oca MMM (2013) Functional properties and dietary fiber characterization of mango processing by-products (*Mangifera indica* L., cv Ataulfo and Tommy Atkins). Plant Foods Hum Nutr 68(3):254–258. https://doi.org/10.1007/s11130-013-0364-y
5. Ajila CM, Aalami M, Leelavathi K, Rao UJSP (2010) Mango peel powder: a potential source of antioxidant and dietary fiber in macaroni preparations. Innov Food Sci Emerg Technol 11(1):219–224. https://doi.org/10.1016/j.ifset.2009.10.004
6. Ajila CM, Prasada Rao UJS (2013) Mango peel dietary fibre: composition and associated bound phenolics. J Funct Foods 5(1):444–450. https://doi.org/10.1016/j.jff.2012.11.017
7. Koubala BB, Kansci G, Mbome LI, Crépeau MJ, Thibault JF, Ralet MC (2008) Effect of extraction conditions on some physicochemical characteristics of pectins from "Améliorée" and "mango" mango peels. Food Hydrocoll 22(7):1345–1351. https://doi.org/10.1016/j.foodhyd.2007.07.005
8. Ajila CM, Bhat SG, Prasada Rao UJS (2007a) Valuable components of raw and ripe peels from two Indian mango varieties. Food Chem 102(4):1006–1011. https://doi.org/10.1016/j.foodchem.2006.06.036
9. Ajila CM, Naidu KA, Bhat SG, Rao UJSP (2007b) Bioactive compounds and antioxidant potential of mango peel extract. Food Chem 105(3):982–988. https://doi.org/10.1016/j.foodchem.2007.04.052
10. Singh RN (1990) Mango. Indian Council of Agricultural Research
11. Pitchaon M (2011) Antioxidant capacity of extracts and fractions from mango (*Mangifera indica* Linn.) seed kernels. Int Food Res J 18(2):523–528
12. Flores-Gallegos AC, Rodriguez-Jasso RM, Aguilar CN (2021) Bioprocessing of Agri-food residues for production of bioproducts. In: Bioprocessing of agri-food residues for production of bioproducts. Apple Academic Press. https://doi.org/10.1201/9781003048305
13. Da Silva LMR, De Figueiredo EAT, Ricardo NMPS, Vieira IGP, De Figueiredo RW, Brasil IM, Gomes CL (2014) Quantification of bioactive compounds in pulps and by-products of tropical fruits from Brazil. Food Chem 143:398–404. https://doi.org/10.1016/j.foodchem.2013.08.001
14. Barreto JC, Trevisan MTS, Hull WE, Erben G, De Brito ES, Pfundstein B, Würtele G, Spiegelhalder B, Owen RW (2008) Characterization and quantitation of polyphenolic compounds in bark, kernel, leaves, and peel of mango (*Mangifera indica* L.). J Agric Food Chem 56(14):5599–5610. https://doi.org/10.1021/jf800738r
15. Abdullah ASH, Mohammed AS, Rasedee A, Saeed Mirghani ME (2015a) Oxidative stress-mediated apoptosis induced by ethanolic mango seed extract in cultured estrogen receptor positive breast cancer MCF-7 cells. Int J Mol Sci 16(2):3528–3536. https://doi.org/10.3390/ijms16023528
16. Abdullah ASH, Mohammed AS, Rasedee A, Mirghani MES, Al-Qubaisi MS (2015b) Induction of apoptosis and oxidative stress in estrogen receptor-negative breast cancer, MDA-MB231 cells, by ethanolic mango seed extract. BMC Complement Altern Med 15(1):45. https://doi.org/10.1186/s12906-015-0575-x
17. Chemat F, Vian MA, Cravotto G (2012) Green extraction of natural products: concept and principles. Int J Mol Sci 13(7):8615–8627. https://doi.org/10.3390/ijms13078615

18. Yadav AS, Singh Yadav A, Pandey (2016) Geographical perspectives of mango production in India. Imp J Interdiscip Res 2(4):257
19. Morton J (1987) Mango: in fruits of warm climates. Miami Florida, pp 221–239
20. Yadav D, Singh Professor S, Deependra Yadav C, Singh S (2017) Mango: history origin and distribution. ~ 1257 ~. J Pharmacog Phytochem 6(6):1257–1262
21. Veeranjaneya Reddy L, Sarathi Reddy Obulam V, Vijaya Sarathi Reddy O, Wee Y-J (2011) Production of ethanol from mango (*Mangifera indica* L.) peel by *Saccharomyces cerevisiae* CFTRI101. Afr J Biotechnol 10(20):4183–4189. http://www.academicjournals.org/AJB
22. Lebaka VR, Wee YJ, Ye W, Korivi M (2021) Nutritional composition and bioactive compounds in three different parts of mango fruit. Int J Environ Res Public Health 18:1–20. https://doi.org/10.3390/ijerph18020741
23. Bhalerao S, Mulmuley G, Anathakrishna S, Potty V (1989) Wash and waste water management in food industry. Fruit and vegetable processing. Indian Food Pack 43:5–11
24. Odunsi AA (2005) Response of laying hens and growing broilers to the dietary inclusion of mango (*Mangifera indica* L.) seed kernel meal. Trop Animal Health Product 37:139
25. Yadav KK, Garg N, Verma A, Kumar S, Trivedi M (2017) Optimization and extraction of oil from mango seed kernel (*Mangifera indica*). Indian J Agric Sci 87(7):943–946. https://doi.org/10.56093/ijas.v87i7.71925
26. Schieber A, Berardini N, Carle R (2003) Identification of flavonol and xanthone glycosides from mango (*Mangifera indica* L. cv. "Tommy Atkins") peels by high-performance liquid chromatography-electrospray ionization mass spectrometry. J Agric Food Chem 51(17):5006–5011. https://doi.org/10.1021/jf030218f
27. Jeevitha GC, Ramamoorthy S, Ahmad F, Saravanan R, Haque S, Capanoglu E (2023) Recent advances in extraction methodologies for the valorization of mango peel wastes. Int J Food Properties 26:3492–3511. https://doi.org/10.1080/10942912.2023.2281255
28. Sogi DS, Siddiq M, Greiby I, Dolan KD (2013) Total phenolics, antioxidant activity, and functional properties of "Tommy Atkins" mango peel and kernel as affected by drying methods. Food Chem 141(3):2649–2655. https://doi.org/10.1016/j.foodchem.2013.05.053
29. Ranganath KG, Shivashankara KS, Roy TK, Dinesh MR, Geetha GA, Pavithra KCG, Ravishankar KV (2018) Profiling of anthocyanins and carotenoids in fruit peel of different colored mango cultivars. J Food Sci Technol 55(11):4566–4577. https://doi.org/10.1007/s13197-018-3392-7
30. Thebaudin JY (1997). Dietary fibers
31. Molina OM, Domínguez-Avila JA, Pareek S, Santana TJM, Aguilar GAG, Lopez-Martínez LX (2023) Valorization of tropical fruit peel powders: Physico-chemical composition, techno-functional properties, and *in vitro* antioxidant and antidiabetic activities. Emirates J Food Agric 35(6):577–587. https://doi.org/10.9755/ejfa.2023.v35.i6.3105
32. Thiruchelvam T, Landahl S, Terry LA (2020) Temporal variation of volatile compounds from Sri Lankan mango (*Mangifera indica* L.) fruit during ripening. J Agric Food Res 2:100053. https://doi.org/10.1016/j.jafr.2020.100053
33. Cheok CY, Mohd Adzahan N, Abdul Rahman R, Zainal Abedin NH, Hussain N, Sulaiman R, Chong GH (2018) Current trends of tropical fruit waste utilization. Crit Rev Food Sci Nutr 58(3):335–361. https://doi.org/10.1080/10408398.2016.1176009
34. Lieb VM, Schuster LK, Kronmüller A, Schmarr HG, Carle R, Steingass CB (2019) Fatty acids, triacylglycerols, and thermal behaviour of various mango (*Mangifera indica* L.) kernel fats. Food Res Int 116:527–537. https://doi.org/10.1016/j.foodres.2018.08.070
35. Luo F, Fu Y, Xiang Y, Yan S, Hu G, Huang X, Huang G, Sun C, Li X, Chen K (2014) Identification and quantification of gallotannins in mango (*Mangifera indica* L.) kernel and peel and their antiproliferative activities. J Funct Foods 8(1):282–291. https://doi.org/10.1016/j.jff.2014.03.030
36. Abdalla AEM, Darwish SM, Ayad EHE, El-Hamahmy RM (2007) Egyptian mango by-product 1. Compositional quality of mango seed kernel. Food Chem 103(4):1134–1140. https://doi.org/10.1016/j.foodchem.2006.10.017

37. Abdel-Aty AM, Salama WH, Hamed MB, Fahmy AS, Mohamed SA (2018) Phenolic-antioxidant capacity of mango seed kernels: therapeutic effect against viper venoms. Rev Bras 28(5):594–601. https://doi.org/10.1016/j.bjp.2018.06.008
38. Puligundla P, Obulam VSR, Oh SE, Mok C (2014) Biotechnological potentialities and valorization of mango peel waste: a review. Sains Malaysiana 43(12):1901–1906. https://doi.org/10.17576/jsm-2014-4312-12
39. Walia N (2013) Production of Bioethanol from Mango Peel. www.ijert.org
40. Muthuswamy M, Arumugam R, Manikandan M (2011) Fermentation of pretreated hydrolyzates of banana and mango fruit wastes for ethanol production. Asian J Exp Biol Sci 2(2):246–256
41. Somda MK, Savadogo A, Ouattara CAT, Ouattara AS, Traore AS (2010) Thermotolerant and alcohol-tolerant yeasts targeted to optimize hydrolyzation from mango peel for high bioethanol production. Asian J Biotechnol 3(1):77–83. https://doi.org/10.3923/ajbkr.2011.77.83
42. Varakumar S, Naresh K, Vijaya O, Reddy S (2012) Preparation of mango (*Mangifera indica* L.) wine using a new yeast-mango-peel immobilised biocatalyst system. Czech J Food Sci 30(6):557
43. Kumar D, Ashfaque M, Ashfaque M, Muthukumar M, Garg N (2012a) Production and characterization of carboxymethyl cellulase from *Paenibacillus polymyxa* using mango peel as substrate. J Environ Biol 33(1):81–84
44. Kumar YS, Kumar PV, Reddy OVS (2012b) Pectinase production from mango peel using *Aspergillus foetidus* and its application in processing of mango juice. Food Biotechnol 26(2):107–123. https://doi.org/10.1080/08905436.2012.670830
45. Tandon DK, Garg N, Kalra SK (1995) Extraction of pectin and fiber from mango peel. Beverage and Food World 2021:20–21
46. Sommano SR, Ounamornmas P, Nisoa M, Sriwattana S, Page P, Colelli G (2018a) Characterisation and physiochemical properties of mango peel pectin extracted by conventional and phase control microwave-assisted extractions. Int Food Res J 25(6):2657
47. Sommano S, Ounamornmas P, Nisoa M, Sriwattana S (2018b) Bioactive functionality of pectin from peels of seven Thai mango cultivars. Acta Hortic 1213:423–428. https://doi.org/10.17660/ActaHortic.2018.1213.62
48. Maran JP, Swathi K, Jeevitha P, Jayalakshmi J, Ashvini G (2015) Microwave-assisted extraction of pectic polysaccharide from waste mango peel. Carbohydr Polym 123:67–71. https://doi.org/10.1016/j.carbpol.2014.11.072
49. Torres-Gallo R, Bayuelo-Bonilla S, Carpio-Ortiz L, Barreto-Rodríguez G, Tirado DF (2022) High-intensity ultrasound-assisted extraction of pectin from mango wastes at different maturity. Int J Food Sci 2022:1. https://doi.org/10.1155/2022/4606024
50. Rehman ZU, Salariya F, Habib F, Shah WH (2004) Utilization of mango peels as a source of pectin. J Chem Soc Pak 26(1):73–76
51. Nguyen HHD, Nguyen HVH, Savage GP (2019) Properties of pectin extracted from Vietnamese mango peels. Food Secur 8(12):629. https://doi.org/10.3390/foods8120629
52. Mugwagwa LR, Chimphango AFA (2019) Box-Behnken design-based multi-objective optimisation of sequential extraction of pectin and anthocyanins from mango peels. Carbohydr Polym 219:29–38. https://doi.org/10.1016/j.carbpol.2019.05.015
53. Aranguren DD, Mantilla-Mantilla M, Durán-Aranguren D, Sierra R (2021) Extraction of pectin from mango (*Mangifera indica*) peels. https://doi.org/10.5071/29thEUBCE2021-3BV.10.2
54. Xia H, Matharu AS (2017) Unavoidable food supply chain waste: acid-free pectin extraction from mango peel *via* subcritical water. Faraday Discuss 202:31–42. https://doi.org/10.1039/c7fd00035a
55. Chen S, Xiao L, Li S, Meng T, Wang L, Zhang W (2022) The effect of sonication-synergistic natural deep eutectic solvents on extraction yield, structural and physicochemical properties of pectins extracted from mango peels. Ultrason Sonochem 86:106045. https://doi.org/10.1016/j.ultsonch.2022.106045

56. Masibo M, He Q (2008) Major mango polyphenols and their potential significance to human health. Compr Rev Food Sci Food Saf 7(4):309–319. https://doi.org/10.1111/j.1541-4337.2008.00047.x
57. Dorta E, Lobo MG, González M (2013a) Improving the efficiency of antioxidant extraction from mango peel by using microwave-assisted extraction. Plant Foods Hum Nutr 68(2):190–199. https://doi.org/10.1007/s11130-013-0350-4
58. Umamahesh K, Sivudu SN, Vijaya O, Reddy S, Sivudu N (2016) Evaluation of antioxidant activity, total phenolics and total flavonoids in peels of five cultivars of mango (*Mangifera indica*) fruit. J Med Plants 4(2):200
59. Jirasuteeruk C, Theerakulkait C (2019) Ultrasound-assisted extraction of phenolic compounds from mango (*Mangifera indica* cv. *Chok Anan*) peel and its inhibitory effect on enzymatic browning of potato puree. Food Technol Biotechnol 57(3):350–357. https://doi.org/10.17113/ftb.57.03.19.5728
60. Safdar MN, Kausar T, Nadeem M, Murtaza M, Sohail S, Mumtaz A, Siddiqui N, Jabbar S, Afzal S (2022) Extraction of phenolic compounds from (*Mangifera indica* L.) and kinnow (*Citrus reticulate* L.) peels for the development of functional fruit bars. Food Sci Technol (Brazil) 42:e09321. https://doi.org/10.1590/fst.09321
61. Pal CBT, Jadeja GC (2020) Microwave-assisted extraction for recovery of polyphenolic antioxidants from ripe mango (*Mangifera indica* L.) peel using lactic acid/sodium acetate deep eutectic mixtures. Food Sci Technol Int 26(1):78–92. https://doi.org/10.1177/1082013219870010
62. Macedo A, Gomes T, Ribeiro C, Moldão-Martins M, Duarte E, Alves VD (2022) Membrane technology for valorization of mango peel extracts. Food Secur 11(17):2581. https://doi.org/10.3390/foods11172581
63. Ojeda GA, Vallejos MM, Sgroppo SC, Sánchez-Moreno C, de Ancos B (2023) Enhanced extraction of phenolic compounds from mango by-products using deep eutectic solvents. Heliyon 9(6):e16912. https://doi.org/10.1016/j.heliyon.2023.e16912
64. Parniakov O, Barba FJ, Grimi N, Lebovka N, Vorobiev E (2016) Extraction assisted by pulsed electric energy as a potential tool for green and sustainable recovery of nutritionally valuable compounds from mango peels. Food Chem 192:842–848. https://doi.org/10.1016/j.foodchem.2015.07.096
65. Oroian M, Escriche I (2015). Antioxidants: Characterization, natural sources, extraction and analysis. Food Res Int. 74:10–36. https://doi.org/10.1016/j.foodres.2015.04.018
66. Dorta E, Lobo MG, González M (2012) Using drying treatments to stabilise mango peel and seed: effect on antioxidant activity. LWT-Food Sci Technol 45(2):261–268. https://doi.org/10.1016/j.lwt.2011.08.016
67. Sultana B, Hussain Z, Asif M, Munir A (2012) Investigation on the antioxidant activity of leaves, peels, stems bark, and kernel of mango (*Mangifera indica* L.). J Food Sci 77(8):C849. https://doi.org/10.1111/j.1750-3841.2012.02807.x
68. Kaur B, Panesar PS, Thakur A (2021) Extraction and evaluation of structural and physicochemical properties of dietary fiber concentrate from mango peels by using green approach. Biomass Convers Biorefinery 15:1815. https://doi.org/10.1007/s13399-021-01740-2
69. Freitas CMP, Coimbra JSR, Souza VGL, Sousa RCS (2021) Structure and applications of pectin in food, biomedical, and pharmaceutical industry: a review. Coatings 11(8):922. https://doi.org/10.3390/coatings11080922
70. Geerkens CH, Nagel A, Just KM, Miller-Rostek P, Kammerer DR, Schweiggert RM, Carle R (2015) Mango pectin quality as influenced by cultivar, ripeness, peel particle size, blanching, drying, and irradiation. Food Hydrocoll 51:241–251. https://doi.org/10.1016/j.foodhyd.2015.05.022
71. Awolu OO, Manohar B (2019) Quantitative and qualitative characterization of mango kernel seed oil extracted using supercritical CO_2 and solvent extraction techniques. Heliyon 5(12):e03068. https://doi.org/10.1016/j.heliyon.2019.e03068
72. Gurjar HR, Raj JD (2022) Extraction of mango kernel (*Magnifera indica*) oil. Asian J Dairy Food Res 41(3):361–364. https://doi.org/10.18805/ajdfr.DR-1855

73. Zahoor S, Anwar F, Qadir R, Soufan W, Sakran M (2023) Physicochemical attributes and antioxidant potential of kernel oils from selected mango varieties. ACS Omega 8(25):22613–22622. https://doi.org/10.1021/acsomega.3c01155
74. Krishnamoorthy R, Hai A, Banat F (2023) Subcritical water extraction of mango seed kernels and its application for cow ghee preservation. PRO 11(5):1379. https://doi.org/10.3390/pr11051379
75. Dorta E, Lobo MG, González M (2013b) Optimization of factors affecting extraction of antioxidants from mango seed. Food Bioprocess Technol 6(4):1067–1081. https://doi.org/10.1007/s11947-011-0750-0
76. Bernal-Mercado AT, Acevedo-Hernandez C, Silva-Espinoza BA, Reynaldo Cruz-Valenzuela M, Gonzalez-Aguilar GA, Nazzaro F, Siddiqui MW, Fernando Ayala-Zavala J, Fratianni F, Javier Vazquez-Armenta F (2018) Antioxidant and antimicrobial capacity of phenolic compounds of mango (*Mangifera indica* L.) seed depending upon the extraction process. J Med Plants by-products 2(46):209. https://doi.org/10.22092/JMPB.2018.118149
77. Shahrim NA, Sarifuddin N, Ismail H (2018) Extraction and characterization of starch from mango seeds. J Phys Conf Ser 1082(1):012019. https://doi.org/10.1088/1742-6596/1082/1/012019
78. Thakare VB, Jadeja GC, Desai MA (2023) Extraction of starch and phenolic compounds from *Mangifera indica* L. var. Kesar seeds and its characterization. Indian J Chem Technol 30(6):822–831. https://doi.org/10.56042/ijct.v30i6.6547
79. Saleh NA, Nordin R, Ahmad AA, Fazial FF (2023) Physiochemical characterization of starch from mango seed kernel extracted using different methods. In: The proceeding of the 1st international conference of chemical science, engineering and technology, vol 2703. AIP Publishing, p 080006. https://doi.org/10.1063/5.0117134
80. Mieles-Gómez L, Quintana SE, García-Zapateiro LA (2023) Ultrasound-assisted extraction of mango (*Mangifera indica*) kernel starch: chemical, techno-functional, and pasting properties. Gels 9(2):136. https://doi.org/10.3390/gels9020136
81. Nayak P, Rayaguru K (2018) Citation: Parv Nayak and Kalpana Rayaguru (2018) Studies on extraction of starch from dried and fresh mango seed kernel. Int J Agric Sci 10(18):7192–7195. https://www.bioinfopublication.org/jouarchive.php?opt=andjouid=BPJ0000217
82. Kittiphoom S (2012) Utilization of mango seed. Int Food Res J 19(4):1
83. Menon L, Mjumdar SD, Ravi U (2014) Mango kernel flour as a potential ingredient in the development of composite flour bread. Indian J Nat Prod Resour 5(1):75–82
84. Jahurul MHA, Zaidul ISM, Norulaini NNA, Sahena F, Jaffri JM, Omar AKM (2014) Supercritical carbon dioxide extraction and studies of mango seed kernel for cocoa butter analogy fats. CYTA—J Food 12(1):97–103. https://doi.org/10.1080/19476337.2013.801038
85. Muchiri DR, Mahungu SM, Gituanja SN (2012) Studies on mango (*Mangifera indica* L.) kernel fat of some Kenyan varieties in Meru. JAOCS. J Am Oil Chem Soc 89(9):1567–1575. https://doi.org/10.1007/s11746-012-2054-6
86. Diarra SS (2014) Potential of mango (*Mangifera indica* L.) seed kernel as a feed ingredient for poultry: a review. World's Poultry Sci J 70(2):279–288. https://doi.org/10.1017/S0043933914000294
87. Franca AS, Oliveira LS, Saldanha SA, Santos PIA, Salum SS (2010) Malachite green adsorption by mango (*Mangifera indica* L.) seed husks: kinetic, equilibrium and thermodynamic studies. Desalin Water Treat 19(1–3):241–248. https://doi.org/10.5004/dwt.2010.1105
88. Yatnatti S, Vijayalakshmi D (2018) Study of soup mix incorporated with starch extract from mango "*Mangifera indica*" seed kernels. Current Research in Nutrition and Food Science 6(3):816–825. https://doi.org/10.12944/CRNFSJ.6.3.24
89. Kanchana S., Raman R. (2024) Utilization of mango processing industrial waste into value added products. https://www.researchgate.net/publication/378698021
90. De La J, Medina C. and García HS (2002). MANGO post-harvest operations-post-harvest compendium. http://www.itver.edu.mx

91. Eitex TT (2017) Valorisation of mango fruit by-products: physicochemical characterisation and future prospect, vol 50. IISTE. Online. www.iiste.org
92. Ziogas V, Tanou G, Molassiotis A, Diamantidis G, Vasilakakis M (2010) Antioxidant and free radical-scavenging activities of phenolic extracts of olive fruits. Food Chem 120(4):1097–1103. https://doi.org/10.1016/j.foodchem.2009.11.058
93. Müller-Maatsch J, Bechtold L, Schweiggert RM, Carle R (2016) Co-pigmentation of pelargonidin derivatives in strawberry and red radish model solutions by the addition of phenolic fractions from mango peels. Food Chem 213:625–634. https://doi.org/10.1016/j.foodchem.2016.06.097
94. Ruiz C, Ramirez C, Pineres C, Angulo M, Hedreira J (2011) Obtaining and characterization of mango peel powder and its use as a source of fiber and a functional ingredient in natural yoghurt. In: Proceedings of the 11th international congress on engineering and food. ICEF11, pp 22–26
95. Masoud MR, El-Hadidy EM (2017) Mango, orange and mandarin peels oleoresins to prepare natural and healthy instant flavor drinks. Suez Canal Univ J Food Sci 4(1):11
96. Noor Aziah AA, Lee Min W, Bhat R (2011) Nutritional and sensory quality evaluation of sponge cake prepared by incorporation of high dietary fiber containing mango (*Mangifera indica* var. *Chokanan*) pulp and peel flours. Int J Food Sci Nutr 62(6):559–567. https://doi.org/10.3109/09637486.2011.562883
97. Sayed MA, Kumar J, Rahman MR, Noor F, Alam MA (2022) Effect of extraction parameters on the yield and quality of pectin from mango (*Mangifera indica* L.) peels. Discov Food 2(1):28. https://doi.org/10.1007/s44187-022-00029-1
98. Galanakis CM (2016) Ultrafiltration for the separation of functional macro-and micromolecules. https://doi.org/10.13140/RG.2.1.5167.2565
99. Ngwasiri PN, Ambindei WA, Adanmengwi VA, Ngwi P, Mah AT, Ngangmou NT, Fonmboh DJ, Ngwabie NM, Ngassoum MB, Aba ER (2022) A review paper on agro-food waste and food by-product valorization into value added products for application in the food industry: opportunities and challenges for Cameroon bioeconomy. Asian J Biotechnol Bioresou Technol 32–61:32. https://doi.org/10.9734/ajb2t/2022/v8i330128
100. Ilyasoglu H, El SN (2014) Nanoencapsulation of EPA/DHA with sodium caseinate-gum arabic complex and its usage in the enrichment of fruit juice. LWT-Food Sci Technol 56(2):461–468. https://doi.org/10.1016/j.lwt.2013.12.002

Part III
Beverage Crops

Chapter 7
Potential and Sustainable Utilization of Tea Processing Waste: A Review

Prangan Duarah and Mihir Kumar Purkait

Abbreviations

CQDs	Carbon quantum dots
EAE	Enzyme-assisted extraction
HPH	High-pressure homogenization
LiS	Lithium sulfur
MAE	Microwave-assisted extraction
MCC	Microcrystalline Cellulose
PFAD	Palm Fatty Acid Distillate
PVA	Poly (vinyl alcohol)
RH	Rice husk
RhB	Rhodamine B
SFE	Supercritical fluid extraction
SSE	Subcritical solvent extraction
STW	Spent tea waste
TLR	Tea leaf residue
TW	Tea waste
UAE	Ultrasound-assisted extraction
WTR-CDS	Waste tea residues carbon dots

P. Duarah
Centre for the Environment, Indian Institute of Technology, Guwahati, India

M. K. Purkait (✉)
Department of Chemical Engineering, Indian Institute of Technology, Guwahati, Assam, India
e-mail: mihir@iitg.ac.in

C. Trois, R. C. Ray (eds.), *Plantation Crop Wastes: Valorization for Economic Sustainability*, Waste as a Resource,
https://doi.org/10.1007/978-3-031-96805-1_7

1 Introduction

Tea, an acknowledged second most popularly consumed beverage in the world, after water, is quite a historical beverage that affects the culture, economy, and society as a whole [1]. Its origins were back in ancient China around 2737 BC. The legend says that the leaves were blown by the wind beneath a tea (*Camellia sinensis* L.) plant and dropped into the boiling water of a Chinese emperor, creating an aromatic and delicious drink. This accidental discovery gradually bloomed into a significant part of Chinese culture, symbolizing refinement, hospitality, and tranquillity. From China, the journey of tea spread far and wide, first reaching Japan in the late eighth century, where it became deeply ingrained in the fabric of Japanese tradition, most notably through the tea ceremony. This was a defining moment for Europe as the United Kingdom embraced tea as a cultural icon and embedded it in everyday life and social rituals, such as afternoon tea. In the Middle East, tea symbolized hospitality and togetherness, adding to its cultural value. The global appreciation of tea is also reflected in its numerous kinds, divided into green, yellow, oolong, and black teas based on their oxidation level. Darjeeling and Assam in India, Ceylon in Sri Lanka, and Turkish tea are some of the most known varieties that underline regional differences and expertise surrounding the production of tea. Apart from its universal appeal as a beverage, tea holds deep meaning to heritage, community, and identity; its story is one of historical richness and contemporary relevance [2].

Global production and consumption of tea have experienced growth over the decades, driven by widespread popularity and cultural significance, but at a rate never seen before. On the other hand, such fast expansion has triggered major environmental and sustainability issues concerning the substantial waste generated in the tea production and consumption chain. Factory tea manufacturing creates enormous quantities of waste products: leaf residues, stems, and fine particulate dust, often wasted or burnt. In both respects, such practices contribute to environmental pollution while simultaneously underusing a byproduct that is so rich in potential. Figure 7.1a–c shows three different types of tea waste which are factory tea waste, tea seeds and domestic tea waste. Millions of tons of tea waste are annually managed ineffectively, thus promoting ecological degradation and losing opportunities for utilization of these materials. Such a situation emphasizes the urgency for

Fig. 7.1 (**a**) Factory tea waste, (**b**) Tea seed, and (**c**) Domestic tea waste

sustainable and innovative management practices, such as waste valorization, to transform the byproducts of the tea industry into high-value products such as biochar, activated carbon, or bioactive compounds. Adopting these practices will help the tea industry reduce its footprint on the environment while generating economic opportunities, hence working hand in glove with sustainability and resource efficiency. This can be pivotal in ensuring that the future of the tea industry stays sustainable and contributes to overall global goals toward environmental protection and circular economy practices.

This chapter aims to explore the research and development initiatives focused on the value addition of tea waste, addressing both the challenges and opportunities associated with its sustainable utilization. Tea waste is a promising area for innovation, given the increasing focus on environmental preservation and circular economy principles. The chapter explores advanced techniques and methodologies for converting byproducts in the tea industry into valuable materials, including biochar, activated carbon, nanomaterials, and biofuels, and their applications in the pharmaceuticals, agriculture, and energy sectors. This chapter aims to provide a comprehensive understanding of the potential for valorization of tea waste by highlighting recent advancements and identifying gaps in current practices, thereby promoting sustainable solutions that are aligned with global efforts toward resource efficiency and waste minimization.

2 Global Production and Composition of Tea Waste

2.1 Worldwide Tea Production Statistics

The global tea market depicts strong growth and a sizeable impact on economic scales, as revenue generated for 2024 is approximated to stand at a collective figure of US$277.30 billion. Revenues are shared between the two major segments: on-premise, at a sum of US$135.60 billion, and off-premise, through restaurants, cafes, and bars with an estimated revenue of US$141.80 billion. The at-home segment is going to grow every year with a compound annual growth rate of 5.79% from 2024 to 2029, reflecting a growing consumer demand for tea in household settings. In global terms, China is the biggest market, accounting for the highest revenue at home: US$48.96 billion in 2024 [3]. In terms of volume, the tea market is estimated to produce 7.54 billion kilograms in 2024. Of this, 7.19 billion kg will be consumed at home, and 351.40 million kg will be consumed out of home. On average, the per capita at-home tea consumption volume is expected to be 0.93 kg, with per capita revenue averaging US$17.50. The at-home segment is expected to grow in volume terms, with a projected 2.8% increase in 2025. Such statistics show that tea continues to be popular and is likely to continue growing, based on both traditional consumption patterns and new opportunities in premium and health-focused tea products.

In 2023, global tea production highlighted the supremacy of a few players, where China was the largest by a huge margin. China produced around 3,250,000 metric tons of tea, which further reestablished it as the world's largest tea producer. This supremacy is not only due to the vast cultivated areas but also because of deep cultural and economic integration in tea production, including the most diverse types of teas: green, black, and specialty teas. India is the second-largest tea-producing country, with 1,367,700 metric tons in global supply. The tea industry is significant for the economy of the country, especially in the regions of Assam and Darjeeling, where high-quality black teas are produced. Kenya is the third-largest producer, with 570,260 metric tons. The country is famous for its strong and fragrant black teas and is one of the major exporters of tea, mainly to the international market for blending teas. Sri Lanka is another important player, with 256,040 metric tons of tea produced in 2023. Famous for its high-quality Ceylon tea, Sri Lanka remains a significant player in the global market for specialty and high-value teas. Other tea-producing countries collectively contributed 1,159,449 metric tons, showing the increasing involvement of countries outside the historical tea-growing region. It's a reflection of cultural value and economic relevance attributed to tea across the world regions, with China, India, Kenya, and Sri Lanka acting as the pivotal actors in meeting the growing global demand for this most popular beverage [4].

Tea waste (TW) is a byproduct of tea production and consumption generated from various processes such as withering, fermentation, leaf collection, drying, soaking, and sorting. Along with an increasing number of tea production and consumption, the amount of waste created at all different stages also increases [5, 6]. They can broadly be classified under categories such as Spent tea leave, industrial byproducts, processing waste, and pruning residues. A large proportion of spent tea leaves are generated from tea preparation in households and restaurants through steeping in hot or steamed water. Industrial-scale waste often comprises large quantities of fibers during the manufacturing processes involved in preparing products that require bottled tea beverages, instant teas, and other canned drinks in which tea leaves are exposed to hot boiling or steaming [6]. In addition, tea refining processes generate processing waste through the removal of excess leaves and other constituents to enhance the quality of the final product.

2.2 *Chemical and Physical Properties of Tea Waste*

TW has distinct physical properties due to its lignocellulosic nature which contains cellulose, hemicellulose, and lignin. The fibrous nature of the material gives it a large specific surface area and porosity, especially after heat or chemical treatment, and is thus useful for adsorption purposes. The bulk density of TW is very low, which makes handling and transportation easy; however, this can increase with compaction or processing. The moisture content in TW changes with the source and its processing stage, thereby varying its storage and degradation behaviors. Morphologically, it is made of fine powder to coarse fragments with rough surfaces and irregular shapes enhancing adsorption capacity. Lignin and cellulose result in moderate thermal stability being important for processes such as pyrolysis and gasification. Usually, dark brown or greenish in color with a rough texture, TW's physical characteristics make it a very versatile material for biochar, activated carbon, and other applications.

TW is also chemically rich and diverse in its composition and varies depending on the processing stage of the tea plant and the type of plant material. Being a source of lignocellulosic biomass with cellulose, hemicellulose, and lignin, these macromolecules play important roles in the structural integrity and are critical to TW, which is of considerable importance in bio-based material production. It is also characterized by various phytochemicals that have shown the presence of bioactive compounds like polyphenols, tannins, flavonoids, saponins, and alkaloids like caffeine [1].

Tea waste is a good source of bioactive compounds and essential nutrients, making it suitable for diverse industrial applications. The high polyphenol content in TW, comprising catechins, theaflavins, and rutin, contributes significantly to its strong antioxidant properties. These polyphenols are widely utilized in pharmaceuticals, cosmetics, and nutraceuticals due to their ability to combat oxidative stress and promote health. Tannins in TW possess protein-binding properties to form protein-bound complexes. Protein-bound complexes are most valuable in animal feed formulations due to their enhancement of protein utilization in ruminants. Alkaloids theobromine and caffeine are also present in TW, which are commercially useful because of their stimulatory action. Caffeine is one of the most desirable components in beverages, medicines, and cosmetic products. TW is also abundant in essential amino acids, omega-3, and omega-6 fatty acids, providing it with great nutritional content and potential for use in feedstock production and dietary supplements [7]. Secondary metabolites such as saponins have surfactant properties and impart antimicrobial activity, allowing them to be used in detergents and biopesticides. The carbohydrate content in TW, especially soluble polysaccharides, has great potential in fermentation processes, thereby facilitating the production of bioethanol and other bio-based chemicals. Minerals such as calcium (Ca), phosphorus (P), potassium (K), magnesium (Mg), and manganese (Mn) further enhance its value, enabling its use in agriculture and soil amendment. This diverse chemical composition makes TW a versatile and sustainable resource for various industrial and environmental applications.

High lignin content is matched by chemical complexity, making the TW highly suitable for the production of biochar and functionalized activated carbon. It is due to this combination of lignin, cellulose, and hemicellulose that gives rise to carbonaceous material with good thermal and chemical stability. Additionally, because TW contains essential minerals and trace elements, such as potassium, magnesium, and calcium, it is suitable for use in soil amendments and fertilizers. These rich, diverse chemical properties underscore TW as a sustainable resource, thereby providing a wide utility basis for different industries-from agriculture to energy, from pharmaceuticals to environmental remediation.

3 Sustainable Utilization of Tea Waste

The sustainable use of TW aims at utilizing its structural richness in lignocellulose and bioactivity to create green products with applications in several sectors. Due to the high polyphenol and tannin content, it can be converted through thermochemical processes into such products as biochar and activated carbon for environmental clean-up purposes, such as water purification and pollutant adsorption. It is, therefore, a good resource for obtaining antioxidants, pharmaceuticals, and animal feed supplements. Extraction of caffeine and other bioactive compounds from TW also supports the development of nutraceuticals and cosmetics. Moreover, TW can be composted or used as a soil amendment to improve soil fertility and promote sustainable agriculture. Innovative applications, such as bioplastics and biofuels, further emphasize its potential in reducing dependency on nonrenewable resources. This transforms TW into high-value products and, therefore, minimizes the environmental footprint, supports a circular economy, and aids in sustainable waste management practices.

3.1 Bioactive Extraction Methods

The extraction of bioactive compounds from TW is a critical step in its sustainable utilization, leveraging its rich composition of polyphenols, tannins, flavonoids, caffeine, and other phytochemicals. Various methods, both conventional and nonconventional, are employed to efficiently extract these compounds, each with distinct advantages and limitations. Traditional techniques like solvent extraction and Soxhlet extraction are widely used for bioactive compound recovery from TW. These methods frequently utilize organic solvents like ethanol, methanol, and acetone in water or aqueous solutions to dissolve and separate the bioactive compounds of interest. Maceration, which involves soaking TW in solvent for a long time, is also widely used and provides convenient and inexpensive extraction. Despite being simple, conventional methods are more frequently associated with large-volume solvent usage, longer time consumption, and degradation due to heat-sensitive compounds. Advanced extraction techniques aimed at increased efficiency, reduced environmental impact, and preservation of the integrity of thermolabile bioactive. Ultrasound-assisted extraction (UAE) utilizes ultrasonic waves that disrupt cell walls, which allows for better penetration by the solvent, thus highly enhancing extraction efficiency while cutting time and solvent usage. Microwave-assisted extraction (MAE) uses microwave energy to heat the solvent and sample to rapidly release bioactive compounds. Supercritical Fluid Extraction (SFE), particularly using supercritical CO_2, is an eco-friendly approach to extract caffeine, polyphenols, and essential oils with minimal solvent residues. Enzyme-assisted extraction (EAE) uses enzymes such as cellulase or pectinase that break down plant cell walls and release intracellular bioactives, thus being an ideal method for extracting polysaccharides and phenolics [8]. Hydrothermal methods, including pressurized hot water extraction, offer selective extraction without the need for organic solvents and, therefore, fit the green chemistry principles. The method has its specific benefits and can be chosen for specific bioactive compounds targeted and for specific purity levels.

Recent advances in extraction technologies have greatly improved the recovery of bioactive compounds from TW, focusing on sustainability and efficiency. For instance, there is a very illustrative work that highlighted the use of microencapsulation technology through spray drying, in which the extracts from TW were sprayed and encapsulated using a thin carbohydrate coating. It maintained freshness and antioxidant properties while making it suitable as a system for drug delivery. Similarly, high-pressure homogenization (HPH) has been developed as a solvent-free, nonthermal extraction method. It effectively breaks down plant tissue cells without subjecting bioactive compounds to high temperatures, thereby being a suitable technique for preserving temperature-sensitive components. Its scalability and environmental benefits make HPH an excellent technology for industrial applications involving agricultural residues such as TW.

Subcritical Solvent Extraction (SSE) further improves the efficiency in the recovery of bioactive compounds from TW. The system operates with a pressurized liquid below its critical point but above the boiling point, thus preserving the chemical integrity of the extracted compound. SSE using an aqueous ethanol–water solvent increases polyphenol content and antioxidant activity much greater compared with conventional methods such as hot water extraction. Integrative approaches, among which ultrasound enzyme-assisted extraction has been developed to overcome the inherent limitations of singular techniques, have also been explored in this regard. UAE, especially when combined with natural deep eutectic solvents, improves tannin, flavonoid, and terpenoid recovery rates significantly. The key parameters involved herein include ultrasonic power where an optimal range has been identified toward balancing efficient cavitation-induced tissue disruption without yield drop due to bubble agglomeration. These cutting-edge methods are effective, though they must be optimized so that issues such as slow-paced processes, risks of potential hazards in high-pressure and high-temperature operations, and degradation of sensitive bioactive compounds are overcome. These results demonstrate the change that innovative methods can introduce for TW valorization in a more sustainable direction with further improvement being expected in the future.

3.2 Conversion into Value-Added Materials

TW holds immense potential for conversion into different value-added products by overcoming environmental and economic constraints. One of the well-known pathways is thermochemical conversion, in which pyrolysis and hydrothermal carbonization convert tea waste into carbon-rich materials, including biochar and hydrochar [9]. Biochar, rich in porosity and functionalized surface, is extensively used in the environment for water purification, improvement of soil, and carbon sequestration. Hydrothermal carbonization carried out under moderate temperature and pressure, yields hydrochar that possesses favorable structural properties to be used in advanced applications such as adsorption and as a precursor for activated carbon production. Carbonaceous materials produced in this process align with sustainability goals in terms of low-cost, effective solutions for environmental remediation.

Advancements in nanotechnology have allowed for the conversion of tea waste into high-value carbon-based nanomaterials such as carbon quantum dots (CQDs) and graphene-like structures. Derived from the intrinsic carbon content of tea waste, using methods such as hydrothermal synthesis and microwave carbonization, these nanomaterials are useful for their photoluminescent properties [5]. Similarly, graphene-like materials derived from tea waste have electrical conductivity and mechanical strength and are suitable for applications in energy devices like supercapacitors and batteries. This conversion pathway not only adds value to tea waste but also propels the advancement of material science and technology.

Tea waste is also a promising feedstock for renewable energy production. Its high organic content makes it suitable for anaerobic digestion, which produces biogas as a clean energy source, and fermentation, which yields bioethanol for fuel applications. These processes reduce reliance on fossil fuels and contribute to decentralized energy systems. Dried tea waste also has considerable thermal energy potential, making it possible to use it as a direct combustion fuel in industrial processes. These energy conversion pathways highlight the role of tea waste in addressing global energy challenges through sustainable bioenergy solutions.

Composting tea waste utilizes its nutrient content for agricultural purposes. The decomposition of tea waste into organic fertilizer enhances the fertility, microbial activity, and overall health of the soil. This is a sustainable recycling process that fits well with organic farming and reduces reliance on chemical fertilizers to produce eco-friendly agriculture. In addition, the fibrous nature of tea waste enables its use in biodegradable products such as paper and packaging. These applications are decreasing environmental pollution by replacing traditional plastics and aligning with the global trend of embracing renewable and biodegradable alternatives.

The conversion of tea waste into value-added products well explains circular economy principles—turning the waste into resource use. Green technologies can combine the innovation of converting this process to make it a stepping stone to sustainable development, reduce the ecological footprint of the tea industry, and drive further resource efficiency in resource-poor sectors.

4 Applications of Tea Waste

Table 7.1 highlights the diverse applications of tea waste as a source material, showcasing its potential in various fields. From bioenergy production to extracting high-value chemicals and the treatment of wastewater, various novel utilization strategies of this abundant byproduct are covered briefly in the section and discussed in detail.

Table 7.1 Applications of tea waste as a starting material

Type of tea waste	Category	Product	Output	Reference
CTC tea Waste	Bioactive	Probiotic formulation	Xanthosine and sucrose	Majumder et al. [10]
Factory tea Waste		Total polyphenol	TPC was 2635.36 ± 52.07 mg GAE/L	Duarah et al. [11]
Compost tea		Biofertilizer	Improved production and plant growth	Zaccardelli et al. [12]
Tea waste		Feed for livestock alternative to the dietary protein source	Goats given 5–10% tea waste diets had lower crude protein digestibility, better nitrogen retention, and no methane emissions compared to the control diet	Sundod et al. [13]
Tea leaf waste	Environmental Remediation	Biochar for fabrication of hybrid nanocomposite	Dye adsorption studies showed excellent removal efficiency (RhB = 95.89%, CR = 94.10%) at 300 K for 25 mg L^{-1} RhB and 22.5 mg L^{-1} Congo red concentrations	Shaikh et al. [9]
Waste black tea		Adsorbent	The adsorption capacity for Cr^{6+} is 9.14 mg/g	Çelebi [14]
Waste green tea		Adsorbent	The adsorption capacity for Cr^{6+} is 8.56 mg/g	
Waste rooibos tea		Adsorbent	The adsorption capacity for Cr^{6+} is 5.12 mg/g	
Domestic tea waste		Magnetite nanoparticles loaded tea waste	The adsorption capacity of prepared composite for the adsorption of cationic dyes was higher than anionic dyes	Madrakian et al. [15]
Spent tea waste	Energy	Biogas	Adding 40% STW with rice husk increases producer gas hydrogen composition and calorific value by 10.29–16.25% and 5.33–6.18 MJ/Nm^3, respectively, and cold gas efficiency by 12–14.5%	Augustine and Sekhar [16]
Factory tea waste		Hydrochar	The optimal hydrochar was obtained at 220 °C for 55 min, achieving a maximum higher heating value of 22.26 ± 1.10 MJ/kg	Duarah et al. [5]
Factory tea waste	Polymer composites	Microcrystalline cellulose	Suitable for filler material	Debnath et al.[17]

(continued)

Table 7.1 (continued)

Type of tea waste	Category	Product	Output	Reference
Tea waste	Energy storage	Hierarchical porous carbon	Showed exceptional supercapacitor performance with 490 F/g capacitance and 99.7% capacitance retention after 1000 cycles at 8 A/g	Ratnaji and Kennedy [18]
Factory tea waste	Food packaging	MCC/PVA film	The addition of 3 wt % MCC doubled the tensile strength of PVA film (from 9.11 ± 1.15 to 18.5 ± 1.93 MPa). PVA composite film with 5 wt b % TWMCC preserved grapes for 18 days, although unpacked grapes lost marketability after a week	Debnath et al. [19]

GAE Gallic acid equivalent, *MCC* microcrystalline cellulose, *PVA* Poly (vinyl alcohol), *RhB* Rhodamine B, *STW* Spent tea waste, *TPC* Total phenolic content

4.1 Biomedical and Pharmaceutical

TW is increasingly recognized for its biomedical and pharmaceutical potential due to its rich content of bioactive compounds, including polyphenols, catechins, tannins, caffeine, and amino acids. These compounds exhibit diverse biological activities such as antioxidant, antimutagenic, anticarcinogenic, and cardiovascular-protective properties, making tea waste a valuable resource for health-related applications. Studies have highlighted that green tea waste possesses higher antioxidant activity compared to black tea and older leaves, primarily due to its catechin content. Catechins are well-documented for their ability to neutralize free radicals, reduce oxidative stress, and prevent mutagenesis and carcinogenesis, contributing to the prevention of chronic diseases such as cancer and cardiovascular disorders [20]. As mentioned earlier, tea contains a diverse range of bioactive compounds, including polysaccharides, in addition to polyphenols. A study by Inbaraj et al. [21] investigated the antioxidant properties of polysaccharides extracted from green tea leaves. The findings demonstrated that these polysaccharides possess significant free radical scavenging activity, which helps mitigate oxidative stress in both humans and animals. Furthermore, the polysaccharides were shown to improve body weight and enhance antioxidant capacity in chickens, highlighting their potential as natural antioxidants [22]. Based on these findings, the study concluded that green tea polysaccharides could be promising candidates for applications in the feed industry. Since tea waste, typically composed of discarded tea leaves, stems, and buds, is abundantly generated, it holds potential as a valuable source of bioactive polysaccharides for further exploration and utilization.

TW also demonstrates utility in probiotic and antimicrobial applications. For example, Kombucha, a probiotic beverage prepared by fermenting leftover tea, contains higher levels of tannins, coumarins, and terpenoids compared to regular tea infusions, enhancing its health-promoting properties [10]. Additionally, the antimicrobial activities of tea polyphenols make them suitable for use as natural food preservatives and antimicrobial agents in pharmaceutical applications. Beyond therapeutic uses, aromatic compounds extracted from TW have applications in fragrance and cosmetics, supporting the development of eco-friendly personal care products.

TW is also valuable in sustainable agricultural practices. It can be valuable in terms of bio-fertilizer and composting materials. Bio-fertilizers derived from tea waste exhibit increased nutritional content, higher moisture retention, and elevated oxygenation capability. The quality of compost is determined by its consistency, maturity, and the chemical content of TW. Tea compost substantially enhances populations of nitrogen-fixing bacteria and nutrient concentrations. It comprises N-triacontanol, which is particularly efficacious in enhancing plant development. Biofertilizers derived from tea waste can improve crop growth and productivity, resulting in significantly increased yields of diverse vegetable and fruit crops [12, 23].

TW contains a considerable quantity of caffeine. Several research investigations indicate the feasibility of extracting caffeine from tea waste, demonstrating that residual caffeine from tea byproducts may be efficiently recovered. The energy beverage and pharmaceutical businesses are deriving greater advantages from caffeine extraction [24]. Further, decaffeinated tea waste can be repurposed as poultry feed, enhancing immune function and productivity when used in moderation. Studies have shown that replacing up to 10% of protein sources in animal diets with tea waste has no significant impact on food intake or health and has even improved the immune response in livestock [13]. TW has also been successfully used in fish feed, contributing to sustainable aquaculture practices by promoting fish growth, health, and well-being. This aligns with circular bioeconomy principles, offering an environmentally sustainable and cost-effective approach to managing agricultural byproducts. By leveraging the bioactive potential of TW, its biomedical and pharmaceutical applications extend beyond conventional uses, presenting an opportunity for sustainable and innovative approaches in healthcare, nutrition, and therapeutic solutions.

4.2 Environmental Remediation Techniques

The byproducts of tea industries possess significant applicability in environmental contexts, particularly as efficient, environmentally sustainable, and cost-effective adsorbents for wastewater treatment. They have significant promise for eliminating heavy metals, dyes, antibiotics, and aromatic chemicals such as phenol and benzene from wastewater. Shaikh et al. reported the synthesis of hybrid nanocomposite biochar aimed at the removal of rhodamine B (RhB) and Congo red. The prepared nanocomposite demonstrated high adsorption efficiency due to the presence of various active surface functional groups, including hydroxyl groups, amino groups, and methylene groups [9]. Madrakian et al. developed magnetic nanoparticle-loaded tea waste to remove seven organic dyes which exhibited rapid removal of dyes with excellent efficiency. The rapid adsorption rate is primarily due to the structural characteristics of the tea leaves and the functional groups present on the tea leaf shell, which contribute to a large surface area and a strong affinity for the efficient and rapid adsorption of dye molecules [15]. Various researchers have also developed absorbents for the removal of heavy metals, similar to the removal of dyes. Yang et al. [25] assessed the adsorption capability of the modified green tea waste for As^{3+} and Ni^{2+}. The estimated maximum adsorption capacities for As^{3+} and Ni^{2+} were 0.4212 mg/g and 0.3116 mg/g, respectively [25]. Çelebi et al. (2020) conducted a study assessing the experimental results regarding the behavior of different tea wastes in the removal of Cr^{6+}, considering both internal and external factors that influence adsorption properties. The adsorption process indicated that waste black tea, waste green tea, and waste rooibos tea were effective, spontaneous, and endothermic for Cr^{6+} removal. The maximum removal efficiencies for Cr^{6+} were achieved under optimal conditions at 88%, 83%, and 73% [14]. Joshi et al. [26] fabricated SiO_2 and tea waste-based nanocomposites for the sequestration of Pb^{2+} and Cd^{2+} ions. The SiO_2@TW nanocomposites exhibited a maximum adsorption efficiency of 153 mg/g for Pb^{2+} and 222 mg/g for Cd^{2+}. In contrast, the residual tea waste demonstrated a maximum adsorption efficiency of 125 mg/g for Pb^{2+} and 142.9 mg/g for Cd^{2+} [26]. Tea waste has also exhibited excellent efficiency in removing aromatic hydrocarbons from aquatic ecosystems. Inbaraj et al. [21] prepared a magnetically activated carbon nanocomposite from green tea leaf waste for evaluation of the adsorption efficiency of 4 priority polycyclic aromatic hydrocarbons. The prepared nanocomposite efficiently removed Benzo[a]anthracene, Chrysene, Benzo[b]fluoranthene, and Benzo[a]pyrene [21]. It was also discovered that the residue from tea waste was useful in eliminating the hazardous organic component pyridine from an aqueous environment. The maximum pyridine removal and uptake in traditional batch adsorption were 33.7 mg/g and 92.3%, respectively, whereas the maximum removal in ultrasound-mediated adsorption was 98.2% and the adsorption capacity was 37.4 mg/g. Therefore, the TW's adsorptive performance was greatly enhanced by the ultrasonic treatment, which also lowered the equilibrium time (90 min) compared to the traditional method (160 min) [27]. TW biochar activated with noncorrosive agents, $KHCO_3$ and CH_3COOK, showed high efficiency in removing sulfadiazine from wastewater. The activated biochars exhibited specific surface areas of 717.6 m^2/g and 648.4 m^2/g, resulting in maximum adsorption capacities of 77.5 mg/g and 58.1 mg/g, respectively [28]. These highlight their potential as effective adsorbents for wastewater treatment.

4.3 Energy Generation from Tea Waste

Bioenergy is rapidly emerging as a cornerstone of sustainable energy solutions in the modern world, driven by the urgent need to reduce reliance on fossil fuels and mitigate environmental challenges. In this scenario, tea waste presents a viable opportunity for the extraction of bioenergy. Numerous studies have been conducted and published on the subject of the generation of bioenergy from waste tea through the utilization of a variety of techniques, including pyrolysis, gasification, combustion, and fermentation procedures. Augustine et al. (2019) demonstrated the potential of spent tea waste (STW) as a sustainable biomass for blending with rice husk (RH) in a downdraft gasifier, achieving notable enhancements in producer gas quality and system efficiency. Blending up to 40% STW with RH increased the calorific value of producer gas from 5.32 to 6.18 MJ/Nm^3 and improved cold gas efficiency from 73.33% to 83.93%, with optimal results observed at an airflow rate of 0.004 kg/s. The addition of STW also elevated the operating temperature from 900 °C to 1010 °C and enhanced the combustion rate, all without requiring modifications to the gasifier. While higher proportions of STW may reduce performance, pretreatment techniques such as briquetting or pelletization could mitigate these limitations. The study underscores STW's utility as a cost-effective and efficient biomass resource for improving gasification systems and its potential for integration with other low-gasification potential biomasses to meet future energy demands [16]. Duarah et al. (2024) investigated the impact of holding time and temperature on the hydrothermal carbonization of factory tea waste to produce solid fuel with enhanced properties. The optimal hydrochar was obtained at 220 °C for 55 min, achieving a maximum higher heating value of 22.26 ± 1.10 MJ/kg. Combustion behavior and kinetics revealed that the activation energy of the prepared hydrochar aligned with the typical range for coal, highlighting its potential as a coal-like biofuel [5]. Pyrolysis of tea waste in a fixed-bed reactor demonstrated a maximum bio-oil yield of 26.80% at 500 °C with a heating rate of 40 °C/min, producing bio-oil with a calorific value of 29.11 MJ/kg and biochar with potential applications in carbon sequestration and soil amendment. Sulfonated tea waste has also been developed as a solid acid catalyst for palm fatty acid distillate conversion, achieving 97% conversion at low temperatures (65 °C) with excellent reusability and acid density, making it feasible for commercial applications [29, 30]. Additionally, seawater-based systems have been explored for bioethanol production from black tea waste, with *Zygosaccharomyces bailii* showing a sixfold higher ethanol yield using immobilized cells compared to free cells, highlighting its potential for reducing freshwater consumption in bioethanol industries [31]. These studies underline the versatility of tea waste and its promise for sustainable energy production and value-added applications.

4.4 Polymer Composites Fabrication

Natural fibers as reinforcing agents in polymer matrices can provide green biodegradable polymer composites with good performance and reduced environmental impact. Synthetic fiber composites are strong but less recyclable, while natural fibers are plentiful, affordable, hydrophobic, and have high specific properties. Thus, mixing both fibers in a polymer matrix yields a hybrid composite with the greatest properties of each. A hybrid polymer composite was developed using tea waste, sisal, and glass fibers in an epoxy resin matrix. Tea waste and sisal fibers were treated with 5 wt% NaOH to remove hydrophilic lignin and hemicellulose before being hybridized with glass fibers. The composite containing 10 wt% tea waste and 20 wt% sisal fibers exhibited the highest tensile (75.6 MPa) and flexural strength (218 MPa), while the composition with 15 wt% tea waste and 15 wt % sisal had the maximum impact strength (96.2 kJ/m^2). Additionally, a composition with 20 wt% tea waste and 5 wt% sisal fibers showed a high sound absorption coefficient of 0.9. The hybridization significantly enhanced tensile (6.3%), flexural (22.4%), impact strength (42.0%), and sound absorption capacity (35%) compared to neat glass fiber composites, making it suitable for automotive applications like bumpers, dashboards, and brake pads [32]. Debnath et al. [17] demonstrated the potential of black tea waste as a renewable source for microcrystalline cellulose (MCC) extraction using microwave-assisted delignification and bleaching. This method, avoiding conventional acid hydrolysis, significantly accelerated the process, producing MCC with 89.77% crystallinity and good thermal properties. Characterization confirmed the removal of amorphous compounds and the isolation of cellulose with a fibrous structure (~23.06 μm), highlighting its suitability as a filler for polymer composites [17]. In another study, used tea dust powder was effectively recycled as reinforcement in epoxy-based polymer composites, fabricated using the hand-layup method with varying tea dust content (10–60 vol%). Mechanical testing revealed that 40 vol% tea dust in the epoxy matrix achieved optimal tensile, flexural, and impact strengths, while higher additions reduced performance due to weaker bonding. This approach underscores the potential of tea waste as a green material for lightweight engineering applications [33].

4.5 Advancements in Energy Storage Devices

With the growing focus on energy and environmental challenges, biomass utilization has emerged as a promising field. Tea waste, rich in cellulose and lignin, offers potential for diverse applications, including energy storage, electrochemical devices, precious metal nanoparticles, recyclable plastics, and sustainable construction materials. Its derivatives, such as microcrystalline and nanocrystalline cellulose, enable the development of high-performance materials like supercapacitors and batteries. Khan et al. [34] reported the synthesis of highly porous carbon with a surface area of 1610 m^2 /g from tea waste through pre-carbonization, acid washing, and KOH activation. The material exhibited excellent electrochemical performance with a specific capacitance of 332 F/g at 1 A/g and remarkable cycling stability, retaining 97.8% capacitance over 100,000 cycles, making it a promising candidate for green supercapacitors [34]. In another study, waste tea-based porous carbons were synthesized through hydrothermal carbonization followed by KOH and $ZnCl_2$ activation and used as host materials for sulfur cathodes in lithium-sulfur (LiS) batteries. The optimized composite with 60 wt% sulfur loading delivered an initial specific capacity of 1048 mAh/g at 0.05 C and retained 627 mAh/g after 100 cycles. It also exhibited good capacity retention of 428 mAh/g at 0.5 C after 100 cycles, highlighting its potential in LiS battery applications [35]. Ratnaji and Kennedy [18] prepared hierarchical porous carbon from tea waste for supercapacitor electrodes through a two-stage process involving pre-carbonization and KOH activation at temperatures ranging from 600 °C to 900 °C. The carbon derived at 800 °C (WTC8) demonstrated exceptional performance, with a high capacitance of 490 F/g and 99.7% capacitance retention after 1000 cycles at 8 A/g, attributed to its well-defined micro-meso porosity and Brunauer-Emmett-Teller surface area of up to 1029 m^2 /g. These studies highlight the potential of converting waste into valuable resources, providing a sustainable approach to energy storage material development [18].

4.6 Other Applications

In addition to these applications that have been discussed, including the utilization of TW in pharmaceuticals, biomedicine, environmental remediation, energy production, and agriculture, there are so many emerging and unconventional applications which illustrate the versatility of this type of material. Applications cover such innovative areas as developing aerogels and hydrogels, based on the lightweight and porous character of the materials derived from TW for use in insulation, water purification, biomedical devices, etc. Similarly, the integration of TW with nanocomposites promises novel possibilities in an advanced material formulation based on further improvements in mechanical, thermal, and functional properties. Jia et al. [36] investigated the use of discarded tea leaf residue as a photothermal material at an interfacial, low-cost, and eco-friendly scale for solar-driven evaporation. Using tea leaf residue (TLR) along with polyvinyl alcohol hydrogels, a high-performance material was obtained, which yielded an evaporation rate of 1.35 kg/m^2 /h with an efficiency of 93.35% under a 1.0 kW/m^2 level of solar irradiation.

The system facilitated water-electricity cogeneration and served as a very efficient dye adsorbent for wastewater treatment. The study shows the TLR's recycling potential and position in sustainable desalination and pollution management, using biomass waste at a minimum cost [36]. Aksu et al. (2024) investigated the isolation of cellulose from tea stem pruning waste in the form of a light-weight (0.22 g·cm^{-3}), eco-friendly hybrid aerogel made from the combination of pure cellulose with homophthalic acid. The obtained aerogel showed very good absorbency, having the capability to hold up to 85 g of water per gram and presented a surface area of 221 m^2/g [37]. In another study, a macroporous (~150 μm) double network hydrogel (TR/PAA) from TW and acrylic acid, showing ultrafast adsorption (~10 min) and high capacities for Cr^{3+}, Pb^{2+}, and Fe^{3+} (206.19, 253.16, and 94.88 mg/g, respectively). The key to the adsorption mechanism lies in the functional groups containing nitrogen and oxygen. Fixed-bed column experiments treated up to 1400 bed volumes (21.6 L) of wastewater efficiently [38]. Another recent study developed waste tea residue carbon dots (WTR-CDs) incorporated into alginate-based hydrogel beads (HB-Alg) for controlled release applications. WTR-CDs were synthesized via a simple carbonization process (λem = 450 nm, WTR − CDs = 18.45%), and hydrogel beads were prepared using an extrusion method. Three loading strategies, ex situ/in situ, were studied, and in situ loaded WTR-CDs showed sustained release due to ionic interactions, wet-beads showed burst release, and dry-beads showed the lowest release [39].

5 Challenges and Future Perspectives

The utilization of TW for the development of value-added products is challenging and promising for future research. One of the major challenges is to scale-up the research focus and production to commercial levels. Most of the current studies are confined to lab-scale experiments, and translating these findings into industrial applications requires addressing scalability issues, ensuring cost-effectiveness, and maintaining the quality of the derived products. Another promising approach is to use TW for biofuel production and reduction of emissions. However, it is confronted with technical barriers such as optimizing conversion processes and regulatory standards. Biofuel production pathways from TW may be an efficient way of reducing dependency on fossil fuels and preventing environmental pollution.

Future prospects in TW utilization involve new frontiers and innovative applications, mainly in agriculture and food production. TW can serve as a viable source of bioactive compounds, soil conditioners, and organic fertilizers that aid in the practice of more sustainable farming. Moreover, the incorporation of TW into food production as a functional ingredient or preservative may open fresh avenues to enhance food quality and shelf life. Advanced technologies, accompanied by interdisciplinary collaboration between research and industry, will release the yet untapped potential of TW to lead to a circular bioeconomy.

6 Summary and Conclusion

Increased demand in the global market for tea has not only increased the production levels but also complicated the management of TW. This chapter throws light on the enormous potential of TW as a valuable biomass resource that can be transformed into high-value products across various industries. By considering its physical and chemical properties, bioactive compounds, and global production metrics, the chapter points out the versatility and environmental significance of TW. The innovative applications of TW in pharmaceuticals, biomedicine, environmental remediation, energy production, and agriculture reflect its significant role in advancing sustainable development. Its full potential can be realized only if challenges related to scaling up production processes, optimizing biofuel conversion technologies, and expanding research into novel applications are overcome. The integration of TW into a circular bioeconomy helps industries reduce environmental burdens and promote sustainable industrial innovation. This chapter argues for further exploration and cooperation to realize the multifaceted benefits of TW, making it a cornerstone for ecological and industrial progress.

References

1. Debnath B, Haldar D, Purkait MK (2021) Potential and sustainable utilization of tea waste: a review on present status and future trends. J Environ Chem Eng 9(5):106179
2. Statista (2024c) Volume of tea consumption worldwide from 2018 to 2029, vol 2024. Statista
3. Statista (2024a) Tea—Worldwide, vol 2024. Statista
4. Statista (2024b) Tea production worldwide from 2010 to 2023, by leading country, vol 2024. Statista
5. Duarah P, Debnath B, Purkait MK (2024a) Synthesis of antibacterial fluorescent carbon dots and green coal-like hydrochar from tea industry byproducts via hydrothermal carbonization. Ind Crop Prod 221:119364
6. Wang Z, Ahmad W, Zhu A, Zhao S, Ouyang Q, Chen Q (2024) Recent advances review in tea waste: high-value applications, processing technology, and value-added products. Sci Total Environ 946:174225
7. Negi T, Kumar Y, Sirohi R, Singh S, Tarafdar A, Pareek S, Awasthi MK, Sagar NA (2022) Advances in bioconversion of spent tea leaves to value-added products. Bioresour Technol 346:126409
8. Joardar S, Duarah P, Purkait MK (2023) Recent advances in myo-inositol recovery and purification from agricultural sources as potential dietary supplements: a review. Sustain Chem Pharm 36:101331
9. Shaikh WA, Kumar A, Chakraborty S, Islam RU, Bhattacharya T, Biswas JK (2022) Biochar-based nanocomposite from waste tea leaf for toxic dye removal: from facile fabrication to functional fitness. Chemosphere 291:132788
10. Majumder S, Ghosh A, Saha S, Acharyya S, Chakraborty S, Sarkar S, Bhattacharya M (2022) Valorization of CTC tea waste through kombucha production and insight into GC-MS based metabolomics. Canrea J Food Technol Nutrit Culinary J 5:38–56
11. Duarah P, Joardar S, Debnath B, Purkait MK (2024b) Optimized extraction of polyphenols from tea factory waste and cost-effective drying methods for sustainable utilization. Bioresour Technol Rep 26:101833
12. Zaccardelli M, Pane C, Villecco D, Palese AM, Celano G (2018) Compost tea spraying increases yield performance of pepper (*Capsicum annuum* L.) grown in greenhouse under organic farming system. Ital J Agron 13(3):229–234
13. Sundod R, Poothong S, Nuengjamnong C (2023) Substitution effect of protein source with tea waste on productive performances and ruminal fermentation in crossbred Saanen lactating goats. Ital J Anim Sci 22(1):453–462

14. Çelebi H (2020) Recovery of detox tea wastes: usage as a lignocellulosic adsorbent in Cr^{6+} adsorption. J Environ Chem Eng 8(5):104310
15. Madrakian T, Afkhami A, Ahmadi M (2012) Adsorption and kinetic studies of seven different organic dyes onto magnetite nanoparticles loaded tea waste and removal of them from wastewater samples. Spectrochim Acta A Mol Biomol Spectrosc 99:102–109
16. Augustine MA, Sekhar SJ (2019) Improvement in the calorific value of producer gas from rice husk with addition of spent tea waste as secondary fuel. Energy and Fuels 33(12):12492–12498
17. Debnath B, Duarah P, Purkait MK (2023) Microwave-assisted quick synthesis of microcrystalline cellulose from black tea waste (*Camellia sinensis*) and characterization. Int J Biol Macromol 244:125354
18. Ratnaji T, Kennedy LJ (2020) Hierarchical porous carbon derived from tea waste for energy storage applications: waste to worth. Diam Relat Mater 110:108100
19. Debnath B, Duarah P, Purkait MK (2024) Sustainable utilization of tea factory waste-derived microcrystalline cellulose as filler material for food packaging films. ACS Sustain Resour Manag 1(12):2518–2529
20. Serdar G, Demir E, Sökmen M (2017) Recycling of tea waste: simple and effective separation of caffeine and catechins by microwave assisted extraction (MAE). Int J Secondary Metabol 4(2):78–89
21. Inbaraj BS, Sridhar K, Chen B-H (2021) Removal of polycyclic aromatic hydrocarbons from water by magnetic activated carbon nanocomposite from green tea waste. J Hazard Mater 415:125701
22. Li X, Chen S, Li J-E, Wang N, Liu X, An Q, Ye X-M, Zhao Z-T, Zhao M, Han Y (2019) Chemical composition and antioxidant activities of polysaccharides from Yingshan cloud mist tea. Oxidative Med Cell Longev 2019(1):1915967
23. Pane C, Palese AM, Spaccini R, Piccolo A, Celano G, Zaccardelli M (2016) Enhancing sustainability of a processing tomato cultivation system by using bioactive compost teas. Sci Hortic 202:117–124
24. Keerthanan S, Bhatnagar A, Mahatantila K, Jayasinghe C, Ok YS, Vithanage M (2020) Engineered tea-waste biochar for the removal of caffeine, a model compound in pharmaceuticals and personal care products (PPCPs), from aqueous media. Environ Technol Innov 19:100847
25. Yang S, Wu Y, Aierken A, Zhang M, Fang P, Fan Y, Ming Z (2016) Mono/competitive adsorption of arsenic(III) and nickel(II) using modified green tea waste. J Taiwan Inst Chem Eng 60:213–221
26. Joshi S, Kataria N, Garg VK, Kadirvelu K (2020) Pb^{2+} and Cd^{2+} recovery from water using residual tea waste and SiO_2@TW nanocomposites. Chemosphere 257:127277
27. Daware GB, Gogate PR (2021) Removal of pyridine using ultrasound assisted and conventional batch adsorption based on tea waste residue as biosorbent. Environ Technol Innov 21:101292
28. He X, Li J, Meng Q, Guo Z, Zhang H, Liu Y (2021) Enhanced adsorption capacity of sulfadiazine on tea waste biochar from aqueous solutions by the two-step sintering method without corrosive activator. J Environ Chem Eng 9(1):104898
29. Basumatary V, Saikia R, Narzari R, Bordoloi N, Gogoi L, Sut D, Bhuyan N, Kataki R (2018) Tea factory waste as a feedstock for thermo-chemical conversion to biofuel and biomaterial. Materials Today Proc 5(11, Part 2):23413–23422
30. Indira D, Das B, Bhawsar H, Moumita S, Johnson EM, Balasubramanian P, Jayabalan R (2018) Investigation on the production of bioethanol from black tea waste biomass in the seawater-based system. Bioresour Technol Rep 4:209–213
31. Rashid U, Ahmad J, Ibrahim ML, Nisar J, Hanif MA, Shean TYC (2019) Single-pot synthesis of biodiesel using efficient sulfonated-derived tea waste-heterogeneous catalyst. Materials 12(14):2293
32. Prabhu L, Krishnaraj V, Gokulkumar S, Sathish S, Ramesh M (2019) Mechanical, chemical and acoustical behavior of sisal–tea waste–glass fiber reinforced epoxy based hybrid polymer composites. Materials Today: Proc 16:653–660

33. Kushwanth Theja K, Bharathiraja G, Sakthi Murugan V, Muniappan A (2023) Evaluation of mechanical properties of tea dust filler reinforced polymer composite. Materials Today: Proc 80:3208–3211
34. Khan A, Senthil RA, Pan J, Osman S, Sun Y, Shu X (2020) A new biomass derived rod-like porous carbon from tea-waste as inexpensive and sustainable energy material for advanced supercapacitor application. Electrochim Acta 335:135588
35. Arie AA, Kristianto H, Cengiz EC, Demir-Cakan R (2020) Waste tea-based porous carbon-sulfur composite cathodes for lithium-sulfur battery. Ionics 26(1):201–212
36. Jia T, Han F, Zhang R, Zhao M, Chen Z, Wang Z, Li L, Qi S, Xi G, Li B (2024) Biomass photothermal hydrogel based on tea leaves residue for cogeneration of clean water and electricity. Adv Sustain Syst 8(6):2300531
37. Aksu K, Kaya M (2024) Cellulose ester-based aerogel: lightweight and highly water-absorbent. ACS Omega 9(3):3701–3708
38. Zhang S, Liu C, Yuan Y, Fan M, Zhang D, Wang D, Xu Y (2020) Selective, highly efficient extraction of Cr(III), Pb(II) and Fe(III) from complex water environment with a tea residue derived porous gel adsorbent. Bioresour Technol 311:123520
39. Nille OS, Patil AS, Vibhute AA, Shendage SS, Tiwari AP, Anbhule PV, Sohn D, Gore AH, Kolekar GB (2024) Route-dependent tailoring of carbon dot release in alginate hydrogel beads (HB-Alg@WTR-CDs): a versatile platform for biomedical applications. Int J Biol Macromol 257:128126

Chapter 8
Valorization of Coffee Bean Processing Wastes and Grounds for Effective Recovery of Value-Added Bio-Based Products

G. Divyashri, O. Sindhu, and M. Anagha

Abbreviations

AD	Anaerobic digestion
GHG	Greenhouse gases
IBD	Inflammatory bowel disease
MAP	Modified atmosphere packaging
OECD	Organisation for Economic Co-operation and Development
ROS	Reactive oxygen species
SCGs	Spent coffee grounds
TBARS	Thiobarbituric reactive oxygen species
t-BOOH	tert-Butyl hydroperoxide

1 Introduction

Coffee is a global industry involving 80 countries and employing around 100 million people, highlighting its worldwide significance. Coffee is a vital export crop that boosts national economies in many developing nations by creating foreign exchange profits and promoting rural development [1]. From planting and

G. Divyashri (✉)
Bioproducts Research, Iom Bioworks Pvt., Ltd.,
Bengaluru, Karnataka, India
e-mail: divyashri@iombio.com

O. Sindhu · M. Anagha
Department of Biotechnology, M S Ramaiah Institute of Technology,
Bengaluru, Karnataka, India

C. Trois, R. C. Ray (eds.), *Plantation Crop Wastes: Valorization for Economic Sustainability*, Waste as a Resource,
https://doi.org/10.1007/978-3-031-96805-1_8

harvesting to processing, marketing, and retail, the production and distribution of coffee create jobs at many phases of the industry (International Trade Centre 2020). Worldwide, coffee consumption has a significant impact on social interactions and cultural customs because it is ingrained in people's everyday lives and traditions [2]. In India, the southern states of Karnataka, Kerala, and Tamil Nadu are the main locations for coffee farming; Karnataka alone produces around 70% of the nation's coffee [3]. India is among the top ten global coffee exporters, and its exports provide a large portion of its foreign exchange earnings (International Trade Centre 2020). In addition, Indian coffee is well known for its distinct flavors and superior quality; several varieties, such as Mysore Nuggets Extra Bold and Monsooned Malabar, are becoming well-known worldwide [4]. Due to the growth of urban consumers' coffee culture and the expansion of coffee shops, India's domestic coffee market is likewise expanding [5].

The waste and by-products produced along the whole value chain, from agricultural production to industrial processing and final consumer use, are a direct result of the world's coffee consumption (Fig. 8.1). These large volumes of waste produced present serious environmental and toxicological problems [6]. The chemical composition of coffee waste varies significantly based on the coffee variety, processing methods, and type of by-products involved [8]. Numerous studies have documented the variations in the chemical components of coffee waste according to the species and the specific processing steps utilized [9].

There are more than 120 species in the genus *Coffea*, which belongs to the family *Rubiaceae*. However, only two varieties, arabica, and robusta, derived from *Coffea arabica* and *C. canephora*, respectively, are commercially produced and distributed globally [10, 11]. About 45% of the weight of coffee cherries (or berries) is made up of pulp, husk, and peel. These components are removed from the coffee beans using

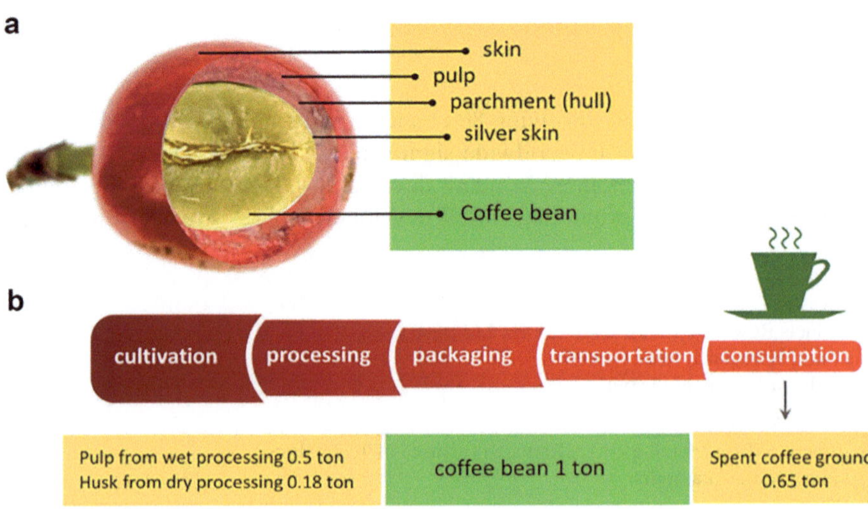

Fig. 8.1 Coffee cherry and its main constituents (**a**) and relevant by-products generated along the value chain of the coffee industry (**b**) [7]

a wet or dry processing method to obtain green coffee beans [12]. Given that the coffee bean makes up almost 50% of the coffee cherry, the primary by-products of the entire coffee value chain are spent coffee grounds (SCGs), which are generated either by the industrial production of soluble coffee or by the final consumption of roasted coffee. For every ton of green coffee (or dried coffee cherries), the by-products include approximately 500 kg of coffee pulp, 180 kg of coffee husk, and 650 kg of SCGs, along with other minor by-products viz., coffee silverskin, coffee parchment, and mucilage [13]. Valorization helps mitigate pollution and reduce greenhouse gas (GHG) emissions and aligns with the principles of the circular economy, where resources are continuously reused and recycled into value-added products.

2 Geographical Origin, Distribution, Production, Productivity, and Description

Coffee is considered the cash crop in many countries, and it stands out to be the second most valuable commodity after oils [14, 15]. Coffee is a perennial crop belonging to the family *Rubiaceae* [16], and genus *Coffea* and has about 103 species worldwide of which mainly three species (*Coffea arabica, C. robusta*, and *C. liberica*) are cultivated for commercial consumption [17]. Ethiopia is the world's largest coffee-producing country and majorly produces and exports the Arabica coffee variety [18, 19]. Coffee production systems can be classified into four types i.e., forest, semi-forest, garden, and plantation type. Naturally grown coffee plants under canopy layers or dense forests with minimal management while harvesting are classified as forest coffee. In semi-forest type, wild coffees are grown under natural environmental conditions after clearing the shrubs and weeds making it possible to grow diverse varieties of coffee. In contrast, the coffees grown with intensive management are garden-type coffee, and the application of modern cultivation techniques and cultural practices to grow the improved varieties are considered in plantation-type coffees [20]. The quality of the coffee bean and its composition varies depending on the climatic conditions, rainfall pattern, soil composition, moisture content, light, humidity, altitudinal variation, temperature, and coffee processing and production procedures [21].

3 Coffee Processing Techniques

Processing coffee is an essential step in turning coffee cherries into roasted and brewed beans to create the popular beverage that is consumed worldwide. The technique employed affects the quantity and quality of by-products produced during processing in addition to the coffee's flavor characteristic [22]. The main techniques for processing coffee are illustrated in Fig. 8.2.

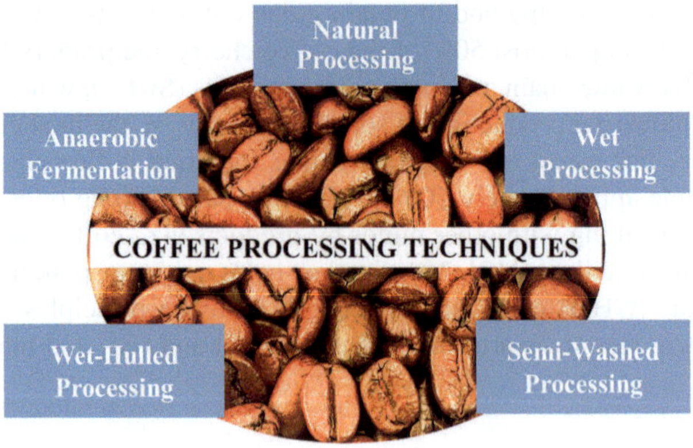

Fig. 8.2 Overview of coffee bean processing techniques

3.1 Natural Processing (Dry Processing)

One of the earliest and most conventional techniques for preparing coffee is dry processing. Using this procedure, after harvesting, the entire coffee cherry is sun-dried [23]. To achieve even drying, the cherries are spread out on broad surfaces like patios or raised beds and turned frequently. To extract the green coffee beans, the peel, pulp, and parchment are mechanically removed from the cherries after they have completely dried. The principal by-products consist of dehydrated pulp and parchment, which are commonly known as coffee husks together. These waste materials can be turned into compost, animal feed, or biomass fuel. This technique of coffee processing is widely utilized in Ethiopia and some portions of Brazil that have little access to water.

3.2 Wet Processing (Washed Processing)

The more recent and water-intensive technique is called wet processing. The coffee cherries are pulped to remove part of the pulp and the outer skin after they are harvested. After that, the beans are fermented in water for 12–48 h to remove any leftover mucilage. The beans are dried, either in the sun or with the aid of machine dryers, after being fermented and the mucilage is removed with a water wash. For many specialty coffees, wet processing yields a desirable flavor profile that is brighter, cleaner, and more acidic. In regions of Africa and Latin America, it is commonly utilized. Wet pulp, mucilage, and wastewater are among the by-products of the wet processing method that need to be carefully managed to avoid environmental contamination [23].

3.3 Semi-Washed Processing (Pulped Natural/ Honey Processing)

A hybrid technique that incorporates aspects of both wet and dry processing is called semi-washed processing. The mucilage remains on the beans after the cherries are pulped, and the beans are then partially dried while the mucilage is still there. Different honey levels, viz., white, yellow, red, or black honey, refer to the degree of mucilage removal and drying time depending on the amount of mucilage left on the beans. Coffee made with this method is typically sweeter and more complex than wet processed coffee, but cleaner than naturally processed coffee. Compared to wet processing, semi-washed processing produces less waste and yields comparable by-products, such as pulp and mucilage, albeit in smaller amounts [24].

3.4 Wet-Hulled Processing (Giling Basah)

The unique technique known as wet-hulled processing is predominantly employed in Indonesia. The beans are pulped, fermented for a brief while, and then partly dried before being partially hulled while still damp. To get the right moisture content, the beans are then dried once more. A characteristic of many Indonesian coffees is their earthy, full-bodied flavor, which is brought about by the wet-hulled processing method. With a focus on pulp and wet parchment, the technique yields by-products that are comparable to those from wet processing [24]. The unusual hulling technique necessitates slightly different waste management requirements, particularly concerning moisture content.

3.5 Anaerobic Fermentation

This is an experimental, relatively new way of preparing coffee. The coffee beans are pulped and then anaerobically fermented in sealed tanks without air. Specialty coffee manufacturers that want to develop standout products are drawn to anaerobic fermentation because it can yield extremely complex and distinctive flavors [22]. Because of the regulated fermentation environment, the anaerobic process produces by-products that are comparable to those produced by wet processing, but frequently in different amounts.

4 Coffee Waste Composition and Generation

Coffee beans have a complex chemical composition that includes a wide range of components that give them their distinct flavors, aroma, and health advantages [25]. Numerous elements contribute to this complexity, including minerals, proteins, lipids, carbohydrates, phenolic compounds, alkaloids (like caffeine), and a variety of volatile and non-volatile substances [26]. During coffee production, a significant portion of the coffee cherry, which encloses the bean, becomes waste material. This includes the parts that are removed during processing, including the pulp, husk, and peel. Most coffee waste is made up of these by-products and the discarded SCGs that are produced after brewing. The chemical composition of these waste products is highly variable, depending on factors such as the coffee species (e.g., *C. arabica* or *C. canephora*), the processing methods used (wet or dry), and the specific part of the coffee cherry or bean.

4.1 Coffee Pulp and Husk

Combined, the pulp and husk account for roughly 45% of the weight of the coffee cherry. Particularly, coffee pulp has high concentrations of cellulose, pectin, and soluble sugars, which makes it a viable raw material for the production of biofuels, bioplastics, and other goods with added value. Nutraceuticals and medicines can benefit from the antioxidant qualities of the phenolic chemicals found in pulp and husk [27].

4.2 Spent Coffee Grounds (SCGs)

Produced by both the final consumption of roasted coffee and the industrial manufacturing of soluble coffee, SCGs are among the most common by-products of coffee. Approximately, 650 kg of SCGs are created for each ton of green coffee that is processed. Organic chemicals found in abundance in SCGs include lipids, proteins, carbohydrates, and bioactive substances such as diterpenes, polyphenols, and caffeine. Because of these elements, SCGs are a useful resource for several applications, including the production of bioenergy, composting, and the extraction of bioactive chemicals for use in medicines and cosmetics [26].

4.3 *Coffee Silverskin and Parchment*

Coffee parchment is the hard, fibrous layer that surrounds the bean and is removed during the hulling process. Coffee silverskin is a thin layer of material that is removed during the roasting process. These by-products are produced in smaller amounts than pulp, husk, and SCGs, but they nevertheless include essential ingredients that can be used in the culinary and nutraceutical industries, including proteins, dietary fiber, and antioxidants [28].

4.4 *Coffee Mucilage*

The sticky, sugary material that envelops the coffee beans within the cherry is called coffee mucilage. It is usually eliminated in the course of wet processing and is a substantial source of organic matter. It is possible to ferment the sugars, pectin, and other polysaccharides found in mucilage to produce bioethanol and other biobased products [29].

5 Valorization Techniques

Valorizing coffee waste such as coffee grounds, coffee husk, coffee parchment, and coffee silver skin reduces the waste disposal cost, generates new revenue streams, and promotes a sustainable coffee production process. There are various techniques through which coffee waste can be utilized to produce value-added products.

5.1 *Bioconversion Methods*

The various bioconversion methods are discussed below.

5.1.1 Composting

Composting involves the spontaneous biological conversion of waste into organic material in the aerobic environment [30]. The coffee waste can be composted alone or co-composted with other degradable municipal solid waste. The compost obtained will be rich in iron and potassium (K) with other micronutrients that are beneficial for plant growth. Degefe et al. [31] reported the use of coffee husk along with cow dung in vermicomposting process. Co-composting coffee pulp and husk with municipal solid waste, cow dung, and poultry manure produced a well-aged organic fertilizer that significantly improved crop growth and yield [32]. It was reported that the co-composting of green waste with coffee by-products reduces GHG emissions [33].

5.1.2 Enzymatic Hydrolysis

Enzymatic hydrolysis offers a sustainable solution for valorizing coffee waste into valuable compounds and can be employed to obtain phenolic compounds with antioxidant properties, dietary fibers with prebiotic potential, and fermentable sugars for biofuel production. The pretreatment followed by enzymatic hydrolysis can increase the extraction of reducing sugars from various coffee wastes [34, 35]. Coffee husk was subjected to enzymatic hydrolysis during the production of fermentable sugars [36]. Chiyanzy et al. [37] reported the use of a cocktail of enzymes, i.e., mannanase and cellulase, for the recovery of soluble solids from the SCGs. Similarly, Jomnonkhaow et al. [38] reported the use of the enzymatic hydrolysis method to valorize SCGs for the production of fermentable sugars, volatile fatty acids single cell protein, and biofuels.

5.1.3 Fermentation

Fermentation offers a versatile and sustainable approach to coffee waste valorization, generating a wide range of valuable products such as biofuel, organic acids, antioxidants, and enzymes. This procedure reduces cellulose crystallinity, removes lignin and hemicelluloses, and increases the surface area available for subsequent biological breakdown. Considering the chemical structure of SCGs, submerged and solid-state fermentation are used as a pretreatment method for the production of short-chain organic acids in the presence of fungi [39]. It was discovered that the combination of enzyme hydrolysis and fermentation was the most efficient way to produce a product with increased antioxidant activity, reducing sugars, free amino acids, and reduction in anti-nutrient compounds [40].

5.1.4 Anaerobic Digestion

Anaerobic digestion (AD), which is the biological process that occurs in the absence of oxygen, is the most preferred solution for the valorization of high moisture substrates like wastewater sludge, organic residues, and food waste for the production of biogas and biofertilizers [41]. AD is a viable alternative for post-hydrothermal liquefaction wastewater due to its acceptability to its energy valorization potential by hydrogen and methane production [42]. SCGs that are rich in organic matter are reported to be potential substrates for biogas production through AD [43]. It has been demonstrated that mixing fresh cow dung with the SCG liquid fraction has enhanced the production of biogas with high methane concentration [44].

5.1.5 Biochar Production

Coffee waste can be used as a raw material for the production of carbon-based products. The residual compounds in used coffee grounds after brewing could influence the coffee's natural structure during the carbonization process. Biochar is the product of the pyrolysis process wherein biomass is heated at a temperature from 200 to 700 °C in the absence of oxygen. Biochar has a wide range of applications based on its chemical and physical characteristics such as an agent for soil improvement [45], reducing GHG emissions [46], and removal of pollutants from wastewater. SCGs may be used to create a carbonaceous adsorbent that can be used for the treatment of wastewater due to its great availability and inexpensive cost [47]. The residue obtained from SCGs during biodiesel production was further valorized for smokeless biochar production [48]. Biochar obtained from spent coffee powder was used for the production of electrically conductive-reinforced plastics [49].

5.1.6 Bioethanol Production

Bioethanol production enhances energy security by reducing reliance on fossil fuels and can reduce GHG emissions. The reducing sugars present in coffee residue waste are a potential substrate for bioethanol production during the fermentation process [50, 51]. Acid or acid and enzymatic hydrolysis are currently being explored for the synthesis of bioethanol from various coffee waste fractions. The coffee husk hydrolysis using acid and cellulolytic enzymes followed by fermentation with lignocellulosic yeast yields the highest ethanol yield compared to other coffee waste samples [52]. The bioethanol production from coffee husk with encapsulated cells is greatly influenced by internal mass transfer effects [53].

5.1.7 Production of Bioplastics

Coffee by-products are being explored for use in the development of bioplastics. Bioplastics developed from coffee by-product-derived compounds may offer surface hydrophobicity, gas impermeability, mechanical resilience, and antioxidant, and antibacterial properties, making them appropriate for the creation of active food packaging. There are two primary approaches for using coffee by-products in the development of sustainable plastic formulations for food packaging. One is the use of unprocessed coffee by-products as functional additives for plastics and the other is the use of extracts made from coffee by-products that can form films or other functional characteristics. Improved tensile strength films are produced by incorporating SCG powder into corn starch-based formulations [54]. Similarly, the films with decreased light transmission, with potential for use in vegetable packaging were developed by incorporation of crude SCG powder into cellulose-based formulations [55]. Pectic polysaccharides obtained from coffee mucilage can be used in the development of biodegradable films with rigid and hydrophobic bioplastics [56].

5.2 *Extraction of Bioactive Compounds*

Various extraction methods like solvent extraction, supercritical fluid extraction, ultrasound-assisted extraction, enzyme-assisted extraction, and membrane separation can be used to extract and purify bioactive compounds from coffee wastes such as polyphenols, caffeine, chlorogenic acid, trigonelline, and dietary fibers. These extracted compounds can be used in food and beverages, pharmaceuticals, cosmetics, and in animal feed.

5.2.1 Extraction of Antioxidants and Polyphenols

The bioactive potential of caffeine, flavonoids, and phenolic chemicals found in coffee and its by-products has drawn attention to their potential uses in pharmaceutical and food applications [57]. Coffee by-products are rich in bioactive compounds such as polyphenols, which exhibit antioxidant properties [58]. The phenolic compounds such as caffeic acid and chlorogenic acid are well known for their antidiabetic, antiviral, anticancer antipyretic, and antilipidemic properties [59]. The conventional water bath extraction method was found to be a better approach than ultrasound-assisted extraction for the extraction of polyphenols from coffee husks [60]. Coffee pulp can also be used as a potential substrate for the extraction of polyphenols with better antioxidant properties, which can be used as an active ingredient in cosmetics and pharmaceutical applications [61]. Coffee silver skin can be proposed to be a low-cost and stable plant material that contains bioactive chemicals with significant antioxidant activity [62].

5.2.2 Caffeine Extraction

Industrial production of synthetic caffeine is underway to meet the growing demand for caffeine and includes laborious steps such as methylation of theobromine under different conditions. However, the cost of such synthesis may increase, and the extraction of caffeine from renewable plant sources might be a more cost-effective option. In line with this, SCGs are a potential source to be considered as a renewable source for caffeine extraction [63]. Studies have also examined the potential of coffee silver skin extracts, which primarily comprise caffeine and chlorogenic acid to lower inflammation-related illnesses and adipogenesis [64]. Caffeine extracted from SCGs finds application in the cosmetic, pharmaceutical, and nutraceutical industries [65]. Green extraction methods based on the use of deep eutectic solvents were developed for the extraction of caffeine from coffee pulp with high antioxidant properties [66].

6 Value-Added Products and Applications

Once thought to be waste, coffee by-products are now recognized as a significant source of bioactive chemicals with applications in food, medicine, and cosmetics, among other industries.

6.1 Bioactive Compounds in Food, Pharmaceuticals, and Cosmetics

Coffee waste is a rich source of bioactive compounds with various applications. It enhances food nutrition, offers pharmaceutical benefits, and augments cosmetic applications. Various applications of coffee waste are listed in Table 8.1 and are elaborated in the below section.

6.1.1 Food Industry

The use of bioactive chemicals derived from coffee by-products in the formulation of nutraceuticals and functional foods is growing. Due to their potent antioxidant qualities, these substances, viz., polyphenols and flavonoids in particular, are useful for raising the nutritional content of food items. Polyphenols and flavonoids, in particular, are known for their strong antioxidant activity, which can help neutralize free radicals and reduce oxidative stress. For example, coffee silverskin has been utilized in the production of enriched biscuits and cookies, where it replaces a portion of wheat flour to enhance the phenolic content and antioxidant capacity of the final product [67]. The potential of coffee silverskin as an ingredient in functional foods has also been demonstrated in studies where it was used to produce cookies with enhanced fiber content and antioxidant properties. These studies showed that while the sensory acceptability of the cookies was maintained, their nutritional profile was significantly improved, making coffee silverskin a promising functional ingredient for the food industry [81]. These bioactive compounds can also be added to beverages and dietary supplements, offering significant health benefits, such as reducing inflammation and improving cardiovascular health [82]. The antimicrobial properties help extend shelf life by inhibiting the growth of spoilage organisms, making them an attractive option for developing clean-label products that meet consumer demands for natural ingredients [83]. Furthermore, the inclusion of SCGs in functional foods is gaining attention. SCGs contain bioactive compounds similar to those found in brewed coffee, such as caffeine, chlorogenic acids, and melanoidins, which have been associated with reduced risks of cardiovascular diseases, cancers, and metabolic disorders [69]. The use of SCGs in food products can enhance their antioxidant properties and potentially moderate chronic human diseases while reducing the environmental impact of coffee waste. Furthermore, the use of coffee

Table 8.1 Application of various coffee waste in food, pharmaceutical, and cosmetic industries

Industry	Type of waste	Application	Property	Reference
Food	Coffee silver skin	Cookies and biscuits	Increases antioxidant activity	Cantele et al. [67]
	Coffee silver skin	Chicken patty	Increases antioxidant activity	Özhamamcı [68]
	Spent coffee grounds	Functional foods	Reduces risks of cardiovascular diseases, cancers, and metabolic disorders	Bevilacqua et al. [69]
	Coffee husk	Pork sausage formulation	Increases antioxidant activity	Araya-Morice et al. [70]
	Parchment	Dietary fiber	Increases antioxidant property	Iriondo-DeHond et al. (2018)
Pharmaceutical	Coffee pulp	Not applicable	Downregulates IL-8 production	Magoni et al. [71]
	Spent coffee grounds	Topical creams, oral supplements, injectables	Reduces inflammation in disorders like psoriasis, rheumatoid arthritis, and other autoimmune disorders	Abbasi-Parizad et al. [72]
	Coffee silverskin Spent coffee grounds	Not applicable	Reduces oxidative stress indicators	Zengin et al. [73]
	Coffee leaf extract	Not applicable	Antibacterial activity	Mesquita Júnior et al. [74]
	Spent coffee grounds	Not applicable	Antifungal and anti-mycotoxigenic properties	Badr et al. [75]
Cosmetic	Coffee silverskin	Skincare product	Increased antioxidant properties	Hayes et al. [76]
	Coffee silverskin	Skincare product	Increased antioxidant properties	Iriondo-DeHond et al. [77]
	Coffee cherry extract	Skincare product	Increased antioxidant properties and decreased hyperpigmentation	Kiattisin et al. [78]
	Coffee pulp extract	Serum	Increased antioxidant properties	Tseng et al. [79]
	Coffee berry extract	Skincare and hair care products	Reduced oxidative stress, increased anti-aging properties	Saewan et al. [80]

husks in fresh pork sausage formulations has shown promising results, particularly in enhancing antioxidant properties and reducing lipid oxidation. A study by Araya-Morice et al. [70] explored the antioxidant potential of coffee husks in fresh pork sausage. The results indicated that the inclusion of coffee husks effectively reduced lipid oxidation, particularly in sausages packaged under modified atmosphere packaging (MAP), where the coffee husk addition helped maintain product quality by inhibiting oxidative processes. Interestingly, the study also found that the addition of coffee husks did not negatively impact the sensory liking of the sausages, suggesting that coffee husks can serve as a natural antioxidant in meat products without compromising consumer satisfaction. Additionally, the potential of coffee silverskin as a functional ingredient has been explored in various meat products. The use of coffee silverskin as a fat replacer in chicken patties was evaluated [68]. The study found that substituting animal fat with coffee silverskin resulted in changes to the physicochemical properties of the patties, such as increased thiobarbituric acid reactive substances (TBARS) values, which are indicative of enhanced antioxidant activity. Iriondo-DeHond et al. [84] conducted a study aimed at confirming silverskin, parchment, and coffee husk as food ingredients that promote health. The results indicated that parchment was found to be a natural source of antioxidant dietary fiber and that coffee husk and silverskin could be fully valorized into two different food ingredients.

6.1.2 Pharmaceutical Industry

Coffee by-products, especially coffee pulp, and SCGs have anti-inflammatory qualities that are worth exploring in the creation of pharmaceuticals intended to treat long-term inflammatory diseases. The study by Magoni et al. [71] showed that quinic acid and procyanidin-rich coffee pulp extracts can successfully downregulate the production of IL-8, a crucial chemokine involved in inflammatory responses. Similarly to this, Abbasi-Parizad et al. [72] showed that SCGs' high flavonoid concentration gives them a potent anti-inflammatory effect. The creation of topical creams, oral supplements, or injectable formulations targeted at lowering inflammation in disorders like psoriasis, rheumatoid arthritis, and other autoimmune diseases could benefit from the ability of SCGs to reduce inflammation.

It has been demonstrated that coffee silverskin and SCGs, which are high in phenolic compounds and chlorogenic acids, lower oxidative stress indicators [73]. This antioxidant capability can be used to create pharmacological formulations that prevent diseases like Parkinson's and Alzheimer's, as well as some forms of cancer and neurodegenerative disorders like cardiovascular disease and neurodegenerative disorders linked to oxidative stress. Coffee leaf extracts have been shown to have antibacterial activity [74]. These extracts are effective against both gram-positive and gram-negative bacteria. These extracts may be transformed into naturally occurring antibacterial agents, which would lessen the need for artificial antibiotics and aid in the fight against antibiotic resistance. Badr et al. [75] demonstrated the antifungal and anti-mycotoxigenic properties of SCG extracts. Furthermore,

Díaz-Hernández et al. [85] investigated the pro-apoptotic and antiproliferative characteristics of SCG extracts, roasted coffee, and green coffee beans. According to the study, the ethanolic extracts of SCGs had the highest total phenolic content and antiradical activity. These findings were directly associated with the powerful antiproliferative effects that the extracts had on cervical cancer (C33A) and lung carcinoma (A549) cells. These cancer cells' viability was greatly reduced by the SCG extracts, which also caused apoptosis, as shown by a rise in the number of apoptotic cells and cell cycle arrest. These results imply that SCGs, which are frequently seen as trash, can be recovered and used as a valuable source of anticancer drugs, offering a long-term solution for cancer treatment. The anthocyanins found in coffee husks, specifically cyanidin-3-glucoside and cyanidin-3-rutinoside, were the subject of research conducted by Fu et al. [86]. The study found that these substances had potent antioxidant qualities, which are vital in reducing oxidative stress, a major contributor to the development of cancer. In human umbilical vein endothelial cells exposed to oxidative stress, the study showed that coffee husk anthocyanins could significantly increase the activity of superoxide dismutase and reduce markers of oxidative damage, such as malondialdehyde and lactate dehydrogenase. These results demonstrate the coffee husks' potential as a natural source of anticancer compounds. The strong anti-inflammatory, anti-proliferative, and antioxidant properties of these chemicals make coffee by-products a prospective tool for the creation of natural anticancer treatments.

6.1.3 Cosmetic Industry

The cosmetic industry has embraced the use of bioactive compounds from coffee by-products, particularly for their antioxidant and anti-aging properties. Coffee silverskin, for instance, is rich in polyphenols and flavonoids, which are incorporated into skincare products to help reduce the signs of aging by neutralizing free radicals and improving skin elasticity [76]. Iriondo-DeHond et al. [77] reported that the extract from coffee silverskin exhibited exceptional antioxidant properties, shielding skin cells from oxidative damage caused by tert-butyl hydroperoxide (t-BOOH). Similarly, the cosmetic potential of coffee cherry extract was investigated. The high phenolic content and antioxidant activity of Arabica coffee cherry extracts, especially those made by ethyl acetate extraction, were emphasized in a study conducted by Kiattisin et al. [78]. Strong anti-tyrosinase activity was demonstrated by these extracts, which is important for lightening skin and decreasing hyperpigmentation. Beyond coffee cherries and silverskin, coffee pulp is another valuable by-product that has several uses. Tseng et al. [79] investigated the effects of coffee pulp extract, which is utilized as a cosmetic ingredient in serums. According to the study, applying coffee pulp extract topically and consuming it regularly greatly increased antioxidant status, decreased tyrosinase activity, and improved several skin characteristics, including elasticity, brightness, moisture content, and collagen production.

Novel delivery technologies, like nanoliposomes, have been developed to improve the stability and efficacy of bioactive compounds found in coffee by-products. Coffee berry extract was investigated by Saewan et al. [80] as a way to encapsulate coffee berry extract and increase the stability and bioavailability of the active ingredients. In human dermal fibroblasts, the study showed that liposomal coffee berry extracts improved protection against oxidative stress and improved skin flexibility and lightening in clinical trials. The anti-aging potential of coffee berry extract was assessed in human dermal fibroblast and hair follicle dermal papilla cells [87]. The findings showed that coffee berry extract increased the activity of the superoxide dismutase and decreased the activity of collagenase and nitric oxide synthesis, two important processes in delaying the aging process of the skin. These results highlight the benefits of coffee berry extract in skincare and hair care products that aim to prevent aging. All this research on coffee silverskin, cherry, pulp, and berry extracts has repeatedly shown their benefits for antioxidant defense against aging and hair growth, which makes them useful components for the development of skincare and hair care products of the future. The importance of coffee by-products in the cosmetics sector is expected to grow as customer demand for eco-friendly and efficacious cosmetics rises, hence improving consumer health and environmental sustainability.

6.2 Biofuels for Renewable Energy

The production of biofuel from coffee by-products is a promising way to improve renewable energy sources and solve the environmental issues related to coffee waste. Due to their high content of carbohydrates, lipids, and other organic materials, these by-products, viz., coffee husk, SCGs, and coffee pulp, can be converted into a variety of biofuels, including biodiesel, bioethanol, biogas, and solid biofuels like briquettes and pellets. Because of their high lipid content, SCGs have been thoroughly researched as a possible feedstock for biodiesel production. According to Battista et al. [88], biodiesel yield can be greatly increased by optimizing the transesterification process, especially by modifying the ratio of coffee oil to methanol and utilizing the right catalyst. By utilizing 2% NaOH as a catalyst and a coffee oil to methanol ratio of 1:8, their study produced a high biodiesel output of 86%. To reduce water consumption and support a more sustainable production process, this procedure also involved the creative processing of SCGs using glycerol, a by-product of the production of biodiesel. Further investigation on the technical, financial, and environmental feasibility of producing biodiesel from SCGs was conducted [89].

Coffee pulp and coffee husk, the primary wastes generated during coffee processing, are rich in lignocellulosic biomass, making them ideal substrates for bioethanol production. To maximize sugar release, Do Viet et al. [90] studied the production of bioethanol from Robusta coffee pulp using enzymatic hydrolysis and alkaline pretreatment. The pretreated coffee pulp was effectively turned into fermentable sugars, and *Saccharomyces cerevisiae* was then used to ferment the sugars to make bioethanol. Similarly, Divyashri et al. [53] investigated the valorization of coffee husk for bioethanol production, comparing different hydrolysis methods. In terms of sugar yield, they found that enzymatic hydrolysis performed better than thermochemical techniques. Production of biogas from coffee waste is a feasible waste-to-energy route, especially from coffee husk and SCGs. Coffee husk and SCGs have a great potential for producing biogas with methane outputs ranging from 225.45 to 329.50 m^3/kg of fresh matter [91].

To optimize the value derived from these coffee by-products and materials, the idea of an integrated biorefinery for coffee by-products has drawn considerable interest. A comprehensive biorefinery model that produces value-added products including biopolymers, antioxidants, and biosurfactants in addition to biofuels like biodiesel, bioethanol, and biogas was proposed [92]. This strategy highlights how crucial it is to create multi-stage biorefineries that can transform coffee by-products into a wide range of beneficial products, improving their economic feasibility and lessening their environmental impact.

6.3 Agricultural Applications

Coffee husk, SCGs, and other coffee by-products can be used in agriculture as an environmentally beneficial and sustainable way to improve soil quality, encourage plant development, and efficiently manage waste. These materials can be converted into useful soil additives, compost, biochar, and biofertilizers, all of which can enhance soil health and agricultural output. SCGs can be added to soil and used in composting operations since they are a good source of organic matter, nitrogen, and other vital elements. Cervera-Mata et al. [93] reported that SCGs can be processed into a variety of by-products, including hydrochar and biochar, each of which has a unique effect on the dynamics of soil nitrogen and carbon. According to the study, biochar produced at higher temperatures (270–400 °C) showed a more stable carbon structure, resulting in lower CO_2 emissions and higher carbon retention in soil, while untreated SCGs and hydrochar increased CO_2 emissions and decreased carbon retention in soil. The potential of SCGs in composting, where they function as organic material rich in nutrients, was further highlighted by Bomfim et al. [94]. The study examined the advantages of employing SCGs in composting, emphasizing how it might increase plant growth and soil fertility without adding harmful substances like heavy metals. They did, however, also highlight the necessity of addressing possible toxicity concerns, especially at high concentrations, which can be remedied by using appropriate composting techniques that break down dangerous materials like caffeine and tannins. The application of coffee-processing leftovers as a potassium (K) fertilizer was investigated [95]. Their research revealed that the high K content of coffee husk and other wastes can be efficiently released into the soil to improve soil fertility. The gradual breakdown of these leftovers guarantees a continuous supply of potassium, which makes them a good substitute for traditional mineral fertilizers.

SCGs have been studied for their potential as a natural bioherbicide in addition to their use as a soil amendment. The usefulness of several agricultural wastes, such as SCGs, as weed control agents in crops was evaluated [96]. According to the study, SCGs effectively decreased the biomass of weeds that naturally emerged while fostering crop development. Because SCGs serve as both a soil supplement and a weed control agent, it has the potential to lessen the need for synthetic pesticides and promote more resilient and sustainable farming methods. It has also been demonstrated that applying composted coffee waste benefits soil microbial communities, especially soil nematodes. The above-mentioned applications of coffee by-products and wastes support sustainable agricultural practices, increase soil fertility, promote plant development, and aid in the recycling of agricultural waste.

7 Summary and Conclusion

Valorizing coffee bean manufacturing wastes and grounds has become an important way to improve the coffee industry's sustainability and lessen its environmental impact. The manufacture of coffee, an industry of worldwide importance, results in a considerable number of solid wastes, including husk, pulp, SCGs, and other by-products. If these wastes are not appropriately managed, they cause serious environmental problems. These by-products, however, are now acknowledged as important resources that can be used to create a range of bio-based goods, such as biofuels and bioactive substances for use in food, medicine, cosmetics, and agriculture. Silverskin, husk, and pulp are coffee by-products that are high in bioactive substances such as flavonoids, polyphenols, and chlorogenic acid. These substances are useful for the food, medicine, and cosmetic industries because of their strong anti-inflammatory, anti-cancer, and antioxidant characteristics. One viable method for producing renewable energy is turning coffee by-products into biofuels. Particularly, SCGs have proven to be an efficient source of biodiesel, and coffee pulp and husk are perfect for producing bioethanol. By converting waste into useful energy resources, the integration of these processes into a biorefinery model improves the coffee industry's economic competitiveness and environmental sustainability. Coffee by-products offer a lot of potential in agriculture as natural bioherbicides, biochar, and soil additions. They provide sustained weed control, increased microbial activity, and better soil fertility. The potential of coffee by-products will be further enhanced by the integration of cutting-edge technologies and creative processing techniques, establishing them as essential resources in the creation of a robust and sustainable global economy.

References

1. World Bank (2019) Global Coffee Market Report. https://documents.worldbank.org/en/publication/documents-reports/documentdetail/689681468337179379/coffee-production-consumption-and-trade
2. Smithsonian Magazine (2018) Coffee here, and coffee there: How different people serve the world's favorite hot drink. https://www.smithsonianmag.com/travel/coffee-here-and-coffee-there-how-different-people-serve-the-worlds-favorite-hot-drink-3214272/
3. Coffee Board of India (2020) Annual Coffee Statistics Report. https://coffeeboard.gov.in/AnnualReportsandAnnualAccounts.html
4. Specialty Coffee Association (2019) US Coffee Market Overview 2018. https://sca.coffee/sca-news/announcements/us-coffee-market-overview#:~:text=Overall%3A,consumption%20was%20US%2414%20billion*
5. International Coffee Organization (2021) Annual Review of the Coffee Market 2021. https://www.ico.org/documents/cy2022-23/annual-review-2021-2022-e.pdf
6. Divyashri G, Murthy TPK, Ragavan KV, Sumukh GM, Sudha LS, Nishka S, Venkataramanaiah RA (2023) Valorization of coffee bean processing waste for the sustainable extraction of biologically active pectin. Heliyon, 9(9)
7. Stufano P, Perrotta A, Labarile R, Trotta M (2022) The second life of coffee can be even more energizing: circularity of materials for bio-based electrochemical energy storage devices. MRS Energy Sustain 9(2):443–460
8. Lee YG, Cho EJ, Maskey S, Nguyen DT, Bae HJ (2023) Value-added products from coffee waste: a review. Molecules 28(8):3562
9. Blinová L, Sirotiak M (2019) Utilization of spent coffee grounds for removal of hazardous substances from water: a review. Research Papers Faculty of Materials Science and Technology Slovak University of Technology 27(44):145–152
10. Campos-Vega R, Loarca-Piña G, Vergara-Castañeda HA, Oomah BD (2015) Spent coffee grounds: a review on current research and future prospects. Trends Food Sci Technol 45(1):24–36
11. Klingel T, Kremer JI, Gottstein V, Rajcic de Rezende T, Schwarz S, Lachenmeier DW (2020) A review of coffee by-products including leaf, flower, cherry, husk, silver skin, and spent grounds as novel foods within the European Union. Food Secur 9(5):665
12. Murthy PS, Naidu MM (2012) Sustainable management of coffee industry by-products and value addition—a review. Resour Conserv Recycl 66:45–58
13. Janissen B, Huynh T (2018) Chemical composition and value-adding applications of coffee industry by-products: a review. Resour Conserv Recycl 128:110–117
14. Bolka M, Emire S (2020) Effects of coffee roasting technologies on cup quality and bioactive compounds of specialty coffee beans. Food Sci Nutr 8(11):6120–6130
15. Tassew AA, Yadessa GB, Bote AD, Obso TK (2022) Location, production systems, and processing method effects on qualities of Kafa biosphere reserve coffees. Agrosyst Geosci Environ 5(2):e20270
16. Arya SS, Venkatram R, More PR, Vijayan P (2022) The wastes of coffee bean processing for utilization in food: a review. Journal of Food Science and Technology, 59(2), 429–444
17. Tsegay G, Redi-Abshiro M, Chandravanshi BS, Ele E, Mohammed AM, Mamo H (2020) Effect of altitude of coffee plants on the composition of fatty acids of green coffee beans. BMC Chem 14:1–11
18. Firdissa E, Mohammed A, Berecha G, Garedew W (2022) Coffee drying and processing method influence quality of arabica coffee varieties (*Coffee arabica* L.) at Gomma I and limmu Kossa, Southwest Ethiopia. J Food Qual 2022(1):9184374
19. Mengistu MW, Workie MA, Mohammed AS (2020) Biochemical compounds of Arabica coffee (*Coffea arabica* L.) varieties grown in northwestern highlands of Ethiopia. Cogent Food Agric 6(1):1741319

20. Tassew AA, Yadessa GB, Bote AD, Obso TK (2021) Influence of location, elevation gradients, processing methods, and soil quality on the physical and cup quality of coffee in the Kafa biosphere reserve of SW Ethiopia. Heliyon 7(8):e07790
21. Alemayehu Tassew A, Berecha Yadessa G, Debela Bote A, Obso TK (2022) The coffee processing method had a more pronounced effect than location and production systems on the overall quality of Kaffa biosphere reserve coffees. Int J Agron 2022(1):9913578
22. Febrianto NA, Zhu F (2023) Coffee bean processing: emerging methods and their effects on chemical, biological and sensory properties. Food Chem 412:135489
23. Pereira LL, Júnior DB, de Sousa LHBP, dos Santos Gomes W, Cardoso WS, Guarçoni RC, ten Caten CS (2021) Relationship between coffee processing and fermentation. In: Quality determinants in coffee production. Springer, Cham, pp 255–301
24. Wulandari S, Ainuri M, Sukartiko AC (2021) Biochemical content of Robusta coffees under fully-wash, honey, and natural processing methods. IOP Conf Ser Earth Environ Sci 819:012067
25. Sunarharum WB, Williams DJ, Smyth HE (2014) Complexity of coffee flavor: a compositional and sensory perspective. Food Res Int 62:315–325
26. Makiso MU, Tola YB, Ogah O, Endale FL (2024) Bioactive compounds in coffee and their role in lowering the risk of major public health consequences: a review. Food Sci Nutr 12(2):734–764
27. Torres-Valenzuela LS, Serna-Jiménez JA, Martínez K (2020) Coffee by-products: nowadays and perspectives. In: Coffee—production and research. IntechOpen, pp 1–18
28. Cao X, Wu H, Viejo CG, Dunshea FR, Suleria HA (2023) Effects of postharvest processing on aroma formation in roasted coffee–a review. Int J Food Sci Technol 58(3):1007–1027
29. Yadav KC, Subba R, Shiwakoti LD, Dhungana PK, Bajagain R, Chaudhary DK, Dahal RH (2021) Utilizing coffee pulp and mucilage for producing alcohol-based beverages. Fermentation 7(2):53
30. Corrêa CL, Penha EM, Freitas-Silva O, Luna AS, Gottschalk LM (2021) Enzymatic technology application on coffee co-products: a review. Waste Biomass Valor 12:3521–3540
31. Degefe G, Mengistou S, Mohammed S (2016) Physico chemical evaluation of coffee husk, wastes of enset (*Enset ventricosum*), vegetable and khat (*Catha edulis*) through vermicomposting employing an epigeic earthworm *Dendrobaena Veneta* (Rosa, 1886). Afr J Biotechnol 15(20):884–890
32. Dadi D, Daba G, Beyene A, Luis P, Van der Bruggen B (2019) Composting and co-composting of coffee husk and pulp with source-separated municipal solid waste: a breakthrough in valorization of coffee waste. Int J Recycl Org Waste Agric 8:263–277
33. San Martin Ruiz M, Reiser M, Kranert M (2020) Enhanced composting as a way to a climate-friendly management of coffee by-products. Environ Sci Pollut Res 27:24312–24319
34. Niglio S, Procentese A, Russo ME, Sannia G, Marzocchella A (2019) Investigation of enzymatic hydrolysis of coffee silverskin aimed at the production of butanol and succinic acid by fermentative processes. Bioenergy Res 12:312–324
35. Ravindran R, Jaiswal S, Abu-Ghannam N, Jaiswal AK (2017) Two-step sequential pretreatment for the enhanced enzymatic hydrolysis of coffee spent waste. Bioresour Technol 239:276–284
36. Silva NC, da Fonseca YA, de Camargos AB, Lima AL, Ribeiro MC, Gurgel LV, Lobo Baêta BE (2022) Pretreatment and enzymatic hydrolysis of coffee husk for the production of potentially fermentable sugars. J Chem Technol Biotechnol 97(3):676–688
37. Chiyanzy I, Brienzo M, García-Aparicio M, Agudelo R, Görgens J (2015) Spent coffee ground mass solubilisation by steam explosion and enzymatic hydrolysis. J Chem Technol Biotechnol 90(3):449–458
38. Jomnonkhaow U, Plangklang P, Reungsang A, Peng CY, Chu CY (2024) Valorization of spent coffee grounds through integrated bioprocess of fermentable sugars, volatile fatty acids, yeast-based single-cell protein and biofuels production. Bioresour Technol 393:130107

39. Pereira J, Cachinho A, de Melo MM, Silva CM, Lemos PC, Xavier AM, Serafim LS (2022) Enzymatic potential of filamentous fungi as a biological pretreatment for acidogenic fermentation of coffee waste. Biomol Ther 12(9):1284
40. Milić MD, Buntić AV, Mihajlovski KR, Ilić NV, Davidović SZ, Dimitrijević-Branković SI (2023) The development of a combined enzymatic and microbial fermentation as a viable technology for the spent coffee ground full utilization. Biomass Convers Biorefinery 13(8):6747–6759
41. Ribeiro IC, Teixeira RA, Rabello LL, Bringhenti JR, Nicolau Korres AM (2020) Composting and anaerobic digestion as biotechnological alternatives for the valorization of used coffee grounds on university campus. In: Universities as living labs for sustainable development: supporting the implementation of the sustainable development goals, pp 789–802
42. Dias ME, Oliveira GHD, Couto PT, Dussán KJ, Zaiat M, Ribeiro R, Tommaso G (2021) Anaerobic digestion of hydrothermal liquefaction wastewater from spent coffee grounds. Biomass Bioenergy 148:106030
43. Mahmoud E, Atabani AE, Badruddin IA (2022) Valorization of spent coffee grounds for biogas production: a circular bioeconomy approach for a biorefinery. Fuel 328:125296
44. Luz FC, Cordiner S, Manni A, Mulone V, Rocco V (2017) Anaerobic digestion of coffee grounds soluble fraction at laboratory scale: evaluation of the biomethane potential. Appl Energy 207:166–175
45. Tangmankongworakoon N (2019) An approach to produce biochar from coffee residue for fuel and soil amendment purpose. Int J Recycl Org Waste Agric 8:37–44
46. Tsai WT (2017) The potential of pyrolysing exhausted coffee residue for the production of biochar. In: Handbook of coffee processing by-products. Academic Press, pp 299–322
47. Al-Awadhi YM, Pradhan S, McKay G, Al-Ansari T, Mackey HR (2022) Coffee waste biochar: a widely available and low-cost biomass for producing carbonaceous water treatment adsorbents. Chem Eng Trans 92:319–324
48. Tongcumpou C, Usapein P, Tuntiwiwattanapun N (2019) Complete utilization of wet spent coffee grounds waste as a novel feedstock for antioxidant, biodiesel, and bio-char production. Ind Crop Prod 138:111484
49. Giorcelli M, Bartoli M (2019) Development of coffee biochar filler for the production of electrical conductive reinforced plastic. Polymers 11(12):1916
50. Banu JR, Kavitha S, Kannah RY, Kumar MD, Atabani AE, Kumar G (2020) Biorefinery of spent coffee grounds waste: viable pathway towards circular bioeconomy. Bioresour Technol 302:122821
51. Nguyen QA, Cho E, Trinh LTP, Jeong JS, Bae HJ (2017) Development of an integrated process to produce d-mannose and bioethanol from coffee residue waste. Bioresour Technol 244:1039–1048
52. Dadi D, Beyene A, Simoens K, Soares J, Demeke MM, Thevelein JM, Van der Bruggen B (2018) Valorization of coffee byproducts for bioethanol production using lignocellulosic yeast fermentation and pervaporation. Int J Environ Sci Technol 15:821–832
53. Divyashri G, Tulsi NP, Murthy TK, Shreyas S, Kavya R, Jaishree IK (2024) Valorization of coffee bean processing waste for bioethanol production: comparison and evaluation of mass transfer effects in fermentations using free and encapsulated cells of Saccharomyces cerevisiae. Bioprocess Biosyst Eng 47(2):169–179
54. Gazonato EC, Maia AAD, Moris VADS, Paiva JMFD (2019) Thermomechanical properties of corn starch based film reinforced with coffee ground waste as a renewable resource. Mater Res 22(2):e20190416
55. Thiagamani SMK, Nagarajan R, Jawaid M, Anumakonda V, Siengchin S (2017) Utilization of chemically treated municipal solid waste (spent coffee bean powder) as reinforcement in cellulose matrix for packaging applications. Waste Manag 69:445–454
56. Valdespino-León M, Calderón-Domínguez G, De La Paz Salgado-Cruz M, Rentería-Ortega M, Farrera-Rebollo RR, Morales-Sánchez E, Terrazas-Valencia F (2021) Biodegradable elec-

trosprayed pectin films: an alternative to valorize coffee mucilage. Waste Biomass Valorization 12:2477–2494
57. Bondam AF, da Silveira DD, dos Santos JP, Hoffmann JF (2022) Phenolic compounds from coffee by-products: extraction and application in the food and pharmaceutical industries. Trends Food Sci Technol 123:172–186
58. Chatzimitakos T, Athanasiadis V, Kotsou K, Palaiogiannis D, Bozinou E, Lalas SI (2023) Optimized isolation procedure for the extraction of bioactive compounds from spent coffee grounds. Appl Sci 13(5):2819
59. Haider K, Haider MR, Neha K, Yar MS (2020) Free radical scavengers: an overview on heterocyclic advances and medicinal prospects. Eur J Med Chem 204:112607
60. Silva MDO, Honfoga JNB, Medeiros LLD, Madruga MS, Bezerra TKA (2020) Obtaining bioactive compounds from the coffee husk (*Coffea arabica* L.) using different extraction methods. Molecules 26(1):46
61. Myo H, Khat-Udomkiri N (2022) Optimization of ultrasound-assisted extraction of bioactive compounds from coffee pulp using propylene glycol as a solvent and their antioxidant activities. Ultrason Sonochem 89:106127
62. Brzezińska R, Wirkowska-Wojdyła M, Piasecka I, Górska A (2023) Application of response surface methodology to optimize the extraction process of bioactive compounds obtained from coffee silverskin. Appl Sci 13(9):5388
63. Mitraka GC, Kontogiannopoulos KN, Batsioula M, Banias GF, Assimopoulou AN (2021) Spent coffee grounds' valorization towards the recovery of caffeine and chlorogenic acid: a response surface methodology approach. Sustain For 13(16):8818
64. Prandi B, Ferri M, Monari S, Zurlini C, Cigognini I, Verstringe S, Tedeschi T (2021) Extraction and chemical characterization of functional phenols and proteins from coffee (*Coffea arabica*) by-products. Biomol Ther 11(11):1571
65. Vandeponseele A, Draye M, Piot C, Chatel G (2021) Study of influential parameters of the caffeine extraction from spent coffee grounds: from brewing coffee method to the waste treatment conditions. Clean Technol 3(2):335–350
66. Loukri A, Sarafera C, Goula AM, Gardikis K, Mourtzinos I (2022) Green extraction of caffeine from coffee pulp using a deep eutectic solvent (DES). Appl Food Res 2(2):100176
67. Cantele C, Tedesco M, Ghirardello D, Zeppa G, Bertolino M (2022) Coffee silverskin as a functional ingredient in vegan biscuits: physicochemical and sensory properties and *in vitro* bioaccessibility of bioactive compounds. Food Secur 11(5):717
68. Özhamamcı İ (2024) Coffee silverskin as a fat replacer in chicken patty formulation and its effect on physicochemical, textural, and sensory properties. Appl Sci 14(15):6442
69. Bevilacqua E, Cruzat V, Singh I, Rose'Meyer RB, Panchal SK, Brown L (2023) The potential of spent coffee grounds in functional food development. Nutrients 2023(15):994
70. Araya-Morice A, Araya-Quesada Y, Cortés N, Caamaño J, Arroyo L (2023) Antioxidant potential of coffee husks in fresh pork sausage. J Food Sci Technol 60(9):2423–2432
71. Magoni C, Bruni I, Guzzetti L, Dell'Agli M, Sangiovanni E, Piazza S, Labra M (2018) Valorizing coffee pulp by-products as anti-inflammatory ingredient of food supplements acting on IL-8 release. Food Res Int 112:129–135
72. Abbasi-Parizad P, De Nisi P, Scaglia B, Scarafoni A, Pilu S, Adani F (2021) Recovery of phenolic compounds from agro-industrial by-products: evaluating antiradical activities and immunomodulatory properties. Food Bioprod Process 127:338–348
73. Zengin G, Sinan KI, Mahomoodally MF, Angeloni S, Mustafa AM, Vittori S, Caprioli G (2020) Chemical composition, antioxidant and enzyme inhibitory properties of different extracts obtained from spent coffee grounds and coffee silverskin. Food Secur 9(6):713
74. Mesquita Júnior GAD, da Costa YFG, Mello VD, Costa FF, Rodarte MP, Costa JDCD, Vilela FMP (2022) Chemical characterisation by UPLC-Q-ToF-MS/MS and antibacterial potential of *Coffea arabica* L. leaves: a coffee by-product. Phytochem Anal 33(7):1036–1044

75. Badr AN, El-Attar MM, Ali HS, Elkhadragy MF, Yehia HM, Farouk A (2022) Spent coffee grounds valorization as bioactive phenolic source acquired antifungal, anti-mycotoxigenic, and anti-cytotoxic activities. Toxins 14(2):109
76. Hayes C, Nurkolis F, Laksemi DAAS, Chung S, Park MN, Choi M, Kim B (2023) Coffee silverskin phytocompounds as a novel anti-aging functional food: a Pharmacoinformatic approach combined with in vitro study. Molecules 28(20):7037
77. Iriondo-DeHond A, Martorell P, Genovés S, Ramón D, Stamatakis K, Fresno M, Del Castillo MD (2016) Coffee silverskin extract protects against accelerated aging caused by oxidative agents. Molecules 21(6):721
78. Kiattisin et al (2019) Antioxidant, anti-tyrosinase, anti-aging potentials and safety of arabica coffee cherry extract. Chiang Mai J Sci 46(5):930–945
79. Tseng YP, Liu C, Chan LP, Liang CH (2022) Coffee pulp supplement affects antioxidant status and favors anti-aging of skin in healthy subjects. J Cosmet Dermatol 21(5):2189–2199
80. Saewan N, Jimtaisong A, Panyachariwat N, Chaiwut P (2023) In Vitro and in vivo anti-aging effect of coffee berry nanoliposomes. Molecules 28(19):6830
81. Dauber C, Romero M, Chaparro C, Ureta C, Ferrari C, Lans R, Vieitez I (2024) Cookies enriched with coffee silverskin powder and coffee silverskin ultrasound extract to enhance fiber content and antioxidant properties. Appl Food Res 4(1):100373
82. Iriondo-DeHond A, Rodríguez Casas A, Del Castillo MD (2021) Interest of coffee melanoidins as sustainable healthier food ingredients. Front Nutr 8:730343
83. Ballesteros LF, Teixeira JA, Mussatto SI (2014) Chemical, functional, and structural properties of spent coffee grounds and coffee silverskin. Food Bioprocess Technol 7:3493–3503
84. Iriondo-DeHond A, Garcia NA, Fernandez-Gomez B, Guisantes-Batan E, Escobar FV, Blanch GP, del Castillo MD (2019) Validation of coffee by-products as novel food ingredients. Innov Food Sci Emerg Technol 51:194–204
85. Díaz-Hernández GC, Alvarez-Fitz P, Maldonado-Astudillo YI, Jiménez-Hernández J, Parra-Rojas I, Flores-Alfaro E, Ramírez M (2022) Antibacterial, antiradical and antiproliferative potential of green, roasted and spent coffee extracts. Appl Sci 12(4):1938
86. Fu XP, Shen XJ, Yin X, Zhang YH, Wang XF, Han ZH, Fan JP (2021) Antioxidant and pro-apoptosis activities of coffee husk (*Coffea arabica*) anthocyanins. Int Food Res J 28(6):1187–1195
87. Saewan N (2022) Effect of coffee berry extract on anti-aging for skin and hair—in vitro approach. Cosmetics 9(3):66
88. Battista F, Zuliani L, Rizzioli F, Fusco S, Bolzonella D (2021) Biodiesel, biogas and fermentable sugars production from spent coffee grounds: a cascade biorefinery approach. Bioresour Technol 342:125952
89. Kamil M, Ramadan KM, Olabi AG, Al-Ali EI, Ma X, Awad OI (2020) Economic, technical, and environmental viability of biodiesel blends derived from coffee waste. Renew Energy 147:1880–1894
90. Do Viet P, Le Pham TQ, Le Nguyen DD (2019) Production of bioethanol from Robusta coffee pulp (*Coffea robusta* L.) in Vietnam. Foods Raw Mater 7(1):10–17
91. Czekała W, Łukomska A, Pulka J, Bojarski W, Pochwatka P, Kowalczyk-Juśko A, Dach J (2023) Waste-to-energy: biogas potential of waste from coffee production and consumption. Energy 276:127604
92. Atabani AE, Ala'a H, Kumar G, Saratale GD, Aslam M, Khan HA, Mahmoud E (2019) Valorization of spent coffee grounds into biofuels and value-added products: pathway towards integrated bio-refinery. Fuel 254:115640
93. Cervera-Mata A, Delgado G, Fernández-Arteaga A, Fornasier F, Mondini C (2022) Spent coffee grounds by-products and their influence on soil C–N dynamics. J Environ Manag 302:114075
94. Bomfim ASCD, De Oliveira DM, Walling E, Babin A, Hersant G, Vaneeckhaute C, Rodrigue D (2022) Spent coffee grounds characterization and reuse in composting and soil amendment. Waste 1:2–20
95. Zoca SM, Penn CJ, Rosolem CA, Alves AR, Neto LO, Martins MM (2014) Coffee processing residues as a soil potassium amendment. Int J Recycl Org Waste Agric 3:155–165
96. Lorenzo P, Guilherme R, Barbosa S, Ferreira AJ, Galhano C (2022) Agri-food waste as a method for weed control and soil amendment in crops. Agronomy 12(5):1184

Chapter 9
Biotechnological Approaches for Cocoa Bean Processing Waste Valorization: A Review

Alfred Elikem Kwami Afedzi, Mark-Jefferson Buer Boyetey, and Hellie Gonu

Abbreviation

AD	Anaerobic digestion
CBH	Cocoa bean shells
CPH	Cocoa pod husks
GMOs	Genetically modified organisms
ICCO	International Cocoa Organization
LCA	Life cycle assessment
NGOs	Nongovernmental organizations
PHAs	Polyhydroxyalkanoates
PHB	Poly(3-hydroxybutyrate)
SDGs	Sustainable development goals
UNFCCC	United Nations Framework Convention on Climate Change

A. E. K. Afedzi (✉)
Fermentation Technology Research Center, Department of Biotechnology, Faculty of Agro-Industry, Kasetsart University, Bangkok, Thailand
e-mail: alfredelikemkwami.a@ku.th

M.-J. B. Boyetey
Cellulose for Future Materials and Technologies Special Research Unit, Department of Biotechnology, Kasetsart University, Bangkok, Thailand

H. Gonu
Fermentation Technology Research Center, Department of Biotechnology, Faculty of Agro-Industry, Kasetsart University, Bangkok, Thailand

Department of Agricultural Biotechnology, Faculty of Agriculture, Food and Consumer Sciences, University for Development Studies, Tamale, Ghana

Ghana EXIM Bank, Accra, Ghana

C. Trois, R. C. Ray (eds.), *Plantation Crop Wastes: Valorization for Economic Sustainability*, Waste as a Resource,
https://doi.org/10.1007/978-3-031-96805-1_9

1 Introduction

The cocoa industry is a crucial economic sector worldwide, with an estimated annual production of around five million metric tons of cocoa beans. Cocoa beans are obtained from cocoa trees (*Theobroma cacao* L.), which are primarily cultivated in tropical regions, with West African countries like Côte d'Ivoire and Ghana accounting for approximately 60% of the world's production [1]. Cocoa beans are used by the food and beverage industry for products such as chocolate, cocoa powder, and beverages, and in the cosmetic industry cocoa butter—all of which contribute to a multibillion-dollar global market.

However, in addition to the increasing demand for cocoa-based products, the cocoa industry generates around 700,000 tons of waste each year, creating a significant inefficiency and strain on land resources in its present form [2]. The disposal or underutilization of these wastes, including cocoa pod husks, bean shells, and cocoa sweatings, contribute to environmental pollution and lost economic opportunities [3–5]. The inadequacy of waste management infrastructure in numerous cocoa-producing regions significantly exacerbates environmental challenges, undermining sustainable agricultural practices and contributing to issues such as soil degradation, deforestation, and ecosystem pollution [6]. The accumulation of cocoa processing waste contributes to greenhouse gas emissions, particularly carbon dioxide and methane, both of which have a global warming potential. Effectively addressing these environmental challenges is crucial for mitigating climate change and ensuring the sustainability of cocoa production systems. In addition to environmental impacts, cocoa processing waste represents a loss of valuable biomass that could be harnessed for economic gain. The wastes are rich in bioactive compounds, fibers, and nutrients, making them suitable for conversion into value-added products [7]. Valorizing these wastes can create new revenue streams, enhance rural economies, and contribute to poverty alleviation in cocoa-producing communities. For effective waste management, there has been a growing interest in valorizing cocoa processing waste through biotechnological approaches.

Valorization involves converting waste materials into economically viable products, thus fostering a circular economy characterized by waste reduction and effective resource utilization. Biotechnological approaches offer innovative solutions for the valorization of cocoa processing waste. Techniques such as microbial fermentation, enzymatic hydrolysis, and bioconversion enable the transformation of waste materials into biofuels, bioplastics, animal feed, and nutraceuticals. For instance, microbial fermentation can produce bioethanol from cocoa pod husks, providing a renewable energy source and reducing reliance on fossil fuels [8]. Enzymatic treatments can extract valuable antioxidants and polyphenols, which have applications in the food and pharmaceutical industries [9, 10]. Integrating these biotechnological methods aligns with the principles of the circular economy, promoting resource efficiency and waste minimization. By converting waste into valuable products, it is possible to reduce environmental impacts, mitigate climate change, and improve the livelihoods of rural communities [11, 12]. Moreover, such approaches can enhance the resilience of the cocoa sector by diversifying income sources and promoting sustainable practices.

This chapter presents a thorough review of biotechnological approaches for valorizing waste from cocoa bean processing, with a focus on environmental and socioeconomic impacts. It highlights the potential of cocoa processing waste as a resource within a circular economy, as well as the traditional and current innovations in waste utilization. Furthermore, this chapter discusses challenges and opportunities in scaling biotechnological methods, and the significance of policy and stakeholder involvement in ensuring the effectiveness of these initiatives.

2 Geographical Origin, Distribution, and Production of Cocoa Beans

2.1 Geographical Origin and Botanical Varieties

Cocoa beans are native to the tropical rainforests of the Amazon Basin and the Orinoco Valley of South America. Archaeological evidence suggests that cocoa was domesticated by indigenous peoples over 5000 years ago [13]. Indigenous cultures such as the Olmec, Maya, and Aztec civilizations cultivated cocoa and revered it for its ceremonial and economic significance [14]. The beans were used as currency and in rituals, highlighting their value in ancient societies. The term "*Theobroma*" aptly means "food of the gods," reflecting its esteemed status in ancient cultures. The Spanish conquest in the sixteenth century led to the introduction of cocoa to Europe, where it became a luxury commodity among the elite [15]. The increasing European demand prompted the establishment of cocoa plantations in tropical colonies, expanding its cultivation beyond its native range.

Cocoa is classified into three primary botanical varieties:

(a) **Criollo**: Known for its superior flavor and aroma, Criollo is considered the finest cocoa variety. However, it is less disease-resistant and yields lower than other varieties, accounting for less than 5% of global production [16].
(b) **Forastero**: The most widely cultivated variety, Forastero is hardy and disease-resistant, offering higher yields. It represents about 80% of global production and is predominantly grown in West Africa [17].
(c) **Trinitario**: A natural hybrid of Criollo and Forastero. It combines the desirable flavor qualities of Criollo with the robustness of Forastero. It accounts for the remaining share of global production [18].

2.2 Global Distribution and Major Producing Regions

Cocoa cultivation is confined to a geographical belt between 20° north and south of the equator, where climatic conditions of temperatures between 21 °C and 32 °C, high humidity, and annual rainfall of 1500–2000 mm are optimal [19].

(a) **West Africa** is the leading cocoa-producing region, contributing approximately 70% of global production. Côte d'Ivoire and Ghana are the top producers, accounting for around 2.2 million tons and 1.1 million tons, respectively, in the 2022/2023 season [20]. Nigeria and Cameroon also contribute significantly.
(b) **Latin America**, the birthplace of cocoa, accounts for about 17% of global production. Ecuador leads in this region, producing around 337,149 tons annually, and is renowned for its fine-flavor Arriba Nacional cocoa [20]. Other contributors include Peru, Brazil, and the Dominican Republic.
(c) **Asia and Oceania** contribute roughly 13% to global cocoa production. Indonesia is the primary producer here, with an output of about 667,296 tons [20]. Papua New Guinea, Malaysia, and the Philippines are also notable producers.

2.3 Cocoa Production: Its Socioeconomic and Environmental Roles

Global cocoa production has been on an upward trend due to increasing demand for chocolate and cocoa-derived products. According to the International Cocoa Organization (ICCO), the 2022/23 cocoa year recorded a production of approximately 5.1 million metric tons although it was forecasted to decline to about 4.3 million metric tons as presented in Table 9.1. This decline is attributed to cocoa prices reaching record levels as production falters due to climate change, disease, lack of investment, and the illegal mining crisis known as "galamsey" in countries such as Ghana [21].

Cocoa farming is predominantly undertaken by smallholder farmers, who account for 80–90% of global production [22]. Farm sizes typically range from 2 to 5 ha.

(a) **Agroforestry Systems**: Traditional cocoa cultivation involves agroforestry practices, intercropping cocoa with shade trees like banana, plantain, and timber species. This system enhances biodiversity, improves soil fertility, and aids in carbon sequestration. Shade trees also provide additional income through the sale of fruits and timber.
(b) **Monoculture Plantations**: To maximize yields, some farmers adopt full-sun monoculture systems. While this increases short-term productivity, it can lead

Table 9.1 Production of cocoa beans (thousand tons) (data from International Cocoa Organization (ICCO), https://www.icco.org/statistics/)

Region	2021/22		Estimates 2022/23		Forecasts 2023/24	
Africa	3589	74%	3719	73.7%	3042	70.2%
Cameroon	295		270		300	
Côte d'Ivoire	2121		2241		1740	
Ghana	683		654		450	
Nigeria	280		315		300	
Others	210		240		252	
Americas	973	20%	1077	21.3%	1035	23.9%
Brazil	220		220		200	
Ecuador	365		454		430	
Others	388		403		405	
Asia and Oceania	265	5%	253	5.0%	255	5.9%
Indonesia	180		160		160	
Papua New Guinea	42		43		45	
Others	43		50		50	
World total	4826	100%	5050	100%	4332	100%

to soil degradation, increased pest and disease incidence, and loss of biodiversity [23].

2.3.1 Socioeconomic Role

Cocoa cultivation is a cornerstone of rural economies in producing countries, supporting the livelihoods of over five million farming households worldwide accounting for 80–90% of global production [22]. In West Africa, smallholder farmers cultivate about 90% of the cocoa produced, typically on farms less than 5 ha. The role of cocoa production in rural economies encompasses the following aspects:

(a) **Employment:** Cocoa farming and related activities provide employment opportunities in rural areas, reducing urban migration.
(b) **Income Generation:** Cocoa is a major export commodity, contributing significantly to national revenues. For example, it accounts for about 15% of Côte d'Ivoire's Gross Domestic Product [24].
(c) **Community Development:** Revenues from cocoa production fund infrastructure projects, education, and healthcare services.

However, cocoa farmers face challenges such as price volatility, pests and diseases, aging tree stocks, and limited access to financing and technology. These issues affect productivity and income stability, highlighting the need for diversification strategies like waste valorization.

2.3.2 Environmental Considerations

The expansion of cocoa cultivation has environmental implications, including deforestation and loss of biodiversity, particularly when forests are cleared for new plantations [25]. Sustainable farming practices, such as agroforestry systems that integrate shade trees with cocoa plants, are being promoted to mitigate these impacts. Waste generated from cocoa processing also poses environmental challenges. Accumulation of cocoa pod husks and other residues can lead to methane emissions and pollution if not managed properly. Valorization of this waste through biotechnological approaches offers a solution that addresses environmental concerns while providing economic benefits.

3 Characteristics and Composition of Cocoa Bean Processing Waste

Cocoa bean processing generates substantial waste during various stages, thereby posing environmental challenges if not properly managed. The primary waste materials include cocoa pod husks (CPH), cocoa bean shells (CBS), and cocoa sweatings (Fig. 9.1), while other by-products include cocoa dust and rejected beans. Understanding the characteristics and composition of these wastes is crucial for developing effective valorization strategies. Cocoa pod husks constitute about 70–75% of the whole fruit's weight and are the outer shells discarded after the extraction of beans for fermentation [26]. Cocoa bean shells are the outer coverings of cocoa beans, removed during the roasting and winnowing processes, accounting for approximately 12–20% of the bean's weight [27]. Cocoa sweatings are liquid by-products rich in sugars, obtained during the fermentation process of cocoa beans.

3.1 Proximate Composition

The proximate composition of cocoa processing wastes (Table 9.2) varies depending on the type and processing conditions. These wastes are rich in organic matter, making them suitable candidates for valorization through biotechnological approaches.

Cocoa pod husks are characterized by high fiber content, primarily composed of cellulose, hemicellulose, and lignin, which make up the structural components of the husk. The high fiber content presents opportunities to produce dietary fibers as a substrate in fermentation processes [35]. Cocoa bean shells, on the other hand, have higher protein and lipid contents compared to pod husks, along with significant amounts of theobromine and other alkaloids [36]. These characteristics make

Fig. 9.1 Product and waste generation along the cocoa processing chain (from farm to final product)

Table 9.2 Proximate composition of cocoa processing wastes

Component	Cocoa pod husks (%)	Cocoa bean shells (%)
Moisture	7–14.4	5–8
Ash	8–2	5–7.6
Crude fiber	11.2–40	25–37.5
Crude protein	6–16.3	15–18
Fat	0.9–5.4	3–19.7
Carbohydrates	Predominantly cellulose, hemicellulose, and lignin (41.9)	Contains theobromine and other alkaloids (14.1%)

Note: Data retrieved from Adeyi [28], Adjin-Tetteh et al. [29], Akinola et al. [30], Dos Anjos Lopes et al. [31], Forero-Nuñez et al. [32], Maliki et al. [33], and Vargas-Arana et al. [34]

CBS a valuable resource for extracting bioactive compounds and developing functional food ingredients.

3.2 Nutritional Values and Bioactive Compounds

Cocoa processing wastes are not only rich in macronutrients but also contain numerous bioactive compounds with potential health benefits. Cocoa pod husks are a good source of minerals such as potassium, calcium, magnesium, and phosphorus [37]. They also contain phenolic compounds with antioxidant properties, which can be harnessed for nutraceutical applications [38, 39]. The high mineral content makes CPH suitable for use as organic fertilizers or soil amendments, contributing to sustainable agricultural practices.

Cocoa bean shells are particularly rich in proteins and dietary fibers. They contain essential amino acids like leucine, phenylalanine, and valine, which are important for human nutrition. Additionally, CBS is a valuable source of methylxanthines such as theobromine and caffeine, and polyphenols like catechins and epicatechins, which have antioxidant and anti-inflammatory properties [40]. These compounds have potential applications in the food, pharmaceutical, and cosmetic industries. Cocoa sweatings, the liquid exudate from cocoa beans during fermentation, are rich in fermentable sugars such as glucose, fructose, and sucrose, making them suitable substrates for microbial fermentation processes [41]. The high sugar content can be utilized to produce bioethanol, vinegar, and other fermented products, adding value to this otherwise discarded by-product.

4 Bio-Valorization of Cocoa Bean Processing Waste in Circular Economy

The concept of a circular economy has emerged as a sustainable alternative to the traditional linear model of "take-make-dispose." In a circular economy, resources are kept in use for as long as possible, extracting maximum value before recovering and regenerating products at the end of their service life [42]. Cocoa production generates substantial amounts of waste, and biotechnological approaches offer innovative solutions to transform these wastes into valuable products, aligning with the principles of a circular economy and contributing to climate change mitigation and rural economic development. Biotechnology utilizes living organisms or their components to develop products and processes for specific uses. In the context of cocoa processing waste, biotechnological methods harness the metabolic capabilities of microorganisms and enzymes to convert organic waste into biofuels, biochemicals, and biomaterials. Furthermore, understanding the comprehensive composition of cocoa processing wastes is essential for designing effective biotechnological processes for their valorization. The rich nutritional profile and presence of bioactive compounds make these wastes valuable resources for producing a wide range of products, from biofuels to functional foods and cosmetics. Addressing the challenges associated with their utilization will enhance the sustainability of the cocoa industry and contribute to environmental conservation and economic development in cocoa-producing regions.

4.1 Microbial Fermentation

Microbial fermentation is a metabolic process carried out by microorganisms such as bacteria, yeasts, or fungi, in which they convert organic compounds, typically carbohydrates like sugars and starches into simpler substances like alcohols, gases, or organic acids [43]. This transformation usually occurs under anaerobic (oxygen-free) conditions. Cocoa processing wastes are rich in carbohydrates, proteins, and other nutrients, making them suitable substrates for microbial growth and fermentation.

4.1.1 Bioethanol Production from Cocoa Pod Husks

Bioethanol production from lignocellulosic biomass such as cocoa pod husks represents a promising avenue for renewable energy generation. The process begins with:

(a) **Pretreatment:** The complex lignocellulosic structure CPH is broken down, enhancing the accessibility of cellulose and hemicellulose. Pretreatment methods include dilute acid hydrolysis, steam explosion, and alkaline treatments, each aiming to disrupt the rigid matrix and increase enzyme accessibility.
(b) **Enzymatic Hydrolysis:** This employs cellulases and hemicellulases to convert cellulose and hemicellulose into fermentable sugars like glucose and xylose. The efficiency of this step is critical, as it directly impacts the yield of fermentable sugars. Factors such as enzyme concentration, temperature, pH, and incubation time must be optimized to maximize sugar release [44].

(c) **Fermentation:** This utilizes microorganisms, typically yeast strains like *Saccharomyces cerevisiae*, to convert the sugars into ethanol under anaerobic conditions. However, *S. cerevisiae* primarily ferments hexose sugars and has limited ability to ferment pentose sugars derived from hemicellulose. To address this limitation, co-fermentation strategies using pentose-fermenting yeasts such as *Pichia stipitis* or genetically engineered strains have been explored [45].

Studies have demonstrated the feasibility of producing bioethanol from CPH with promising yields [46, 47]. Yadav et al. [48] reported significant ethanol production using optimized pretreatment and fermentation conditions, presenting the potential of this approach with a high fermentation efficiency (67.17%). Despite its potential, several challenges hinder the large-scale implementation of bioethanol production from CPH. The presence of inhibitory compounds formed during pretreatment, such as furfural and hydroxymethylfurfural (HMF), can impede microbial fermentation. Mitigation strategies include detoxification processes and the development of inhibitor-tolerant yeast strains [49]. Additionally, the high cost of enzymes necessitates the exploration of cost-effective enzyme production methods and the use of enzyme recycling techniques to enhance economic viability.

4.1.2 Biogas Production Through Anaerobic Digestion

Anaerobic digestion (AD) of cocoa processing waste offers an efficient method for biogas production, converting organic matter into a mixture of methane and carbon dioxide. The AD process involves a consortium of microorganisms working synergistically in four stages: hydrolysis, acidogenesis, acetogenesis, and methanogenesis [50]. Cocoa processing waste, with its high organic content, serves as an excellent substrate for AD. However, the lignin-rich nature of CPH can limit biodegradability. Pretreatment methods, such as mechanical grinding or thermal treatment, can enhance the digestibility of the biomass. Co-digestion with nitrogen-rich substrates like animal manure has also been shown to improve nutrient balance, enhance microbial activity, and increase biogas yields [51].

Biogas produced from cocoa processing waste can be utilized for cooking, heating, or electricity generation, providing an alternative energy source for rural communities. The digestate, a by-product of AD, is rich in nutrients and can be used as a biofertilizer, promoting sustainable agriculture. Implementation of small-scale biogas plants in cocoa-producing regions like Ghana [52], Brazil [53], and Nigeria [54] has demonstrated the dual benefits of waste management and energy production. Challenges in AD include maintaining process stability, as fluctuations in pH, temperature, or substrate composition can affect microbial activity. Inhibitory substances, such as ammonia or volatile fatty acids, can accumulate and impede methanogenesis. Addressing these issues requires careful monitoring and control of operating conditions, as well as further research into robust microbial consortia capable of withstanding process perturbations.

4.1.3 Production of Organic Acids and Other Metabolites

Microbial fermentation of cocoa processing waste can also yield valuable organic acids, such as lactic acid, which has applications in the food industry and biodegradable plastics production. Cocoa sweatings, rich in fermentable sugars, serve as an ideal substrate for lactic acid bacteria (*Lactobacillus* spp.). Bobiles et al. [55] successfully produced lactic acid from cocoa husk hydrolysates, highlighting the potential for industrial-scale production. Similarly, the production of citric acid, enzymes, and single-cell proteins from cocoa waste has been explored [56]. Fermentation using *Aspergillus niger* increased total crude protein and antioxidant properties while reducing the theobromine content [57]. These metabolites have diverse applications, ranging from food additives to animal feed supplements, and contribute to the diversification of products derived from cocoa processing waste.

4.2 Enzymatic Treatments and Extraction of Bioactive Compounds

Enzymatic treatments offer a selective and environmentally friendly approach to extracting valuable compounds from cocoa processing waste. Enzymes catalyze specific biochemical reactions under mild conditions, reducing the need for harsh chemicals and minimizing environmental impact.

4.2.1 Extraction of Pectin from Cocoa Pod Husks

Pectin, a polysaccharide used as a gelling agent in the food industry, can be extracted from CPH. The extraction process typically involves acid hydrolysis, where dilute acids solubilize pectic substances. Enzymatic treatments using pectinases can enhance extraction efficiency by breaking down complex polysaccharides and facilitating the release of pectin [58]. The extracted pectin from CPH has been found to possess comparable gelling properties to commercial pectin derived from citrus peels or apple pomace [59, 60]. Utilizing CPH as a pectin source not only adds value to cocoa waste but also offers an alternative raw material for the pectin industry, potentially reducing costs and supply constraints. Challenges in pectin extraction include variability in pectin yield and quality due to differences in cocoa varieties, growing conditions, and processing methods [61]. Optimizing extraction parameters, such as pH, temperature, and enzyme concentration, is essential to achieve consistent product quality suitable for industrial applications.

4.2.2 Recovery of Antioxidants and Polyphenols from Cocoa Bean Shells

Cocoa bean shells are rich in bioactive compounds, including polyphenols, flavonoids, theobromine, and caffeine, which exhibit antioxidant properties. Enzymatic treatments can enhance the extraction of these compounds by degrading cell wall components and releasing bound phenolics. Enzymes such as cellulases, hemicellulases, and pectinases are employed to facilitate this process [62]. The recovered antioxidants have potential applications in the food, pharmaceutical, and cosmetic industries as natural additives and nutraceutical ingredients. The use of environmentally friendly solvents like water or ethanol aligns with green chemistry principles and ensures the safety of the extracted compounds for human consumption. One of the main challenges is the stability of phenolic compounds, which can be affected by factors such as light, temperature, and pH [63]. Developing appropriate formulation and encapsulation techniques can enhance the stability and bioavailability of these compounds, expanding their commercial potential.

4.3 Genetic Engineering and Synthetic Biology Innovations

Advancements in genetic engineering and synthetic biology offer opportunities to enhance biotechnological processes for cocoa waste valorization. By modifying the genetic makeup of microorganisms, it is possible to improve their metabolic capabilities, substrate utilization, and product yields.

4.3.1 Engineered Microorganisms for Improved Fermentation

Traditional microbial strains used in fermentation may have limitations in substrate range, tolerance to inhibitors, or product yields. Genetic engineering allows for the development of strains with enhanced characteristics:

(a) **Recombinant Yeasts:** *Saccharomyces cerevisiae* engineered to express heterologous cellulases can directly ferment cellulose to ethanol, reducing process steps. A report indicates that engineered *S. cerevisiae* expressing genes for xylose utilization enables the fermentation of both hexose and pentose sugars from lignocellulosic hydrolysates, improving ethanol yields [64]. Similarly, modifications to increase tolerance to inhibitors formed during pretreatment processes can enhance fermentation efficiency. The overexpression of efflux pumps or detoxification enzymes helps microorganisms withstand the presence of compounds like furfural and 5-hydroxymethylfurfural (HMF) [65].

(b) **Bacteria with Expanded Substrate Range:** *Escherichia coli*, *Corynebacterium glutamicum*, and *Pseudomonas putida* modified to metabolize pentose sugars enhance the utilization of hemicellulose-derived sugars. The *P. putida* strain co-utilized glucose or cellobiose with xylose to completely deplete the sugars [66], while engineered *C. glutamicum* assimilated xylose and xylooligosaccharides [67].

While genetically engineered microorganisms offer significant advantages, regulatory hurdles and public acceptance remain challenges. Strict regulations govern the use of GMOs, particularly in food-related applications, necessitating rigorous safety assessments and compliance with legal frameworks.

4.3.2 Synthetic Biology for Novel Bioproducts

Synthetic biology combines principles of engineering and biology to design and construct new biological systems or redesign existing ones for useful purposes. In the context of cocoa waste valorization, synthetic biology can be applied to create microbial cell factories capable of producing high-value products like bioplastics, specialty chemicals, or pharmaceuticals from waste substrates.

Metabolic pathway engineering involves redirecting metabolic fluxes to increase the production of target metabolites. For example, metabolic pathways can be engineered to channel substrates toward the production of polyhydroxyalkanoates (PHAs)—biodegradable polymers with applications in packaging and biomedical devices [68]. By optimizing gene expression, enzyme activity, and metabolic fluxes, it is possible to enhance PHA production from cocoa processing waste. Engineered *Cupriavidus necator* DSM 545 and *Saccharomyces cerevisiae* Fm17 produced PHA from CPH [69], and *Bacillus firmus* CECT 14 produced poly(3-hydroxybutyrate) (PHB) from CBS [70].

The application of synthetic biology extends to the development of biosensors for monitoring fermentation processes or detecting contaminants in waste streams. Such innovations improve process control, efficiency, and safety in biotechnological applications. The challenges in synthetic biology include the complexity of biological systems, potential unintended consequences of genetic modifications, and ethical considerations. Ensuring biosafety and addressing public concerns are critical for the acceptance and success of synthetic biology applications.

Over the years, biotechnological approaches have advanced significantly in the utilization of the main waste components from the cocoa bean processing chain (cocoa pod husk, sweating, and bean shell) to produce value-added products, as summarized in Fig. 9.2.

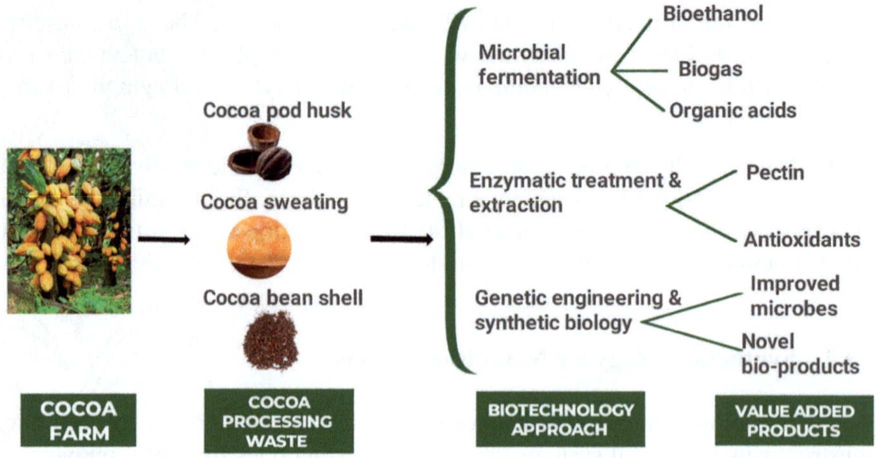

Fig. 9.2 Biotechnological approaches for cocoa processing waste valorization into value-added products

5 Applications and Products Derived from Cocoa Processing Waste

The valorization of cocoa processing waste not only addresses environmental concerns but also creates economic opportunities through the development of various applications and products. This section explores both traditional and innovative uses of cocoa waste, highlighting their significance in promoting sustainability and enhancing rural economies.

5.1 Traditional Uses of Cocoa Processing Waste

Historically, cocoa processing waste has been utilized in several conventional ways, primarily at a local or small-scale level. These traditional uses include animal feed, fertilizers, and compost, which contribute to waste reduction and resource recycling within agricultural systems.

5.1.1 Animal Feed

Cocoa pod husks (CPH) and cocoa bean shells (CBS) have been used as feed ingredients for livestock due to their availability and nutrient content. The fibrous nature of CPH makes it suitable as a roughage source in ruminant diets [71]. Studies have shown that incorporating CPH into animal feed can be beneficial when properly processed to reduce anti-nutritional factors such as theobromine, which can be toxic to animals at high levels [72].

Processing methods like ensiling, fermentation, and supplementation with enzymes can improve the nutritional value and digestibility of cocoa waste-based feeds. However, limitations exist due to the presence of compounds like theobromine and tannins, which require careful management to avoid adverse effects on animal health.

5.1.2 Fertilizers and Compost

Cocoa processing waste is rich in organic matter and essential nutrients such as potassium, calcium, and magnesium, making it suitable for use as organic fertilizer and compost. Composting cocoa waste enhances soil fertility, improves soil structure, and promotes sustainable agricultural practices [73, 74]. Composting involves the aerobic decomposition of organic matter by microorganisms, resulting in a stable humus-like product. The application of composted cocoa waste to cocoa plantations and other crops can reduce the reliance on chemical fertilizers, lower production costs, and mitigate environmental pollution associated with waste accumulation.

5.2 Innovations in Product Development

Advancements in technology and increased emphasis on sustainability have led to innovative applications of cocoa processing waste. These innovations harness the potential of cocoa waste to produce biofuels, biochar, food and feed additives, nutraceuticals, and cosmetics.

5.2.1 Biofuels and Bioenergy

The conversion of cocoa processing waste into biofuels presents a promising avenue for renewable energy production. As discussed in Sect. 4, bioethanol and biogas can be produced from CPH and cocoa sweatings through microbial fermentation and anaerobic digestion, respectively.

In addition to liquid and gaseous biofuels, direct combustion of cocoa waste for heat and power generation has been explored. Cocoa shells, with a higher calorific value due to their lipid content, can be used as a biomass fuel in boilers and furnaces [75]. Pyrolysis of cocoa waste to produce bio-oil and syngas is another area of interest, contributing to energy diversification and reducing dependence on fossil fuels.

5.2.2 Biochar and Soil Amendments

Biochar production from cocoa processing waste involves the thermal decomposition of biomass under limited oxygen conditions (pyrolysis). Biochar is a carbon-rich material that can be used as a soil amendment to improve soil fertility, water retention, and carbon sequestration [76]. The application of biochar derived from cocoa waste has been shown to enhance soil properties and increase crop yields [77, 78]. Furthermore, biochar production contributes to climate change mitigation by stabilizing carbon in soils for extended periods, reducing greenhouse gas emissions.

5.2.3 Food and Feed Additives

The extraction of valuable compounds from cocoa processing waste has led to the development of food and feed additives. Cocoa bean shells are a source of dietary fibers, antioxidants, and bioactive compounds that can be incorporated into food products to enhance nutritional value and functional properties [7]. In the feed industry, cocoa waste-derived additives can serve as natural growth promoters and health enhancers for livestock and aquaculture species [79, 80]. The incorporation of polyphenol-rich extracts from cocoa waste in animal diets has been associated with improved antioxidant status and immune function [81, 82].

5.2.4 Nutraceuticals and Cosmetics

The presence of bioactive compounds such as polyphenols, flavonoids, theobromine, and caffeine in cocoa processing waste opens opportunities in the nutraceutical and cosmetic industries. Extracts from cocoa waste exhibit antioxidant, anti-inflammatory, and anti-aging properties, making them suitable for use in dietary supplements and skin care products [83]. Cocoa polyphenols have been linked to cardiovascular health benefits, neuroprotective effects, and improved metabolic function [84, 85]. The utilization of cocoa waste in the development of nutraceuticals aligns with consumer trends favoring natural and sustainable ingredients. In cosmetics, cocoa butter extracted from cocoa bean shells can be used in lotions, creams, and soaps, providing moisturizing and emollient effects [86, 87]. The valorization of cocoa waste in these industries adds economic value and promotes waste minimization. Figure 9.3 illustrates the diverse applications of cocoa processing waste, categorized into conventional and novel uses.

5.3 Potential Challenges

While cocoa processing wastes offer numerous opportunities for valorization, various issues require attention. The presence of lignin and tannins in high amounts can affect the digestibility of these materials when used in animal feed, necessitating

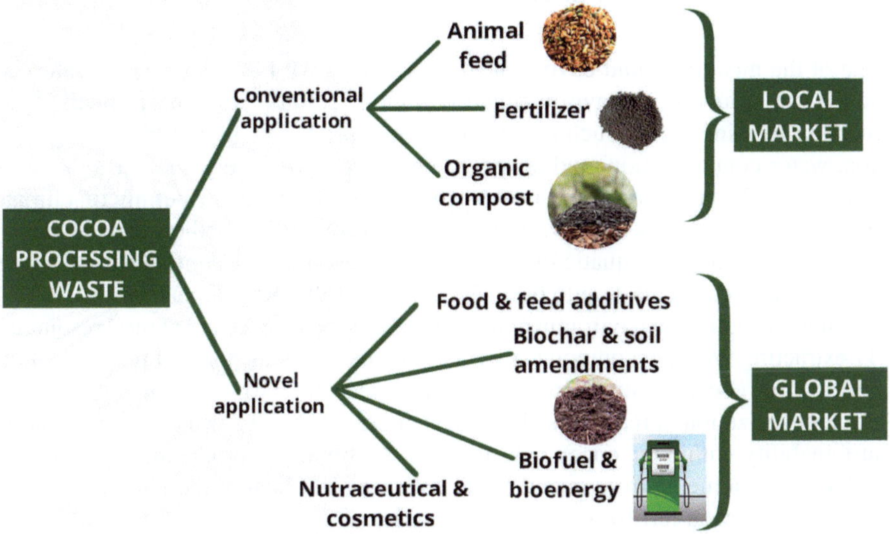

Fig. 9.3 Applications of cocoa processing waste: conventional uses for the local market and novel uses for the global market

pretreatment processes to reduce antinutritional factors [7]. There is also the risk of mycotoxin contamination due to fungal growth during improper storage, which poses health risks and requires careful monitoring. Additionally, the potential accumulation of heavy metals such as cadmium and lead, depending on environmental conditions, necessitates regular testing to ensure safety in applications involving human consumption [88–90].

6 Environmental and Economic Impacts of Biotechnological Valorization

The biotechnological valorization of cocoa processing waste offers a transformative approach to mitigating environmental challenges and stimulating economic growth in cocoa-producing regions. By converting waste materials into valuable products through advanced biotechnological methods, significant environmental benefits can be realized, including pollution reduction, resource conservation, and climate change mitigation. Economically, these innovations open new markets, create employment opportunities, and enhance the resilience of rural economies. This section examines the specific environmental advantages and economic impacts of biotechnological valorization, providing a comprehensive analysis of its potential to contribute to sustainable development.

6.1 Environmental Benefits

One of the most profound environmental advantages of biotechnological valorization is the reduction of environmental pollution. Traditional disposal methods of cocoa processing waste, such as open dumping and burning, lead to soil degradation, water contamination, and air pollution. Decomposing organic waste releases methane and carbon dioxide, potent greenhouse gases contributing to climate change [91]. Leachates from waste piles can infiltrate water bodies, causing eutrophication and harming aquatic life. Biotechnological processes mitigate these issues by converting cocoa waste into biofuels, biochemicals, and biomaterials.

Another significant environmental benefit is the conservation of natural resources. By extracting valuable compounds such as pectin, antioxidants, and bioactive molecules from cocoa waste, there is a reduced need for virgin raw materials. This efficient utilization of resources aligns with circular economy principles, promoting sustainability within the cocoa industry [92]. Additionally, producing biofertilizers and soil amendments from cocoa waste enhances soil health, reducing the dependency on chemical fertilizers that can have adverse environmental impacts.

Climate change mitigation is a critical global priority, and biotechnological valorization contributes to this effort. The reduction of greenhouse gas emissions through methane capture and the substitution of fossil fuels with biofuels from cocoa waste supports national and international commitments to combat climate change [93]. Moreover, the production and application of biochar from cocoa waste sequester carbon in soils, acting as a long-term carbon sink and enhancing soil fertility.

6.2 Economic Impacts

Economically, biotechnological valorization opens new avenues for income generation and diversification in cocoa-producing regions. By transforming waste into high-value products such as biofuels, bioplastics, and nutraceuticals, farmers and local enterprises can access new markets and revenue streams. This diversification reduces dependence on cocoa bean sales alone, enhancing economic resilience against market fluctuations and price volatility in the global cocoa market.

The establishment of biotechnological facilities stimulates employment opportunities, contributing to job creation and skill development. Roles in production, quality control, maintenance, research, and development expand employment prospects beyond traditional agriculture [94]. The need for technical expertise promotes education and training initiatives, building human capital and fostering innovation in rural areas. Entrepreneurship is also encouraged, with individuals and cooperatives establishing businesses centered around waste valorization.

Infrastructure development is another economic benefit, as biotechnological initiatives often require improvements in transportation, energy supply, and communication networks. Enhanced infrastructure not only supports the efficient operation of valorization processes but also benefits the wider community by improving access to services and markets. For example, bioenergy production from cocoa waste can provide electricity to rural areas, supporting electrification efforts and enabling other economic activities.

Moreover, biotechnological products derived from cocoa waste have the potential to enhance trade balances and foreign exchange earnings for cocoa-producing countries. High-value products like bioplastics and nutraceuticals meet the growing global demand for sustainable and eco-friendly goods [95]. Exporting these products generates foreign exchange earnings and strengthens national economies, positioning countries as leaders in sustainable innovation.

Strengthening rural economies and promoting social equity are also significant economic impacts. By engaging local communities in value addition activities, economic benefits are distributed more widely, reducing inequality and enhancing social well-being. Cooperative models and inclusive business practices ensure that smallholder farmers and marginalized groups can participate effectively and reap the rewards of biotechnological advancements. Empowering women through involvement in processing and entrepreneurial activities further contributes to gender inclusion and community development.

7 Policy Implications, Sustainability, and Ethical Considerations

The biotechnological valorization of cocoa processing waste presents a multifaceted opportunity to enhance environmental sustainability, economic development, and social well-being. To fully realize this potential, it is essential to consider the policy frameworks, sustainability, and ethical considerations that influence the adoption and success of these initiatives.

7.1 Policy and Regulatory Frameworks

One of these important frameworks is international agreements and standards. Global environmental challenges such as climate change, resource depletion, and waste management have prompted international bodies to establish agreements and standards that influence national policies. Instruments like the Paris Agreement under the United Nations Framework Convention on Climate Change (UNFCCC) emphasize the reduction of greenhouse gas emissions and the promotion of sustainable practices [96]. The Sustainable Development Goals (SDGs) set by the United Nations outline objectives related to responsible consumption and production (Goal 12), climate action (Goal 13), and partnerships for the goals (Goal 17). These international commitments encourage cocoa-producing countries to adopt policies that support waste valorization. By aligning national strategies with global agreements, countries can access financial mechanisms such as the Green Climate Fund and benefit from technology transfer and capacity-building initiatives.

National policies and government support play a critical role in creating an enabling environment for the biotechnological valorization of cocoa waste. Policies that support research and development, provide financial incentives, and establish regulatory frameworks are essential. For example, governments can offer tax incentives, grants, or low-interest loans to encourage investment in biotechnological facilities [97]. Implementing feed-in tariffs or renewable energy mandates can stimulate the market for bioenergy produced from cocoa waste.

Regulatory frameworks should address standards for product quality, environmental protection, and the use of genetically modified organisms (GMOs) where applicable [98]. Clear regulations reduce uncertainty for investors and ensure that products meet safety and environmental criteria. In Ghana, for instance, the government has developed policies to promote bioenergy and support sustainable agriculture, which can facilitate cocoa waste valorization initiatives [99].

Role of institutions and stakeholders: Collaboration among various stakeholders including government agencies, research institutions, industry players, nongovernmental organizations (NGOs), and local communities is vital for the success of waste valorization projects. Institutions can facilitate partnerships, coordinate efforts, and provide platforms for knowledge exchange. Establishing innovation hubs or centers of excellence focused on biotechnology and sustainable agriculture can drive research and development [100]. Engaging stakeholders in policy formulation ensures that diverse perspectives are considered, promoting policies that are practical and socially acceptable. Public–private partnerships can leverage the strengths of different sectors, combining resources, expertise, and networks to implement projects effectively. Engaging stakeholders in policy formulation ensures that diverse perspectives are considered, promoting policies that are practical and socially acceptable. Public–private partnerships can leverage the strengths of different sectors, combining resources, expertise, and networks to implement projects effectively.

7.2 Sustainability and Ethical Considerations

Life cycle assessment (LCA) and environmental impact ensure the environmental sustainability of biotechnological valorization processes. LCA is a systematic method for assessing the environmental aspects associated with a product, process, or service from cradle to grave [101]. Applying LCA to cocoa waste valorization helps identify areas where environmental performance can be improved, such as reducing energy consumption, minimizing emissions, or optimizing resource use. Strategies to enhance environmental sustainability include adopting renewable energy sources for processing facilities, implementing waste minimization practices, and utilizing environmentally friendly reagents and materials [102]. Continuous monitoring and reporting of environmental performance promotes transparency and accountability, ensuring that valorization initiatives contribute positively to sustainability goals.

Social responsibility and equity through ethical considerations are paramount in ensuring that cocoa waste valorization contributes to social well-being. Equitable distribution of benefits, respect for local cultures, and engagement with communities are critical for social sustainability. Inclusive business models, such as cooperative structures and community-based enterprises, facilitate equitable participation and benefit-sharing among stakeholders [103]. Empowering local communities through education and training enhances their ability to engage effectively in value addition activities. Addressing potential concerns related to the use of genetically modified organisms (GMOs) requires transparent communication and adherence to regulatory requirements. Ethical guidelines for biotechnology emphasize the importance of biosafety, biosecurity, and the precautionary principle [104]. Engaging stakeholders in dialogue and education can build trust and acceptance, ensuring that valorization initiatives are culturally appropriate and socially beneficial.

Consumer perception and market acceptance: Consumer attitudes toward products derived from waste or involving biotechnology significantly influence market success. Concerns about safety, quality, and environmental impacts need to be addressed proactively. Marketing strategies should highlight the benefits of cocoa waste-derived products, such as sustainability, natural origin, and social impact, to enhance consumer appeal. Certification schemes and eco-labels can enhance credibility and appeal to environmentally conscious consumers. Providing clear information about product attributes, production processes, and ethical standards helps build consumer trust and drives market acceptance [105]. Educating consumers about the environmental and economic benefits of valorization can further support the adoption and popularity of bioproducts derived from cocoa waste.

8 Conclusion and Future Perspectives

The biotechnological valorization of cocoa processing waste emerges as a transformative strategy that addresses critical environmental challenges while fostering economic growth in cocoa-producing regions. By converting waste materials into valuable bioproducts such as biofuels, bioplastics, and nutraceuticals, this approach significantly mitigates issues like soil contamination, water pollution, and greenhouse gas emissions. Economically, it diversifies income streams for farmers and local enterprises, enhances market competitiveness, and stimulates employment and skill development in rural communities. Effective policy frameworks, technological innovations, and strategic implementation are essential to maximize these benefits, ensuring that valorization initiatives are both sustainable and economically viable. Ultimately, leveraging biotechnological advancements to valorize cocoa waste not only promotes environmental stewardship but also contributes to the socioeconomic resilience of cocoa-dependent communities.

Looking forward, the biotechnological valorization of cocoa processing waste holds immense potential for further advancements and broader implementation. Continued research and development are crucial to refine bioprocessing technologies, enhance efficiency, and reduce costs, making valorization more accessible and scalable. Innovations in synthetic biology and genetic engineering can lead to the development of more robust microbial strains and optimized enzymes, increasing the yield and quality of bioproducts. Additionally, the integration of digital technologies such as artificial intelligence and the Internet of Things can streamline operations, improve process monitoring, and enable predictive maintenance, thereby enhancing overall efficiency. International collaboration and partnerships will play a pivotal role in accelerating the adoption of valorization practices, facilitating knowledge exchange, and providing access to funding and technical expertise. Governments and industry stakeholders must continue to develop supportive policies and incentives that encourage investment in biotechnological ventures and promote sustainable practices. Moreover, fostering education and capacity-building initiatives will ensure that local communities are equipped with the necessary skills and knowledge to participate effectively in valorization projects. By embracing these future directions, the cocoa industry can achieve greater environmental sustainability, economic prosperity, and social well-being, positioning itself as a leader in sustainable agricultural practices.

References

1. Gyan JK, Bajan B (2023) Market analysis on cocoa beans export: the case of Ghana and Cote d'Ivoire in West Africa. J Agribusiness Rural Dev 66:375. https://doi.org/10.17306/J.JARD.2022.01646
2. RSI W. R. I (2024) How cacao can do so much more than make chocolate. Abidjan. https://www.wri.org/insights/hidden-benefits-cacao-waste
3. Anoraga SB, Shamsudin R, Hamzah MH, Sharif S, Saputro AD (2024) Cocoa by-products: A comprehensive review on potential uses, waste management, and emerging green technologies for cocoa pod husk utilization. Heliyon 10:e35537. https://doi.org/10.1016/j.heliyon.2024.e35537
4. Sánchez M, Laca A, Laca A, Díaz M (2023a) Cocoa bean shell: A by-product with high potential for nutritional and biotechnological applications. Antioxidants 12:1028. https://doi.org/10.3390/antiox12051028
5. Vásquez ZS, de Carvalho Neto DP, Pereira GVM, Vandenberghe LPS, de Oliveira PZ, Tiburcio PB, Rogez HLG, Góes Neto A, Soccol CR (2019) Biotechnological approaches for cocoa waste management: A review. Waste Manag 90:72–83. https://doi.org/10.1016/j.wasman.2019.04.030
6. Anokye J, Abunyewa AA, Jørgensen U, Kaba JS, Twum-Ampofo K, Dawoe E et al (2024) Mitigation of greenhouse gas emissions through shade systems and climate-smart soil fertility interventions in cocoa landscapes in the semi-deciduous ecological zone of Ghana. Soil Adv 1:100001. https://doi.org/10.1016/j.soilad.2024.100001

7. Afedzi AEK, Obeng-Boateng F, Aduama-Larbi MS, Zhou X, Xu Y (2023) Valorization of Ghanaian cocoa processing residues as extractives for value-added functional food and animal feed additives—a review. Biocatal Agric Biotechnol 52:102835. https://doi.org/10.1016/j.bcab.2023.102835
8. Dahunsi SO, Adesulu-Dahunsi AT, Izebere JO (2019) Cleaner energy through liquefaction of cocoa (*Theobroma cacao*) pod husk: Pretreatment and process optimization. J Clean Prod 226:578–588. https://doi.org/10.1016/j.jclepro.2019.04.112
9. Disca V, Jaouhari Y, Carrà F, Martoccia M, Travaglia F, Locatelli M et al (2024) Effect of carbohydrase treatment on the dietary fibers and bioactive compounds of cocoa bean shells (CBSs). Food Secur 13:1. https://doi.org/10.3390/foods13162545
10. Mellinas AC, Jiménez A, Garrigós MC (2020) Optimization of microwave-assisted extraction of cocoa bean shell waste and evaluation of its antioxidant, physicochemical and functional properties. LWT-Food Sci Technol 127:109361. https://doi.org/10.1016/j.lwt.2020.109361
11. Jensch C, Schmidt A, Strube J (2022) Versatile green processing for recovery of phenolic compounds from natural product extracts towards bioeconomy and cascade utilization for waste valorization on the example of cocoa bean shell (CBS). Sustain For 14:3126. https://doi.org/10.3390/su14053126
12. Oussou KF, Guclu G, Kelebek H, Selli S (2023) Valorization of cocoa, tea, and coffee processing by-products-wastes. In: Capanoglu E, Navarro-Hortal MD, Forbes-Hernández TY, Battino M (eds) Valorization of wastes/by-products in the design of functional foods/supplements, vol 107. Academic Press, pp 91–130
13. Zarrillo S, Gaikwad N, Lanaud C, Powis T, Viot C, Lesur I et al (2018) The use and domestication of *Theobroma cacao* during the mid-Holocene in the upper Amazon. Nat Ecol Evol 2(12):1879–1888. https://doi.org/10.1038/s41559-018-0697-x
14. Coe SD, Coe MD (2013) The true history of chocolate. Thames & Hudson, London
15. Presilla ME (2009) The new taste of chocolate: a cultural and natural history of cacao with recipes. Random House Digital, Inc.
16. Motamayor JC, Lachenaud P, da Silva e Mota J, W, Loor R, Kuhn DN, Brown JS, Schnell RJ (2008) Geographic and genetic population differentiation of the Amazonian chocolate tree (*Theobroma cacao* L). PLoS One 3:e3311.
17. ICCO, O. I. C (2021) Quarterly bulletin of cocoa statistics, vol 1. International Cocoa Organization. https://www.icco.org
18. Afoakwa EO (2016) Chocolate science and technology. Wiley, New York
19. Wood GAR, Lass RA (2008) Cocoa. John Wiley & Sons
20. WCF, W. C. F (2024) Cocoa producing countries:2024. https://worldpopulationreview.com/country-rankings/cocoa-producing-countri
21. Donkor P, Siabi EK, Frimpong K, Frimpong PT, Mensah SK, Vuu C et al (2024) Impacts of illegal artisanal and small-scale gold mining on livelihoods in cocoa farming communities: A case of Amansie West District. Ghana Resour Policy 104879(91):104879
22. Scudder M, Wampe N, Waviki Z, Applegate G, Herbohn J (2022) Smallholder cocoa agroforestry systems is increased yield worth the labour and capital inputs? Agric Syst 196(103350):103350. https://doi.org/10.1016/j.agsy.2021.103350
23. Armengot L, Barbieri P, Andres C, Milz J, Schneider M (2016) Cacao agroforestry systems have higher return on labor compared to full-sun monocultures. Agron Sustain Dev 36:70. https://doi.org/10.1007/s13593-016-0406-6
24. IDH, the S. T. I (2023) Côte d'Ivoire Cocoa & Forests Initiative 2022 Annual Report. https://www.idhsustainabletrade.com/publication/cote-divoire-cocoa-forests-initiative-2022-annual-report/
25. Sassen M, van Soesbergen A, Arnell AP, Scott E (2022) Patterns of (future) environmental risks from cocoa expansion and intensification in West Africa call for context specific responses. Land Use Policy 119:106142. https://doi.org/10.1016/j.landusepol.2022.106142

26. Ouattara LY, Kouassi EKA, Soro D, Soro Y, Yao KB, Adouby K et al (2021) Cocoa pod husks as potential sources of renewable high-value-added products: A review of current valorizations and future prospects. Bioresources 16(1):1988–2020
27. Botella-Martínez C, Lucas-Gonzalez R, Ballester-Costa C, Pérez-Álvarez JÁ, Fernández-López J, Delgado-Ospina J et al (2021) Ghanaian cocoa (*Theobroma cacao* L.) bean shells coproducts: effect of particle size on chemical composition, bioactive compound content and antioxidant activity. Agronomy 11:401. https://doi.org/10.3390/agronomy11020401
28. Adeyi O (2010) Proximate composition of some agricultural wastes in Nigeria and their potential use in activated carbon production. J Appl Sci Environ Manag 14(1):55. https://doi.org/10.4314/jasem.v14i1.56490
29. Adjin-Tetteh M, Asiedu N, Dodoo-Arhin D, Karam A, Amaniampong PN (2018) Thermochemical conversion and characterization of cocoa pod husks a potential agricultural waste from Ghana. Ind Crop Prod 119:304–312. https://doi.org/10.1016/j.indcrop.2018.02.060
30. Akinola AO, Eiche JF, Owolabi PO, Elegbeleye AP (2018) Pyrolytic analysis of cocoa pod for biofuel production. Niger J Technol 37(4):1026–1031
31. Dos Anjos Lopes SM, Martins MV, de Souza VB, Tulini FL (2023) Evaluation of the nutritional composition of cocoa bean shell waste (*Theobroma cacao*) and application in the production of a phenolic-rich iced tea. J Culin Sci Technol 21:818–828. https://doi.org/10.1080/15428052.2021.2016531
32. Forero-Nuñez CA, Jochum J, Sierra FE (2015) Effect of particle size and addition of cocoa pod husk on the properties of sawdust and coal pellets. Ingeniería e Investigación 35(1):17–23
33. Maliki M, Ikhuoria EU, Aluyor P (2023) Proximate analysis of selected agricultural waste for their nutritional potential. North Afr J Food Nutr Res 7(15):117–125
34. Vargas-Arana G, Merino-Zegarra C, Tang M, Pertino MW, Simirgiotis MJ (2022) UHPLC–MS characterization, and antioxidant and nutritional analysis of cocoa waste flours from the Peruvian Amazon. Antioxidants 11:595. https://doi.org/10.3390/antiox11030595
35. Soares TF, Oliveira MBPP (2022) Cocoa by-products: characterization of bioactive compounds and beneficial health effects. Molecules 27:1625. https://doi.org/10.3390/molecules27051625
36. Djali M, Santasa K, Indiarto R, Subroto E, Fetriyuna F, Lembong E (2023) Proximate composition and bioactive compounds of cocoa bean shells as a by-product from cocoa industries in Indonesia. Foods 12:3316. https://doi.org/10.3390/foods12173316
37. Kone K, Akueson K, Norval G (2020) On the production of potassium carbonate from cocoa pod husks. Recycling 5:23. https://doi.org/10.3390/recycling5030023
38. Adi-Dako O, Ofori-Kwakye K, Frimpong Manso S, Boakye-Gyasi MEL, Sasu C, Pobee M (2016) Physicochemical and antimicrobial properties of cocoa pod husk pectin intended as a versatile pharmaceutical excipient and nutraceutical. J Pharm 2016:7608693. https://doi.org/10.1155/2016/7608693
39. Van NT, Thi PD, Long VB, Van NH, Ngoc TL (2021) Microwave-assisted extraction for maximizing the yield of phenolic compounds and antioxidant capacity from cacao pod husk (*Theobroma cacao* L.). Curr Nutr Food Sci 17(2):225–237. https://doi.org/10.2174/1573401316999200503032017
40. Cantele C, Rojo-Poveda O, Bertolino M, Ghirardello D, Cardenia V, Barbosa-Pereira L, Zeppa G (2020) *In vitro* bioaccessibility and functional properties of phenolic compounds from enriched beverages based on cocoa bean shell. Food Secur 9:715. https://doi.org/10.3390/foods9060715
41. Tejeda JF, Arango-Angarita J, Cuervo JL (2023) Effect of solar pre-drying and yeast starter inoculation treatments on the chemical composition of cocoa (*Theobroma cacao* L.) beans from southwestern Colombia. Foods 12:4495. https://doi.org/10.3390/foods12244455
42. Blomsma F, Tennant M (2020) Circular economy: preserving materials or products? Introducing the resource states framework. Resour Conserv Recycl 156:104698. https://doi.org/10.1016/j.resconrec.2020.104698

43. Soto B (2019) Fermentation processes. Scientific e-Resources
44. Yusof F, Khanahmadi S, Amid A, Mahmod SS (2016) Cocoa pod husk, a new source of hydrolase enzymes for preparation of cross-linked enzyme aggregate. Springerplus 5:1–18
45. Subudhi S, Mudgil D, Saha K, Kumar Sarangi P, Pal P (2024) Yeast as a cell factory for fermentative production of ethanol from xylose. J Taiwan Inst Chem Eng 105616:105616. https://doi.org/10.1016/j.jtice.2024.105616
46. Hernández-Mendoza AG, Saldaña-Trinidad S, Martínez-Hernández S, Pérez-Sariñana BY, Láinez M (2021) Optimization of alkaline pretreatment and enzymatic hydrolysis of cocoa pod husk (T*heobroma cacao* L.) for ethanol production. Biomass Bioenergy 154:106268. https://doi.org/10.1016/j.biombioe.2021.106268
47. Shet VB, Sanil N, Bhat M, Naik M, Mascarenhas LN, Goveas LC et al (2018) Acid hydrolysis optimization of cocoa pod shell using response surface methodology approach toward ethanol production. Agric Nat Resour 52(6):581–587. https://doi.org/10.1016/j.anres.2018.11.022
48. Yadav A, Dong C-D, Sharma D, Tsai M-L, Sun P-P, Nargotra P et al (2024) Integrated choline chloride/citric acid-microwave pretreatment for efficient nanolignin extraction and bioethanol production from cocoa pod husk waste. Energy Environ 0958305X241270269:09583 05X241270269
49. Santana NB, Dias JCT, Rezende RP, Franco M, Oliveira LKS, Souza LO (2018) Production of xylitol and bio-detoxification of cocoa pod husk hemicellulose hydrolysate by *Candida boidinii* XM02G. PLoS One 13(4):e0195206. https://doi.org/10.1371/journal.pone.0195206
50. Kabaivanova L, Petrova P, Hubenov V, Simeonov I (2022) Biogas production potential of thermophilic anaerobic biodegradation of organic waste by a microbial consortium identified with metagenomics. Life 12:702. https://doi.org/10.3390/life12050702
51. Mosquera J, Rangel C, Thomas J, Santis A, Acevedo P, Cabeza I (2021) Biogas production by pilot-scale anaerobic co-digestion and life cycle assessment using a real scale scenario: independent parameters and co-substrates influence. Processes 9:1875. https://doi.org/10.3390/pr9111875
52. Awafo EA, Owusu PA (2022) Energy and water mapping of the cocoa value chain in Ghana. Sustain Product Consumpt 29:341–356. https://doi.org/10.1016/j.spc.2021.10.027
53. Bandeira FJS, Ribeiro JAS, Mesquita ALA, Mesquita ALA, Torres EA (2023) Potential use of palm oil and cocoa waste biomasses as sources of energy generation by gasification system in the state of Pará, Brazil. Engenharia Agrícola 43:e20220151
54. Anyanwu CN, Ibeto CN, Eze IS, Ezeoha SL (2013) Present and prospective energy use potentials of selected agricultural wastes in Nigeria. J Renew Sustain Energy 5:32703. https://doi.org/10.1063/1.4808046
55. Bobiles SC, Elegado FB, Millena CG, Merca FE (2022) Small-scale cacao (*Theobroma cacao* L.) fermentation process utilizing cacao pod husk. Food Res 6(4):236–245
56. Perwitasari U, Agustina NT, Pangestu R, Amanah S, Saputra H, Andriani A et al (2023) Cacao pod husk for citric acid production under solid state fermentation using response surface method. Biomass Convers Biorefinery 13(8):7165–7173. https://doi.org/10.1007/s13399-021-01690-9
57. Sianoun N, Pongyeela P, Chungsiriporn J (2024) *Aspergillus niger* fermentation of cocoa pod husk for increased total crude protein utilizing response surface methodology. Biomass Convers Biorefinery 15:5313. https://doi.org/10.1007/s13399-024-05303-z
58. Rozendo AS, Vandenberghe LP, de Mattos PB, Rogez HL, Soccol CR (2024) Pectinase production from cocoa pod husk in submerged fermentation and its application in the clarification of apple juice. Fermentation 10:337. https://doi.org/10.3390/fermentation10070337
59. Hennessey-Ramos L, Murillo-Arango W, Vasco-Correa J, Paz Astudillo IC (2021) Enzymatic extraction and characterization of pectin from cocoa pod husks (*Theobroma cacao* L.) using celluclast® 1.5 L. Molecules 26:1473. https://doi.org/10.3390/molecules26051473
60. Yapo BM, Koffi KL (2013) Extraction and characterization of gelling and emulsifying pectin fractions from cacao pod husk. Nature 1(4):46–51

61. Haile S, Ayele A (2022) Pectinase from microorganisms and its industrial applications. Sci World J 2022:1881305. https://doi.org/10.1155/2022/1881305
62. Huynh GH, Van Pham H, Hong Nguyen HV (2023) Effects of enzymatic and ultrasonic-assisted extraction of bioactive compounds from cocoa bean shells. J Food Meas Charact 17(5):4650–4660. https://doi.org/10.1007/s11694-023-01986-6
63. Albuquerque BR, Heleno SA, Oliveira MBPP, Barros L, Ferreira ICFR (2021) Phenolic compounds: current industrial applications, limitations and future challenges. Food Funct 12(1):14–29. https://doi.org/10.1039/D0FO02324H
64. Ko JK, Um Y, Woo HM, Kim KH, Lee S-M (2016) Ethanol production from lignocellulosic hydrolysates using engineered *Saccharomyces cerevisiae* harboring xylose isomerase-based pathway. Bioresour Technol 209:290–296. https://doi.org/10.1016/j.biortech.2016.02.124
65. Brandt BA, Jansen T, Görgens JF, van Zyl WH (2019) Overcoming lignocellulose-derived microbial inhibitors: advancing the *Saccharomyces cerevisiae* resistance toolbox. Biofuels Bioprod Biorefin 13:1520–1536. https://doi.org/10.1002/bbb.2042
66. Dvořák P, de Lorenzo V (2018) Refactoring the upper sugar metabolism of *pseudomonas putida* for co-utilization of cellobiose, xylose, and glucose. Metab Eng 48:94–108. https://doi.org/10.1016/j.ymben.2018.05.019
67. Imao K, Konishi R, Kishida M, Hirata Y, Segawa S, Adachi N et al (2017) 1,5-Diaminopentane production from xylooligosaccharides using metabolically engineered displaying beta-xylosidase on the cell surface. Bioresour Technol 245:1684–1691. https://doi.org/10.1016/j.biortech.2017.05.135
68. Dalton B, Bhagabati P, De Micco J, Padamati RB, O'Connor K (2022) A review on biological synthesis of the biodegradable polymers polyhydroxyalkanoates and the development of multiple applications. Catalysts 12:319. https://doi.org/10.3390/catal12030319
69. Ramos LH, Cisneros-Yupanqui M, Santisteban Soto DV, Lante A, Favaro L, Casella S, Basaglia M (2023) Exploitation of Cocoa pod residues for the production of antioxidants, polyhydroxyalkanoates, and ethanol. Fermentation 9:843. https://doi.org/10.3390/fermentation9090843
70. Sánchez M, Laca A, Laca A, Díaz M (2023b) Cocoa bean shell as promising feedstock for the production of poly(3-hydroxybutyrate) (PHB). Appl Sci 13:975. https://doi.org/10.3390/app13020975
71. Anya MI, Ozung PO, Igwe PA (2018) Blood profile of west African dwarf (WAD) bucks fed raw and processed cocoa pod husk meal based–diets in the humid high rainforest zone of Nigeria. Global J Pure Appl Sci 24(2):125–135
72. Figueroa KHN, García NVM, Vega RC (2020) Cocoa by-products. In: Food wastes and by-products: nutraceutical and health potential. Wiley, New York, pp 373–411. https://doi.org/10.1002/9781119534167.ch13
73. Atere CT, Olayinka A (2013) Soil chemical properties and growth of maize (*Zea mays* L.) as affected by cacao pod compost based N and P fertilizer. Nigerian. J Soil Sci 23(1):83–93
74. Fidelis C, Rajashekhar Rao BK (2017) Enriched cocoa pod composts and their fertilizing effects on hybrid cocoa seedlings. Int J Recycl Org Waste Agric 6:99–106. https://doi.org/10.1007/s40093-017-0156-8
75. Nelson N, Darkwa J, Calautit J (2021) Prospects of bioenergy production for sustainable rural development in Ghana. J Sustain Bioenergy Syst 11:227. https://doi.org/10.4236/jsbs.2021.114015
76. Venkatesh G, Gopinath KA, Reddy KS, Reddy BS, Prabhakar M, Srinivasarao C et al (2022) Characterization of biochar derived from crop residues for soil amendment, carbon sequestration and energy use. Sustain For 14:2295. https://doi.org/10.3390/su14042295
77. Ogunlade MO, Odesola IF, Akappo OA, Ige EO (2016) Effects of temperature on some chemical properties and yield of biochar derived from cocoa pod husks for soil amendment. In: Proceedings of the 2nd annual conference of biochar initiative of Nigeria, pp 74–78
78. Pouangam Ngalani G, Dzemze Kagho F, Peguy NNC, Prudent P, Ondo JA, Ngameni E (2023) Effects of coffee husk and cocoa pods biochar on the chemical properties of an acid soil

from West Cameroon. Arch Agron Soil Sci 69(5):744–758. https://doi.org/10.1080/03650340.2022.2033733
79. Kuo H-W, Chang C-C, Cheng W (2022) Pectin from dry cacao pod husk mediates growth performance, immune resistance responses and carbohydrate metabolism of *Litopenaeus vannamei* through dietary administration. Aquaculture 548:737613. https://doi.org/10.1016/j.aquaculture.2021.737613
80. Reiche AM, Tretola M, Eichinger J, Munger A, Eggerschwiler L, Hutten A et al (2023) Influence of supplementing bakery by-products and cocoa bean shells to a grass-based diet on feed intake, milk production and ruminal fermentation of early-lactating dairy cows. Ital J Anim Sci 22:242
81. Carta S, Nudda A, Cappai MG, Lunesu MF, Atzori AS, Battacone G, Pulina G (2020) Short communication: cocoa husks can effectively replace soybean hulls in dairy sheep diets—effects on milk production traits and hematological parameters. J Dairy Sci 103:1553–1558. https://doi.org/10.3168/jds.2019-17550
82. Hikmah H, Alam G, Syamsu JA, Salengke S, Nahariah N (2020) The digestive and physiological visceral organs of male Bali cattle were fed with cocoa bean shells. IOP Conf Ser Earth Environ Sci 492:12063. https://doi.org/10.1088/1755-1315/492/1/012063
83. Agudelo C, Bravo K, Ramírez-Atehortúa A, Torres D, Carrillo-Hormaza L, Osorio E (2021) Chemical and skincare property characterization of the main cocoa byproducts: extraction optimization by RSM approach for development of sustainable ingredients. Molecules 26:7429. https://doi.org/10.3390/molecules26247429
84. Barrera-Reyes PK, de Lara JC-F, González-Soto M, Tejero ME (2020) Effects of cocoa-derived polyphenols on cognitive function in humans. Systematic review and analysis of methodological aspects. Plant Foods Hum Nutr 75:1–11. https://doi.org/10.1007/s11130-019-00779-x
85. Grassi D, Ferri C, Desideri G (2016) Brain protection and cognitive function: cocoa flavonoids as nutraceuticals. Curr Pharm Des 22:145–151
86. Akinrotohun HO, Oyedele AO, Igbeneghu OA (2021) Formulation development of topical antibacterial lotion with Theobroma cacao pod husk ash extract for treatment of shaving bumps. J Anal Pharm Res 10(1):14–24
87. Shakoor R, Hussain N, Bashir H, Ashraf MH (2024) Plant biomass materials in cosmetic applications. In: Jawaid M, Khan A, Asiri A, Abdullah M (eds) Plant biomass applications. Academic Press, pp 185–208. https://doi.org/10.1016/B978-0-443-15465-2.00011-2
88. Barraza F, Schreck E, Lévêque T, Uzu G, López F, Ruales J et al (2017) Cadmium bioaccumulation and gastric bioaccessibility in cacao: A field study in areas impacted by oil activities in Ecuador. Environ Pollut 229:950–963. https://doi.org/10.1016/j.envpol.2017.07.080
89. Dewayani W, Syamsuri R, Septianti E, Suriany Halil W, Nurlaila, Rahman A (2022) Good handling practice study to reduce the level of contamination in cocoa beans in east Luwu. IOP Conf Ser Earth Environ Sci 1024:12067. https://doi.org/10.1088/1755-1315/1024/1/012067
90. Mohamadi S, Mahmudiono T, Zienali T, Sadighara P, Omidi B, Limam I, Fakhri Y (2024) Probabilistic health risk assessment of heavy metals (Cd, Pb, and As) in cocoa powder (*Theobroma cacao*) in Tehran, Iran market. Int J Environ Health Res 34(1):257–272. https://doi.org/10.1080/09603123.2022.2146070
91. Wang N, He Y, Zhao K, Lin X, He X, Chen A et al (2024) Greenhouse gas emission characteristics and influencing factors of agricultural waste composting process: a review. J Environ Manag 354:120337. https://doi.org/10.1016/j.jenvman.2024.120337
92. Rojas LF, Zapata P, Ruiz-Tirado L (2022) Agro-industrial waste enzymes: perspectives in circular economy. Current opinion in green and sustainable. Chemistry 34:100585. https://doi.org/10.1016/j.cogsc.2021.100585
93. Victor-Ekwebelem MO, Ugwuocha CS, Yakimov MM (2024) The way forward for bioenergy technology for developing countries. In: Maddela NR, Aransiola SA, Ezugwu CI, Winkelstroter Eller LK, Scalvenzi L, Meng F (eds) Microbial biotechnology for bioenergy. Elsevier, pp 347–376. https://doi.org/10.1016/B978-0-443-14112-6.00017-1

94. Boadi SA, Olwig MF, Asare R, Bosselmann AS, Owusu K (2022) The role of innovation in sustainable cocoa cultivation: moving beyond mitigation and adaptation. In: Coromaldi M, Auci S (eds) Climate-induced innovation: mitigation and adaptation to climate change. Springer International Publishing, Cham, pp 47–80. https://doi.org/10.1007/978-3-031-01330-0_3
95. Ogbu CC, Okey SN (2023) In: Ahmad F, Sultan M (eds) Agro-industrial waste management: the circular and bioeconomic perspective. IntechOpen, Rijeka. https://doi.org/10.5772/intechopen.109181
96. Naser MM, Pearce P (2022) Evolution of the international climate change policy and processes: UNFCCC to Paris agreement. Oxford University Press. https://doi.org/10.1093/acrefore/9780199389414.013.422
97. Vadlani PV (2020) Financing strategies for sustainable bioenergy and the commodity chemicals industry. Green Energy to Sustainability, In, pp 569–586. https://doi.org/10.1002/9781119152057.ch23
98. Rozas P, Kessi-Pérez EI, Martínez C (2022) Genetically modified organisms: adapting regulatory frameworks for evolving genome editing technologies. Biol Res 55:31. https://doi.org/10.1186/s40659-022-00399-x
99. Agyekum EB (2020) Energy poverty in energy rich Ghana: A SWOT analytical approach for the development of Ghana's renewable energy. Sustain Energy Technol Assess 40:100760. https://doi.org/10.1016/j.seta.2020.100760
100. Kunert KJ, Botha A-M, Oberholster PJ, Yocgo R, Chimwamurombe P, Vorster J, Foyer CH (2020) Factors facilitating sustainable scientific partnerships between developed and developing countries. Outlook Agric 49(3):204–214. https://doi.org/10.1177/0030727020939592
101. Andrews ES (2009) Guidelines for social life cycle assessment of products: social and socio-economic LCA guidelines complementing environmental LCA and life cycle costing, contributing to the full assessment of goods and services within the context of sustainable development. UNEP/Earthprint
102. Chen T-L, Kim H, Pan S-Y, Tseng P-C, Lin Y-P, Chiang P-C (2020) Implementation of green chemistry principles in circular economy system towards sustainable development goals: challenges and perspectives. Sci Total Environ 716:136998. https://doi.org/10.1016/j.scitotenv.2020.136998
103. Louman B, Girolami ED, Shames S, Primo LG, Gitz V, Scherr SJ et al (2022) Access to landscape finance for small-scale producers and local communities: a literature review. Land 11:1444. https://doi.org/10.3390/land11091444
104. Singh K, Zeeshan Z, Karna S (2023) The regulation of biotechnology: ethical, legal, and social implications. Interdiscip Stud Soc Law Politics 2(3):44–50
105. Kumar P, Polonsky M, Dwivedi YK, Kar A (2021) Green information quality and green brand evaluation: the moderating effects of eco-label credibility and consumer knowledge. Eur J Mark 55(7):2037–2071. https://doi.org/10.1108/EJM-10-2019-0808

Part IV
Masticatory and Nut Crops

Chapter 10
Functional Uses of Unmarketable Betel Leaf (*Piper betle* L.) in Food and Other Sectors: A Review

Alaka Mohanty and Sujata Mohanty

Abbreviations

ABTS	2,2′-azino-bis(3-ethylbenzothiazoline-6-sulfonic acid)
DPPH	2,2-diphenyl-1-picrylhydrazyl
GC-MS	Gas chromatography–mass spectrometry

1 Introduction

The *Piperaceae* family, comprising small trees, shrubs, or lianas, encompasses over 10 genera and more than 1500 species [1]. These *Piper* species are distributed across various regions, with approximately 700 species found in tropical America, 300 in the Asian tropics, and another 150 in Africa [2]. Renowned for their medicinal properties, these plants have been integral to traditional medicine practices in China, Indonesia, India, and Korea [3]. Many members of the *Piper* genus are commercially predominant in several agro-climatic zones of some countries. *Piper nigrum*, *P. betle*, *P. aduncum*, *P. longum*, and *P. cubeba* are the customary members of the family *Piperaceae. P. betle* is the most studied species, cultivated for its fresh leaves and essential oil rich in vital bioactive constituents with various therapeutic potentials [4]. Despite its great economic potential, the crop faces serious spoilage due to disease and pest attacks. The unavailability of proper marketing, storage facilities, and changing prices are the key factors in the huge wastage of surplus betel leaves. The shelf life of betel leaves depends on the storage conditions but can range from a few days to over 40 days. These unmarketable leaves could be utilized

A. Mohanty · S. Mohanty (✉)
Department of Biotechnology, Rama Devi Women's University, Bhubaneswar, Odisha, India
e-mail: sujatamohanty@rdwu.ac.in

C. Trois, R. C. Ray (eds.), *Plantation Crop Wastes: Valorization for Economic Sustainability*, Waste as a Resource,
https://doi.org/10.1007/978-3-031-96805-1_10

for the extraction of essential oils and the development of value-added products like bars, tonics, perfumes, and food additives.

P. betle L., commonly known as betelvine, is a perennial, dioecious, and tropical shade-loving creeper. Betelvine is known as "Green Gold" for its medicinal and economic importance. It is mainly cultivated for its heart-shaped green leaves and used for human consumption and also for herbal formulations and products. In India, betel leaf is referred to by various names in several local languages that include "Tambula" (Sanskrit), "Vettilai" (Tamil), "Vettilakkoti" (Malayalam), "Nagarbel" (Gujurati), "Villayadela" (Kannada), "Pan" (Hindi and Bengali), "Videch-pan" (Marathi), and "Tamalapaku" (Telugu). In addition to India, it is also known as "Burg-e-Tanbol" and "Tanbol" in Persian and Arabic, respectively [5, 6]. Worldwide, there are approximately 100–150 variabilities of betel leaf, with 40 of them found in India [7]. The taxonomic classification of *Piper betle* L. is as follows:

Kingdom	Plantae
Division	Magnoliophyta
Class	Magnoliopsida
Order	Piperales
Family	Piperaceae
Genus	*Piper*
Species	*Betle*

2 Botanical Description

The betel leaf is characterized by its heart-shaped dorsiventral green foliage, though variations exist in shades ranging from yellow-tinged greens to deep, blackish hues. Its vine bears a weak cylindrical stem initially green, which matures into a semi-woody texture displaying earthy brown or yellow tones. In its early stage, the stem may exhibit a vibrant green hue adorned through parallel red lines. Adventitious roots sprout individually from the node, facilitating the vine's ascent along host plants or support structures, reaching impressive heights of 15–50 feet annually through prolific branching. Periodic pruning is conducted to enhance leaf quality. Leaves are petiolate and glabrous, alternate along the stem, varying in 5–20 cm length and 3–15 cm width, and are influenced by genetic factors. The foliage features a prominent midrib along with 4–8 additional ribs emanating from the base, converging toward the apex, which is either acute or acuminate. The leaf margin remains entire, with the stout petiole. On dioecious plants, male and female flowers manifest separately, characterized by spiky, dense, cylindrical structures lacking sepals or petals and measuring 3–10 cm or more. Fruit production is rare under Indian agroclimatic conditions (Fig. 10.1).

Fig. 10.1 *Piper betle* vines in Baraja (plantation)

3 Origin and Distribution

Piper species grow well in the high altitudes like cloud forests along with lowland tropical rainforests. Over 700 species of *Piper* have been identified across both the hemispheres (Northern and Southern), cultivated extensively in countries such as India, Philippines, Malaysia, Sri Lanka, Indonesia, and other East African and Southeast Asian nations [8]. *Piper betle* originates in Malaysia and is native to Central and Eastern Malaysia. These plants have a history dating back around 2500 years, with cultivation spreading from the Malaysian region and tropical Asia to Madagascar and East Africa. This species is cultivated broadly in China and India owing to its well-established roles in time-worn practices and its enormous medicinal properties [9–13]. However, there are approximately 100 varieties of betelvine found all over the world; among them, 40 are found in India [14].

4 Traditional Uses of Betelvine

Piper betle is a highly significant plant globally due to its medicinal value and its role in traditional practices [15]. The plant is popular in India because of its important role in various aspects of life, including social, cultural, and religious activities. It is widely recognized as a medicinal plant across Asia, with its leaves being the most extensively used and studied part. Chewing betel leaves is a common practice in many countries, believed to offer benefits such as freshening breath, strengthening gums, preserving dental health, and aiding digestion [16, 17].

Betel leaves are mostly used as a mouth freshener, but they are also renowned for treating many diseases such as cold, cough, rheumatism, bronchial asthma, and stomachalgia. This plant is also used traditionally for curing several diseases such as cuts, conjunctivitis, pyorrhoea, injuries, leucorrhoea, boils, constipation, otorrhoea, mastoiditis, mastitis, swelling of gum, and abscesses [18]. In traditional medicine, betel leaves find diverse applications, including vaginal douching in Indonesia [19], mouthwash for gargling in India and Thailand [20], and being utilized in the treatment of dental issues, joint pain, arthritis and headache in Malaysia. Moreover, in Sri Lanka, the juice of betel leaf is employed to address skin diseases [21, 22]. Betel leaves contain a volatile essential oil with antiseptic, cardiotonic, blood pressure-lowering, and both cardiac and respiratory depressant effects. The antiseptic quality of this oil comes from chavicol, a bioactive component. Betel leaf essential oil also has anti-bacterial, anti-fungal, and anti-protozoan properties [23]. Betel leaves, along with its fruits, flowers, and roots, have carminative, antiseptic, and stimulant properties, making them useful for treating malaria. Additionally, betel leaves can be employed to address various ailments both externally and internally. External applications include treating boils, ulcers, bruises, nasal ulcerations, night blindness, eye injuries, respiratory issues, throat infections, headaches, hoarseness of voice, and inflamed joints. The internal uses include treating bad breath, gastritis, stomach disorders, filaria, throat irritation, and impaired digestion. Additionally, the essential oil in the leaves contains volatile compounds like chavibetol, eugenol, chavicol, and hydroxyl chavicol (Fig. 10.2), which contribute to its anticancer, antifungal, and antioxidant properties.

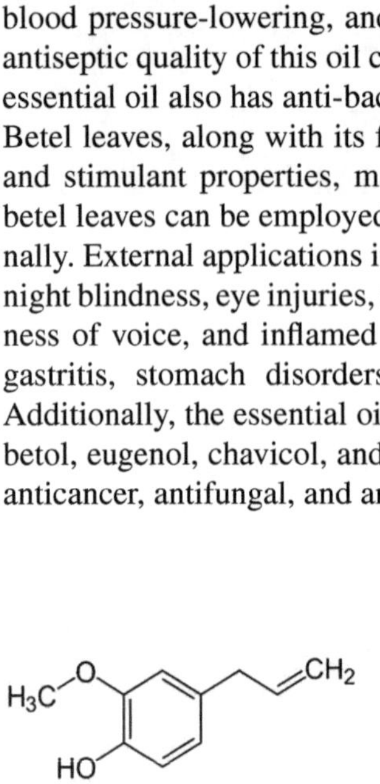

Eugenol
($C_{10}H_{12}O_2$)

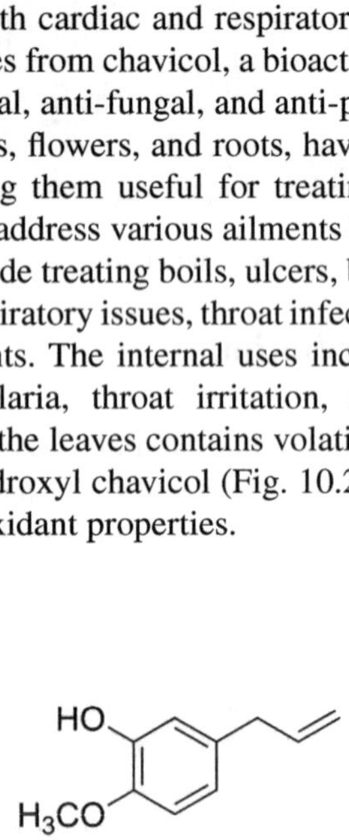

Chavibetol
($C_{10}H_{12}O_2$)

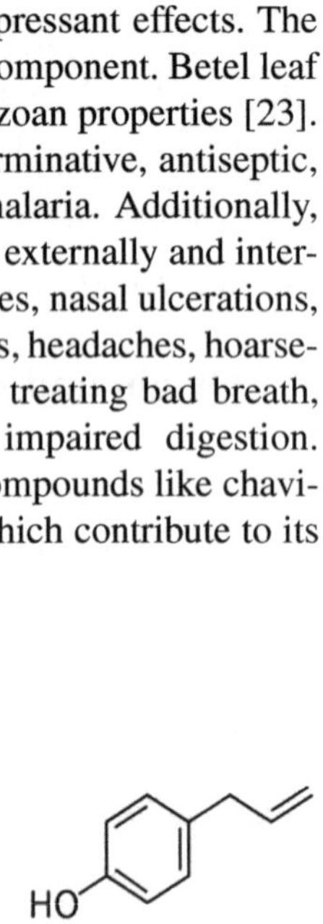

Chavicol
($C_9H_{10}O$)

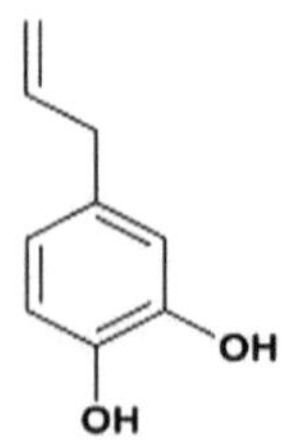

Hydroxychavicol
($C_9H_{10}O_2$)

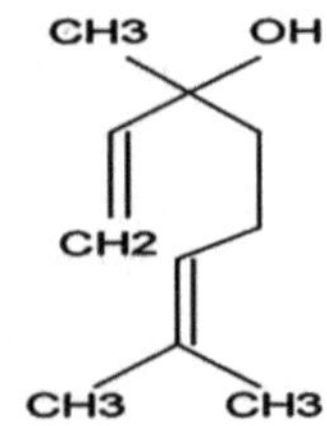

Linalool
($C_{10}H_{18}O$)

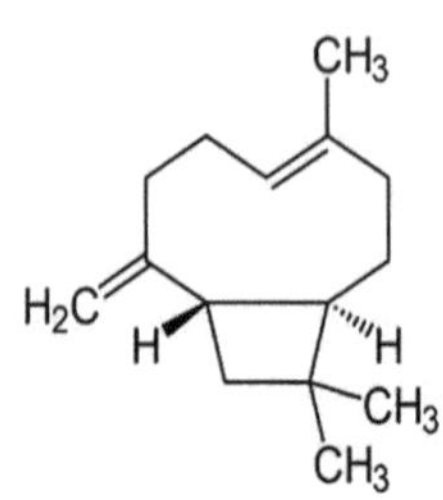

β - Caryophyllene
($C_{15}H_{24}$)

Fig. 10.2 Structure of some bioactive compounds from *Piper betle*

5 Commercial Application of Betelvine Leaves in the Food and Beverages Industry

Several commercial products containing betel leaves are available, including mouthwash, cosmetic, and personal care goods like toothpaste, face cream, soap, shampoo, antiseptic lotions, perfumes, and medicinal items [24]. As a result, the future creation of medicinal products from betelvine leaves shows potential for treating dermatophyte infections and preventing oral ailments. These nanoparticles have promising antibacterial activity and could be investigated further for use in medical devices and biomedical and pharmaceutical industries [25]. This betel leaf oil is a food additive that shows promise in the food industry for extending and protecting product shelf life during processing and storage. This essential oil not only aids in preventing food product spoilage but also ensures consumer health safety, notably by suppressing aflatoxin production. Investigations have shown that betel leaf essential oil in apple juice might be able to kill spores or prevent them from germinating, which are crucial processes for reducing fungal infections and the production of mycotoxins [22, 26].

6 Bioactivity Studies of Betelvine

Bioactivity studies in betelvine have been studied and reported by many workers (Table 10.1). Antibacterial activities of betelvine essential oil and extract have been examined against a wide range of plants and animal pathogenic bacteria. Saxena et al. [27] tested the antibacterial and antioxidant prospects of leaf essential oil of two betelvine landraces—Bangladeshi and Desawari—against a group of pathogenic microbes. The essential oils of these two landraces were reported to exhibit antibacterial potential against *Staphylococcus aureus*, *Klebsiella pneumonia*, and *Staphylococcus epidermidis*. Deshpande and Kadam [28] studied the antibacterial activity of betelvine against some oral pathogens, which are the main causative agents of dental caries and oral cancer like *Streptococcus mutans*. Methanol extract exhibited greater inhibitory activity followed by ethyl acetate extract of betelvine leaves. These results suggest that betelvine could be employed as a natural anticarcinogenic agent in oral care products. GC-MS (gas chromatography–mass spectrometry) analysis of these extracts revealed the presence of antibacterial and antioxidant compounds, which form the primary platform for further pharmacological studies [28]. Betel leaf oil is reported as a very effective antimicrobial agent [38]. The antimicrobial activity of the Vellaikodi landraces betel oil was reported as opposed to some dental pathogens like *Lactobacillus acidophilus*, *Staphylococcus aureus*, *Saccharomyces cerevisiae*, *Streptococcus mutans*, and *Candida albicans*. Based on the antimicrobial activity assay data, the Vellaikodi landraces of betelvine leaf oil exhibited notable effectiveness, attributed to its high content of 5-(2-propenyl)-1,3-benzodioxole. The antimicrobial potential against tested pathogens and the

Table 10.1 Various studied bioactivities of *Piper betle* L.

S. No.	Plant part	Active substance	Studied bioactivity	References
1	Leaves of Bangladeshi and Desawari landraces	Essential oil	Antibacterial activities against *staphylococcus aureus*, *klebsiella pneumonia*, *staphylococcus epidermidis*	Saxena et al. [27]
2	Leaves	Methanol extract	Antibacterial activities against *streptococcus mutans*, the causative agent of dental caries and oral cancer	Deshpande and kadam [28]
3	Leaves of Vellaikodi landraces	5-(2-propenyl)-1,3-benzodioxole from essential oil	Antibacterial activities against dental pathogens like *lactobacillus acidophilus*, *Staphylococcus aureus*, *Saccharomyces cerevisiae*, *Streptococcus mutans*, and *Candida albicans*	Sugumaran et al. [29]
4	Cured leaves	Essential oil	Antimicrobial activity against *mycobacterium smegmatis*	Madhumita et al. [8]
5	Leaves	Methanol and aqueous extracts	Strong inhibitory activity against the yeasts *M. Pachydermatis* and *C. Albicans*.	Rowa and Hob [30]
6	Leaves	Essential oil	Antimicrobial activity against multidrug-resistant bacteria like *Escherichia coli*, *Citrobacter freundii*, *Acinetobacter baumannii*, *Klebsiella pneumoniae*, *Enterobacter cloacae*, and *Pseudomonas aeruginosa*	Das et al. [31]
7	Leaves	Hydroxychavicol from chloroform extract	Reduce the growth of biofilm by *C. Albicans*	Ali et al. [32]
8	Leaves of Bangladeshi and Desawari landraces	Essential oil	Inhibitory activity against *Saccharomyces cerevisiae*, *Candida albicans*, *Candida rugosa*, *aspergillus flavus*, and *aspergillus Niger*	Saxena et al. [27]
9	Leaves	Chavibetol of essential oil	Potent antioxidant activity against DPPH and ABTS free radical	Das et al. [31]
10	Leaves	Essential oil	Antioxidant activity through DPPH free radical scavenging assay, reducing power and chelating effect of ferrous ions	Rowa and Hob [30]
11	Leaves	4-Allylphenolic components of essential oil	Mushroom tyrosinase inhibition activity with IC50 value 126 ppm	Rowa and Hob [30]

(continued)

Table 10.1 (continued)

S. No.	Plant part	Active substance	Studied bioactivity	References
12	Leaves	Essential oil	Larvicidal activity against the causative agent of dengue fever *Aedes aegypti* L.	Wahyuni [33]
13	Leaves	Methanol extract	Mosquito larvicidal activity	Rowa and Hob [30]
14	Leaves	Essential oil	Insecticidal activity on adult *M. Domestica*	Mohottalage et al. [34]
15	Leaves of Bangla Mahoba and Kapoori Vellaikodi landraces	Methanolic extracts	In vitro anti-leishmanial activity	Misra et al. [35]
16	Leaves	Ethanolic extract and essential oil	Antihistaminic activity tested using chlorpheniramine maleate as standard in Guinea pig	Hajare et al. [36]
17	Leaves	Methanolic extract	Hypolipidemic activity in high-fat diet-induced hyperlipidemia rat	Thirumalai et al. [37]

presence of the bioactive constituent of betelvine leaf oil suggested it is a source of antimicrobial components for dental caries in commercial toothpaste [29]. The antibacterial activity of essential oil of both fresh and cured betelvine leaves was evaluated against a multidrug-resistant bacterium, i.e., *Mycobacterium smegmatis*, a gram-positive bacterium, *Staphylococcus aureus*, and a gram-negative bacteria, *Pseudomonas aeruginosa.* The results obtained from these studies showed that the essential oil of cured leaf sample had significantly higher antimicrobial activity against *M. smegmatis* than the essential oil of fresh leaf samples [8].

The antioxidant, and antimicrobial activity of *P. betle* leaf essential oil of 30 different landraces in India was reported. Maghai landrace with a high percentage of chavibetol possesses higher antioxidant activity than others, followed by Utkalsudama and Nagar Pan landraces. Similarly, the leaf essential oil of all tested 30 landraces showed potent antibacterial activity against multidrug-resistant strains such as *Escherichia coli*, *Acinetobacter baumannii*, *Klebsiella pneumoniae*, *Citrobacter freundii*, *Enterobacter cloacae*, and *Pseudomonas aeruginosa* with inhibition zone ranges from 10 to 25 mm (diameter) [31].

The antioxidant activity, antimicrobial activity, and tyrosinase inhibition potential of the betelvine leaf essential oil and leaf aqueous and methanol extracts were studied by Rowa and Hob [30]. The leaf extracts obtained using both methanol and water as solvent displayed strong activity against the yeasts *Malassezia pachydermatis* and *Candida albicans*. The betelvine oil also exhibited positive antimicrobial properties against all the tested pathogens. Among them, the strongest activity was noticed against *C. albicans* followed by *S. aureus* and *Malassezia pachydermatis*. The antioxidant activity of the betel leaf extracts and essential oil was evaluated by using different techniques like DPPH(2,2-diphenyl-1-picrylhydrazyl) free radical scavenging activity, reducing power, and chelating effect of ferrous ions. It was also reported that the leaf essential oil showed mushroom tyrosinase inhibition activity in a concentration-dependent manner with an IC50 value of 126 μg/ml. The authors reported the major act in the inhibition of tyrosinase is due to the presence of 4-allyl phenolic components present in the essential oil [30]. Foo et al. [39] suggested the aqueous extraction method for antimicrobial activities of betelvine leaves to be a cost-effective and green process as water is non-toxic and eco-friendly, and the yield of aqueous extract was also high as compared to ethanol. Hydroxychavicol was isolated from the chloroform extract of betelvine leaves. The presence of this active compound in betel leaf was considered for its antifungal activity against 124 fungal strains. This compound was also noticed to reduce the growth of biofilm by *C. albicans*. Therefore, the antifungal activity expressed by this hydroxychavicol permits its use as a potent antifungal agent in mouthwash against oral *Candida* infections and also in treating topical infections [32].

The antifungal activity of two landraces of betelvine leaves, namely Bangladeshi and Desawari, was evaluated against some pathogenic microbial strains including *Saccharomyces cerevisiae*, *C. albicans*, *C. rugosa*, *Aspergillus flavus*, and *A. niger*. The antimicrobial potential of the oils of both these two landraces was found to be active against fungal strains. According to them, these essential oils could be utilized in creating antimicrobial agents as a natural remedy, serving as an alternative to synthetic antimicrobial products [27]. The larvicidal activity of the leaf essential oil from Indonesian betelvine was reported against *Aedes aegypti* L., which is the major vector of dengue fever in Indonesia [33]. The betelvine essential oils from Indonesia against *A. aegypti* showed potential alternatives to new bioinsecticide against *A. aegypti*, replacing insect resistance to insecticidal *Bacillus thuringiensis* toxins expressed in a transgenic plant. Rowa and Hob [30] also reported that the essential oil of betelvine exhibited mosquito larvicidal activity. The betelvine leaf extract prepared using methanol as solvent showed more larvicidal activity as compared to the aqueous extract.

The betelvine leaf essential oil obtained from Sri Lanka exhibits insecticidal activity on adult *Musca domestica*. This leaf oil was found to be a highly potent insecticide than Ceylon citronella oil as both the eugenol (6.4%) and safrole (52.7%) themselves exhibit high insecticidal potential; they may be responsible for the insecticidal activity of leaf essential oil of betelvine. Safrole has been tested to be more toxic than eugenol when compared with the positive control citronellal with an exposure time of 24 and 28 h [34]. Widowati et al. [40] determined the anticancer, antioxidant, and apoptosis-promoting properties of the leaf extracts of the Piperaceae family members (*P. betle* L., *P.fragile* Benth, *P. pellucidum* L., *P. bellatum* L., *P. aduncum* L.) on cervical cancer cells, particularly HeLa cell line. The result obtained from the research demonstrated that the entire extracts exhibit high anticancer activity, which increases along with the incubation time, i.e., longer the incubation time, the higher the anticancer activity. Among all the extracts tested, *P. betle* extract exhibited the highest antioxidant potential as revealed by SOD (superoxide dismutase) activities and DPPH scavenging assays and is also found to be the active inducer of apoptosis in HeLa cells at an incubation period of 24 and

48 h. From the overall study, it was reported that this extract shows limited anticancer activity in HeLa cells, with the most potent anticancer effects observed in *P. pellucidum* and *P. fragile* extracts.

Misra et al. [35] reported the comparative in vitro anti-leishmanial activity of methanolic extracts from two betelvine landraces (i.e., Bangla Mahoba and Kapoori Vellaikodi). The experiment stated that Bangla Mahoba was able to inhibit both the stages of *Leishmania* parasites selectively. This is initiated by stimulating the apoptotic actions of the production of reactive oxygen species, intending the mitochondria without any cytotoxicity toward macrophages. Based on the differences in the chemical composition of the essential oil assayed through GC-MS, further research was continued to evaluate the bioactivity of the methanolic extract of Kapoori Vellaikodi and Bangla Mahoba. As in betelvine (var. Bangla Mahoba), essential oil eugenol content was higher than the eugenol content of Kapoori Vellaikodi landrace essential oil, and methanolic extract of Bangla Mahoba essential oil showed the anti-leishmanial efficacy mediated through apoptosis. Therefore, betelvine leaves, being directly consumed by many people all over the world, can be used as folk medicine having a high therapeutic potential.

Hajare et al. [36] reported that the betelvine plant is effective in treating histaminic-related diseases due to the presence of its antihistaminic potential. The pharmacological evaluation of betelvine leaf ethanolic extract and essential oil has been conducted on guinea pigs using chlorpheniramine maleate as standard, which reveals their antihistaminic activity. Thirumalai et al. [37] evaluated and concluded that methanolic leaf extract of betelvine exhibited a significant hypolipidemic activity in high-fat diet-induced hyperlipidemia rats.

The effect of betel leaf ethanolic extract at doses of 100 mg/kg and 200 mg/kg body weight was evaluated in guinea pigs with asthma and compared with a standard antihistaminic drug, chlorpheniramine. The results obtained from this study revealed that betelvine leaf has significant anti-asthmatic activity [41].

Betel leaves also possess radioprotective properties as reported by Bhattacharya et al. [42]. This activity has been demonstrated in betelvine ethanolic extract using two in vitro models: rat liver mitochondria and pBR 322 plasmid DNA. The radioprotective activity of this extract can be ascribed to the superoxide and hydroxyl radicals scavenging potential of this extract together with its lymphoproliferative activity. It was found that allyl pyrocatechol and chavibetol are the phenolic constituents responsible for the radical scavenging property in the betel leaf extract.

7 Bioactive Constituents of Betelvine Leaf Essential Oils

The oil yield percentage of betel leaf essential oil ranges from 0.1%–2.0%. Betelvine is rich in many secondary metabolites including flavonoids, phenolic acids, alkaloids, terpenoids, and saponins [43]. The essential oil has cardiac, hypersensitive, and respiratory relaxant effects together with antibacterial and antifungal potential. It also has varied pharmacological potentials due to the occurrence of chavibetol in its essential oil [44–46]. Besides chavibetol, another active chemical constituent hydroxychavicol, which belongs to the phenol group, has also been detected in betel leaves. Hydroxychavicol is utilized in the pharmaceutical industry as it possesses anticarcinogenic activity [47]. Rawat et al. [45] studied five betelvine landraces and identified those based on the presence of volatile chemical components in their essential oil. A total of 51 components were identified, and eugenol was found to be the major constituent found in almost all the landraces. Similarly, eugenol, linalool, estragole, α-copaene, caryophyllene, and chavicol were found to be the most predominant components of essential oil of betelvine leaves. A total of 33 and 30 volatile compounds were isolated from fresh leaves and cured samples of betel representing a total of 98.41% and 97.34% of total oil, respectively [8]. Furthermore, the essential oil yield percentage and bioactive constituents of many available varieties of betelvine were studied by many workers [29, 48–50]. The yield was reported to be different in various betel landraces such as Mitha, Sanchi, and Bangla with 2.0%, 1.70%, and 0.80% of essential oil, respectively [14]. Caburian and Osi [51] reported 1.44% as the average volatile oil yield from betelvine leaves. The essential oil yield of 0.31%, 0.2%, and 0.08%, in volume by weight basis, was reported [29]. The yield of essential oil from the fresh new leaves was much higher than that of the stored leaves [52]. Gas chromatography coupled with mass spectroscopy (GC-MS) analysis is considered to be one of the best techniques to identify the volatile constituents, long chain, and branched chain hydrocarbons, esters, acids, and alcohols as it provides more precise and accurate information in qualitative analysis [28]. GC-MS analysis of Bangladeshi and Deswari cultivars of betelvine from India resulted in the identification of 35 and 25 compounds representing 86.11% and 85.41%, respectively. The major constituents recognized were eugenol, α-selinene, β-selinene, methyl eugenol, α-farnesene, and germacrene-D, which are found to be similar in both the oils except safrole, which was absent in the leaf oil of Bangladesi landrace [27]. The major components of betelvine leaf essential oil obtained from Sri Lanka were safrole (52.7%), allyl pyrocatechol diacetate (15.4%), eugenol (6.4%), and eugenyl acetate (5.8%) [34]. A wide variation was observed in the oil components of Indian, Vietnamese, Malaysian, and Philippine betel leaves. From the literature survey, it was found that leaf essential oil of Philippine betelvine contains chavibetol (53.1%) and chavibetol acetate (15.5%). These are phenolic compounds that serve as the major constituents. Besides the major components, betel leaf essential oil also consists of other phenolic compounds like eugenol, allylpyrocatechol monoacetate, methyl eugenol, safrole, allylpyrocatechol diacetate, camphene, p-cymene, terpenes, 1,8-cineole, alpha-pinene, caryophyllene, Beta-pinene, and limonene [53]. Betel oil from southern India contains safrole (39.9%) as one of the major constituents, and other components include eugenol (9.0%), terpinene-4-ol (6.3%), and allo-pyrocatechol monoacetate (8.5%). Eugenol was found to be the major compound of *Piper betle* cultivars named Ramtek Bangla and Desi Bangla of India with 90.5% and 82.2%, respectively, and methyl eugenol was also found at 4.1% and 6.9%, respectively [54]. Other minor components like terpineol, p-cymene, and terpinyl acetate were also present in the betel oil. Eugenol, at 26.65% and

28.29%, and terpenyl acetate, at 44.93% and 45.9%, were found in Desi Desawari and Mahoba cultivars of India, respectively.

8 Conclusion and Future Prospects

Betelvine is an economically and industrially important medicinal plant. The plant provides economic independence to thousands of betelvine growers in Asia. Despite so much economic value and abundant medicinal properties, the plant still lacks proper exploration in the global market due to restricted cultivation, the existence of a synonym, and unstable market demand. The use of unmarketable betelvine leaves for essential oil extraction would enhance the export potential of the plants in the national and international markets. Furthermore, future research is highly essential in bioactivity studies of betelvine essential oil. Development of elite planting material, creating a local market, and proper storage of unused betelvine leaves are the major necessities to meet the demand in the future.

References

1. Alves HDS, Rocha WRVD, Braz-Filho R, Chaves MCDO (2017) Isolation of monoterpene dihydrochalcones from *Piper montealegreanum* Yuncker (Piperaceae). Molecules 22:874
2. Atiya A, Sinha BN, Ranjan Lal U (2018) New chemical constituents from the *Piper betle* Linn. (Piperaceae). Nat Prod Res 32:1080–1087
3. Jeon H, Kim K, Kim YD, Lee SE (2019) Naturally occurring piper plant amides potential in agricultural and pharmaceutical industries: perspectives of piperine and piperlongumine. Appl Biol Chem 62:1–7
4. Preethy TT, Aswathi KK, Mannambeth RJ, Pillai AV (2016) Betelvine leaves—a green treasure house of useful chemicals. Int J Recent Sci Res 7(3):9216–9221
5. Das S, Parida R, Sandeep IS, Nayak S et al (2016) Biotechnological intervention in betelvine (*Piper betle* L.): a review on recent advances and future prospects. Asian Pac J Trop Med 9(10):938–946
6. Thomas UC, Chandini S, Thomas A (2013) Integrated nutrient management studies in betelvine (*Piper betle* L.). Int J Appl Res Stud 48:2–7
7. Ravindran PN, Johny AK, Nirmal BK (2002) Spices in our daily life. Satabdi Smaranika 2:102–105
8. Madhumita M, Guha P, Nag A (2019) Extraction of betel leaves (*Piper betle* L.) essential oil and its bio-actives identification: process optimization, GC-MS analysis and anti-microbial activity. Ind Crop Prod 138:111578
9. Dwivedi V, Tripathi S (2014) Review study on potential activity of *Piper betle*. J Pharmacog Phytochem 3(4):93–98
10. Periyanayagam K, Mubeen M, Saleem M, Basha SS (2011) Phytochemical studies and GC/MS analysis on the isolated essential oil from the leaves of *Piper betle* L. var. sirugamani1 (SGM1). J Pharm Res 4:2411–2413
11. Punuri JB, Sharma P, Sibyala S, Tamuli R et al (2012) Piper betle-mediated green synthesis of biocompatible gold nanoparticles. Int Nano Lett 2:18

12. Ranade SA, Soni A, Kumar N (2011) In: Grillo O (ed) SPAR profiles for the assessment of genetic diversity between male and female landraces of the dioecious betelvine plant (*Piper betle* L.). Ecosystems Biodiversity, Rijeka, pp 443–464
13. Thanh L, Dung NX, Luu HV, Leclercq PA (2002) Chemical composition of the leaf oil from *Piper betle* L. cultivated in Vietnam. J Essent Oil-bearing Plants 5(1):38–42
14. Guha P (2006) Betel leaf: the neglected green gold of India. J Hum Ecol 19(2):87–93
15. Hossain F, Anwar M, Akhtar S, Numan SM (2017) Uses impact of betel leaf (*Piper betle* L.) on public health. Md Sci J Public Health 5(6):408–410
16. Fazal F, Mane PP, Rai MP, Thilakchand KR et al (2014) The phytochemistry, traditional uses and pharmacology of *Piper betle*. linn (Betel Leaf): a pan-asiatic medicinal plant. Chin J Integr Med:1–11
17. Kaypetch R, Thaweboon S (2018) Antifungal property of *Piper betle* leaf oil against oral *Candida* species. In: MATEC web of conferences, vol 242. EDP Sciences, p 01021
18. Agarwal T, Singh R, Shukla AD, Waris I et al (2012) Comparative analysis of antibacterial activity of four *Piper betle* varieties. Adv Appl Sci Res 3(2):698–705
19. Joesoef MR, Sumampouw H, Linnan M, Schmid S et al (1996) Douching and sexually transmitted diseases in pregnant women in Surabaya, Indonesia. Am J Obstet Gynecol 174(1):115–119
20. Chowdhury U, Baruah PK (2020) Betelvine (*Piper betle* L.): a potential source for oral care. Curr Botany 11:87–92
21. Arambewela LS, Arawwawala ML, Withanage D, Kulathunga S (2010) Efficacy of betel cream on skin ailments. J Complement Integrat Med 7(1):48
22. Nayaka NMDMW, Sasadara MMV, Sanjaya DA, Yuda PESK et al (2021) *Piper betle* (L): recent review of antibacterial and antifungal properties, safety profiles, and commercial applications. Molecules 26(8):2321
23. Shah SK, Garg G, Jhade D, Patel N (2016) *Piper betle*: phytochemical, pharmacological and nutritional value in health management. Int J Pharm Sci Rev Res 38(2):181–189
24. Madhumita M, Guha P, Nag A (2020) Bio-actives of betel leaf (*Piper betle* L.): a comprehensive review on extraction, isolation, characterization, and biological activity. Phytother Res 34(10):2609–2627
25. Rashida MMO, Islam MS, Haque MA, Rahman MA et al (2016) Antibacterial activity of polyaniline coated silver nanoparticles synthesized from *Piper betle* leaves extract. Iranian J Pharm Res 15(2):591
26. Basak S, Guha P (2017) Use of predictive model to describe sporicidal and cell viability efficacy of betel leaf (*Piper betle* L.) essential oil on *Aspergillus flavus* and *Penicillium expansum* and its antifungal activity in raw apple juice. LWT 80:510–516
27. Saxena M, Khare NK, Saxena P, Syamsundar KV et al (2014) Antimicrobial activity and chemical composition of leaf oil in two varieties of *Piper betle* from northern plains of India. J Sci Indust Res 2(1):95–99
28. Deshpande SN, Kadam DG (2013) Evaluation of antibacterial activity of extracts of *P. Betle* [Linn.] leaves against *Streptococcus mutans*. J Pharm Pharm Sci 4(11):1040 1050
29. Sugumaran M, Suresh GM, Sankarnarayanan K, Yokesh M et al (2011) Chemical composition and antimicrobial activity of Vellaikodi variety of *Piper betle* Linn. leaf oil against dental pathogens. Pharm Technol 3:2135–2139
30. Rowa LCM, Hob JC (2009) The antimicrobial activity, mosquito larvicidal activity, antioxidant property and tyrosinase inhibition of *Piper betle*. J Chin Chem Soc 56:653–658
31. Das S, Sandeep IS, Mohapatra P, Kar B et al (2022) A comparative study of essential oil profile, antibacterial and antioxidant activities of thirty *Piper betle* landraces towards selection of industrially important chemotypes. Ind Crop Prod 187:115289
32. Ali I, Khan FG, Suri KA, Gupta BD et al (2010) In vitro antifungal activity of hydroxychavicol isolated from *Piper betle* L. Ann Clin Microbiol Antimicrob 9(1):7
33. Wahyuni D (2012) Larvicidal activity of essential oils of *Piper betle* from the Indonesian plants against *Aedes aegypti* L. J Appl Environ Biol Sci 2(6):249–254

34. Mohottalage S, Tabacchi R, Guerin PM (2007) Components from Sri Lankan *Piper betle* L. leaf oil and their analogues showing toxicity against the housefly, *Musca domestica*. Flavour Fragr J 22(2):130–138
35. Misra P, Kumar A, Khare P, Gupta S et al (2009) Pro-apoptotic effect of the landrace Bangla Mahoba of *Piper betle* on *Leishmania donovani* may be due to the high content of eugenol. J Med Microbiol 58(8):1058–1066
36. Hajare R, Darvhekar VM, Shewale A, Patil V (2011) Evaluation of antihistaminic activity of *Piper betle* leaf in Guinea pig. Afr J Pharm Pharmacol 5:113–117
37. Thirumalai T, Tamilselvan N, David E (2014) Hypolipidemic activity of *Piper betle* in high fat diet induced hyperlipidemic rat. J Acute Dis 3(2):131–135
38. Adeltrudes BC, Marina OO (2010) Characterization and evaluation of antimicrobial activity of the essential oil from leaves of *Piper betle* L. Int Educ Sci Res J 2:2–13
39. Foo LW, Salleh E, Mamat SNH (2015) Extraction and qualitative analysis of *Piper betle* leaves for antimicrobial activities. Int J Eng Technol 2:1–8
40. Widowati W, Wijaya L, Wargasetia TL, Bachtiar I et al (2013) Antioxidant, anticancer, and apoptosis-inducing effects of *Piper* extracts in HeLa cells. J Exp Integrat Med 3(3):225–230
41. Misra KH, Ramu BK, Ranjita N, Bandyopadhyay M (2014) Evaluation of anti-asthmatic effect of ethanol extract of *Piper betle* Linn. Against histamine induced bronchospasm in Guinea pigs. Int J Basic Appl Chem Sci 4(1):67–73
42. Bhattacharya S, Subramanian M, Bauri A, Kamat JP (2005) Radioprotecting property of the ethonolic extract of the *Piper betel* leaf. J Radiat Res 46:165–171
43. Valle DL, Puzon JJM, Cabrera EC, Rivera WL (2016) Thin layer chromatography bioautography and gas chromatography-mass spectrometry of antimicrobial leaf extracts from Philippine *Piper betle* L. against multidrug-resistant bacteria. Evid Based Complement Altern Med 2016:1–7
44. Lakshmi BS, Naidu KC (2010) Comparative morphoanatomy of *Piper betle* L. cultivars in India. Annal Biol Res 1(2):128–134
45. Rawat AK, Tripathi RD, Khan AJ, Balasubrahmanyam VR (1989) Essential oil components as markers for identification of *Piper betle* L. cultivars. Biochem Syst Ecol 17:38–55
46. Thurairajah N, Rahim ZH (2007) The crude aqueous extract of *Piper betle* L. and its antibacterial effect towards *Streptococcus mutans*. Am J Biochem Biotechnol 3:10–15
47. Bhide SV, Zariwala MB, Amonkar AJ, Azuine MA (1991) Chemopreventive efficacy of a betel leaf extract against benzopyrene induced for stomach tumors in mice. J Ethnopharmacol 34:207–213
48. Ahmad FS, Mohamed SM, Jamil M, Jani NA (2024) Chemical composition, antibacterial activity, and ADME studies of leaf essential oil of *Piper betle*. Malaysian J Chem 26(3):456–464
49. Garg SC, Jain R (1996) Chavicol rich essential oil of *Piper betle* L., Cultivar Sagar Bangla. Euro Cosmetics 5:27–28
50. Ramalakshmi K, Sulochanamm G, Rao LJM, Borse BB et al (2002) Impact of drying on quality of betel leaf (*Piper betle* L.). J Food Sci Technol 39(6):619–622
51. Caburian AB, Osi MO (2010) Characterization and evaluation of antimicrobial activity of the essential oil from the leaves of *Piper betle* L. E-Int Sci Res J 2(1):2–13
52. Pradhan D, Suri KA, Pradhan DK, Biswasroy P (2013) Golden heart of the nature—*Piper betle* L. J Pharmacog Phytochem 1(6):147–152
53. Rimando AM, Han BH, Park JH, Cantoria MC (1986) Studies on the constituents of Philippine *Piper betle* leaves. Arch Pharm Res 9:93–97
54. Sharma ML, Rawat SKS, Balusubrahmanyam VR, Singh A (1983) Studies on essential of betelvine leaf *Piper betle* part II, Vars: Desi Bangla, Ramtek Bangla, Calcutta Bangla. Indian Perfumer 27:91–93

Chapter 11
Areca Nut Husk: Valorization into Bioproducts, Biomaterials and Biofuels

Arigo R. W. Sangma and S. R. Joshi

Abbreviation

ACE	Angiotensin-I-converting enzyme
ANH	Areca nut husk
ANHA	Areca nut husk ash
HDL	High-density lipoproteins
LDL	Low-density lipoproteins

1 Introduction

Global waste has increased due to population growth and a rise in food demand. Extensive research has been done on turning biowaste into biomaterials, with an exploration of different strategies [1]. India stands as one of the foremost producers of food crops globally, yielding substantial quantities of agricultural residues. The inherent bulkiness of these residues, coupled with the absence of organized markets for their commercialization, deterioration of quality during storage, and the urgency to ready fields for the cultivation of successive crops, compels farmers to resort to the practice of burning these residues. This method of disposal contributes significantly to the emission of particulate matter and greenhouse gases [2]. The environmentally friendly conversion of biomass into biofuels, different chemicals, and biomaterials is receiving more and more focus to safeguard the health and prosperity of current and upcoming generations. In the realm of biomass valorization technologies, hydrothermal conversion assisted by microwave has emerged as a

A. R. W. Sangma · S. R. Joshi (✉)
Microbiology Laboratory, Department of Biotechnology and Bioinformatics, North-Eastern Hill University, Shillong, Meghalaya, India

C. Trois, R. C. Ray (eds.), *Plantation Crop Wastes: Valorization for Economic Sustainability*, Waste as a Resource,
https://doi.org/10.1007/978-3-031-96805-1_11

Fig. 11.1 Pictorial depiction of the areca nut plant and the cross-section of the nut (fruit)

cutting-edge option, able to provide various reaction products for the energy, pharmaceutically important products, and chemical industries [3].

The consumption and utilization of areca nut (*Areca catechu* L.) are notably important in the Northeast part of India, particularly in Meghalaya. Specific types of raw nuts undergo fermentation to alter their flavor and taste. This process, traditionally carried out, has not been researched, and the variety and behavior of microbes present during the process remain unexamined [4]. The areca nut (Fig. 11.1), with a 60–70-year lifespan and tropical adaptation, remains vulnerable to various abiotic and biotic factors. Thus, investigating its adaptation strategies over time is essential [5].

2 Geographical Origin, Distribution, Production, and Productivity

2.1 Geographical Origin

The areca nut plant is native to China, Malaysia, and Indonesia although it is exotic to India, Fiji, Madagascar, Japan, Kenya, Pakistan, Philippines, Papua New Guinea, Solomon Islands, Samoa, Tanzania, and Sri Lanka [6]. Areca nut is originally found in Southeast Asia, especially in the tropical zones of Malaysia and the Philippines. It is believed that genuine wild populations do not exist in archipelagos [7].

2.2 Distribution

The areca nut thrives in tropical, consistently humid climates with annual rainfall ranging from 1500 to 5000 mm (60–200 inches) and is best suited for altitudes between 0 and 900 m (0 to 2950 ft). While it is suited to different types of soil types, it necessitates good drainage with higher moisture retention capacity. The growth rate is average, approximately 0.5 m annually (20 annually) [8, 9]. Even though the areca nut has fewer trees and leaves compared to other palms like coconut and oil palm, it still generates a similar amount of waste biomass (12 ‑18 t/ ha for coconut and 14–15 t/ ha for oil palm) because of the higher density of population per unit area. Areca nut generates a significant amount of recyclable biomass annually with leaf, leaf sheath, bunch wastes, and husk contributing to a total of 10–12 t/ha per year. The palm tree produces more recyclable biomass from its leaves compared to husk and other organic waste materials. The biomass from the husk makes up about 16.5–17% of the overall weight of dried areca nuts, yet it is mostly unused as a fuel source [10–12].

2.3 Production

Coconut and areca nut palms are primarily utilized for economic reasons in the palm tree-populated areas [13]. The tropical plant areca nut shows the best growth in temperatures range of 21 °C to 32 °C, in areas with high humidity and an average rainfall of 1500 to 5000 mm annually. The species is affected by cold temperatures, needs shielding from powerful gusts of wind and lack of water, and thrives in different types of tropical forests at elevations of up to 900 m. It favors well-drained soil with a pH range of 5–8, particularly deep clay loams that are high in organic material. Full exposure to sunlight is ideal, while young plants may benefit from some shade. Irrigation and manure are needed to make light sandy soil suitable for use [7].

2.4 Productivity

In a seven-year study on areca nut productivity and nutrient uptake, consistent nutrient application is crucial for boosting yields in laterite soils. Sustainable yields of approximately 2600 kg/ ha can be achieved through organic waste recycling. Chemical fertilizers showed a significant productivity increase of 73–85% with nitrogen, phosphate, and potassium (NPK) when compared to 48–59% with vermicompost and 46–63% with integrated treatments. Utilizing organic waste meets the N and P requirements, while alternative K sources should be investigated for better yields [14].

3 Description, Cultivation, Proximate Composition, Nutritional Values, and Uses

3.1 Description

The areca nut tree is tall with a single trunk and circular scars from fallen leaves. The large, pinnate leaves have many long, dark green leaflets fused with widened petioles. The flower cluster is a compact spadix within a spathe, with fragrant male flowers on thread-like branches on top and female flowers below. The fruits are oval, greenish-yellow turning orange when ripe, with a fleshy outer layer and a hard seed inside (Figs. 11.2 and 11.3). The tree can reach 15–25 meters in height, with leaves up to 2.5 meters long. Male flowers have greenish-yellow sepals, white petals, and six stamens, while female flowers have a perianth and a three-celled ovary [7]. A fibrous, oval drupe measuring 5 to 10 × 3 to 5 cm (2 to 4 × 1.2 to 2 in) varies in color from yellow to orange or turns reddish upon ripening. Seeds are typically ovoid, ellipsoidal, or even globular (3 to 4 × 2 to 4 cm [1.2 to 1.6 × 0.8 to 1.6 in]), a flattened base is occasionally visible, and the endosperm is a ruminate (accompanied by rigid reddish tissue from internal integument extending horizontally into pale brown endosperm); the embryo is conical and situated at the seed's base [8]. In Asia, areca nuts are peeled, dried, pickled, and sold in markets for consumption. The hard nuts are cracked, sliced, or shredded. Cracked nuts are ground into powder

Fig. 11.2 Ripe areca nut fruits

Fig. 11.3 Section of ripe fruits

for chewing, sliced nuts are taken in whole and chewed, and shredded nuts are used to make "gutkha" with tobacco or "paan masala" with fennel seeds [7]. Crop residues, in addition to being primarily utilized as livestock feed, are also employed as raw materials in the production of bioenergy for producing pulp in the paper sector packaging items, organic matter for composting, fertilizers, roofing material, and material for covering the soil [2].

3.2 Cultivation

During the early stages, areca nut is intercropped with yearly and every other year crops, while in the bearing stage, it is grown alongside crops such as banana, cocoa, pepper, clove, and acid lime in a high-density multi-species cropping system. Therefore, there is a vast opportunity to enhance organic waste and nutrient recycling within areca nut-dependent agricultural systems [12, 15, 16]. Only organic nutrition (vermicompost made from recyclable biomass) lowers areca nut kernel production during 11 years when compared to inorganic nutrition. The application of NPK has been shown to increase areca nut yield by 18%, while drip fertigation of 75% NPK has been shown to increase yield by 25% over vermicompost application [12].

3.3 Proximate Composition

The husk fibers of the areca nut are mostly made up of cellulose, along with various amounts of hemicellulose, lignin, pectin, and proto-pectin [17]. Areca nut contains over 50 compounds that include alkaloids, flavonoids, tannins, triterpenes, fatty acids, and other types of compounds. The fiber content of areca but husk (ANH) was 80.20% (ranging from 67.39 to 86.11%). In comparison to other fruit fibers [cotton (32%), coir (40%)], this fiber content is quite significant [18]. The seed of the areca nut has alkaloids including arecaine, arecoline, arecaidine, guvacine, guvacoline, and choline. It also contains tannin, gum, gallic acid, and a variety of minerals like calcium, copper, iron, and phosphorus (Fig. 11.4) [8].

3.4 Nutritional Values

The nutrient content found in these areca nut waste materials ranges from 0.62% to 1.59% N, 0.07% to 0.16% P, and 0.75% to 1.25% K. Areca nut produces around 95 g of N, 10 g of P, and 110 g of K per palm annually from recyclable biomass. Since husks have a high K content, they can serve as a viable K source in organic farming because areca nut requires a lot of K for growth. The effective reusing of these waste materials provides a significant amount of nutrients and largely fulfills the nutritional needs of crops [10–12].

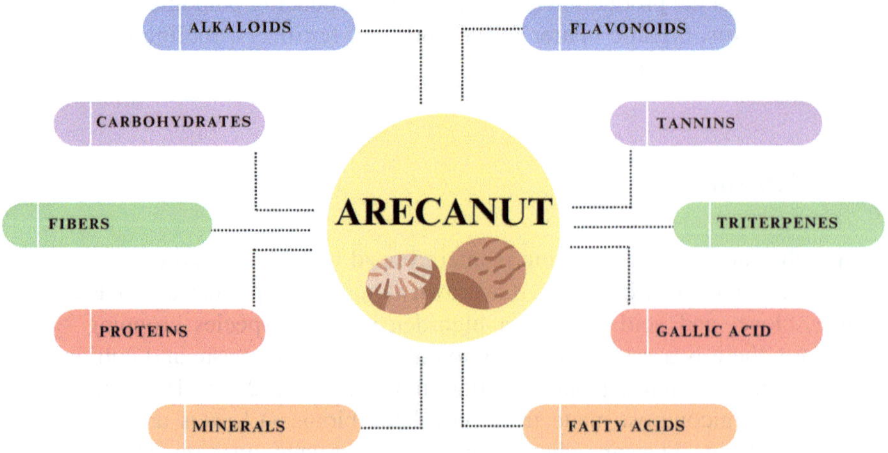

Fig. 11.4 Composition and nutritional values of areca nut

3.5 Uses

Areca nut has been utilized in traditional medicine since ancient times, especially in countries of South and Southeast Asia. Its application in ancient times was often referenced in the Sanskrit medical texts and subsequently in Hindu and Buddhist writings [19, 20]. Four elements underpin the widespread popularity of chewing areca nut and betel quid: social acceptance, religious practices, perceived health advantages, and dependence. It serves as an astringent, mouth freshener after eating; a flavor enhancer, laxative, and intoxicant; and for issues like indigestion, impotence, and gynecological concerns, parasitic intestinal infections, and to prevent pregnancy-related morning sickness [21, 22].

Agro-waste such as ANH can act as an appropriate medium in the solid-state fermentation procedure with mushrooms. Mushrooms break down lignocellulosic materials by producing lignocellulosic enzymes and use the resulting products to generate their fruiting bodies. Areca palm leaves serve as an excellent substrate for maximum oyster mushroom yield, complemented by topsoil. For the production of the spawn of two oyster mushrooms, four substrates including the coffee husk, chopped areca nut leaf sheath, paddy grains, and sorghum were utilized as media, which all facilitated *Pleurotus eous* achieving 100% colonization in 18,10, 16, and 10 days, respectively [23]. The uses and processing for areca nut products are shown in Fig. 11.5.

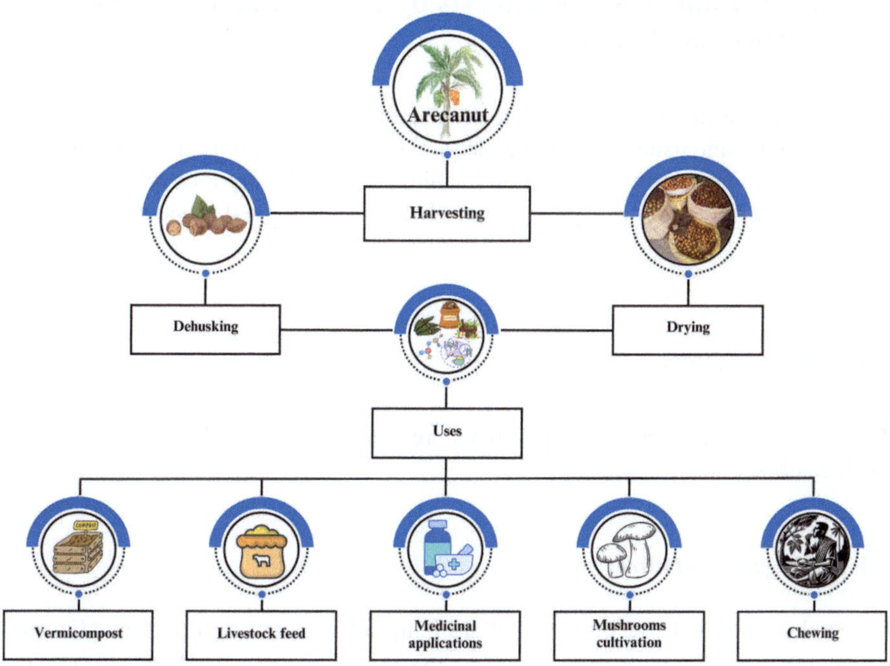

Fig. 11.5 Uses and processing for areca nut products

4 Processing for Food and Beverages, and Residue and by-Product Generation

In India, the biomass that can be recycled from areca nut is estimated at 4.5 to 5.4 million tons per year. Using them for making vermicompost and cultivating oyster mushrooms, as a source of livestock feed and other valuable by-products, are both sustainable and economically viable (Fig. 11.5). The isolation of cellulose nanofibrils from ANH allows effective use of waste. Chemical treatments eliminate non-cellulosic materials, while mechanical methods like grinding and homogenization individualize fibers to the nano dimension, enhancing their potential applications [24]. The nuts contain 8–12% fat that can be extracted using organic solvents such as chloroform and hexane. Fat can be made edible by refining it with alkalis. In making confectionary products, the refined fats can be mixed with cocoa fat and can be used in various products as an extender to cocoa butter. In the making of sweets and biscuits, this refined fat can be used as a substitute for the traditionally used vanaspati [25]. Creating nonwoven samples from areca nut fibers adds new functionality to the agri-waste fibers in the Meghalaya region [26]. Areca husk-activated carbons exhibited the highest killing effect and inhibition when tested against a waterborne pathogen that is widely accessible in *Escherichia coli* [27].

5 Valorization of Nut Waste, Its by-Products, and Prospects in a Circular Economy

Areca nut waste can be valorized in a circular economy by extracting valuable compounds from it and using them in a variety of applications. Valorization of waste can help reduce environmental impact and achieve a closed material cycle. It can also help generate new economic possibilities. The list of valorized products is shown in Table 11.1.

5.1 Bioproducts

The various bioproducts developed from areca nut wastes are discussed in the following sub-sections.

5.1.1 Organic Fertilizer

Researchers studied the potential of utilizing an African nightcrawler (*Eudrilus eugeniae*) to transform organic wastes in areca nut and cocoa pod waste into vermicompost. The findings showed that organic wastes can be effectively transformed into vermicompost, achieving a recovery rate of 74.65–87.75% within a 3-month

Table 11.1 Valorized products from areca nut husk

S. No.	Products	References
1.	Bioplastic	Dalimunthe et al. [28]
2.	Substrate for mushroom	Thulasi et al. [23]
3.	Textile fibers	Vasudeva et al. [17]
4.	Pyrolytic oil	Mishra et al. [29]
5.	Composites	Lazim et al. [30]
6.	Organic fertilizer	Annamalai et al. [31]
7.	Biomolecules • Manganese peroxidase • Citric acid • Ethanol • Cellulase production • Fungal pectinase • Xylose	Nidhi et al. [32], Narayanamurthy et al. [33], Naveen Kumar et al. [34], Sreena and Sebastian [35], Jayanna Naveenkumar et al. [36], Vardhan et al. [37]
8.	Bioadsorbents • Pb^{2+} removal • Brilliant green dye removal. • Methylene blue adsorption • Iron removal • Phenol removal • Lead and copper removal • Oil spill cleanup	Ramesh et al. [38], Baidya and Kumar [39], Tabassum et al. [40], Subramani et al. [41], Girish (2017); Nagarajappa et al. (2020), Nik Ab Lah et al. [42]
9.	Feed additives • Chicken feed • Fish feed	Azis et al. [43], Murwani et al. [44], Mahata et al. [45], Ifedi et al. [46]
10.	Biofuels	Das et al. [27]
11.	Biochar	Vijayanand et al. [47]
12.	Natural dyes	Kabir et al. [48]
13.	Paper making	Setty et al. [49]

composting period. The microbial population in vermicompost was significantly greater than in traditional compost [31]. A study was conducted where the nutritional value of ANH and its decomposition with microbial consortia over three years. The husk's C: N ratio saw a notable decrease following composting with the consortia. Levels of nutrients such as N, P, K, Zn, Mn, Fe, and Cu in compost rose throughout 90 days. As compost aged, nutrient levels increased when treated with microbial consortia, cow dung, and glyricidia [50]. Using areca palm waste for growing oyster mushrooms and then turning the remaining material into organic fertilizer through composting or vermicomposting is a practical way to recycle lignocellulosic biomass. Areca leaf sheath shows potential as a suitable medium for large-scale cultivation of oyster mushrooms (*Pleurotus sajor-caju*) [12, 51, 52].

5.1.2 Soil Amendment

Composting is a successful way to handle biodegradable solid waste. In a study done on composting areca nut waste, 8 kg of materials were utilized in a pit of precise measurements. Microbial organisms were added to the layers of waste to assist in decomposition. The findings demonstrated a decline in organic carbon levels and a rise in nitrogen levels, along with a significant decrease in C:N ratios in comparison to untreated waste. Areca nut waste treated with various microorganisms showed a notable decrease in C:N ratios, suggesting enhanced decomposition [53]. When ANH fiber was compared to fine fiber composites, coarse fiber composites were found to have a higher impact strength composed of two layers of glass fiber encased in a coarse layer of ANH fiber [54]. The response of soil microbiology and enzymes to the recycling of organic matter and the use of chemical fertilizers in an areca nut-based cropping system on laterite soil was assessed. Treatments consisted of organic matter recycling along with its combination with various grade concentrations of NPK. Notable differences in microbial populations, microbial biomass carbon, microbial quotient, and phosphatase and dehydrogenase activities were observed as a result of nutrient management in the basins of different component crops within the areca nut-based cropping system [55]. The pyrolytic transformation of an abundantly found agricultural leftover—the ANH—into biochar as a sustainable strategy for residue management was studied where the use of areca nut husk biochar along with vetiver (*Chrysopogon zizanioides* L.) was employed as an innovative agrotechnological method to evaluate enhancements in soil quality. The amendment of areca nut husk biochar in the soil resulted in a rise in both plant growth and relative water content that may be linked to the rise in cation exchange capacity as well as organic carbon and soil nutrients [56].

5.1.3 Biomolecules

Fungal strains were tested for manganese peroxidase production through the plate assay method by using ANH as a substrate by solid-state fermentation and found that *Fusarium* species was the most effective [32] and even *Phanerochaete*, a type of white-rot fungus was also used to produce enzymes that modify lignin, such as manganese peroxidase [57]. Areca nut husk was found to be a suitable material for citric acid production by *Aspergillus niger* through the solid-state fermentation method, offering a solution to environmental pollution caused by disposing of the residue [33]. The efficient use of ANH as an economical carbon source for the synthesis of low-cost endoglucanase can subsequently be utilized for the development of second-generation biofuel [58]. The fermentation of ANH to produce ethanol demonstrates potential as a biofuel supplement for addressing global warming and pollution. Microbial pretreatment with cellulolytic fungi and fermentation with *Zymomonas mobilis* NCIM 2915 resulted in the highest ethanol production where ANH was found to be a suitable substrate, indicating the effectiveness of fungal pretreatment for biofuel production [34]. Agro waste such as ANH was used for cellulase production as a carbon source in production media and the activity was comparable to that of what is carboxy methyl cellulose activity [35] and even *Trichoderma viride* in solid-state fermentation utilized the agricultural husk of areca nut as a feeding base for the synthesis of the cellulase enzyme [59]. *Penicillium canescens* exhibited the highest pectinase activity by utilizing ANH waste to create fungal pectinase using submerged fermentation for industrial purposes [36]. The most neglected lignocellulosic substance—the ANH—was pretreated and characterized as a source of inexpensive xylose because it contains a substantial quantity of xylose, a type of pentose sugar [37]. Research was conducted on the anticoccidial properties of areca nut extract (ANE) regarding coccidiosis. ANE and diclazuril reduced the mucosal harm induced by coccidial infection. The supplementation of diet ANE elevated the levels of interleukin 2 (IL-2) in infected chickens compared to the negative control group and the results showed ANE had anticoccidial properties and a beneficial effect on intestinal mucosa damage in broiler chicks with coccidiosis [60]. The valorization of areca nut waste into bioproducts is shown in Figs. 11.6 and 11.7.

5.2 *Biomaterials*

The various biomaterials developed from areca nut wastes are discussed in the following sub-sections.

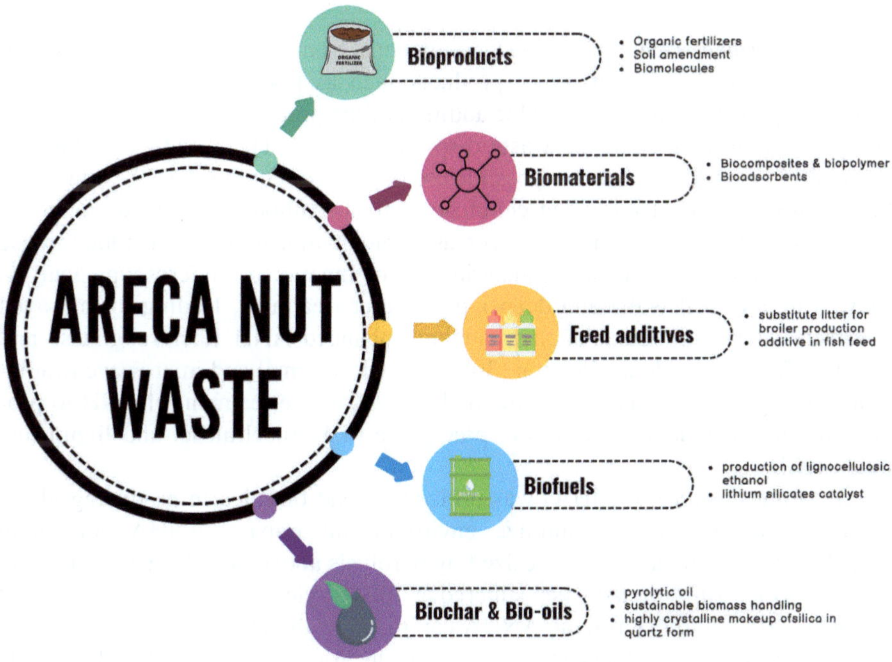

Fig. 11.6 Valorization of areca nut waste

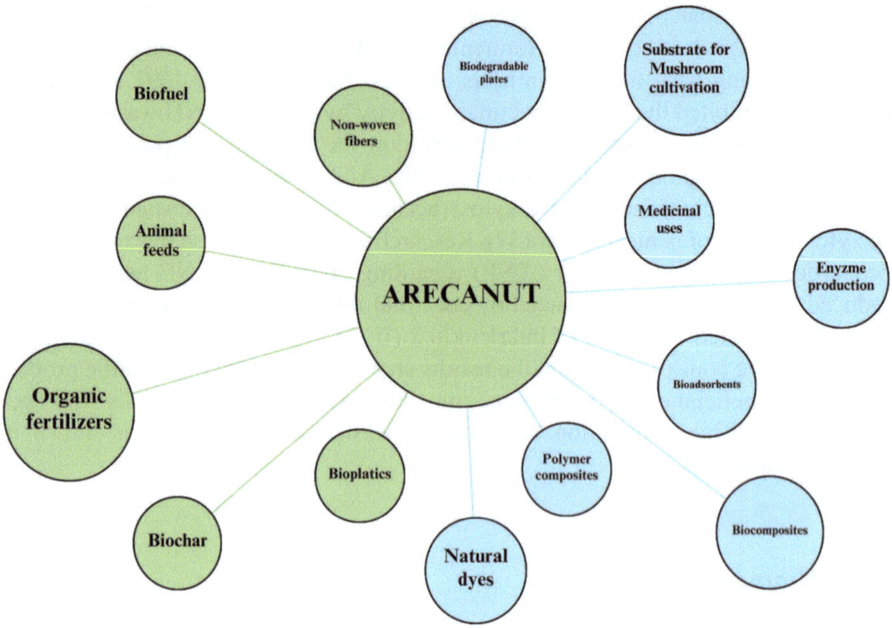

Fig. 11.7 Areca nut substrate and its products

5.3 Biocomposites and Biopolymer

Anaerobic fermentation of areca husk components from urban solid waste and rural residues can produce a variety of products such as fiber, biogas, compost, and digester liquids that can lead to value addition of the products and improve the sustainability and viability of the waste disposal system [61]. A study was done by examining waste ANH fibers for possible use in polymer composites. Treating ANH fibers with 5% NaOH increased cellulose content. Although tensile strength and Young's modulus were found to be decreased, elongation at break was found to rise due to the removal of cementing materials. Scanning electron microscopy analysis showed thinner cell walls and deeper pores after treatment, leading to improved interfacial shear strength. Alkali treatment enhanced ANH fiber roughness and potential for composite application [30], and hydrothermal and microwave irradiation were also utilized as heating methods for alkali pretreatment of ANH to produce residue abundant in cellulose, precipitate of hemicellulose, and lignin-rich liquor [62].

Areca nut husk ash (ANHA) was utilized instead of silica in producing glass-ceramic zinc silicate to minimize environmental impact. ANHA containing 29.17%–45.43% silica was synthesized by pyrolysis above 700 °C. Zinc oxide and ANHA formed zinc silicate when sintered at 700 °C, yielding superior outcomes at elevated temperatures. Utilizing ANHA can result in the inexpensive and environmentally friendly manufacturing of zinc silicate for phosphors [63]. The tensile strength of ANH fiber is comparable to that of coir and kenaf fiber, making ANH fibers ideal for reinforcing polymer composites [64]. Evaluating the tensile properties of alkali-treated maize fiber and areca hybrid composites compared to untreated fiber hybrid composites showed that treated fibers achieve higher quality than untreated ones. An increase in stress was noticed in composites of natural fibers treated with 10% NaOH and reinforced with tamarind fruit fibers and ANH fibers, which made the epoxy hybrid. The mechanical properties such as tensile strength, bending stress, and impact of polypropylene composites filled with areca nut short fiber at various compositions (3, 5, 10, 20, and 30 wt.%) decreased as the percentage of filler increased. Composites containing 10% areca nut fiber demonstrated superior mechanical strength [65–67].

5.3.1 Bioadsorbents

ANH hydrochar can act as an adsorbent to eliminate Pb^{2+} due to its demonstrated effectiveness. The highest elimination rate of 95.08% was observed at a concentration of 25 mg/L in the sewage water [38]. ANH treated with 0.1 M NaOH has been effectively utilized as a reliable adsorbent for brilliant green dye removal. Chemically treated ANH from a water solution can serve as a cost-effective alternative material for effectively removing brilliant green dye from the water solution [39]. Activated ANH hydrochar using NaOH also served as an adsorbent for methylene blue adsorption [40]. The process of pyrolyzing ANH to produce biochar can help in the removal of iron from water solutions. Adsorption research indicated that the adsorption ability heightened as the initial iron concentration and contact duration rose, yet declined with greater adsorbent dosage. The adsorption process showed it to be spontaneous, exothermic, and involving both physisorption and chemisorption mechanisms. Analyses and successful elimination of iron using pyrolyzed areca husk indicate its potential as a viable alternative adsorbent [41]. The use of ANH, an agricultural by-product, as an adsorbent for phenol removal from wastewater was investigated. The adsorbent underwent chemical treatment with phosphoric acid at an impregnation ratio of 1:4. The chemically treated adsorbent demonstrated a surface area of 678.78 m^2/g, indicating good potential. Optimization was performed using response surface methodology for factors like dosage and temperature. Adsorption followed the Langmuir Isotherm and pseudo-second-order kinetics, with a capacity of 27.746 mg/g, highlighting the ANH's efficacy in wastewater treatment [68]. Wastewater effluents significantly contribute to water pollution, particularly from iron and metal industries in India, which release heavy metals. These metals are now recognized as critical environmental pollutants. Soil aquifer treatment is utilized for groundwater recharge, leveraging the unsaturated zone and aquifer as natural filters to diminish pollutant concentrations via various processes. The soil aquifer treatment system study assesses the removal efficiency of lead and copper using an areca nutshell revealing that copper removal is more effective than lead in demonstrating superior adsorption capacity [69].

A scientific research study was conducted to evaluate the ANH as a potential material as an oil adsorbent for oil spill cleanup. The ANH was evaluated for the adsorption of crude oil and diesel, and the observations revealed the oil sorption capacity of ANH for heavy crude oil to be around 14.8%, suggesting it could serve as a potential sorbent material to contemporary options while the use of ANH for diesel absorption is not advisable due to its low effectiveness [42].

5.4 Feed Additives

ANH can serve as a substitute litter material for broiler production without compromising the performance of the chickens where the gizzard weight was greater in broilers raised on rice hulls compared to those raised on dried leaves and ANH [43]. The study evaluated broiler performance when fed fermented versus unfermented areca nutshell. Utilizing this waste for poultry feed improved performance and lowered cholesterol levels. The objective was to ferment the areca nutshell with Indigenous microorganisms to produce cellulase, reducing crude fiber before feeding. The results showed that fermented areca nut shells significantly enhanced daily feed intake, body weight gain, and carcass content while reducing abdominal fat, cholesterol, HDL (high-density lipoproteins), and LDL (low-density lipoproteins) levels. Fermented areca nutshell can be included up to 6% in broiler rations, yielding positive impacts [45].

Utilizing inexpensive substrates instead of synthetic materials is a method to lower expenses. This research investigates the potential for the endoglucanase enzyme production by *Bacillus subtilis* MUS1 utilizing five types of agricultural waste as a source of carbon. The agro-wastes, including ANH, jackfruit outer rind, banana peduncle, pepper waste, and tamarind husk, were evaluated for their potential as carbon sources for the production of the enzyme. For increasing the endoglucanase enzyme production, statistical methods were used for the optimization of the ANH medium, and the wastes ANH demonstrated the best endoglucanase activity. Enhancing the ANH medium resulted in a 2.2-fold rise in endoglucanase activity [58]. A study was conducted on fingerlings (*Clarias gariepinus*) with starting weights from 4.55 ± 0.25 to 5.53 ± 1.00 g to assess the impact of adding areca nut seed flour in varying amounts to the diet formulation on growth performance, nutrient use, body composition, histopathology, and blood parameters. The proximate analysis findings indicated an enhancement in carbohydrate content along with a decrease in protein content as the incorporation level rises indicating that areca nut seed flour may prove to be beneficial as an additive in fish feed because of its rich carbohydrate content [46]. The alternate supplementation of areca nut and *Anredera cordifolia* leaf powder every 3 days for 18 days decreased the fecal associated with endoparasites and enhanced histopathology of tissues affected by endoparasites in layer hens, particularly in the caecum [44].

5.5 Biofuels

With the increasing demand for solutions for increasing sustainable energy worldwide, agro-waste is becoming a key contender for biofuel production. The most effective technique for saccharification of ANH was found to be fungus treatment, which can be used to transform lignocellulosic waste into valuable substances like fermentable sugars for the production of lignocellulosic ethanol [27]. ANH treated with lithium compound is a successful catalyst for converting used cooking oil into biodiesel. The catalyst predominantly consists of lithium silicates as its primary crystalline materials. It exhibits great performance in transesterification with minimal catalyst usage and gentle reaction conditions as it is perfect for creating biodiesel from inexpensive feedstocks with high levels of free fatty acids and is reusable [70].

5.6 Biochar and Bio-Oil

Biochar is a solid substance produced through the process of biomass carbonization via destructive distillation. It can be included in soil to enhance soil functions and lessen emissions from biomass that naturally decay into greenhouse gases. The process of converting biomass such as areca nut waste into biochar through thermal methods could be a viable approach for sustainable biomass handling [47]. Significant advancements are being achieved in utilizing all parts of plant material for extracting energy by converting plant remnants through pyrolysis making ANH, a potential candidate for pyrolysis [71]. Pyrolysis of areca nut sawdust with a catalyst in an optimal setting greatly enhanced the characteristics of pyrolytic oil, which is an intricate mixture of hundreds of components that make up this synthetic liquid fuel [72]. Areca nut shell ash possesses a highly crystalline makeup featuring silica in the quartz form. It also includes sylvite (KCl), iron oxides (FeO and Fe_2O_3), and potassium sulfate (K_2SO_4). Scanning electron microscope images demonstrate the existence of particles with irregular shapes in all ash powders. Thermal analysis findings indicate that the majority of weight loss occurs between 400 °C and 600 °C due to the elimination of volatile as well as residual carbon content. Areca nut ash exhibits weight reduction exceeding 700 °C owing to the vaporization of KCl [73].

6 Techno-Economic Analyses and Industrial Prospects

The areca nut kernel possesses a notably higher protein content of 9.89 to 14.62 g/100 g, but its use is limited. To enhance its applications in pharmaceuticals and food industries, extraction of areca nut kernel globulin was done leading to the identification of inhibitor peptides of angiotensin-I converting enzyme (ACE). A novel ACE-inhibitory heptapeptide (Ala-Pro-Lys-Ile-Glu-Glu-Val) was synthesized, showcasing the areca nut kernel's potential for antihypertensive drugs and functional foods [74]. Tannin and natural dyes represent the most promising products for development due to their significant market demand and relatively high added value [75]. The cell wall composition of ANH is quite loose and finely fibrous even though the aspect ratio of ANH, which was 61, was lower than that of rice straw with 125 aspect ratios. It still satisfied the essential criteria for fiber production in the pulp and paper sector [76].

The areca nut-based multispecies cropping system offers extra income and serves as a safeguard against price fluctuations of the primary crop. It holds the promise of creating job prospects to enhance the quality of life in rural areas. It is widely accepted that a multispecies cropping system is essential for making farming both profitable and sustainable [77]. Areca nut is a vital commercial crop in India, supporting numerous farming families. India leads globally in areca nut area (49%) and production (50%), primarily in the southwestern and northeastern regions, especially in Karnataka, Kerala, Assam, West Bengal, and Meghalaya. Additionally, in certain regions, areca nut leaf sheaths are employed in the fabrication of disposable cups and plates, packing cases, plywood, ornamental veneer panels, and picture frames. The primary source of fuel wood is utilized for culinary purposes, while residues from horticultural activities associated with coconut and areca nut cultivation serve as a resource for water heating applications [78].

An analytical examination of the research reveals significant opportunities within the agribusiness sector arising from the processing of green areca nuts. This potential not only facilitates the generation of substantial income and employment prospects but also fosters a complementary interrelationship between the agricultural and nonagricultural segments of the rural economy [79]. Understanding farmers' multidimensional behavior and socioeconomic characteristics is vital for adopting better agricultural practices and closing the technological gap.

7 Real-Time Applications, Patents

Natural dyes that can emit white light have garnered attention for applications in sensors, display systems, and back-lights. A study was done to examine the energy transfer and luminescence decay in a natural dye from ANH, enhanced with carbon dots. The carbon nanoparticles, produced through a hydrothermal treatment of the dye, measure 4–5 nm. White light emission, characterized by favorable Commission Internationale del' Eclairage (CIE) values (0.28, 0.34), occurs when the areca nut extract is combined with its corresponding carbon dots upon photoexcitation at 380 nm and therefore highlights the potential of using natural dyes with fluorescent nanoparticles for white luminescent displays [80]. Areca nut-based farming is efficient and has the potential to provide significantly important ecosystem services such as yield, economic viability, soil organic carbon enrichment, sequestration of carbon, and water use. Chemisorption significantly influences the adsorption mechanisms of methylene blue and Congo red utilizing biochars derived from pine nut husks. The resultant biochars, when combined with hydroxyapatites, demonstrate a notable efficacy in the removal of methylene blue, a cationic dye attributable to the negatively charged surface properties of the biochars, along with the interactions facilitated by the phosphate and calcium constituents present in the hydroxyapatites [81].

8 Socioeconomic Impacts, Climate Change Mitigation, and Well-being

Areca nut husk offers significant economic benefits by creating employment and income opportunities, especially in developing regions with a prevalent areca nut industry. Converting the husk into value-added products like handicrafts, furniture, and building materials supports local economies and generates new jobs. Biomass fuel exposure is a major cause of respiratory diseases, including chronic obstructive pulmonary disease. The burning of ANH, often overlooked, has not been formally studied for its effects on respiratory health in areca farming areas. The risk of indoor air pollution is significantly high due to poor ventilation, making the smoke from ANH burning a largely unexplored component of environmental tobacco smoke [82]. Numerous studies have demonstrated that low doses of areca nut possess pharmacological effects, including deworming, anti-inflammatory benefits, enhancing gastrointestinal function, reducing blood lipids, preventing atherosclerosis, and exhibiting anti-depressant qualities. The key mechanism at play in these effects is to diminish the production of reactive oxygen species. It disrupts the biochemical metabolic function of bacteria. Prolonged intake of areca nuts in significant amounts can lead to various negative reactions or associated malignant conditions in the human body [83]. Areca nut husk has the potential to provide significant socioeconomic and environmental benefits. Research has shown that the cascade use of biomass can lead to increased carbon storage, employment, and production costs, while also resulting in increased energy use and greenhouse gas emissions [84]. Similarly, the utilization of agricultural waste like oil palm ash can provide environmental and economic advantages [85]. In the case of ANH, its versatility and potential applications make it a valuable resource that could be leveraged to address various challenges.

The utilization of ANH can have a positive impact on climate change mitigation. The use of ANH as a renewable energy source, either through direct combustion or the production of biofuels, can help reduce the carbon footprint of various industries and contribute to the transition toward a more sustainable energy mix. Areca nut cultivation is a significant economic activity in many parts of the world, particularly in South and Southeast Asia. As the effects of climate change continue to manifest, understanding the interplay between areca nut production, waste management, and climate change mitigation strategies becomes increasingly crucial. Researchers have observed the devastating impacts of climate change phenomena, such as erratic rainfall patterns and temperature fluctuations, on other cash crops like cocoa in Malaysia, highlighting the urgent need to investigate the specific vulnerabilities of areca nut cultivation to climate change [86].

The ANH is a versatile and cost-effective material with potential for various applications. It can be used as a reinforcement particulate in aluminum metal matrix composites, improving their mechanical properties compared to traditional alloys. Processed into powdery particulates, the husk enhances the final composite's strength and durability. With desirable traits such as high strength, low density, and environmental friendliness, the husk serves as an eco-friendly alternative to synthetic materials. Its abundance in areca palm-growing regions further supports its use as a readily available resource for composite development [87].

9 Conclusion and Future Perspectives

Areca nut husk valorization has significant prospects across industries, energy, agriculture, handicrafts, and environmental conservation. With proper investment in technology, infrastructure, and awareness programs, it can become a key player in circular economies, promoting sustainability while enhancing socioeconomic benefits in regions where areca nuts are cultivated. The valorization of ANH has garnered significant attention in recent years as researchers and industries explore sustainable and economically viable methods to convert this agricultural waste into valuable products. Therefore developing circular value chains for industrial processing by-products, such as areca nut husk, can help eliminate environmental issues associated with waste disposal and contribute to a more sustainable future.

References

1. Mishra B, Mohanta YK, Reddy CN, Reddy SDM, Mandal SK, Yadavalli R, Sarma H (2023) Valorization of agro-industrial biowaste to biomaterials: an innovative circular bioeconomy approach. Circ Econ 2:100050
2. Bhuyan N, Narzari R, Gogoi L, Bordoloi N, Hiloidhari M, Palsaniya DR, Deb U, Gogoi N, Kataki R (2020) Valorization of agricultural wastes for multidimensional use. In: Pandey A, Negi S, Soccol CR (eds) Current developments in biotechnology and bioengineering. Elsevier, pp 41–78
3. Gao Y, Remón J, Matharu AS (2021) Microwave-assisted hydrothermal treatments for biomass valorisation: a critical review. Green Chem 23:3502–3525
4. Sangma ARW, Joshi SR (2021) Areca nut: traditional processing, uses and products potential of the husk. NEHU J 19:55–65
5. Hebbar KB, Ramesh SV, Bhat R (2024) Predicting current and future climate suitability for arecanut (*Areca catechu* L.) in India using ensemble model. Heliyon 10:e26382

6. Orwa, C., Mutua, A., Kindt, R., Jamnadass, R. and Anthony, S. (2009). *Areca catechu* L. Agroforestry Database: A Tree Reference and Selection Guide 4
7. Lim TK (2012) Areca catechu. In: Lim TK (ed) Edible medicinal and non-medicinal plants. Springer, Cham, pp 260–276
8. Ansari A, Mahmood T, Bagga P, Ahsan F, Shamim A, Ahmad S, Shariq M, Parveen S (2021) *Areca catechu*: A phytopharmacological legwork. Food Front 2:163–183
9. Staples GW, Bevacqua RF (2006) *Areca catechu* (betel nut palm). Species profiles for Pacific Island. Agroforestry 1:1–9
10. Bhat NT, Mohapatra AR (1989) Effect of supplying nutrients through organic manures, inorganic fertilisers and their combination on areca nut crop. J Plant Crop 16:443–447
11. Bhat R, Sujatha S (2012) Influence of biomass partitioning and nutrient uptake on yield of areca nut grown on a laterite soil. Commun Soil Sci Plant Anal 43:1757–1767
12. Sadasivuni S, Bhat R, Pallem C (2015) Recycling potential of organic wastes of areca nut and cocoa in India: a short review. Environ Technol Rev 4:91–102
13. Praveen R, Bandyopadhyay A, Babu S, Anjaneyulu A (2021) Waste to wealth with special reference to coconut, areca nut and oil palm. Adv Hortic 10:139–161
14. Sujatha S, Bhat R (2013) Impact of vermicompost and nitrogen-phosphorus-potassium application on biomass partitioning, nutrient uptake and productivity of areca nut. J Plant Nutr 36:976–989
15. Bhat R, Sujatha S (2011) Arecanut based high density multispecies cropping/farming systems. In: Areca nut based cropping/farming systems. ICAR-CPCRI, pp 27–44
16. Sujatha S, Bhat R, Balasimha D, Apshara ES (2011) Arecanut-based inter/mixed cropping systems. In: Arecanut-based cropping/farming systems. CPCRI, Kasaragod, pp 6–26
17. Vasudeva N, Rangaswamy BE, Tejaswini M, Bhat S, Sahithya G, Annapurna V (2019) Areca husk fibres as agro-waste to value added products in textile sector-A practicability study. Int J Adv Res 7:106–111
18. Deshmukh PS, Shahare PU, Bhanage GB, Dhekale JS, Patil PG (2016) Some engineering properties of areca nut husk (*Areca catechu* L.) useful in waste to wealth conversion. Indian J Arecanut Spices Med Plant 18:42–50
19. Peng W, Liu Y-J, Hu M-B, Yan D, Gao Y-X, Wu C-J (2017) Using the "target constituent removal combined with bioactivity assay" strategy to investigate the optimum arecoline content in charred areca nut. Sci Rep 7:40278
20. Salehi B, Konovalov DA, Fru P, Kapewangolo P et al (2020) *Areca catechu*-from farm to food and biomedical applications. Phytother Res 34:2140–2158
21. Auluck A, Hislop G, Poh C, Zhang L, Rosin MP (2009) Areca nut and betel quid chewing among south Asian immigrants to Western countries and its implications for oral cancer screening. Rural Remote Health 9:1118
22. Strickland SS (2002) Anthropological perspectives on use of the areca nut. Addict Biol 7:85–97
23. Thulasi EP, Thomas PD, Ravichandran B, Madhusudhanan K (2010) Mycelial culture and spawn production of two oyster mushrooms, *Pleurotus florida* and *Pleurotus eous* on different substrates. Int J Biol Technol 1:39–42
24. Chandra J, George N, Narayanankutty SK (2016) Isolation and characterization of cellulose nanofibrils from areca nut husk fibre. Carbohydr Polym 142:58–166
25. Kammardi DTNP (2012) Report of special scheme on cost of cultivation of areca nut in Karnataka. GOI. Department of Agricultural Economics, University of Agricultural Sciences, GKVK, Bangalore
26. Das P (2019) Underutilized Meghalyan areca nut husk waste fiber for development of nonwoven textile material. Int J Pure Appl Biosci 7:563–567
27. Das J, Debnath C, Nath H, Saxena R (2022) Antibacterial effect of activated carbons prepared from some biomasses available in north-East India. Energy Sources Part A: Recovery Util Environ Effects 44:4928–4938

28. Dalimunthe NF, Ambarita ED, Barus LE, Hanief S (2024) Characterization of a bioplastic spoon based on kepok banana (*Musa acuminata balbisiana* C.) starch with the addition of areca nut shells (*Areca catechu* L.) and glycerol. IOP Conf Series Earth Environ Sci 1352:012009
29. Mishra RK, Gariya B, Savvasere P, Dhir D, Kumar P, Mohanty K (2024) Thermocatalytic pyrolysis of waste areca nut into renewable fuel and value-added chemicals. Am Chem Soc Omega 9:25779–25792
30. Lazim Y, Salit MS, Zainudin ES, Mustapha M, Jawaid M (2014) Effect of alkali treatment on the physical, mechanical, and morphological properties of waste betel nut (*Areca catechu*) husk fibre. Bioresources 9:7721–7736
31. Annamalai SJK, Naik R, Jaganathan D, Mathew AC (2017) By-product utilization of coconut, areca nut and cocoa processing. In: Product diversification in plantation crops. Today & Tomorrow's Printers and Publishers, New Delhi, pp 149–203
32. Nidhi KV, Varsha BS, Prajwal M, Sai PSG, Ananda HV, Prakruthi G (2020) A biotechnological approach on screening and production of manganese peroxidase by *Fusarium* sp. from areca nut husk. Euro J Pharm Med Res 7:310–314
33. Narayanamurthy G, Ramachandra YL, Rai SP, Manohara YN, Kavitha BT (2008) Areca husk: an inexpensive substrate for citric acid production by *aspergillus Niger* under solid state fermentation. Indian J Biotechnol 7:99–102
34. Naveen Kumar KJ, Thippeswamy B, Krishnappa M (2016) Bioethanol fermentation from fungal pretreated lignocellulosic areca nut (*Areca catechu* L.) husk using yeasts and *Zymomonas mobilis* NCIM 2915. Int J Curr Microbiol App Sci 5:571–582
35. Sreena CP, Sebastian D (2015) Cost effective cellulase production by *Bacillus subtilis* MUS1 using lignocellulosic biomass residues. In: Biodiversity and evaluation: perspectives and paradigm shifts. Cochin University of Science and Technology, pp 268–270
36. Jayanna Naveenkumar K, Thippeswamy B, Krishnappa M (2014) Production and partial purification of fungal pectinases from areca nut (*Areca catechu* L.) husk waste under submerged fermentation. Int J Adv Biotechnol Res 5:188–201
37. Vardhan H, Mahato RB, Sasmal S, Mohanty K (2022) Production of xylose from pre-treated husk of areca nut. J Nat Fibers 19:131–144
38. Ramesh S, Sundararaju P, Banu KSP, Karthikeyan S, Doraiswamy U, Soundarapandian K (2019) Hydrothermal carbonization of areca nut husk biomass: fuel properties and sorption of metals. Environ Sci Pollut Res 26:3751–3761
39. Baidya KS, Kumar U (2021) Adsorption of brilliant green dye from aqueous solution onto chemically modified areca nut husk. South African J Chem Eng 35:33–43
40. Tabassum M, Bardhan M, Novera TM, Islam MA, Hadi Jawad A, Islam MA (2020) NaOH-activated betel nut husk hydrochar for efficient adsorption of methylene blue dye. Water Air Soil Pollut 231:1–13
41. Subramani BS, Shrihari S, Manu B, Babunarayan KS (2019) Evaluation of pyrolyzed areca husk as a potential adsorbent for the removal of Fe^{2+} ions from aqueous solutions. J Environ Manag 246:345–354
42. Nik Ab Lah NKI, James Ketek JA, Tengku Mohd TA (2021) Characteristics and oil sorption effectiveness of areca nut husk. Malaysian J Chem Eng Technol 4:167–171
43. Azis A, Zudaidah Z, Afriani A (2020) Betel nut husks as an alternative litter materials for broiler production. Int J Poult Sci 19:81–85
44. Murwani R, Kusumanti E, Naumova EN (2022) *Areca catechu* L. and *Anredera cordifolia* (ten) Steenis supplementation reduces faecal parasites and improves caecal histopathology in laying hens. Int J Vet Sci Med 10:52–63
45. Mahata ME, Deni I, Nofrianti A, Rizal Y, Ardi A (2020) Effect of nutritional supplements of fermented and unfermented betel nutshell waste in performance of broiler chickens. J World's Poult Res 10:118–124
46. Ifedi EN, Ajayi IA, Olaifa F, Kakojulo A (2016) Nutritional evaluation, toxicological effect and possible utilization of *Areca catechu* seed flours as an additive in feed formulation for African catfish fingerlings (*Clarias gariepinus*). Annal Food Sci Technol 17:55–67

47. Vijayanand C, Kamaraj S, Sriramajayam S, Ramesh D (2016) Biochar production from areca nut waste. Int J Farm Sci 6:43–48
48. Kabir SMM, Dhar AK, Bhattacharjee M (2020) The use of natural *Areca catechu* dyes for silk and nylon and its halochromic effect. J Textile Inst 111:882–889
49. Setty RTK, Krishnappa RY, Prasanna. and Angadi S. B. (1983) Paper making potentiality of arecanut husk. Indian Pulp and Paper Technical Association 20:74–78
50. Gurumurthy SB, Patil SV, Shankarappa TH, Prashantha A (2018) Bioconversion of areca nut husk; residue to compost *via* microbial consortia. Int J Curr Microbiol App Sci 7:2883–2888
51. Moorthy VK (1991) Production technology of the oyster mushroom *Pleurotus sajor-caju* Fr singer. Ph.D. Thesis, Mangalore University, Mangalagangotri, India
52. Moorthy VK, Mohanan RC (1991) Evaluation of various organic.substrates for the cultivation of *Pleurotus sajor-caju*. J Plant Crop 19:65–69
53. Nagaraja R, Gurumurthy BR, Shivanna MB (2014) Bio softening of areca nut waste areca husk, leaf and leaf sheath for value added compost. Int J Res Appl Nat Soc Sci 2:105–112
54. Muralidhar N, Kaliveeran V, Arumugam V, Srinivasula Reddy I (2019) A study on areca nut husk fibre extraction, composite panel preparation and mechanical characteristics of the composites. J Instit Eng (India) 100:135–145
55. Ravi B, Sujatha S, Thomas GV (2008) Soil microbiological and enzymological response to organic matter recycling and chemical fertilization in areca nut based cropping system. J Plant Crop 36:281–289
56. Gujre N, Mitra S, Agnihotri R, Sharma MP, Gupta D (2022) Novel agrotechnological intervention for soil amendment through areca nut husk biochar in conjunction with vetiver grass. Chemosphere 287:132443
57. Rajan A, Kurup JG, Abraham TE (2010) Solid state production of manganese peroxidases using areca nut husk as substrate. Braz Arch Biol Technol 53:555–562
58. Sreena CP (2020) Arecanut husk: A potential feedstock for enhanced production of endoglucanase by *Bacillus subtilis* MU S1. Environ Technol Innov 19:100989
59. Bhagyamma CV, Sharad S, Gudigar SG, Reshma SV (2011) Bioprocessing of areca husk in solid state fermentation for cellulase production using *Trichoderma viride*. Indian J Appl Res 1:15–17
60. Wang D, Zhou L, Li W, Zhou H, Hou G (2018) Anticoccidial effects of areca nut (*Areca catechu* L.) extract on broiler chicks experimentally infected with *Eimeria tenella*. Exp Parasitol 184:16–21
61. Chanakya HN, Sreesha M (2012) Anaerobic retting of banana and areca nut wastes in a plug flow digester for recovery of fiber, biogas and compost. Energy Sustain Dev 16:231–235
62. Singh A, Singh PK, Wang W, Shrivastava AK (2020) Microbes biology: microbes in wetland and bioprospection of microbes. In: Upadhyay AK, Singh R, Singh DP (eds) Restoration of wetland ecosystem: A trajectory towards a sustainable environment. Springer, Singapore, pp 87–99
63. Anuar MF, Fen YW, Azizan MZ, Rahmat F, Zaid MHM, Khaidir REM, Omar NAS (2021) Sustainable production of areca nut husk ash as potential silica replacement for synthesis of silicate-based glass-ceramics materials. Materials 14:1141
64. Yusriah L, Sapuan SM, Zainudin ES, Mariatti M (2014) Characterization of physical, mechanical, thermal and morphological properties of agro-waste betel nut (*Areca catechu*) husk fibre. J Clean Prod 72:174–180
65. Ashok RB, Srinivasa CV, Basavaraju B (2018) A review on the mechanical properties of areca fiber reinforced composites. Sci Technol Mater 30:120–130
66. Girisha C, Sanjeevamurthy S, Gunti R, Manu S (2012) Mechanical performance of natural fiber-reinforced epoxy-hybrid composites. Int J Eng Res Appl 2:615–619
67. Kumar GCM (2008) A study of short areca fiber reinforced PF composites. Proc World Congress Eng 2:2–4
68. Girish CR, George GM (2017) Phenol removal from wastewater using areca nut husk (*Areca catechu*) as adsorbent. Int J Mech Eng Technol 8:1–9

69. Nagarajappa DP, Keshava Kumar S (2020) Removal of heavy metals by soil aquifer treatment system in conjunction with natural adsorbents. Int Res J Modern Eng Technol Sci 8:1396–1400
70. Vinu V, Band Binitha NN (2020) Lithium silicate based catalysts prepared using areca nut husk ash for biodiesel production from used cooking oil. Materials Today: Proceedings 25:241–245
71. Gokul PV, Singh P, Singh VP, Sawarkar AN (2019) Thermal behavior and kinetics of pyrolysis of areca nut husk. Energy Sources, Part A: Recovery Util Environ Effects 41:2906–2916
72. Mishra RK, Mohanty K (2021) Fuel properties and compositional analysis of *Areca catechu* sawdust over MgO and ZSM-5 catalysts. J Energy Inst 94:252–262
73. Umamaheswaran K, Batra VS (2008) Physico-chemical characterisation of Indian biomass ashes. Fuel 87:628–638
74. Liu X, Li G, Wang H, Qin N, Guo L, Wang X, Shen S (2021) Identification, characterization and antihypertensive effect *in vivo* of a novel ACE-inhibitory heptapeptide from defatted areca nut kernel globulin hydrolysates. Molecules 26:3308
75. Jannah M, Machfud M, Sugiarto S (2021) Potential added value of areca nut products in Aceh. Jurnal Teknologi Industri Pertanian 31:190–197
76. Yuan J, Zhang H, Zhao H, Ren H, Zhai H (2023) Study on dissociation and chemical structural characteristics of areca nut husk. Molecules 28:1513
77. Jaganathan D, Nagaraja NR (2015) Perception of farmers about areca nut based multi-species cropping system. Indian Res J Extension Educ 15:49–54
78. Ramachandra TV, Joshi NV, Subramanian DK (2000) Present and prospective role of bioenergy in regional energy system. Renew Sust Energ Rev 4:375–430
79. Mula G, Sarker SC, Mondal S (2020) Economics of agribusiness in areca nut processing-an analytical study in Terai zone of West Bengal. J Crop Weed 16:66–77
80. Mathew J, Joy J, Kumar SA, Philip J (2019) White light emission by energy transfer from areca nut husk extract loaded with carbon dots synthesized from the same extract. J Lumin 208:356–362
81. Kim HJ, Han JW, Yu JH, Jun BM, Chon K (2024) Adsorption of charged synthetic organic dyes by pine nut husk biochars modified with hydroxyapatite: adsorption mechanisms and reusability. Desalin Water Treat 321:100970
82. Katipalla A, Ballal P, Shivanna NK (2024) Areca nut husk: A burning issue? Indian J Resp Care 13:140–141
83. Wang Z, Guo Z, Luo Y, Ma L, Hu X, Chen F, Li D (2024) A review of the traditional uses, pharmacology, and toxicology of areca nut. Phytomedicine 156005:156005
84. Suominen T, Kunttu J, Jasinevičius G, Tuomasjukka D, Lindner M (2017) Trade-offs in sustainability impacts of introducing cascade use of wood. Scand J For Res 32:588–597
85. Foo KY, Hameed BH (2009) Value-added utilization of oil palm ash: A superior recycling of the industrial agricultural waste. J Hazard Mater 172:523–531
86. Chizari A, Mohamed Z, Shamsudin MN, Seng KWK (2017) The effects of climate change phenomena on cocoa production in Malaysia. Int J Environ Agric Biotechnol 2:2599
87. Joseph OO, Babaremu KO (2019) Agricultural waste as a reinforcement particulate for aluminum metal matrix composite (AMMCs): a review. Fibers 7:33

Chapter 12
Cashew Apple and Nutshell: A Potential Bio-Resource for the Production of Beverages, Bio-Sourced Chemicals, Materials, and Fuels

Sixberth Mlowe, Kennedy J. Ngwira, and James E. Mgaya

Abbreviations

CDA	Cellulose diacetate
CNSL	Cashew nut shell liquid
CNS	Cashew nut shell
DVB	Divinylbenzene
HCPs	Hypercrosslinked polymers
HHV	Higher heating values
HPPDP	4-[(4-hydroxy-2-pentadecenylphenyl)diazenyl]phenol
LHV	Lower heating values
MIPs	Molecularly imprinted polymers
PVC	Polyvinyl chloride
PPAR	Proliferator-activated receptors
SCW	Subcritical water
SC-CO_2	Supercritical carbon dioxide
TAE	Tannic acid equivalent

S. Mlowe (✉) · J. E. Mgaya
Chemistry Department, Dar es Salaam University College of Education, Dar es Salaam, Tanzania
e-mail: sixberth.mlowe@duce.ac.tz

K. J. Ngwira
Molecular Sciences Institute, School of Chemistry, University of the Witwatersrand, Johannesburg, South Africa

C. Trois, R. C. Ray (eds.), *Plantation Crop Wastes: Valorization for Economic Sustainability*, Waste as a Resource,
https://doi.org/10.1007/978-3-031-96805-1_12

1 Introduction

Balancing the industrial production of essential materials with the need for environmental conservation is an ongoing global challenge [1]. Due to climate change, resource depletion, and pollution, harnessing renewable materials has become essential rather than optional. Consequently, there is a strong demand for technological innovations to enhance the efficient use of renewable resources, including catalytic conversion of biomass into industrial fine chemicals and energy materials, as well as protocols to improve the shelf life of perishable products.

The significance of utilizing renewable resources can be understood through several aspects, including sustainable resource management, waste reduction, resource efficiency, and abundant supply [2, 3]. The sustainable nature of renewable resources implies that the resources can be replenished, meeting energy and material needs without depleting finite resources, thus balancing consumption and conservation. Likewise, the use of agro-wastes as feedstocks in production processes enhances resource efficiency by converting them into valuable products like biofuels, bioplastics, and fertilizers, which minimizes the amount that ends up in landfills. The large quantities of wastes generated during various agricultural processes, such as crop residues, fruit peels, and husks make these materials readily available as raw materials for various industrial processes.

This chapter focuses on the valorization of cashew products, specifically, the cashew apple, cashew nut shells, and cashew nut shell liquid. Cashew apples are highly perishable fruits and must be processed into value-added products; otherwise, they are thrown out as wastes [4]. On the other hand, cashew nut shells are produced after extraction of the edible kernel and are often treated as wastes. According to data published by the Food and Agriculture Organization of the United Nations (FAOSTAT 2024), the global production of cashew nuts (in shell) reached 3,852,868 tons in the year 2022, and the major producing countries were (metric tons) Côte d'Ivoire (970,000), India (752,000), Vietnam (341,680.33), Philippines (217,583.2), Tanzania (216,906.6), Benin (215,000), Indonesia (163,083), Brazil (147,137), Burkina Faso (145,246), Mozambique (144,823), Guinea-Bissau (116,484.6), and Ghana (107,678). Globally, there hasn't been a smooth trend in the production of cashew nuts as shown in Fig. 12.1. Despite the up-and-down trend for the major part of the decade, the overall global production of cashew nuts has increased from 3,050,698 tons in the year 2010 to 3,852,868 tons in 2022 (FAOSTAT 2024). On the other hand, the production of cashew apples seemed to decrease from 1,998,640 tons in the year 2010 to 1,329,862 tons in 2022. The decreased cashew apple production over a decade is an indication of the decrease in the popularity of cashew apples among consumers. This highlights the need to publicize the nutritional value of cashew apples and the available opportunities for adding their value.

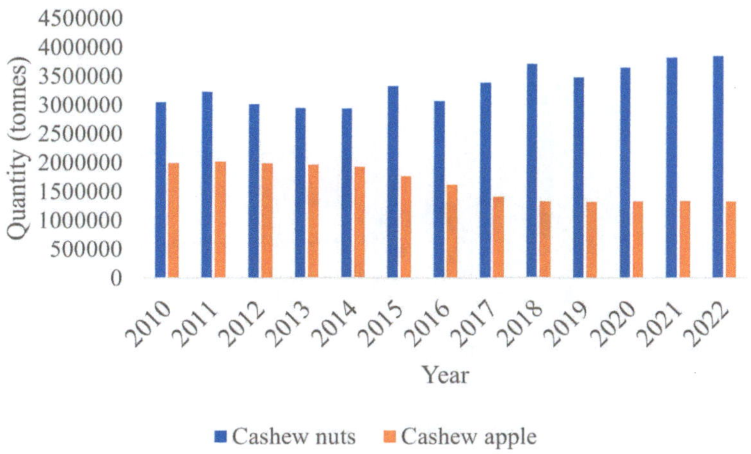

Fig. 12.1 Global production of cashew nuts and cashew apples over a decade (FAOSTAT, 2024)

2 Cashew Apple

The cashew (*Anacardium occidentale* L.) tree typically produces cashew apples and the nuts grow attached to the bottom of the apple (Fig. 12.2). Cashew nuts contain kernels, which are edible seeds and a good source of healthy fats, protein, vitamins, and minerals [5]. Cashew apples are the fleshy and fibrous fruits of the cashew tree, often described as sweet and juicy. Despite the nutritional value and the availability of product value-addition opportunities, cashew apples are still not popular compared to other fruits. This may be attributed to various factors including limited knowledge about their nutritional values and health benefits, high perishability rates, poor post-harvest management, and less affordable processing equipment.

2.1 Nutritional Composition of Cashew Apple

Cashew apples generally contain significant amounts of vitamin C, sugars (fructose and sucrose), amino acids, fibers, carotenoids, polyphenols, and minerals, such as potassium, magnesium, sodium, and iron [6, 7]. These contents make cashew apples a potential resource for producing various products, including health drinks, food supplements, bioethanol, and livestock feed. The high content of vitamin C is associated with various health benefits of cashew apple juice. Research reports have indicated that supplementation with cashew apple juice produces positive impacts

Fig. 12.2 Cashew apple and cashew nut

during high-intensity exercise and also improves immunological mechanisms [8–11]. Cashew apples contain about 10% w/w sugars, predominantly glucose and fructose [12]. The significant sugar content in cashew apples is ideal for fermentation to produce alcoholic beverages [6, 13]. The various essential minerals found in cashew apples are beneficial for human health, including muscle function, energy production, activation of enzymes, membrane function, regulating fluid balance, muscle contractions, nerve signals, strong bones and teeth, repair processes in the body, formation of hemoglobin, regulating the formation of hormones, supporting immune function, and DNA synthesis [7, 14]. The fiber content in cashew apples is helpful in blood sugar control by decreasing the absorption of sugar in the bloodstream. This is particularly beneficial for people with diabetes [15, 16]. Cashew apples are also significant sources of polyphenols. Prominent polyphenolic compounds present in cashew apples include tannins, flavonoids (anthocyanins, myricetin, quercetin, kaempferol), and phenolic acids (caffeic acid, coumaric acid, ferulic acid, and gallic acid) [6]. Polyphenols have several health benefits due to their antioxidant properties [17, 18]. A summary of the nutritional composition of cashew apples is presented in Table 12.1.

2.2 Uses of Cashew Apple

Cashew apples can be consumed raw, however, less commonly compared with other fruits due to their astringency and high perishability rate [21]. To extend the fruit's shelf life and make it more versatile in culinary applications, the fruit is processed into other products such as juice, jelly, candied fruit, pectin, syrup, jam, and fermented drinks [22–24].

Medically, cashew apples contain bioactive compounds that exhibit antioxidant, antimicrobial, antimutagenic, antitumor, anti-fungal, and anti-inflammatory activities [25, 26]. In some parts of the world, cashew apples or their juice are used to treat diseases such as sore throat, dysentery, diarrhea, uterine complaints, dropsy, cholera, tonsillitis, and rheumatism [27–29].

Table 12.1 Chemical and nutritional composition of cashew apples [19, 20]

Component	Amount
g/100 g	
Moisture	86.1
Protein	0.8
Fat	0.2
Carbohydrate	12.6
Total sugars	8.4–10.6
Crude fiber	0.6
Ash	0.3
Mg/100 g	
Calcium (ca)	0.2
Potassium (K)	0.0708
Phosphorus (P)	19
Magnesium	0.0348
Iron (Fe)	0.4
Vitamin B1(thiamin)	0.2
Vitamin B2 (riboflavin)	0.2
Niacin	0.5
Vitamin C	200
Total phenolics	815.3

In industries, cashew apples are potential raw materials to produce bioethanol, biosurfactants, and lactic acid [30–32]. The fermentable sugars contained in cashew apples can be easily converted into ethanol through fermentation processes. Toure and co-workers carried out batch experiments for bioethanol production using cashew apple feedstock under both aerobic and anaerobic conditions [32]. The highest ethanol concentration (75.75 g/L) was obtained at 33° C under aerobic conditions, indicating the great potential of the apples as raw materials for bioethanol production.

In biosurfactant production using cashew apples, the cashew apple juice serves as a source of carbon and energy for the microorganisms involved in the surfactant production [33]. Felix and coworkers isolated various *Bacillus* strains and assessed their capability to produce biosurfactants using cashew apple feedstock [31]. The biosurfactants produced by the selected strains were able to reduce significantly the surface tension of water and the interfacial tension of n-hexadecane. The biosurfactants also showed remarkable stability against changes in pH, temperature, and salinity.

Cashew apples are also good raw materials to produce lactic acid, which involves crushing cashew apples to extract the juice, which is then inoculated with lactic acid bacteria, such as *Lactobacillus* species. These bacteria convert the sugars in the juice into lactic acid under anaerobic conditions (absence of oxygen). The mixture is incubated at a controlled temperature (typically between 30–40 °C) for a specific period, allowing the bacteria to proliferate and produce lactic acid. Silveira and co-workers employed *Lacticaseibacillus casei* B-442 for the production of lactic acid using cashew apples under optimum conditions of $(NH_4)_2SO_4$ (6 g/L) and reducing sugar (50 g/L) obtained from cashew apple juice, at 37 °C in a medium of pH 6.5. Under these conditions, they were able to obtain a 95% yield of lactic acid [13].

2.3 Production of Beverages from Cashew Apple

Transforming cashew apples into juices, wines, or other beverages increases their market value and encourages farmers to harvest and sell them rather than letting them go to waste. The process also extends the shelf life of cashew apples, as the beverages can be processed and preserved, significantly prolonging the fruit's usability [4]. Adding preservatives such as citric acid (0.01%) and sodium benzoate (0.01%) to cashew apple juice extends the shelf-life of the juice satisfactorily up to 90 days [34]. Successful extension of the shelf life of cashew apple juice has also been achieved by treating the juice with optimum thermosonication [35].

2.3.1 Alcoholic Beverages

The production of alcoholic beverages from cashew apples generally involves inoculating cashew apple juice with yeast and allowing the juice to ferment for several days to weeks, depending on the desired beverage. Yeast plays a crucial role in the fermentation of cashew apple juice by converting sugars present in the juice into alcohol. The commonly used strain of yeast for alcohol production is the *Saccharomyces cerevisiae*. This yeast has been extensively used for many years in the brewery and wine industries due to its tolerance to a wide range of pHs. The yeast also gives a high yield of ethanol and can withstand high ethanol concentrations [36]. Despite the high efficiency of *S cerevisiae*, researchers have continued to search for other strains of yeasts to improve the sensory characteristics of fruit wine [37]. Depending on the mode of preparation, the alcoholic beverages that can be produced from cashew apples include cashew wine, cashew feni, and Caju Amigo. Production of wine from cashew apples using *S cerevisiae* results in cashew apple wine with sensory qualities generally acceptable as an alcoholic beverage; however, there is a significant difference in taste, flavor, aroma, and aftertaste observed between cashew apple wine and the commercially available grape wine [38]. The high content of tannins in the cashew apple wine is presumably the reason for the difference observed. According to Mohanty et al. [38], the production of cashew apple wine involves the following general procedure: Clean cashew apples are dipped in brine (5%) for 3 days to reduce tannin content. The apples are then steamed for 15 min at 15 lb. pressure and crushed to obtain cashew apple juice. The juice is filtered and treated with sodium metabisulfite as a preservative. The desired pH of 3.6 and 17 °Bx are archived by adding tartaric acid and cane sugar, respectively, into the juice. The juice is then inoculated with a starter culture such as *S. cerevisiae* var. *bayanus*, and fermentation is allowed to proceed at 32 °C for

6 days. The resulting wine is racked two or three times at 15-day intervals to remove the sediment deposited in the wine. After racking, the wine is clarified by adding 0.04% bentonite, followed by the addition of preservatives before bottling.

Cashew Feni is a popular alcoholic, which originated in Goa, India. For the production of feni, the cashew apple juice is transferred to a fermentation vessel and left to ferment for about 10 days. The fermented juice is then distilled resulting in a spirit with unique flavor. The first distillation often gives a low-strength distillate (10% to 15% alcohol). The twice or triple-distilled feni can have the strength of up to 40% v/v alcohol [39]. After distillation, the feni is aged in glass or wooden barrels to enhance its flavor and aroma, although sometimes is consumed unaged.

Caju Amigo is another type of alcoholic drink made from cashew apples and is one of the popular drinks in Brazil. The drink is made by mixing cashew juice with *cachaça (a popular Brazilian drink prepared by distilling fermented sugar cane juice). During the preparation of* Caju Amigo, *equal parts of the* cashew juice and *cachaça are simply mixed in a glass and served straight (TasteAtlas, 2024).*

2.3.2 Nonalcoholic Beverages

Cashew apples can be processed into juice and be served straight or the juice can be further clarified into Cajuina. Making unclarified cashew juice involves washing the fully ripe cashew apples and cutting them into halves to remove the seeds. The cashew apple pieces are then pressed using a pressing machine to obtain pressed cashew apple juice. Alternatively, the cashew apple pieces are blended in a blender until smooth. The blended cashew apple mixture is then filtered through a fine-mesh strainer or cheesecloth to remove any solids [40]. Cajuína is a processed nonalcoholic drink made from cashew apples, originating from the northeast region of Brazil. It differs from regular cashew juice in preparation and characteristics. While regular cashew juice is typically made by simply pressing or blending fresh cashew apples to extract the juice, the production of cajuina involves stages of clarification and heat treatment. According to the operation guide published by a researcher from the Brazilian Agricultural Research Corporation (EMBRAPA) [41], the production of cajuina involves a series of steps. The clean ripe cashew apples are carefully crushed and pressed aiming to produce good yields and satisfactory juice quality, which is free of excessive amounts of tannins. The juice is then subjected to clarifying agents (commonly gelatin), which produce flocculation when they come in contact with the tannins in the juice. Gelatin must be added in the form of a watery solution (10% concentration), at a temperature of approximately 50 to 60 °C. Heating is necessary to dissolve gelatin in the water because proteins do not dissolve in cold water. After clarification, the juice is filtered using cotton or felt fabric pieces installed in three or four filters placed on top of each other. The final step is filling and sealing whereby the clarified juice is bottled while hot (70–80 °C). The purpose of filling bottles with hot clarified juice is to create a vacuum inside the bottles so that there will be no air when the juice cools inside the sealed bottles. There is a notable difference in quality between clarified and unclarified cashew apple juice

[40]. Treatment of both blended and pressed cashew apple juice with clarifying agents (gelatin) results in an optimum decrease in tannin from 217.6 mg/100 ml TAE (tannic acid equivalent) to 24.6 mg/100 ml TAE for pressed juice and from 258.0 mg/100 ml TAE to 55.0 mg/100 ml TAE for blended juice. Clarification of cashew apple juice with a low concentration (0.2 g/L) of gelatin in a liter of juice yields sweet and less astringent juice with a high-intensity yellow color. On the other hand, there is no significant difference in terms of pH, total soluble solids, and titratable acidity for the cashew apple juice with and without clarification.

3 Cashew Nut Shell (CNS)

The cashew nut consists of a cashew shell of about 70–75% (w/w) and, therefore, is the major by-product of the cashew processing industry [42]. CNSs are globally considered waste; yet, they offer a possibility for alternate applications due to their unique physical, chemical, and thermal properties.

3.1 Composition and Properties of CNS

Despite some disparities, extensive studies have reported significant work on the chemical and physical properties of CNS (Table 12.2). The compositional structure of the CNS consists of several layers: epicarp, mesocarp (which is the primary source of the popular oil cashew nut shell liquid), endocarp, and testa [48]. Information on the composition and physical properties of cashew nut shells is essential in handling, drying, heating, cashew nut shell liquid (CNSL) extraction, and other related processes. In addition, CNSs are often overlooked in agricultural waste management discussions and are frequently discarded or incinerated, leading to environmental concerns.

Cashew nut shells have not yet been fully harnessed, as most continue to be disposed of as agricultural waste. Recent studies however attest CNS as a source of valuable renewable resources with a variety of benefits [42, 49]. This food waste holds considerable potential as a feedstock for making biofuels and value-added chemical products, along with the possible use as an animal feed component owing to its abundance. In this section, appreciable alternative applications will be discussed that focus on energy production and material development for various uses including adsorbents and CNS-derived activated carbon.

3.2 Potential Applications of CNS

The potential applications of CNS are discussed in the following sections.

Table 12.2 Some reported physical properties of cashew nut shells (CNS)

Property	Composition	Reference(s)
Ultimate analysis (wt %)		
Carbon (C)	*45.7–63.20*	Ajith Kumar and Ramesh [43]; Chung et al. [44], de Paiva et al. [45], Zinla et al. [46]
Hydrogen (H)	3.7–7.68	Chung et al. [44], de Paiva et al. [45], Zinla et al. [46]
Nitrogen (N)	0.2–0.94	Chung et al. [44], de Paiva et al. [45], Zinla et al. [46]
Sulfur (S)	0–0.11	de Paiva et al. [45], Zinla et al. [46]
Oxygen (O)	21.9–45.21	Ajith Kumar and Ramesh [43], Chung et al. [44], de Paiva et al. [45], Zinla et al. [46]
Proximate analysis (wt %)		
Moisture content	3.83–9.43	Ajith Kumar and Ramesh [43], Chung et al. [44], de Paiva et al. [45]
Volatile matter	72.93–97.39	Paternina Reyes et al. [47], Chung et al. [44], de Paiva et al. [45]
Fixed carbon	11.98–18.17	Chung et al. [44], de Paiva et al. [45]
Ash	0.87–7.53	Paternina Reyes et al. [47], de Paiva et al. [45]
Heating values (MJ/kg)		
HHV	17.01–24.2	de Paiva et al. [45]
LHV	15.75–23.6	Chung et al. [44], de Paiva et al. [45]
Chemical composition (wt %)		
Hemicellulose	0.70–70.42	Paternina Reyes et al. [47], de Paiva et al. [45]
Cellulose	11.50–64.57	Paternina Reyes et al. [47], de Paiva et al. [45]
Lignin	17.8–24.31	Paternina Reyes et al. [47], de Paiva et al. [45]
Bulk density (kg/ m^{-3})	220.50–478	Ajith Kumar and Ramesh [43], de Paiva et al. [45]

3.2.1 Biofuels

CNS waste offers a promising alternative for sustainable energy production through pyrolysis due to low moisture and ash content. Utilizing CNS as biofuel can reduce reliance on fossil fuels and costly energy sources, providing a viable solution for domestic cooking and heating. Most commonly, CNS in the form of briquettes accounts for often energy source produced from CNS waste, with significant calorific content [50]. Production of CNS briquettes is done by grinding cashew nut shell waste into a fine powder and then compressing the mixture into solid briquettes [50]. To ensure cohesive integrity and withstand compression, mixing with a binder usually improves the stability of the briquettes. As materials made from CNS biochar lack cohesion, a binding agent is used to improve the property. Cassava flour is an example of a commonly used binder in various research projects, valued for its physicochemical properties, accessibility, and affordability [50]. The study by Sawadago and co-workers used cassava starch as a binding agent in the production of briquettes from cashew cake [51]. The formed briquettes gave a calorific value of 25.7 MJ/kg, a density of 0.91 g/cm^3, a compressive strength of 382.89 kPa, and an impact resistance of 61.10 with a combustion performance comparable to wood charcoal. Analysis of the influence of the amount of cassava starch as a binder in the briquette performance has revealed that the physical and mechanical characteristics of the briquettes increase with the binder content, but this lowers the overall calorific value of the briquettes [43]. In another study, bio-briquettes boasting a greater calorific value of 29.49 MJ/kg were produced using tapioca flour (cassava starch) as a binding agent, along with ethanol and warm water in a 50:50 ethanol-to-water ratio to create a slurry or paste [52]. A few reports have also investigated other binding materials in the production of CNS briquettes. For example, sawdust from carpentry work was used as a binder along de-oiled cashew nut shells during briquetting, which resulted in increased burning time of briquettes [53]. Sugarcane bagasse, when used as a binder, results in the formation of CNS briquettes with increased densification and a higher calorific value of 33.02 MJ/kg [54]. The produced biofuel had a density and mechanical strength that was idyllic for use in homes and biomass power plants.

Numerous studies have explored the utilization of various ingredients combined with CNS, based on the local availability of agricultural wastes. For instance, dual agro wastes of CNS and areca nuts were used for fuel briquette production combined with the binder (cassava flour) through the densification method [55]. The study reported that the highest average calorific value of 21 MJ/kg was obtained when the quantity of cashew nut shells used was 65% by weight. Briquettes composed of carbonized cashew nut shells and rice husks in a weight ratio of 7:3 provided high compressive strength and calorific value [50]. The report indicated that the amount of CNS had a significant impact on calorific value and thermal efficiency, while rice husk charcoal mixtures enhanced stability and compressive strength. CNS residue left after extracting CNSL has also been explored for the production of charcoal [56]. The researchers combined the CNS residue and rice husks to improve flammability and utilized the torrefaction process in making charcoal briquettes. Their findings show that a higher ratio of rice husks results in reduced compressive strength. A briquette with a 75:25% mixture ratio exhibited the highest compressive strength, with a higher heating value (HHV) of 6243.81 kcal/kg. In a separate study, mixtures of CNS and sugarcane bagasse were used to produce bio briquettes with the highest calorific value of 31.73 MJ/kg using cassava starch as a binder in the 50:10:40 compositions [57]. The studies show that charcoal produced from CNS not only provides similar HHV to traditional wood charcoals but also offers advantages such as faster heating and reduced fuel consumption. Furthermore, the relatively low cost of production suggests that CNS can play a significant role in sustainable energy solutions, contributing to reduced reliance on fossil fuels while promoting local waste management practices.

3.2.2 Adsorbents

The industrial application of activated carbon adsorbents is often limited by its costly manufacturing and regeneration processes. In response, there has been a growing interest in developing low-cost adsorbents using abundant industrial and agricultural biomass wastes, which not only provide an effective alternative but also contribute to waste reduction [58]. One limitation of using biomass wastes as adsorbents is their limited adsorption capacities; however, this limitation can be addressed through chemical treatments with substances like sulfuric acid (H_2SO_4), nitric acid (HNO_3), phosphoric acid (H_3PO_4), and sodium hydroxide (NaOH) to enhance their structure and surface properties [59, 60]. There are several studies on the use of CNS as sorption materials to remove various metal ions and metalloids, dyes, and phenols from polluted aqueous environments. A study by Nuithitikul and co-workers used CNS sorbent for Pb(II) ions decontamination from water [61]. The treatment of CNS with 1 M H_2SO_4, which was used to modify the surfaces and ensure improved adsorption capacities, positions CNS as a potential low-cost adsorbent for Pb(II) decontamination. Easy preparation and its maximum adsorption capacity results were similar to those of various biosorbents reported elsewhere [62]. Hunaidah and co-workers investigated the influence of particle size and pH variation on the adsorption of lead (Pb) and zinc (Zn) using activated carbon derived from cashew nut waste [63]. Their study showed that as the pH approaches acidic levels, the adsorption ability improves, and the smaller the grain size, the greater the adsorption capacity. Several other studies have documented notable applications of the cashew nut shell as a natural adsorbent in the removal of the metal ions including Cd^{2+}, Pb^{2+}, Ni^{2+}, and Cr^{3+} from contaminated water [64, 65]. The studies showed good linear adjustments for all mathematical models (Langmuir, Freundlich, and Dubinin–Radushkevich).

Dye organic compounds containing chromophores are reported to be harmful to the environment due to various industrial processes. Several studies have been carried out to investigate the utilization of CNS feedstock to remove dye wastes in the field of wastewater treatment. An activated carbon adsorbent derived from CNS produced using orthophosphoric acid as the activating agent demonstrated an adsorption capacity of 520 mg/g and good recyclability for up to five cycles in adsorbing methylene blue for wastewater treatment [66]. A further study by Spagnoli and coworkers reported considerably higher adsorption capacity values for cashew nut shell-derived activated carbon adsorbent with higher surface areas for removing methylene blue dye [67].

There are also recent studies that have a focus on reducing pharmaceutical pollutants, such as antibiotics, which are heavily used in medical therapy and cattle and poultry industries for the treatment of microbial infections as well as for the growth of livestock. In the study by Aljeboree and co-workers, cashew nut shells were used to produce H_3PO_4-activated carbon adsorbents for the removal of amoxicillin (AMX) contaminants from aqueous solutions [68]. Ahmadou and co-workers investigated the adsorption properties of two main families of toxic mycotoxins (aflatoxins and ochratoxin A) by various biochars obtained from CNS produced by the pyrolysis process for use in agriculture [69]. The study showed that varying pyrolysis temperature had an effect on the adsorption rate for Ochratoxin A whereas, for aflatoxin, adsorption rates were unaffected. Other current studies have explored the application of cashew nut shells for potential use in gas sorption. For example, Fonseca-Bermúdez and co-workers synthesized activated carbon derived from cashew nut shells using a potassium carbonate (K_2CO_3) activation process for greenhouse gas capture (CO_2 and CH_4). The synthesized materials demonstrated a remarkable surface area of 1693 m^2/g along with substantial total micropore volumes. In general, cashew nut shells, as biomass waste from post-industrial processes, highlight their importance in addressing environmental challenges [70].

4 Cashew Nut Shell Liquid (CNSL)

CNSL is an important bioresource, obtained as a byproduct of the cashew processing industry.

4.1 Chemical Composition and Structure of CNSL

CNSL consists of a mixture of phenolic compounds with C15–long alkylene chains appended on the *meta* position. Depending on the degree of unsaturation of the side chain, the phenolics are classified as saturated, monoenes, dienes, and trienes [71, 72]. The four major constituents of CNSL are anacardic acids, cardanols, cardols, and 2-methyl cardols (Fig. 12.3).

4.2 Extraction and Purification Methods for CNSL

Three methods are commonly used to extract CNSL, and these are solvent, mechanical, and thermal extractions [73]. The composition of the phenolic compounds in CNSL depends on the extraction method employed. Natural CNSL contains anacardic acids (60–65%), cardanols (10–15%), cardols (20%), and traces of 2-methyl

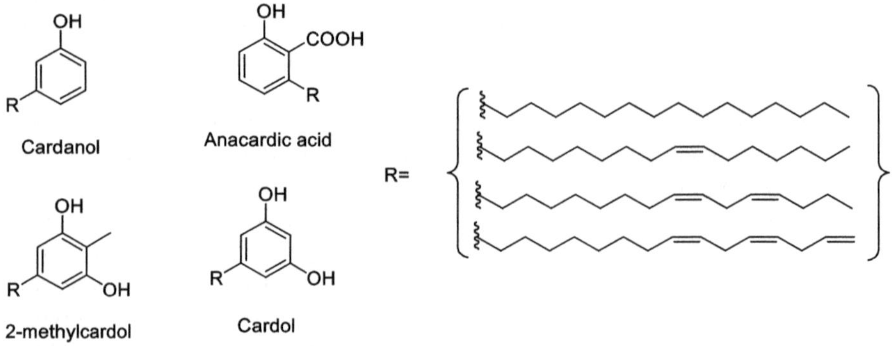

Fig. 12.3 Major constituents of CNSL

cardols and is obtained by cold solvents or mechanical extraction [74]. In contrast, technical CNSL is obtained by roasting the cashew nuts at elevated temperatures. This leads to the decarboxylation of anacardic acids to yield mainly cardanols (60–65%), cardols (15–20%), polymeric materials (10%), and traces of 2-methyl cardols [74].

4.2.1 Solvent Extraction

This method utilizes commonly used organic solvents that are both chlorinated such as dichloromethane, chloroform, and carbon tetrachloride as well as unchlorinated ones, for example, ethyl acetate, diethyl ether, hexane, methanol, ethanol, acetone, pet ether, and toluene. Chlorinated solvents, when used, tend to form emulsions than unchlorinated solvents [75]. In general, polar solvents furnish higher amounts of extracted CNSL than non-polar solvents [76]. Several techniques are employed under solvent extraction, and these include ultrasonic extraction, Soxhlet extraction, static solvent extraction, supercritical carbon dioxide, and subcritical water (SCW) extraction [77]. Solvent extraction methods suffer from long extraction times and the usage of large volumes of solvents [78]. In addition, challenges in the filtration process, the generation of colored impurities, and the persistence of organic solvents in the end products have been reported ([79]). Furthermore, the toxicity of some solvents poses both environmental and health concerns [80, 81]. The use of supercritical carbon dioxide (SC-CO_2) and SCW have drawn much interest as promising eco-friendly and efficient alternatives [77].

4.2.2 Mechanical Extraction

The mechanical extraction method is the oldest and most popular approach for obtaining natural CNSL. The extraction process is simple and effective and results in thicker and high-quality CNSL [82, 83]. The shells are compressed under high pressure using a hydraulic or screw press. The released oil is filtered using a filter press. This approach suffers from low yields as significant amounts of the oil remain trapped in the residue after processing [75, 84].

4.2.3 Thermal Extraction

This method involves subjecting the shells to elevated temperatures, which makes the shells crack releasing the oil. High temperatures, however, facilitate the decarboxylation of anacardic acids to cardanols, producing technical CNSL. This is a method of choice if cardanol-rich CNSL is the target [73, 83]. Thermal extraction techniques used include roasting and hot oil bath methods [76].

4.3 Applications of CNSL-Based Materials

The versatility of CNSL and its phenolic lipid components has led to numerous applications such as the fabrication of bioplastics and biopolymers, the development of coatings, adhesives and resins, and pharmaceutical and cosmetic products [85, 86].

4.3.1 Bioplastics and Biopolymers

CNSL has been described as a versatile monomer for polymers and polyol synthesis [87, 88]. The inherent functional features of CNSL phenolic lipids enable polymerization. For instance, the unsaturation of the 15-carbon side chain could facilitate radical-mediated addition polymerization while the phenolic properties are an excellent feature for condensation polymerization [89, 90].

The synthesis of thermoplastic polyurethane (PU) from cardanol for use in the telecommunications industry has been reported [91]. Cardanol was first converted to 4-[(4-hydroxy-2-pentadecenylphenyl)diazenyl]phenol (HPPDP), which was subsequently used as a monomer to prepare the polyurethane (Fig. 12.4).

Copolyesters were also synthesized by solution polymerization of terephthaloyl chloride with cardanol-derived HPPDP and 1,4-butane diol [92]. The oxidative polymerization of cardanols mediated by iron catalyst has been reported by Kobayashi (Fig. 12.5) [93, 94]. The cross-linked polymers displayed properties similar to commercially available CNSL-formaldehyde resin.

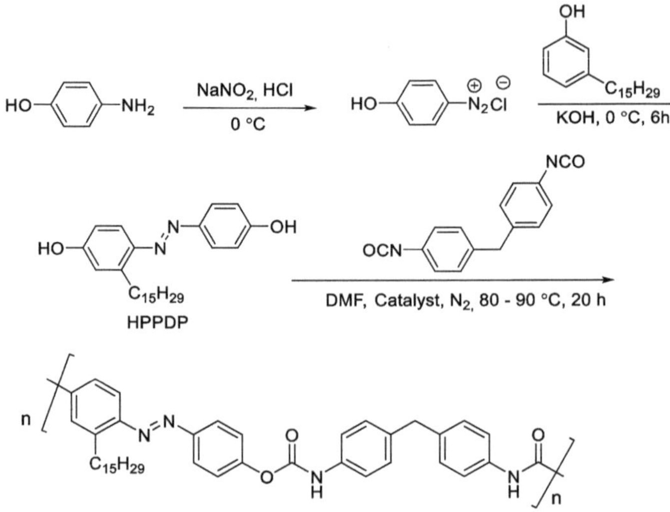

Fig. 12.4 Synthesis of thermoplastic polyurethane from cardanol

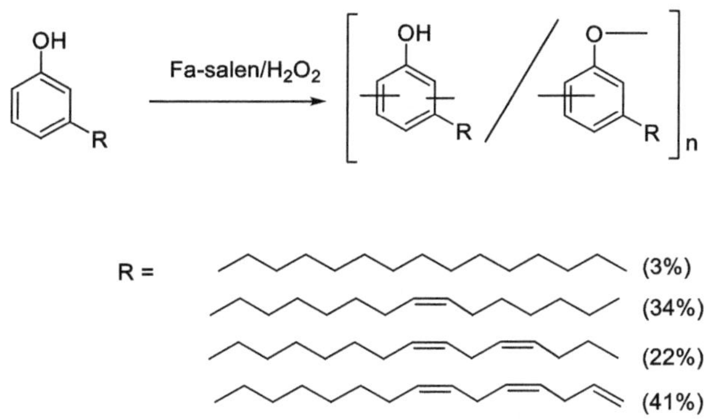

Fig. 12.5 Iron-mediated oxidative polymerization of cardanols

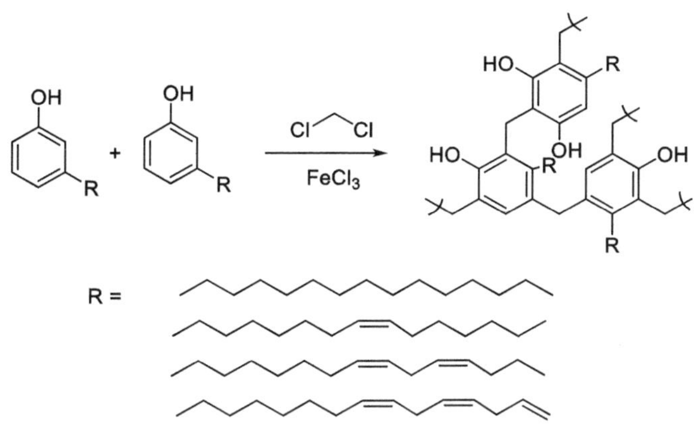

Fig. 12.6 Polymerization of cardanols and cardols to form HCPs

Boron trifluoride diethyl etherate ($BF_3O.Et_2O$) has been reported to initiate the cationic polymerization of cardanols. It polymerized cardanol in a short time at 140 °C [89]. Hypercrosslinked polymers (HCPs) for getting rid of volatile organic compounds have been prepared from CNSL using the Friedel–Crafts reaction (Fig. 12.6). One of the prepared HCPs displayed excellent adsorption of *o*-xylene [95].

Cardanol was utilized in the synthesis of a biopolymer composite with electrical properties by combining it with cassava starch [96]. The resulting biopolymer exhibited better values of electrical resistance. The thermal and mechanical properties of a thermoplastic cellulose diacetate (CDA) appended to hydrogenated cardanol were also explored. The CDA-based bioplastics showed high heat and water resistance [97]. Molecularly imprinted polymers (MIPs) for recognition of propranolol have been prepared using anacardanyl acrylate (AnAcr) and anacardanyl methacrylate (AnMcr) derived from anacardic acids as monomers (Fig. 12.7). These

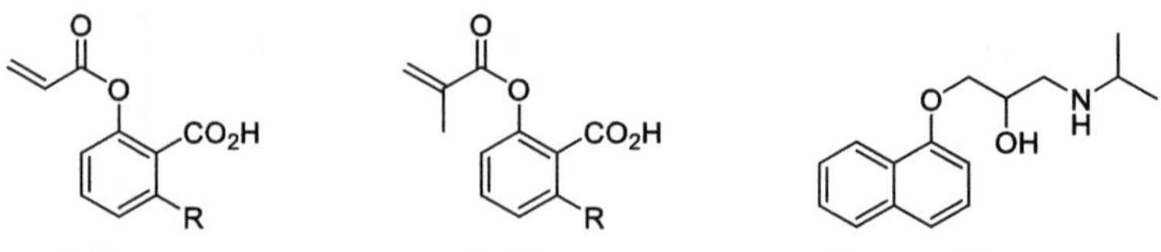

Fig. 12.7 Structures of functional monomers and the propranolol

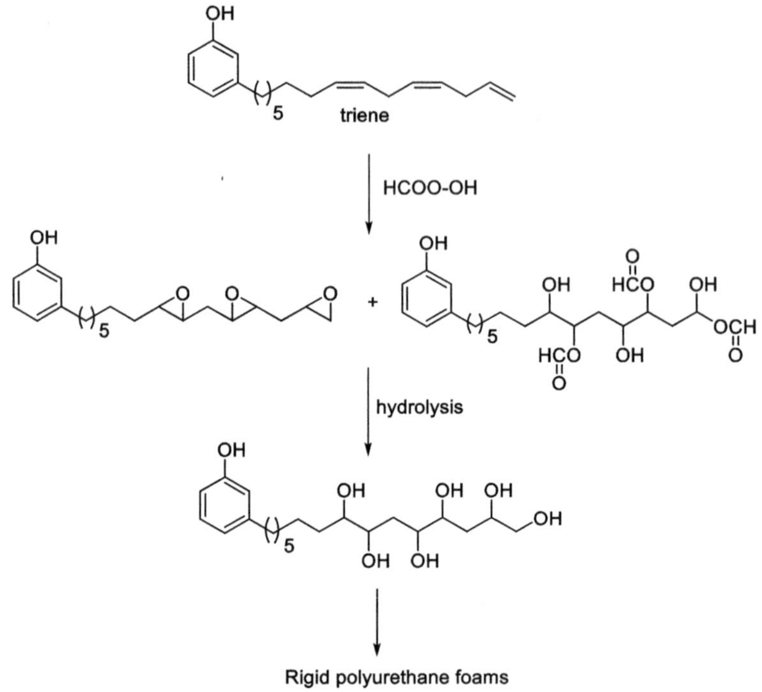

Fig. 12.8 Synthesis of polyols

were cross-linked with divinylbenzene (DVB) and ethylene glycol dimethacrylate (EGDMA) [98]. The DVB-based MIPs showed much improved noncovalent propranolol affinity and binding specificity.

CNSL phenolics have also been functionalized into polyols, which are chemical building blocks for developing surfactants and soft polymers [88]. Bio-based polyols for fabricating rigid polyurethane foams have been synthesized by the epoxidation of the unsaturated side chain followed by hydrolysis (Fig. 12.8) [99].

Low-density foams with good compressive structures were obtained when Mannich-based polyols were used [99]. Huo et al. [100] also synthesized autocatalytic polyols from cardanol, which were then converted to polyurethane. The resultant polyurethane foams had high compression strength and were thermal stable [100]. A sulfonic acid functionalized solid polymer catalyst has been prepared by treating CNSL with formaldehyde and sulfonating the resultant resin with chlorosulfonic acid [101]. The catalyst was used to mediate Friedel–Crafts reactions of indoles and furfuraldehydes.

4.3.2 Coatings, Adhesives, and Resins

CNSL phenolics have been extensively used to fabricate resins, coatings, adhesives, and plasticizers. Yokoyama et al. [102] synthesized bio-based bisphenol epoxy resin from hydrogenated cardanol, which could replace the petroleum-based diglycidyl ether [102]. The bio-based resin was synthesized by treating cardanol with propionaldehyde to give the bisphenol, which, upon exposure to epichlorohydrin, furnished the desired ether (Fig. 12.9).

Cardanol acetate was synthesized and used as a secondary plasticizer for PVC (polyvinyl chloride) owing to its partial miscibility with PVC [103]. Mesoporous superhydrophobic metal (Mn(II)) coordinated bio-polyurethanes were synthesized from CNSL as corrosion-resistant coatings [104]. The films/coatings were reported to be solvent-resistant, had low capacitance, and exhibited high impedance. Furthermore, the films inhibited various strains of bacteria [104]. In another study, Zafar et al. assembled polymeric thin films from Mg(II)-ion-coordinated methylolated-CNSL for antibacterial applications (Fig. 12.10) [105].

The prepared films were reported to inhibit both the Gram-positive and Gram-negative strains of bacteria and displayed excellent thermal stability [105]. Polyphenalkamines as anticorrosive coatings were synthesized from cardanol [106]. Cardanol was exposed to formaldehyde and ethylenediamine, which furnished phenalkamine (PhAA), which was then subsequently polymerized to

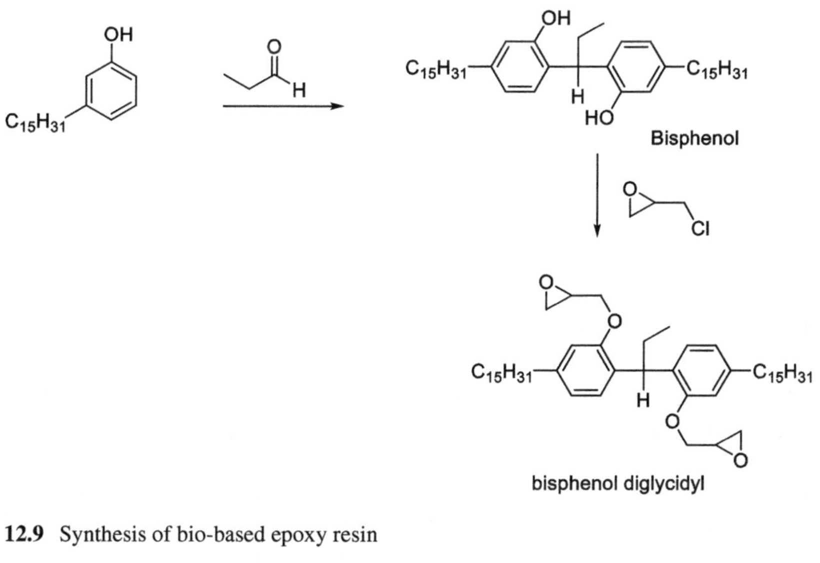

Fig. 12.9 Synthesis of bio-based epoxy resin

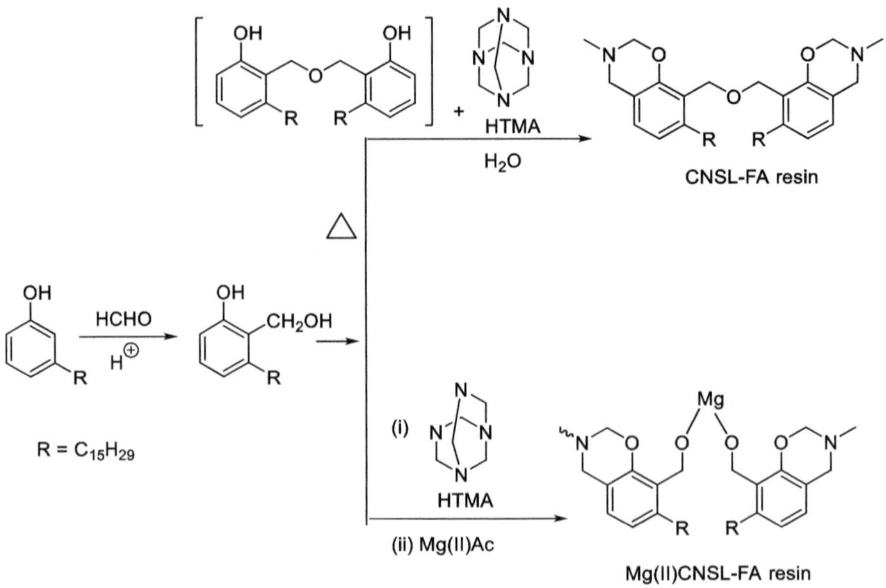

Fig. 12.10 Synthesis of CNSL-derived resins

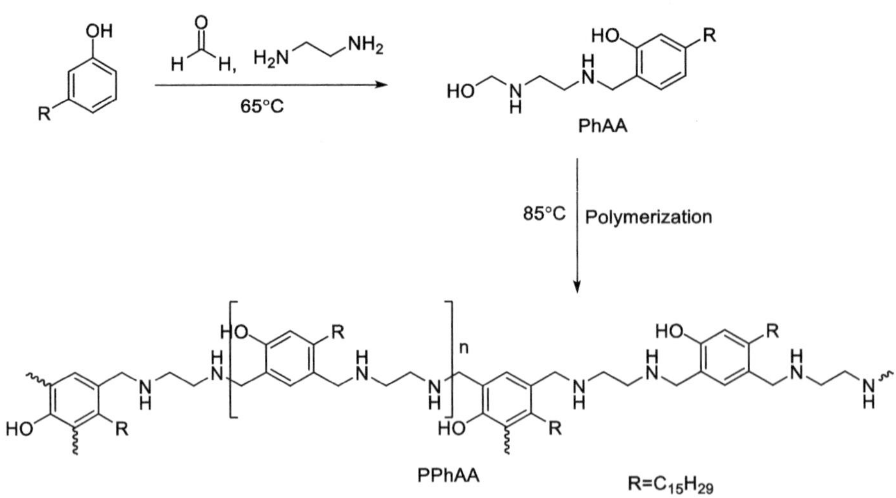

Fig. 12.11 Synthesis of polyphenalkamines

polyphenalkamine (PPhAA), as shown in Fig. 12.11. The coatings were reported to provide good protection to materials against both icing and corrosive environments.

Sodium alginate-based biopolymer films have been prepared by incorporating CNSL at various doses ranging from 0 to 1.5% [107]. Films spiked with CNSL provided better thermal resistance, reduced homogeneity, and increased elasticity.

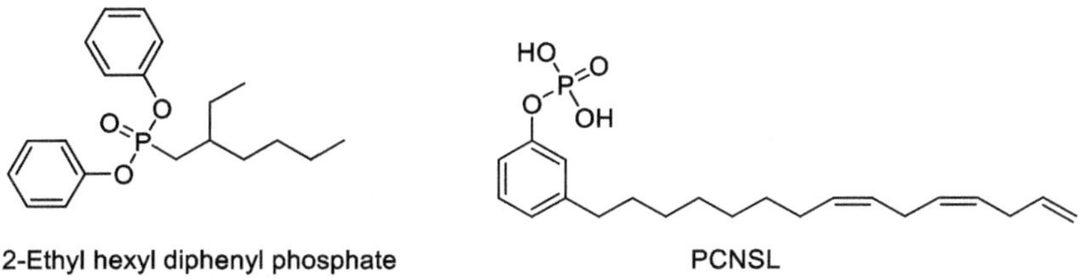

Fig. 12.12 Commercial phosphate and bio-based phosphate (PCNSL)

Fig. 12.13 Synthesis of novolac resins from cardanols

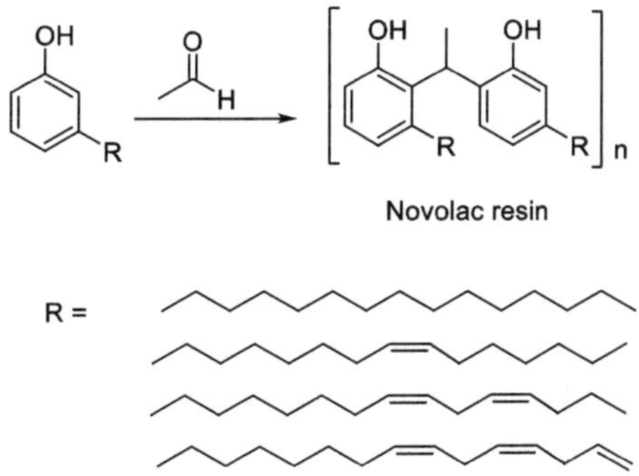

Natural rubber was modified with phosphorated CNSL (PCNSL) as a plasticizer. The rubber had better cure retardation than that modified by 2-ethyl hexyl diphenyl phosphate (Santicizer) (Fig. 12.12). Furthermore, the rubber displayed flame retardancy, better tensile properties, and thermal stability [108].

Cardanol, as a plasticizer, antioxidant, and activator of natural rubber has similar properties such as tensile strength when compared with commercial plasticizers, activators, and antioxidants [109]. Cardanol has also been used as a replacement for commercial phenols to synthesize novolac resins (Fig. 12.13) via condensation polymerization with acetaldehyde [110]. The resin displayed excellent thermal stability.

Gosh and Karak reported on the assembly of anticorrosive self-healable polyurethane coating from CNSL and other additives [111]. The obtained polyurethane displayed good surface hydrophobicity, good corrosion resistance properties, and reasonable mechanical and thermal properties. Interestingly, the polyurethane was able to self-mend after sustaining mechanical damage.

4.3.3 Pharmaceutical and Cosmetic Products

In search of sustainable ways to synthesize drugs and cosmetic products, researchers have turned to CNSL phenolics as alternative and bio-renewable chemical building blocks. Pharmaceutical and cosmetic ingredients have been prepared from CNSL.

Pharmaceutical Products

CNSL hybrid molecules targeting mitochondrial function in *Trypanosoma brucei* have been synthesized [112]. The phenolic and modified phenolic constituents from CNSL were exposed to substituted 2-bromo-1,4-naphthoquinones to generate various CNSL hybrids (Fig. 12.14).

The shorter chain versions of the CNSL hybrids synthesized from ozonolysis intermediate exhibited a tenfold trypanocidal activity compared with the baseline compound 2-phenoxy- naphthalene-1,4-dione. The synthesis of novel sulfonamides as antibacterial agents from anacardic acids has been reported [113].

Methylation of the phenolic and carboxylic acid protons, followed by ozonolysis of the resulting ester and subsequent reduction, yielded a primary alcohol. Several functional group interconversions afforded the desired sulfonamides (Fig. 12.15). Several sulfonamides displayed higher antibacterial activities than ampicillin used as a positive control. Cole-Hamilton and colleagues synthesized several pharmaceutical drugs like norfenefrine, *rac*-phenylephrine, etilefrine, and fenoprofen from cardanol [114]. The key advanced intermediate to access the pharmaceutical drugs was 3-vinylphenol, which was synthesized from cardanol via ethenolysis and isomerizing metathesis (Fig. 12.16).

Treating 3-vinylphenol with iron phthalocyanine gave the desired norfenefrine and *N*-methylation of norfenefrine mediated by methyl triflate furnished phenylephrine. Alkylating using ethyl triflate or reductive amination of norfenefrine gave etilefrine. For fenoprofen, cardanol was exposed to bromobenzene under basic conditions, which gave the phenoxy-cardanol (Fig. 12.17). Ethenolysis, isomerizing metathesis, methoxy carbonylation, and subsequent hydrolysis gave the desired fenoprofen [114]. Ethonolysis and cross-metathesis strategy was also deployed in the assembly of 2-hydroxy-6-tridecylbenzoic acid, a tyrosinase inhibitor, utilizing crude CNSL [115] as depicted in Fig. 12.18.

Analogs of rivastigmine, a drug used to treat dementia and whose structure is related to alkaloid (−)-3-*O*-acetylspectaline (Fig. 12.19), have been assembled from

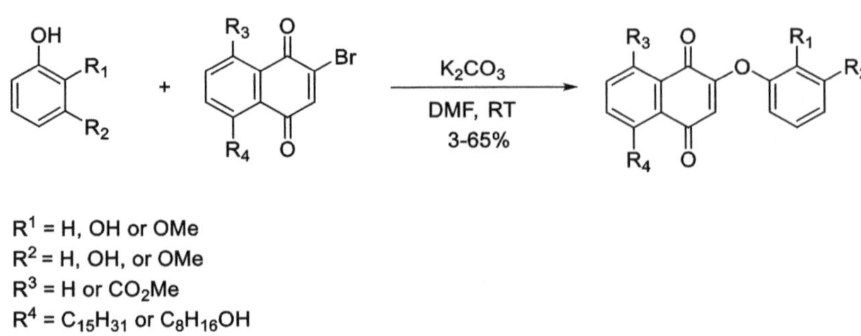

Fig. 12.14 Synthesis of naphthoquinone–CNSL hybrids

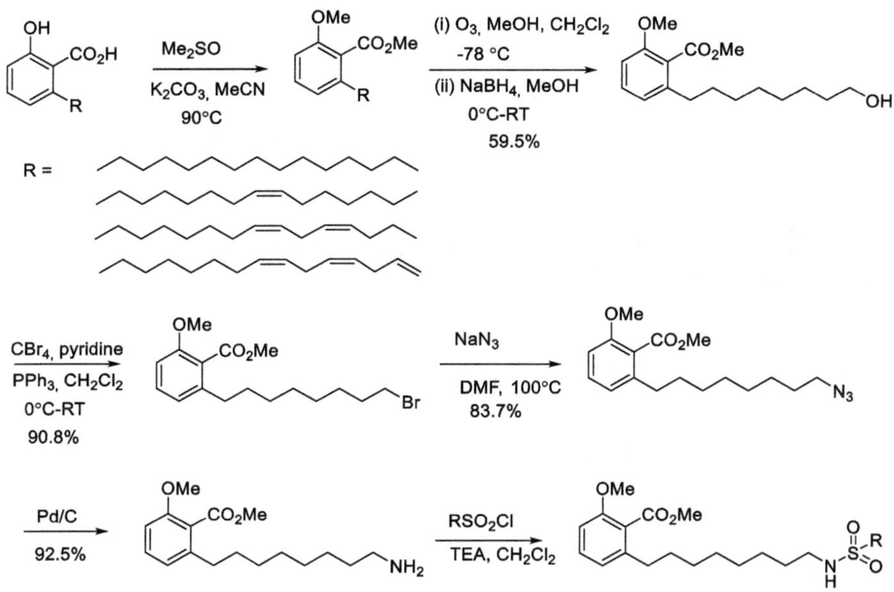

Fig. 12.15 Synthesis of sulfonamides from anacardic acids

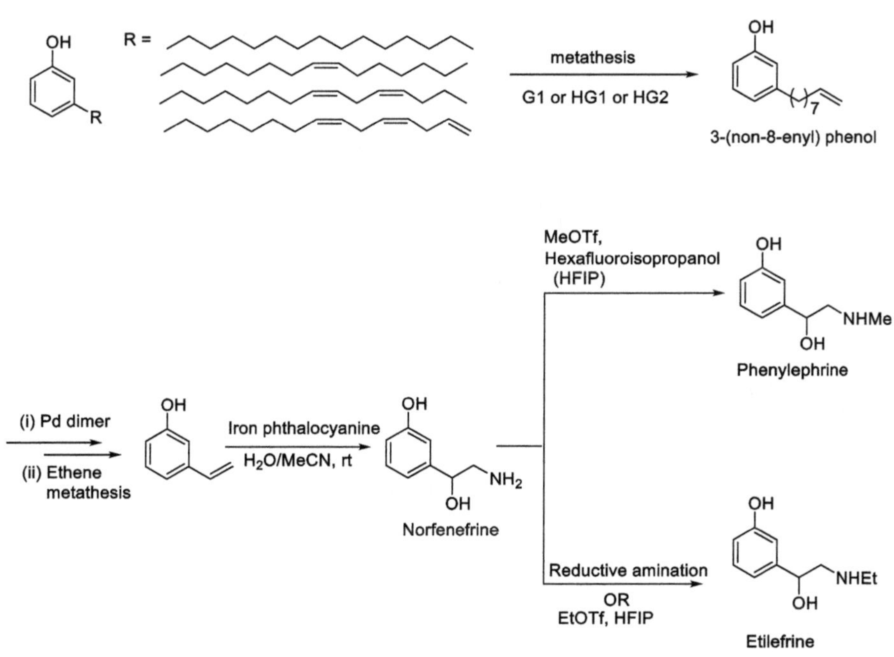

Fig. 12.16 Synthesis of norfenefrine, phenylephrine, and etilefrine

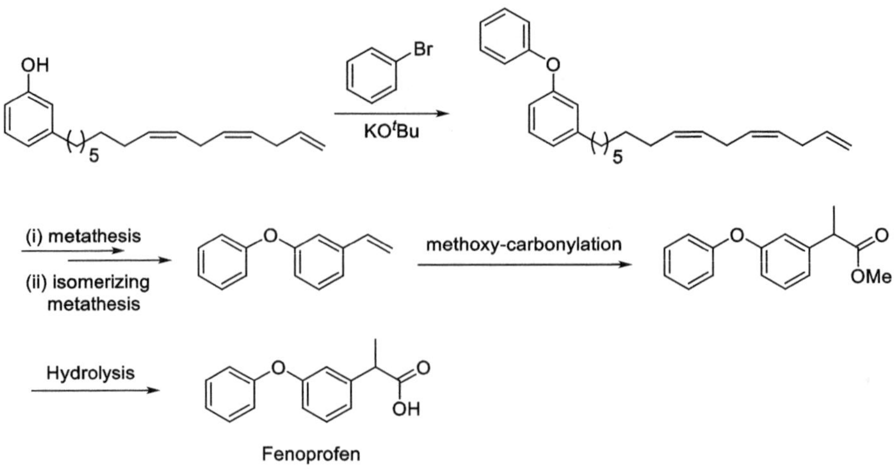

Fig. 12.17 Synthesis of fenoprofen

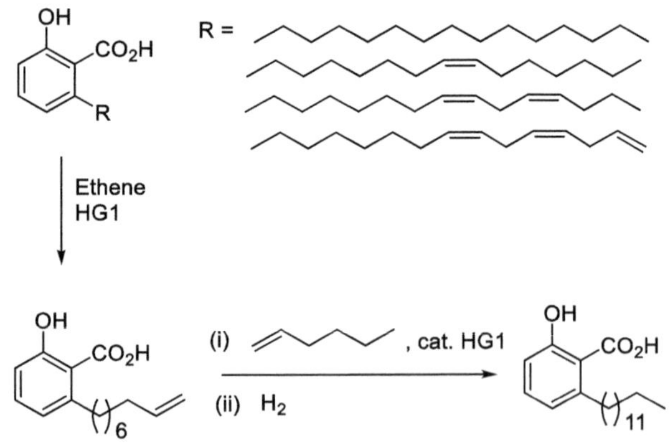

Fig. 12.18 Synthesis of an anacardic acid-derived tyrosine inhibitor

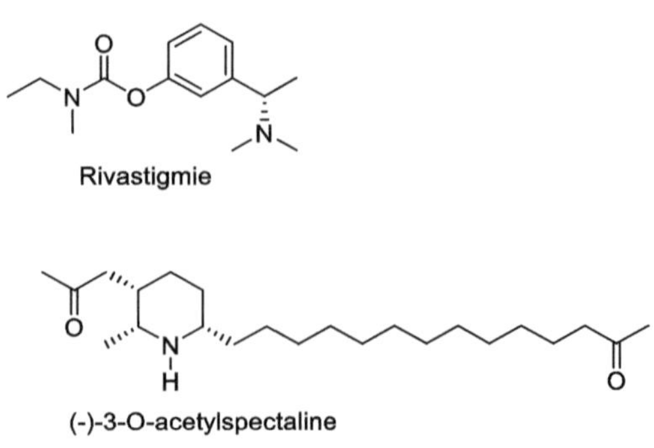

Fig. 12.19 Structure of rivastigmine and (−)-3-O-acetylspectaline

cardanol [116]. Twenty-four dual-binding acetylcholinesterase (AChE) inhibitors were synthesized and a few examples of these are illustrated in Fig. 12.20. Nineteen compounds exhibited at least 50% enzyme inhibition, and most analogs did not display noticeable toxicity. Multifunctional-bearing cholinesterase inhibitors were also synthesized based on cardanol-derived LDT16 (Fig. 12.21; [117]). The piperidine and azepine appended derivatives emerged as good cholinesterase inhibitors and exhibited anti-amyloid and antioxidant activities. The general synthetic route for accessing the inhibitors is shown in Fig. 12.22 [117]. Cardanols were first acetylated followed by ozonolysis and subsequent reduction, which gave the primary alcohol. Exposing the alcohol to iminium ions generated from the reaction of paraformaldehyde and suitable secondary amines furnished the desired targets.

CNSL-derived phenolic lipids to treat metabolic diseases have been synthesized from anacardic acid (Fig. 12.23) and cardanol/isoanacardic acid (Fig. 12.24) [118]. These were tested for the activity of peroxisome proliferator-activated receptors (PPARs). Several compounds exhibited PPAR agonists in the low micromolar

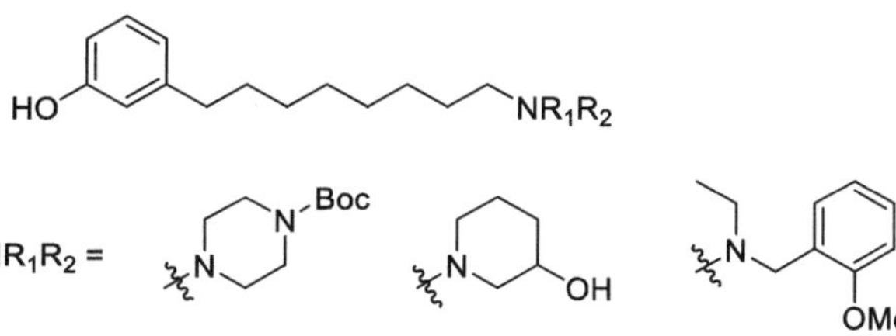

Fig. 12.20 CNSL-derived acetylcholinesterase (AChE) inhibitors

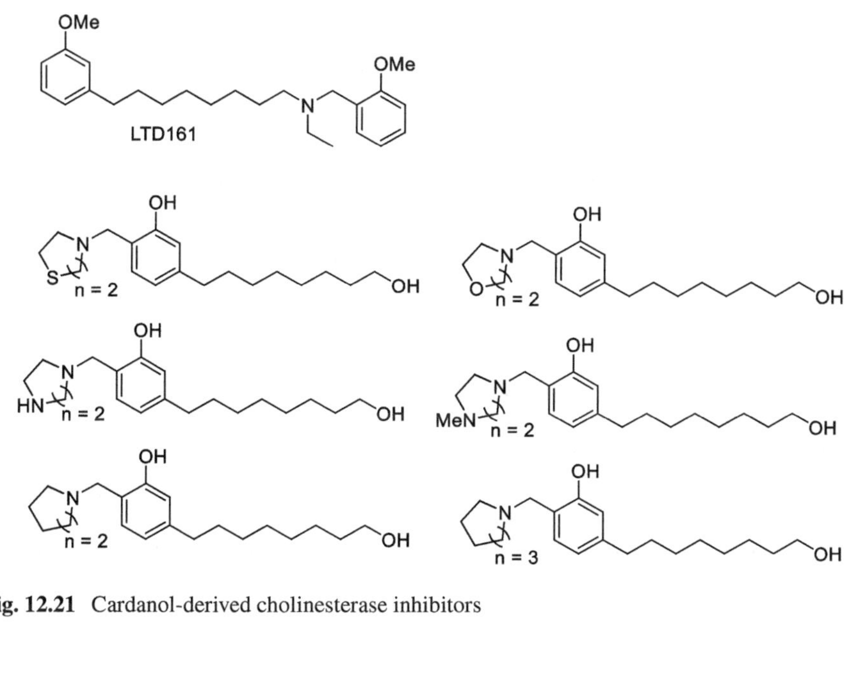

Fig. 12.21 Cardanol-derived cholinesterase inhibitors

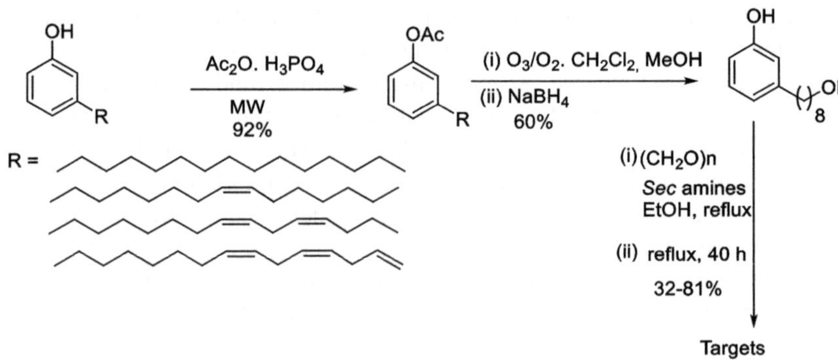

Fig. 12.22 Synthesis of cholinesterase inhibitors

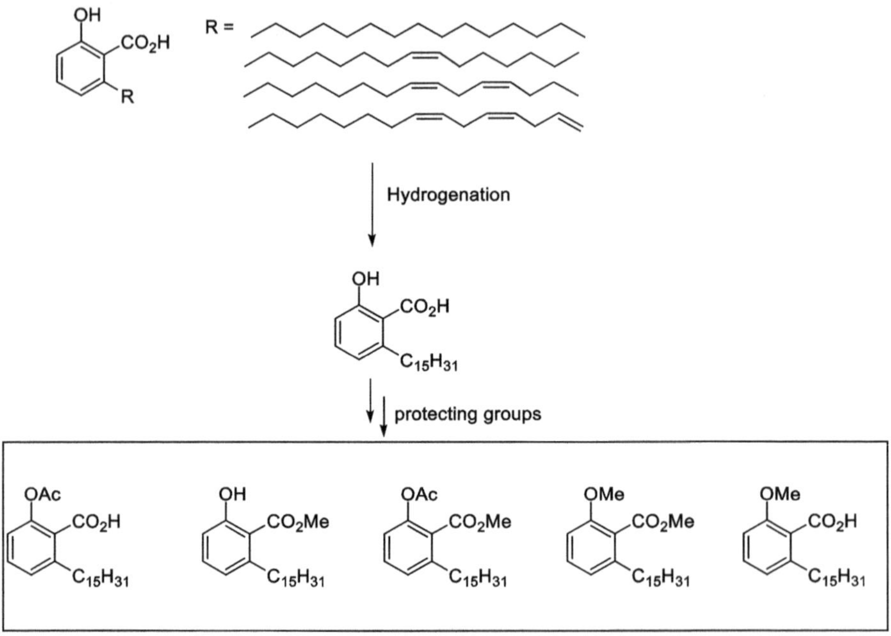

Fig. 12.23 Synthesis of derivatives of anacardic acid

range. One analog from cardanol (LDT409) showed partial agonist activity for PPARα and PPARγ.

Cytogenotoxic and mutagenic profiling of CNSL and cardanol have been reported by Islam and co-workers [119]. In vivo studies revealed that cardanol induced genotoxic effects on bone marrow cells of a female mouse, but this phenomenon was not observed with CNSL. Both cardanol and CNSL exhibited non-mutagenic activity.

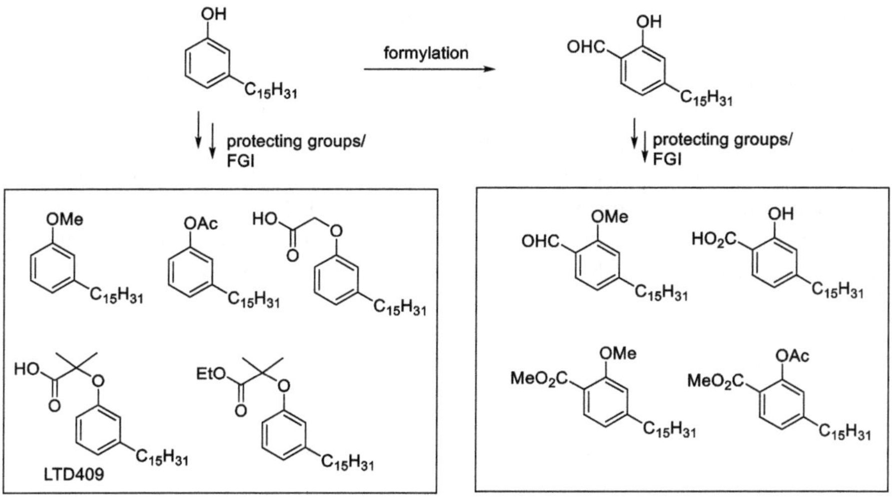

Fig. 12.24 Synthesis of derivatives of cardanol and isoanacardic

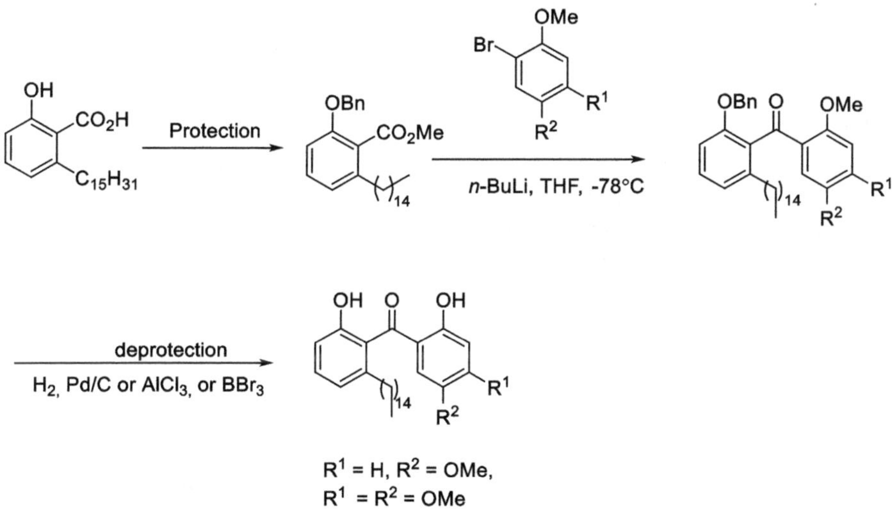

Fig. 12.25 Synthesis of anacardic acid-derived benzophenones

Cosmetic Ingredients

The major drawbacks of current UV protection agents used in cosmetic industries include aqua toxicity, coral bleaching, and poor biodegradability, and they are petrochemically derived among others [120, 121]. Mitigating these drawbacks remains a key challenge. The synthesis of several classes of UV filters from anacardic acid and cardanol has been reported [122]. Anacardic acid-derived benzophenones were synthesized as displayed in Fig. 12.25. Protection of anacardic acid followed by condensation with substituted bromo benzenes and subsequent deprotection afforded hydrogen-bonded benzophenones.

Benzophenones were also synthesized from cardanol (Fig. 12.26). Hydrogenated cardanol was treated with benzoyl chloride, and the resulting ester was subjected to a Fries rearrangement reaction yielding the mono-hydroxy-benzophenones. The second hydroxyl group was introduced via a late-stage oxidation reaction [122].

Triazines have also been synthesized from cardanol (Fig. 12.27). Formylation of cardanol followed by reduction gave a primary alcohol, which was subsequently treated with benzamidine hydrochloride to give unsymmetrical triazine. Another

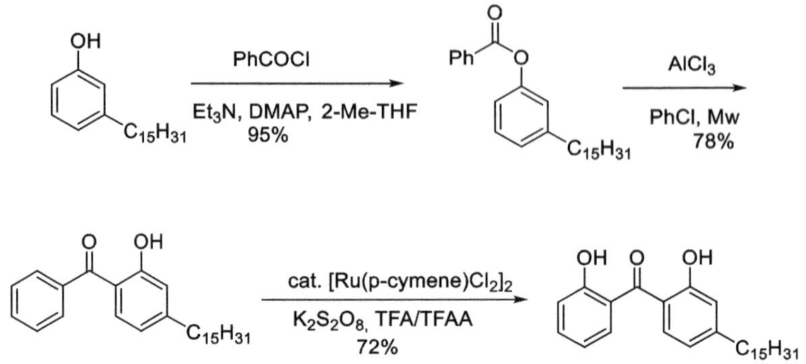

Fig. 12.26 Synthesis of benzophenones from cardanol

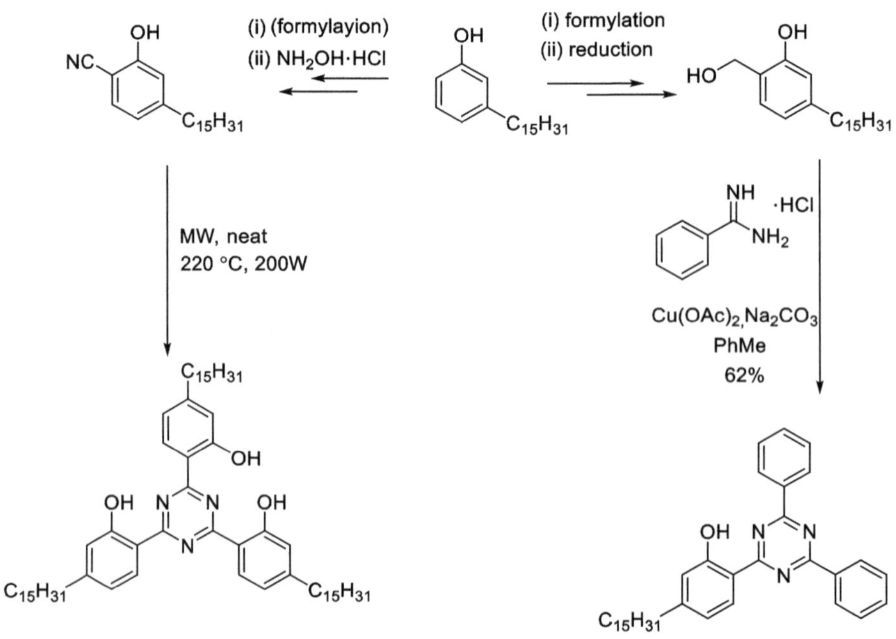

Fig. 12.27 Synthesis of triazines from cardanol

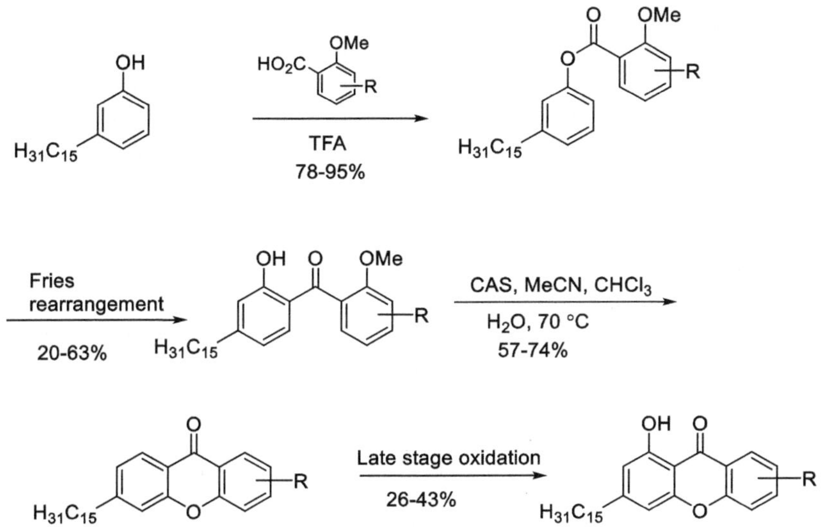

Fig. 12.28 CAS-mediated synthesis of xanthones from cardanol

triazine where all the appended aromatic rings were derived from cardanol was synthesized by converting formylated cardanol to a nitrile, which was subsequently cyclotrimerized to afford the desired symmetrical triazine [122].

Most of the compounds synthesized exhibited good to excellent UV absorbance. The triazines, in particular, exhibited broad-spectrum UV absorption in the UVA and UV region highlighting their potential for commercialization [122]. Jagot et al. also synthesized xanthones derived from cardanol as potential UV absorbers (Fig. 12.28) [123].

Hydrogenated cardanol was subjected to an esterification reaction with substituted 2-methoxybenzoic to yield substituted esters. Fries rearrangement of the esters yielded substituted benzophenones, which upon cyclization afforded the substituted xanthones. Late-stage oxidation furnished the hydrogen-bonded xanthones. Two hydrogen-bonded xanthones displayed good broad-spectrum UV absorption with excellent molar absorption coefficients.

5 Conclusion and Future Perspectives

The development of bio-based materials and technologies derived from bio-renewable resources offers a strategic solution to mitigate the economic challenges posed by the use of conventional resources. By harnessing the potential of cashew nut agro-waste, we can contribute to both environmental sustainability and economic stability, while reducing reliance on non-renewable resources. Extensive work on the utilization of cashew nut agro-waste by-products supports sustainability and the principles of a circular economy. Studies reported so far demonstrate that this type of agro-waste has significant potential industrial applications. Although a plethora of literature is available on the valorization of cashew nut agro-waste, extensive research especially on the utilization of the waste is imperative.

References

1. Liu S (2020) Interactions between industrial development and environmental protection dimensions of sustainable development goals (SDGs): evidence from 40 countries with different income levels. Environ Socio-Economic Stud 8:60–67. https://doi.org/10.2478/environ-2020-0018
2. Dugmore TIJ, Clark JH, Bustamante J et al (2017) Valorisation of biowastes for the production of green materials using chemical methods. Top Curr Chem 375:1–49. https://doi.org/10.1007/s41061-017-0133-8
3. Ning P, Yang G, Hu L et al (2021) Biotechnology for biofuels recent advances in the valorization of plant biomass. Biotechnol Biofuels 1–22:102. https://doi.org/10.1186/s13068-021-01949-3
4. Aluko A, Kassim N, Makule E (2023a) Investigating the optimal treatment to improve cashew apple juice quality and shelf life. J Food Process Preserv 2023:1–12. https://doi.org/10.1155/2023/4155761
5. Jamshidi S, Moradi Y, Nameni G et al (2021) Effects of cashew nut consumption on body composition and glycemic indices: a meta-analysis and systematic review of randomized controlled trials. Diabetes Metab Syndr Clin Res Rev 15:605–613. https://doi.org/10.1016/j.dsx.2021.02.038
6. Akyereko YG, Yeboah GB, Wireko-Manu FD et al (2023) Nutritional value and health benefits of cashew apple. JSFA Rep 3:110–118. https://doi.org/10.1002/jsf2.107
7. Bhakyaraj RKS (2012) Minerals and bioactive compounds in cashew apple (*Anacardium occidentale* L.). J Food Resour Sci 1:32–36
8. Prasertsri P (2017) Cashew apple juice: contents and effects on health. Nutr Food Sci Int J 4:2–5. https://doi.org/10.19080/nfsij.2017.04.555629
9. Prasertsri P, Roengrit T, Kanpetta Y et al (2013) Cashew apple juice supplementation enhanced fat utilization during high-intensity exercise in trained and untrained men. J Int Soc Sports Nutr 10:1. https://doi.org/10.1186/1550-2783-10-13
10. Prasertsri P, Roengrit T, Kanpetta Y et al (2019) Cashew apple juice supplementation enhances leukocyte count by reducing oxidative stress after high-intensity exercise in trained and untrained men. J Int Soc Sports Nutr 16:1–11. https://doi.org/10.1186/s12970-019-0299-2
11. Prasertsri P, Tong-un T, Roengrit T et al (2016) Cashew apple juice supplementation increases endurance and strength performance in cyclists. J Exerc Physiol Online 19:59–70
12. Gutiérrez-Paz C, Rodríguez-Moreno M-C, Hernández-Gómez M-S, Fernández-Trujillo JP (2024) Processing effects on bioactive compounds and potential benefits for human health. Food Secur 13:1–19
13. Silveira RMSCPMLFAAGFANFS (2012) Cashew apple juice as substrate for lactic acid production. Food Bioprocess Technol 5:947–953. https://doi.org/10.1007/s11947-010-0382-9
14. Ali AAH (2023) Overview of the vital roles of macro minerals in the human body. J Trace Elem Miner 4:100076. https://doi.org/10.1016/j.jtemin.2023.100076
15. Weickert MO, Pfeiffer AFH (2018) Impact of dietary fiber consumption on insulin resistance and the prevention of type 2 diabetes. J Nutr 148:7–12. https://doi.org/10.1093/jn/nxx008
16. Weickert MO, Pfeiffer AFH (2008) Metabolic effects of dietary fiber consumption and prevention of diabetes. J Nutr 138:439–442. https://doi.org/10.1093/jn/138.3.439

17. Arts ICW, Hollman PCH (2005) Polyphenols and disease risk in epidemiologic studies. Am J Clin Nutr 81:317S. https://doi.org/10.1093/ajcn/81.1.317s
18. Williamson G (2017) The role of polyphenols in modern nutrition. Nutr Bull 42:226–235. https://doi.org/10.1111/nbu.12278
19. Bhat MG (2008) Nutritive value of Cashew. ICAR-National Research Center for Cashew, Puttur. https://cashew.icar.gov.in/wp-content/uploads/2017/04/Nutritive%20Value%20of%20Cashew.pdf
20. van Walraven N, Stark AH (2024) From food waste to functional component: cashew apple pomace. Crit Rev Food Sci Nutr 64:7101–7117. https://doi.org/10.1080/10408398.2023.2180616
21. Dheeraj, Srivastava A, Mishra A (2023) Mitigation of cashew apple fruits astringency. Environ Sustain 6:319–329. https://doi.org/10.1007/s42398-023-00276-7
22. David CT, Prasad SM (2015) Value added products from cashew apple. Acta Hortic 1080:383–389. https://doi.org/10.17660/ActaHortic.2015.1080.51
23. Lowor S, Yabani D, Winifred K, Agyente-Badu C (2016) Production of wine and vinegar from cashew (*Anacardium occidentale)* "Apple". Braz Biotechnol J 12:1–11. https://doi.org/10.9734/bbj/2016/23366
24. Sobhana A (2019) Cashew apple utilization-generating wealth from waste. Adv Nutr Food Sci 4:1–5. https://www.opastonline.com/
25. Cavalcante AAM, Rubensam G, Picada JN et al (2003) Mutagenicity, antioxidant potential, and antimutagenic activity against hydrogen peroxide of cashew (*Anacardium occidentale*) apple juice and cajuina. Environ Mol Mutagen 41:360–369. https://doi.org/10.1002/em.10158
26. Cruz Reina LJ, Durán-Aranguren DD, Forero-Rojas LF et al (2022) Chemical composition and bioactive compounds of cashew (*Anacardium occidentale*) apple juice and bagasse from Colombian varieties. Heliyon 8:e09528. https://doi.org/10.1016/j.heliyon.2022.e09528
27. Aderiye BI, David OM, Atere VA (2015) Administration of cashew extracts in the treatment of some infections and diseases. Adv Med Plant Res Rev 3:75–86
28. Laxmanaswami B, Urooj A (2018) Phytochemical profile and antimicrobial activity of cashew apple (*Anacardium occidentale* L.) extract. GSC Biol Pharm Sci 5:095–098. https://doi.org/10.30574/gscbps.2018.5.3.0152
29. Salehi B, Gültekin-özgüven M, Kirkin C, Özçelik B (2020) Antioxidant , antimicrobial , and anticancer effects of *Anacardium* plants: an ethnopharmacological perspective. Front Endocrinol 11:1–16. https://doi.org/10.3389/fendo.2020.00295
30. Anoopkumar AN, Gopinath C, Annadurai S et al (2024) Biotechnological valorisation of cashew apple: prospects and challenges in synthesising wide spectrum of products with market value. Bioresour Technol Rep 25:101742. https://doi.org/10.1016/j.biteb.2023.101742
31. Nogueira Felix AK, Martins JJL, Lima Almeida JG et al (2019) Purification and characterization of a biosurfactant produced by *Bacillus subtilis* in cashew apple juice and its application in the remediation of oil-contaminated soil. Colloids Surfaces B Biointerfaces 175:256–263. https://doi.org/10.1016/j.colsurfb.2018.11.062
32. Toure N, Atobla K, Cisse M, Dabonne S (2019) Bioethanol production from cashew (*Anacardium occidentale* L.) apple juice by batch fermentation using *Saccharomyces cerevisiae* E450. Int J Biol Chem Sci 13:2546–2559. https://doi.org/10.4314/ijbcs.v13i6.10
33. Begum W, Saha B, Mandal U (2023) A comprehensive review on production of bio-surfactants by bio-degradation of waste carbohydrate feedstocks: an approach towards sustainable development. RSC Adv 13:25599–25615. https://doi.org/10.1039/d3ra05051c
34. Talasila U, Vechalapu RR, Shaik KB (2014) Clarification, preservation, and shelf life evaluation of cashew apple juice. Food Sci Biotechnol 21:709. https://doi.org/10.1007/s10068-012-0092-3
35. Deli MGEP, Kirit BD, Ağçam E et al (2022) Changes in cashew apple juice treated with optimum thermosonication during storage. Food Chem Adv 1:100120. https://doi.org/10.1016/j.focha.2022.100120

36. Kasavi C, Finore I, Lama L et al (2012) Evaluation of industrial *Saccharomyces cerevisiae* strains for ethanol production from biomass. Biomass Bioenergy 45:230–238. https://doi.org/10.1016/j.biombioe.2012.06.013
37. Rêgo ESB, Rosa CA, Freire AL et al (2020) Cashew wine and volatile compounds produced during fermentation by non-saccharomyces and saccharomyces yeast. Lwt 126:109291. https://doi.org/10.1016/j.lwt.2020.109291
38. Mohanty S, Ray P, Swain MR, Ray RC (2006) Fermentation of cashew (*Anacardium occidentale* L.) "apple" into wine. J Food Process Preserv 30:314–322. https://doi.org/10.1111/j.1745-4549.2006.00067.x
39. Kannan V, Rangarajan V, Manjare SD, Pathak PV (2021) Microbial production of value-added products from cashew apples—an economical boost to cashew farmers. J Pure Appl Microbiol 15:1816–1832. https://doi.org/10.22207/JPAM.15.4.71
40. Aluko A, Makule E, Kassim N (2023b) Effect of clarification on physicochemical properties and nutrient retention of pressed and blended cashew apple juice. Food Sci Nutr 11:1891–1903. https://doi.org/10.1002/fsn3.3222
41. Paiva FFA de (2013) Cashew Juice (Clarified) Original in Portuguese Researcher at Brazilian Agricultural Research Corporation-Embrapa
42. Cruz T, Maranon A, Hernandez C et al (2024) Exploring the potential of cashew nutshells: a critical review of alternative applications. Bioresources 19:6768–6803
43. Ajith Kumar TT, Ramesh ST (2023) Sustainable production of cashew nutshell briquettes: experimental assessment and optimization of factors affecting the physical and fuel characteristics. Biomass Convers Biorefinery 13:16969–16990. https://doi.org/10.1007/s13399-021-02234-x
44. Chung KWY, Blin J, Lanvin C et al (2024) Pyrolysis of cashew nut shells-focus on extractives. J Anal Appl Pyrolysis 179:106452. https://doi.org/10.1016/j.jaap.2024.106452
45. de Paiva EM, Mattos ALA, da Silva JCG et al (2024) Pyrolysis of cashew nutshell residues for bioenergy and renewable chemicals: kinetics, thermodynamics, and volatile products. J Anal Appl Pyrolysis 177:106303. https://doi.org/10.1016/j.jaap.2023.106303
46. Zinla BTD, Koffi EPM, Koua KB et al (2024) Life cycle assessment of cashew nutshell briquettes produced in Côte d'Ivoire. Open J Appl Sci 14:2411–2430. https://doi.org/10.4236/ojapps.2024.149159
47. Paternina Reyes MJ, Unfried Silgado J, Santa Marín JF et al (2023) Cashew nutshells: a promising filler for 3D printing filaments. Polymers (Basel) 15:4347. https://doi.org/10.3390/polym15224347
48. Oliveira NN, Mothé CG, Mothé MG, de Oliveira LG (2020) Cashew nut and cashew apple: a scientific and technological monitoring worldwide review. J Food Sci Technol 57:12–21. https://doi.org/10.1007/s13197-019-04051-7
49. Kyei SK, Eke WI, Nagre RD et al (2023) A comprehensive review on waste valorization of cashew nutshell liquid: sustainable development and industrial applications. Clean Waste Syst 6:100116. https://doi.org/10.1016/j.clwas.2023.100116
50. Pathaveerat S, Pruengam P, Lekrungroenggid N (2024) The potential of cashew nut shell waste in charcoal briquettes after producing cashew nut shell liquid through cold extraction. Results Eng 23:102579. https://doi.org/10.1016/j.rineng.2024.102579
51. Sawadogo M, Tchini Tanoh S, Sidibé S et al (2018) Cleaner production in Burkina Faso: case study of fuel briquettes made from cashew industry waste. J Clean Prod 195:1047–1056. https://doi.org/10.1016/j.jclepro.2018.05.261
52. Ifa L, Yani S, Nurjannah N et al (2020) Techno-economic analysis of bio-briquette from cashew nut shell waste. Heliyon 6:e05009. https://doi.org/10.1016/j.heliyon.2020.e05009
53. Ganesan S, Vedagiri P (2022) Production of sustainable biomass briquettes from de-oiled cashewnut Shell. Mater Today Proc 68:2484–2492. https://doi.org/10.1016/j.matpr.2022.09.179

54. Kadjo BS, Sako MK, Diango KA et al (2024) Production of a high-energy solid biofuel from biochar produced from cashew nut shells. Clean Eng Technol 21:100776. https://doi.org/10.1016/j.clet.2024.100776
55. Chungcharoen T, Srisang N (2020) Preparation and characterization of fuel briquettes made from dual agricultural waste: cashew nut shells and areca nuts. J Clean Prod 256:120434. https://doi.org/10.1016/j.jclepro.2020.120434
56. Yohana E, Arijanto KIE, Lazuardi A (2017) Development of briquette fuel from cashew shells and rice husk mixture. AIP Conf Proc 1788:030017. https://doi.org/10.1063/1.4968270
57. Sarafa OOS, Olafimihan EO et al (2022) Production and characterization of cashew nut shells and sugarcane bagasse composite briquettes as an alternative source of energy. Int J Adv Eng Manag 4:1177. https://doi.org/10.35629/5252-040911771186
58. Papadaki MI, Mendoza-Castillo DI, Reynel-Avila HE et al (2021) Nut shells as adsorbents of pollutants: research and perspectives. Front Chem Eng 3:1–15. https://doi.org/10.3389/fceng.2021.640983
59. Islam MT, Hyder AG, Saenz-Arana R et al (2019) Removal of methylene blue and tetracycline from water using peanut shell derived adsorbent prepared by sulfuric acid reflux. J Environ Chem Eng 7:102816. https://doi.org/10.1016/j.jece.2018.102816
60. Mashkoor F, Nasar A (2019) Preparation, characterization and adsorption studies of the chemically modified *Luffa aegyptica* peel as a potential adsorbent for the removal of malachite green from aqueous solution. J Mol Liq 274:315–327. https://doi.org/10.1016/j.molliq.2018.10.119
61. Nuithitikul K, Phromrak R, Saengngoen W (2020) Utilization of chemically treated cashew-nut shell as potential adsorbent for removal of Pb(II) ions from aqueous solution. Sci Rep 10:1–14. https://doi.org/10.1038/s41598-020-60161-9
62. Ouyang XK, Yang LP, Wen ZS (2014) Adsorption of Pb(II) from solution using peanut shell as biosorbent in the presence of amino acid and sodium chloride. Bioresources 9:2446–2458. https://doi.org/10.15376/biores.9.2.2446-2458
63. Hunaidah UMAA, Fayanto S et al (2019) Activated carbon from cashew nut waste and its application as a heavy metal absorbent. J Phys Conf Ser 1321:022004. https://doi.org/10.1088/1742-6596/1321/2/022004
64. Coelho GF, Gonçalves AC, Tarley CRT et al (2014) Removal of metal ions cd (II), Pb (II), and Cr (III) from water by the cashew nut shell *Anacardium occidentale* L. Ecol Eng 73:514–525. https://doi.org/10.1016/j.ecoleng.2014.09.103
65. Kumar PS, Ramalingam S, Kirupha SD et al (2011) Adsorption behavior of nickel(II) onto cashew nut shell: equilibrium, thermodynamics, kinetics, mechanism and process design. Chem Eng J 167:122–131. https://doi.org/10.1016/j.cej.2010.12.010
66. Anchan HN, Jadhav S, Dutta S (2023) Activated carbon from cashew nut husk and cashew nut shell wastes: synthesis, characterization, and adsorption studies. Chem Select 8:1–7. https://doi.org/10.1002/slct.202302831
67. Spagnoli AA, Giannakoudakis DA, Bashkova S (2017) Adsorption of methylene blue on cashew nut shell based carbons activated with zinc chloride: the role of surface and structural parameters. J Mol Liq 229:465–471. https://doi.org/10.1016/j.molliq.2016.12.106
68. Aljeboree AM, Aljaberi ASOK, Salah OH et al (2024) Adsorption of amoxicillin drug onto activated carbon prepared from cashew nut shell by H_3PO_4 activation: studies on equilibrium isotherm by nonlinear equations. Adv J Chem Sect A 7:347–354. https://doi.org/10.48309/ajca.2024.434307.1474
69. Ahmadou A, Napolı A, Durand N, Montet D (2019) Production of biochar from cashew nut shell to adsorb mycotoxins (aflatoxins and ochratoxin a) under different conditions for use in agriculture. Agric Food Chem 03:37
70. Fonseca-Bermúdez ÓJ, Giraldo L, Sierra-Ramírez R et al (2024) Cashew nut shell biomass: a source for high-performance CO_2/CH_4 adsorption in activated carbon. J CO2 Util 83:102799. https://doi.org/10.1016/j.jcou.2024.102799

71. Bhatia B, Amarnath N, Rastogi SK, Lochab B (2024) Isolation of cardanol fractions from cashew nutshell liquid (cnsl): a sustainable approach. SustainChem 5:68–80
72. Paramashivappa R, Phani Kumar P, Vithayathil PJ, Srinivasa Rao A (2001) Novel method for isolation of major phenolic constituents from cashew (*Anacardium occidentale* L.) nut shell liquid. J Agric Food Chem 49:2548–2551. https://doi.org/10.1021/jf001222j
73. Mubofu EB, Mgaya JE (2018) Chemical valorization of cashew nut shell waste. Top Curr Chem 376:8. https://doi.org/10.1007/s41061-017-0177-9
74. Phani Kumar P, Paramashivappa R, Vithayathil PJ et al (2002) Process for isolation of Cardanol from technical cashew (*Anacardium occidentale* L.) nut shell liquid. J Agric Food Chem 50:4705–4708. https://doi.org/10.1021/jf020224w
75. Garkal DJ, Bhande RS (2014) Review on extraction and isolation of cashew nutshell liquid. Int J Innov Eng Res Technol 1:1–8
76. Senthil Kumar P, Arun Kumar N, Sivakumar R, Kaushik C (2009) Experimentation on solvent extraction of polyphenols from natural waste. J Mater Sci 44:5894–5899. https://doi.org/10.1007/s10853-009-3834-8
77. Yuliana M, Tran-Thi NY, Ju Y-H (2012) Effect of extraction methods on characteristic and composition of Indonesian cashew nut shell liquid. Ind Crop Prod 35:230–236. https://doi.org/10.1016/j.indcrop.2011.07.007
78. Luque de Castro MD, Garcia-Ayuso LE (1998) Soxhlet extraction of solid materials: an outdated technique with a promising innovative future. Anal Chim Acta 369:1–10. https://doi.org/10.1016/S0003-2670(98)00233-5
79. Tyman JHP, Johnson RA, Muir M, Rokhgar R (1989) The extraction of natural cashew nutshell liquid from the cashew nut (*Anacardium occidentale*). J Am Oil Chem Soc 66:553–557. https://doi.org/10.1007/BF02885447
80. Hanmoungjai P, Pyle L, Niranjan K (2000) Extraction of rice bran oil using aqueous media. J Chem Technol Biotechnol 75:348–352. https://doi.org/10.1002/(SICI)1097-4660(200005)75:5<348::AID-JCTB233>3.0.CO;2-P
81. Joshi DR, Adhikari N (2019) An overview on common organic solvents and their toxicity. J Pharm Res Int 28:1–18
82. Himabindu T, Raguram V, Kumar ST (2015) Review on various methods of extraction of cashew nut shell liquid and isolation of anacardic acid. Int J Institutional Pharm Life Sci 5:293–308
83. Mgaya J, Shombe GB, Masikane SC et al (2019) Cashew nutshell: a potential bio-resource for the production of bio-sourced chemicals, materials and fuels. Green Chem 21:1186–1201. https://doi.org/10.1039/C8GC02972E
84. Chaudhari A, Thakor N, Mahale D et al (2012) Physical properties of cashew nut shell liquid. Int J Appl Innov Eng Manag 6:172–175
85. Sharma P, Gaur VK, Sirohi R et al (2020) Valorization of cashew nut processing residues for industrial applications. Ind Crop Prod 152:112550. https://doi.org/10.1016/j.indcrop.2020.112550
86. Telascrêa M, Leão AL, Ferreira MZ et al (2014) Use of a cashew nut shell liquid resin as a potential replacement for phenolic resins in the preparation of panels—a review. Mol Cryst Liq Cryst 604:222–232. https://doi.org/10.1080/15421406.2014.968509
87. Lubi MC, Thachil ET (2000) Cashew nut shell liquid (CNSL)—a versatile monomer for polymer synthesis. Des Monomers Polym 3:123–153. https://doi.org/10.1163/156855500300142834
88. Rojtman E, Denis M, Sirvent C et al (2024) Polyols from cashew nut shell liquid (CNSL): corner-stone building blocks for cutting-edge bio-based additives and polymers. Polym Chem 15:4375–4415. https://doi.org/10.1039/D4PY00851K
89. Antony R, Pillai CKS, Scariah KJ (1990) GPC studies on the cationic polymerization of cardanol initiated by boron trifluoride diethyletherate. J Appl Polym Sci 41:1765–1775. https://doi.org/10.1002/app.1990.070410733
90. Hemalatha IN, Dakshinamurthy H, Aggarwai JS (1968) No Title. Paint Manuf 38:29

91. Bhunia HP, Jana RN, Basak A et al (1998) Synthesis of polyurethane from cashew nut shell liquid (CNSL), a renewable resource. J Polym Sci Part A Polym Chem 36:391–400. https://doi.org/10.1002/(SICI)1099-0518(199802)36:3<391::AID-POLA3>3.0.CO;2-V
92. Bhunia HP, Basak A, Chaki TK, Nando GB (2000) Synthesis and characterization of polymers from cashewnut shell liquid: a renewable resource: V synthesis of copolyester. Eur Polym J 36:1157–1165. https://doi.org/10.1016/S0014-3057(99)00171-8
93. Ikeda R, Tanaka H, Uyama H, Kobayashi S (2000) A new crosslinkable polyphenol from a renewable resource. Macromol Rapid Commun 21:496–499. https://doi.org/10.1002/(SICI)1521-3927(20000501)21:8<496::AID-MARC496>3.0.CO;2-G
94. Ikeda R, Tanaka H, Uyama H, Kobayashi S (2002) Synthesis and curing behaviors of a crosslinkable polymer from cashew nut shell liquid. Polymer (Guildf) 43:3475–3481. https://doi.org/10.1016/S0032-3861(02)00062-9
95. Liu Z, Liu H, Wang Y et al (2022) Preparation of hyper crosslinked polymers with cashew nut shell liquid for removal of volatile organic compounds. Polym Eng Sci 62:1823–1832. https://doi.org/10.1002/pen.25967
96. Arrieta AA, Ducuara JA, Combatt EM (2023) Valorization of cashew nut processing by-product: development of a cardol/starch biopolymer composite with electrochemical properties and technological potential. Eastern-European J Enterp Technol 3:32–41. https://doi.org/10.15587/1729-4061.2023.282208
97. Iji M, Moon S, Tanaka S (2011) Hydrophobic, mechanical and thermal characteristics of thermoplastic cellulose diacetate bonded with cardanol from cashew nutshell. Polym J 43:738–741. https://doi.org/10.1038/pj.2011.57
98. Philip JYN, Buchweishaija J, Mkayula LL, Ye L (2007) Preparation of molecularly imprinted polymers using anacardic acid monomers derived from cashew nut shell liquid. J Agric Food Chem 55:8870–8876. https://doi.org/10.1021/jf0718289
99. Suresh KI (2013) Rigid polyurethane foams from cardanol: synthesis, structural characterization, and evaluation of polyol and foam properties. ACS Sustain Chem Eng 1:232–242. https://doi.org/10.1021/sc300079z
100. Huo S, Jin C, Liu G et al (2019) Preparation and properties of biobased autocatalytic polyols and their polyurethane foams. Polym Degrad Stab 159:62–69. https://doi.org/10.1016/j.polymdegradstab.2018.11.019
101. Kotkar GD, Tilve SG (2024) Sulfonic acid-functionalized solid polymer catalyst from crude cashew nut shell liquid: synthesis of tetra(indolyl)methanes and bis(indolyl)methanes from xylochemicals. ACS Omega 9:47543. https://doi.org/10.1021/acsomega.4c06337
102. Yokoyama Y, Yasui T, Takeda A et al (2023) Novel bio-based flexible bisphenol epoxy resin derived from cashew nut shell liquid. Polym J 55:859–867. https://doi.org/10.1038/s41428-023-00782-9
103. Greco A, Brunetti D, Renna G et al (2010) Plasticizer for poly(vinyl chloride) from cardanol as a renewable resource material. Polym Degrad Stab 95:2169–2174. https://doi.org/10.1016/j.polymdegradstab.2010.06.001
104. Shaily SA, Zafar F et al (2023) Superhydrophobic Mn(II)-coordinated technical cashew nut shell liquid-based bactericidal and corrosion-resistant advanced polyurethane coatings. Mater Today Commun 35:105947. https://doi.org/10.1016/j.mtcomm.2023.105947
105. Zafar F, Khan S, Ghosal A et al (2019) Clean synthesis and characterization of green nanostructured polymeric thin films from endogenous mg (II) ions coordinated methylated-cashew nutshell liquid. J Clean Prod 238:117716. https://doi.org/10.1016/j.jclepro.2019.117716
106. Masood S, Ghosal A, Zafar F et al (2023) Epoxy-free polyphenalkamine one-pack sustainable coating from technical cashew nut shell liquid: fabrication and electrochemical studies for anti-corrosive applications. Mater Today Sustain 22:100362. https://doi.org/10.1016/j.mtsust.2023.100362
107. Vasconcelos L, de Souza M, de Oliveira J et al (2021) Elaboration and characterization of bioactive films obtained from the incorporation of cashew nut shell liquid into a matrix of sodium alginate. Antioxidants 10(9):1378; 10.3390/antiox10091378

108. Menon ARR, Pillai CKS, Nando GB (1998) Modification of natural rubber with phosphatic plasticizers: a comparison of phosphorylated cashew nut shell liquid prepolymer with 2-ethyl hexyl diphenyl phosphate. Eur Polym J 34:923–929. https://doi.org/10.1016/S0014-3057(97)00224-3
109. Alexander M, Thachil ET (2010) The effectiveness of cardanol as plasticiser, activator, and antioxidant for natural rubber processing. Prog Rubber Plast Recycl Technol 26:107–124. https://doi.org/10.1177/147776061002600301
110. Marliyana SD, Wahyuningsih S, Hayati NN (2022) Synthesis of cardanol-based acetaldehyde novolac resin from cashew nut shell liquid (CNSL). J Kim Sains dan Apl 25(9):23. https://doi.org/10.14710/jksa259316-321
111. Ghosh T, Karak N (2020) Cashew nut shell liquid terminated self-healable polyurethane as an effective anticorrosive coating with biodegradable attribute. Prog Org Coatings 139:105472. https://doi.org/10.1016/j.porgcoat.2019.105472
112. Cerone M, Uliassi E, Prati F et al (2019) Discovery of sustainable drugs for neglected tropical diseases: cashew nut shell liquid (CNSL)-based hybrids target mitochondrial function and ATP production in *Trypanosoma brucei*. Chem Med Chem 14:621–635. https://doi.org/10.1002/cmdc.201800790
113. Reddy NS, Rao AS, Chari MA et al (2012) Synthesis and antibacterial activity of sulfonamide derivatives at C-8 alkyl chain of anacardic acid mixture isolated from a natural product cashew nut shell liquid (CNSL). J Chem Sci 124:723–730. https://doi.org/10.1007/s12039-012-0253-1
114. Shi Y, Kamer PCJ, Cole-Hamilton DJ (2019) Synthesis of pharmaceutical drugs from cardanol derived from cashew nut shell liquid. Green Chem 21:1043–1053. https://doi.org/10.1039/C8GC03823F
115. Pollini J, Bragoni V, Gooßen LJ (2018) Synthesis of a tyrosinase inhibitor by consecutive ethenolysis and cross-metathesis of crude cashew nutshell liquid. Beilstein J Org Chem 14:2737–2744. https://doi.org/10.3762/bjoc.14.252
116. Uliassi E, de Oliveira AS, de Nascente LC et al (2021) Cashew nut shell liquid (CNSL) as a source of drugs for Alzheimer's disease. Molecules 26(18):5441. https://doi.org/10.3390/molecules26185441
117. Espro C, Paone E, Mauriello F et al (2021) Sustainable production of pharmaceutical, nutraceutical and bioactive compounds from biomass and waste. Chem Soc Rev 50:11191–11207. https://doi.org/10.1039/D1CS00524C
118. Sahin C, Magomedova L, Ferreira TAM et al (2022) Phenolic lipids derived from cashew nut shell liquid to treat metabolic diseases. J Med Chem 65:1961–1978. https://doi.org/10.1021/acs.jmedchem.1c01542
119. de Leite AS, Islam MT, MFCJ P et al (2019) Cytogenotoxic and mutagenic profiling of cashew nut shell liquids and cardanol. Clin Phytoscience 5:37. https://doi.org/10.1186/s40816-019-0129-8
120. Chatzigianni M, Pavlou P, Siamidi A et al (2022) Environmental impacts due to the use of sunscreen products: a mini-review. Ecotoxicology 31:1331–1345. https://doi.org/10.1007/s10646-022-02592-w
121. Vuckovic D, Tinoco A, Ling L et al (2022) Conversion of oxybenzone sunscreen to phototoxic glucoside conjugates by sea anemones and corals. Science 376:644–648. https://doi.org/10.1126/science.abn2600
122. Ngwira KJ, Kühlborn J, Mgani QA et al (2019) Valorisation of cashew nut shell liquid phenolics in the synthesis of UV absorbers. European J Org Chem 2019:4778–4790. https://doi.org/10.1002/ejoc.201900743
123. Izak J, Rahman A, Ntsimango S, Ngwira KJ, de Koning CBFM (2022) Hydrogen-bonded xanthones as potential uv absorbers: the synthesis of xanthones from bio-renewable cardanol utilizing a ceric ammonium sulfate (CAS)-mediated oxidation reaction. SynOpen 06:58–66. https://doi.org/10.1055/s-0040-1719903

Part V
Oil Yielding Crops

Chapter 13
Coconut Husk and Shell Waste: A Bioresource for the Production of Materials and Bioenergy

Ibrahim Ahmed Abass, Edgar C. Mapunda, Elisante E. Mshiu, and Sixberth Mlowe

Abbreviations

MPa	Mega Pascal
TPCS	Thermoplastic cassava starch
UV	Ultraviolet
W/m·K	Watts per meter per Kelvin

1 Introduction

The coconut tree (*Cocos nucifera* L.) holds significant agricultural importance and is highly cultivated in tropical and subtropical regions, with its cultivation spanning more than 12 million hectares worldwide [1, 2]. Asia-Pacific countries including Indonesia, the Philippines, India, and Sri Lanka are the main producers of coconuts, accounting for about 80% of production. Challenges in coconut production are based on climate changes, which include extreme weather events like droughts,

I. A. Abass (✉) · E. C. Mapunda
Chemistry Department, College of Natural and Applied Sciences, University of Dar es Salaam, Dar es Salaam, Tanzania
e-mail: ibrahim.abass@suza.ac.tz

E. E. Mshiu
Department of Geosciences, School of Mines and Geosciences, University of Dar es Salaam, Dar es Salaam, Tanzania

S. Mlowe
Chemistry Department, Dar es Salaam University College of Education, Dar es Salaam, Tanzania

C. Trois, R. C. Ray (eds.), *Plantation Crop Wastes: Valorization for Economic Sustainability*, Waste as a Resource,
https://doi.org/10.1007/978-3-031-96805-1_13

cyclones, and diseases such as lethal yellowing and pests [3]. However, research efforts are ongoing to develop high-yielding and disease-resistant coconut varieties, improve agricultural practices, and develop sustainable pest management strategies.

The global trend in the generation of coconut husk and shell waste has significantly increased due to the high volume of coconut production, with Asia, Latin America, and Africa as the leading producers [4]. This increase in production generates a large amount of waste, including husks and shells, which can make up to 65% by weight, through which coconut processing industries struggle to manage these wastes [3, 5]. Based on the waste management schemes, husk waste produces coir fiber and pith, while shell waste is utilized in bio-energy production or transformed into biochar [3]. The effort of transforming these wastes into useful products is limited, resulting in a large portion of the waste ending up in landfills, leading to environmental hazards such as groundwater contamination, methane emissions, and mosquito breeding [6]. The better utilization of these wastes aligns with sustainable development goals by fostering a circular economy, lowering emissions, and minimizing the environmental impact of agricultural waste.

2 Coconut Husk and Shell Composition

2.1 Chemical Composition

Coconut husk and shell wastes (Fig. 13.1) are chemically composed of cellulose, hemicellulose, and lignin. On a dry weight, coconut husk contains almost 37.55% cellulose, 31.18% hemicellulose, and 34.15% lignin [7, 8]. Cellulose provides high crystallinity and structural rigidity, making it a potential for paper production, textiles, and nanocellulose-based advanced composites [9]. Hemicellulose contributes to the hydrophilic nature of the husk, enhancing water absorption for agricultural mulching and applied in oil spill mitigation strategies [10]. Lignin helps in

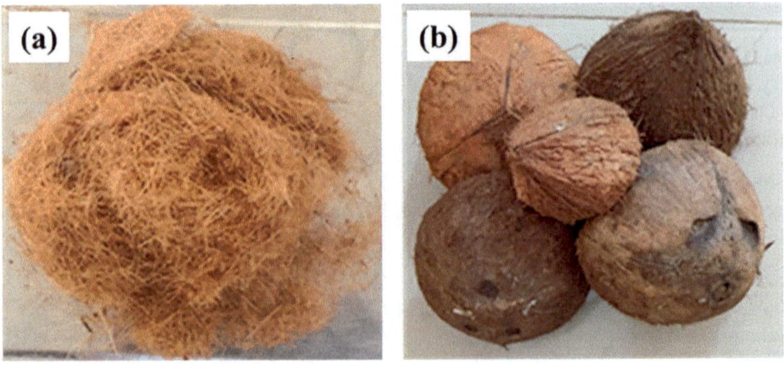

Fig. 13.1 Coconut waste materials used for the production of different materials: (**a**) coconut husk and (**b**) coconut shells

mechanical strength and microbial resistance, acting as a binder between cellulose and hemicellulose. The high content of lignin enhances applications in biochar production, energy generation, and bioplastics [9]. On the other hand, the minimal ash content enriched with minerals like calcium and silica improves the husk's chemical stability and adsorption properties, for a low moisture content (~2.7%), low density, and high porosity [11].

2.2 Physical Properties and Structural Characteristics

Coconut husks and shells exhibit unique physical properties and structural characteristics, making them versatile for several applications. Their high specific gravity (~1.15 g/cm^3) and low bulk density (0.1–0.3 g/cm^3) make them lightweight yet durable, ideal for packaging, insulation, and lightweight composites. The fibrous structure leads to high porosity, allowing significant air and water retention, which benefits water management soil improvement, and water in agriculture [12]. Moisture content is from 8% to 20%, with dried husks having lower levels to prevent decay in industrial applications. Coconut fibers with remarkable water absorption retain 130–180% of their dry weight, making them suitable for oil spill remediation [8]. Mechanically, the fibers reveal tensile strength of up to 200 MPa and elasticity, stretching to 30% before breaking, which enhances durable applications like geotextiles, ropes, and composites [13]. Coconut fibers with low thermal conductivity (0.04–0.06 W/m·K) make them good insulators, while stability up to 200 °C increases their suitability for heat-exposed environments [14]. Structurally, the coconut waste fibers have thick cell walls of lignin, cellulose, and hemicellulose with diameters up to 200 μm, and varying lumen sizes that enhance buoyancy and absorption. The husk's layered composition, with an outer shell, fibrous mesocarp, and pith-rich inner layer, helps in mechanical strength and supports various functional applications [9].

2.3 Comparison with Other Agricultural Wastes

Coconut husk stands out among agricultural wastes due to its unique chemical composition, physical properties, and diverse potential applications. It contains 37.55% cellulose, 31.18% hemicellulose, and 34.15% lignin, making it durable but resistant to degradation [7, 8, 15]. Compared to rice husk, which has 30% cellulose, 20% hemicellulose, and 20% lignin with high silica content (up to 20%), coconut husk offers better tensile strength and water absorption. Wheat straw, with similar cellulose content (~35%), has lower lignin (~10–15%) and is more biodegradable, making it suitable for composting and animal bedding [16]. Sugarcane bagasse, with 40% cellulose, 28% hemicellulose, and 25% lignin, is ideal for bioethanol production and composite reinforcement [15, 16]. Physically, coconut husk's low density

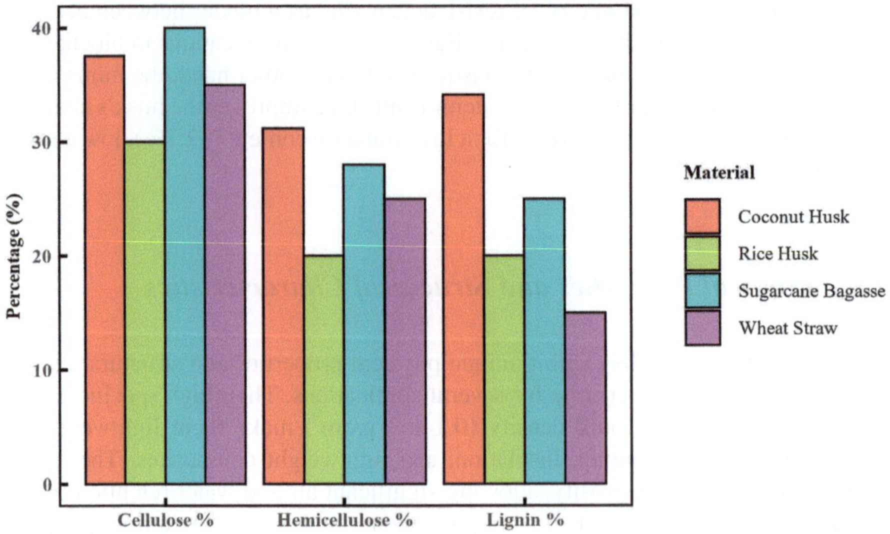

Fig. 13.2 Biomass compositions of coconut husk, rice husk, wheat straw, and sugarcane bagasse

(0.1–0.3 g/cm^3), high porosity, and tensile strength (100–200 MPa) make it effective for water retention, oil absorption, and sorbent applications [15]. Other agricultural wastes like rice husk and sugarcane bagasse are lighter and versatile for bio-composites and bioenergy. Coconut husks are applied in coir production and oil spill cleanup [17], biochar from coconut wastes is used for soil structure enhancement [18], rice husk used for bioenergy and silica extraction [19], wheat straw used for bioplastics and soil amendments [20], and sugarcane bagasse used for bioethanol production [21] and paper production [22]. Figure 13.2 shows a comparison of biomass compositions of coconut husk, rice husk, wheat straw, and sugarcane bagasse [7, 8, 15, 16].

3 Applications of Coconut Husk and Shell

Coconut husk and shell by-products are used in a wide range of applications due to their properties, abundance, and renewability. The fibrous husks are extensively utilized in the production of eco-friendly coir fibers and biodegradable products [16]. The hard and dense shell is used for producing biochar, activated carbon, and construction materials [15]. These by-products support sustainable industrial practices and offer eco-friendly solutions for waste management and resource conservation. These applications underscore the importance of coconut by-products in enhancing environmental sustainability.

3.1 Coconut Husk and Shell in Material Production

Coconut husk and shell are composed of high amounts of lignin, cellulose, and hemicellulose [7]. Due to their easy availability and durability, fibrous coconut husks are primarily processed into coir fibers, which are used in products such as brushes, mats, ropes, and textiles [23]. In addition, coir serves as a raw material for biodegradable goods like eco-friendly packaging and geotextiles, as well as a reinforcement material in composite assemblies. Figure 13.3 shows the utilization of coconut husk and shell wastes for the manufacturing of different products. The high absorbency of coconut shells coir qualifies its use in agriculture for soil conditioning and oil spill cleanup [24]. Given the nature of the coconut shell, it is a vital source of composite materials, biochar, and activated carbon because of the presence of the lignin structure [18]. For example, activated carbon from coconut shells is well known for its use in air and water purification due to its exceptional adsorption capabilities. Coconut shells are also known for their use as a renewable fuel, production of eco-friendly construction materials, and ornamental goods [1].

The husk and shell of coconuts both play a significant part in the production of sustainable materials, lowering waste and promoting environmentally friendly innovation. However, their use comes with several difficulties that may limit their efficacy and scalability. The unpredictability of coconut husk and shell quality and availability is one of the main problems. Because the source material varies naturally, these materials may differ in quality and properties. The consistency of the fibers and shells is affected by several variables, including harvesting techniques, climate, and coconut age. The resulting materials' qualities are difficult to standardize and manage due to this diversity, which also makes using them on a broad scale more difficult [3]. Processing coconut shells and husks presents another barrier because they are both hard materials that need complex processing methods to be broken down into useful forms. It is difficult to remove and modify their lignin, cellulose, and hemicellulose content for use in composite materials or other applications [25]. The coconut shell's rough and brittle texture further complicates its use

Fig. 13.3 Illustration showing applications of coconut wastes

by making it difficult to prepare without specialist equipment. Materials derived from coconut husk and shells are biodegradable; however, in some applications, their longevity may be limited. For instance, a study by Hasan and co-workers (2021) showed that coconut-based composites degrade over time when exposed to moisture or UV light, decreasing their lifespan in outdoor uses like packaging and construction. In the study, the formation of biocomposite panels with improved thermo-mechanical performances, which could be feasible for industrial applications, was successful.

Limited durability can restrict their utilization in areas requiring long-term performance. Processing coconut shells and husks may potentially have negative effects on the environment and human health. When dust and particles are produced by mechanical and chemical processes, workers may be at risk for health problems if appropriate safety precautions are not taken [26]. Furthermore, if not handled appropriately, some chemical treatments intended to enhance material qualities have negative environmental implications, therefore making it difficult to employ these materials in sustainable production [27]. Another barrier to the wide use of coconut husk and shell in the manufacturing of materials is economic viability. To properly break down and handle these materials, special equipment may be required, which can result in a significant initial investment in the infrastructure [27]. This can make coconut husk and shell-based by-products less economically viable compared to other more established raw materials. Finally, narrow knowledge and research in this area pose a significant challenge. While there is growing attention to using coconut husks and shells for material production, the full potential of these materials is not yet fully known. More research is needed to optimize processing techniques, improve material properties, and assess their long-term sustainability and performance in various applications.

3.1.1 Bio-Based Composites

Bio-based composites are environmentally friendly materials created by mixing natural fibers with a binder or matrix that is typically obtained from conventional polymers and renewable resources. For a variety of purposes, these composites are intended to provide environmentally benign substitutes for synthetic, non-biodegradable materials. Bio-based composites resulting from coconut husks and shells are categorized into four types based on composition and application. First, coir fiber-reinforced composites composed of coconut husks are known to be robust, lightweight, and suitable for furniture, automotive parts, and packaging. Alkali treatment often enhances their bonding and performance properties [28]. Second, hybrid fiber composites, combining coir with other natural fibers or fillers like calcium carbonate, exhibit enhanced strength, and reduced water absorption, making them suitable for automotive and construction uses [29]. Third, cement-based composites normally incorporate husk fibers into cement matrices, resulting in thermally insulated, robust materials for construction applications. Finally, coconut shell composites, enriched with carbon, especially are used in activated carbon composites for environmental remediation through water and air filtration systems.

Coconut shells and husks are especially promising because of their remarkable mechanical qualities, availability, and renewability [30]. The potential of coconut husk and shell in bio-based composites has been extensively studied, with an emphasis on its remarkable qualities and sustainability. Coconut husk is suitable for polymer reinforcement because it is high in lignin and cellulose, which give it stiffness, endurance, and resistance to microbial destruction [31]. Similarly, coconut shell powder enhances mechanical strength, toughness, water resistance, and thermal stability, especially in the automotive and construction industries, because of its high density and hydrophobic qualities. These materials, which are abundant as agricultural waste and biodegradable when combined with bio-resins, provide superior acoustic and insulating qualities at a lower cost. Fiber-matrix adhesion is further improved by surface treatment of coir fibers, improving composite performance.

Research conducted by Srivastava [32] on reinforced biochar composites comprising 80% epoxy, 10% biochar, and 5% coconut coir revealed higher flexural modulus, suggesting enhanced resistance to bending forces. Coconut husks are reported to be used in reinforcing concrete materials by increasing tensile strength while reducing cracking [30]. Chemical modification on the surface of coconut waste particles has also been reported to produce suitable bio-fillers in the preparation of epoxy composites featuring good thermal, mechanical, wettability, and flammability properties [33]. To obtain high-quality composites from coconut waste, the process typically begins with soaking the material in water for several days to promote anaerobic fermentation and soften the fibers for easier separation. This is followed by the crushing of coconut shells, which are then ground into fine particles and sieved for use. After separation, the coir fibers are cleaned and dried. Treating the coconut waste with sodium hydroxide (NaOH) improves the surface characteristics of the fibers. Subsequently, bio-epoxy resins are mixed with coir fibers and shell fillers using fabrication techniques to produce the final composite material [34].

Coir fiber composites fabricated with thermoplastic and thermosetting polymers exhibit superior mechanical and thermal properties, making them suitable for use in automotive parts, construction panels, and packaging materials [28, 35]. Several studies have documented coconut husk composites possessing lightweight and thermal insulation properties, making them ideal for partitions and ceiling tiles [36]. Another study conducted by Ichim and co-workers prepared biocomposites by thermoforming using coir fibers as reinforcing material and polypropylene as the matrix [37]. The biocomposites produced showed that tensile and bending strength decreased with increasing coir fiber content. Coconut-based composites' revolutionary potential for green technology and sustainable industries will determine their destiny. Advanced functional composites are being developed for specific applications such as thermal insulation including hybrid and nano-composites. Their affiliation with the circular economy promotes sustainable resource use and waste reduction by turning agricultural waste into valuable products. Nanotechnology and chemical treatment innovations keep making them more competitive with synthetic materials, while environmentally friendly developments like entirely biodegradable composites reduce their negative effects on the environment.

3.1.2 Activated Carbons

A highly porous form of carbon, activated carbon, is produced with a vast surface area, which makes it an exceptional adsorbent for eliminating pollutants, impurities, and poisons from gases, liquids, and solids. The utilization of coconut husks and shells to synthesize activated carbon has recently emanated due to their high carbon content. The porous nature, high surface area, and adsorption capabilities of the activated carbon made from coconut waste make it suitable for a range of uses [38]. Activated carbon has a widespread array of carbonized materials with a high degree of surface area and porosity. Given the excellent performance properties of activated carbon made from agro-wastes, coconut shell activated carbon is applied in domestic and industrial wastewater treatment and purification, desalination, separating gases and refining, odor and pollutant removal, and medical applications. Recently, several industrial by-products derived from agro-waste have been converted into activated carbon materials [34]. Studies have revealed efficacy in numerous contexts; for example, coconut shell-based activated carbon adsorbent can adsorb up to 90% of heavy metals, organic pollutants, and oil spills (up to 5 g/g) from water, as well as pharmaceutical drugs like paracetamol from the gastrointestinal tract [39–41]. Granular activated carbon is used to purify air by absorbing volatile organic compounds and capturing up to 1.5 mmol/g of gases such as CO_2 from industrial flue gases [42, 43]. Activated carbon is used in medical applications for hemoperfusion devices to remove uremic toxins from the blood [44], in agriculture used to decrease pesticide residues such as atrazine and glyphosate from runoff water [45]. Granular activated carbon removes color and impurities from sugarcane syrups without altering their flavor [46], during gold extraction, granular activated carbon is used to recover gold up to 95% of gold cyanide complexes with recovery rates of more than 95% [47], and activated carbon also acts as a catalyst material in hydrogenation and oxidation [48].

Producing activated carbon from coconut husks and shells involves carbonization and activation via physical, chemical, or combined methods [34]. Physical activation involves carbonizing coconut shells at 400–500 °C in an inert atmosphere, followed by steam activation at 800–900 °C, resulting in microporous carbon with a high surface area of 950 m^2/g [49]. Alternatively, activation with CO_2 at 850 °C for 2–3 hr. improves pore structure [50]. Chemical activation using agents such as potassium hydroxide (KOH) at 700 °C or H_3PO_4 at 450 °C results in high surface areas up to 1200 m^2/g and mesoporous structures. Combined methods, such as $ZnCl_2$ pretreatment followed by steam activation or KOH impregnation with subsequent heating, increase pore volume and adsorption efficiency for specific applications like pesticide elimination [34]. Industrial-scale production regularly employs rotary kilns, ensuring consistent quality for water and air treatment applications [51].

Nevertheless, challenges remain in the production process. Variability in feedstock, influenced by factors such as moisture content and maturity, affects the consistency of the final product. Additionally, insufficient drying and cleaning can introduce impurities, thereby reducing the adsorption capacity of the activated carbon [52]. Energy requirements for carbonization and activation, often beyond 800 °C, elevate costs, though alternatives such as solar-assisted pyrolysis are being explored [53]. Chemical activation poses environmental concerns due to water pollution and hazardous waste from washing processes [34]. The energy-intensive nature of production contributes to significant greenhouse gas releases [54]. Moreover, balancing pore development, surface area, and production cost are challenges, with advanced optimization techniques such as response surface methodology being resource-intensive [52]. However, there is a need for creative, effective, and environmentally responsible processes for the production of activated carbon from coconut by-products. Researchers have extensively worked on producing coconut shell powder charcoal for use in agricultural and environmental applications such as fertilizers. For example, the increment of coconut shell charcoal improved the soil characteristics (ex-limestone mining media) in addition to supporting the growth of *Acacia mangium* seedlings [55]. Marlinae and co-workers demonstrated that coconut shells, husks, and water hyacinths improved the quality of dug well water in flood areas [56]. The study showed that dissolved oxygen (DO) levels increased in multiple villages, whereas turbidity and total suspended solids (TSS) decreased.

3.1.3 Construction Materials

Coconut husk-derived materials have commonly been reported to be used in the construction industry due to their thermal qualities, sustainability, durability, and lightweight properties [6]. The use of fibers and powder from coconut wastes in a variety of composites and construction materials has been shown to improve performance and promote environmentally friendly methods. Their use as reinforcements in construction materials such as cement, concrete, and polymer composites has been reported to significantly improve mechanical qualities [17]. For instance, a study on the thermal and mechanical properties of biodegradable thermoplastic cassava starch (TPCS) blended with coconut husk fiber particles showed increased intermolecular interactions and thermal stability of the biocomposites [57]. The produced biodegradable TPCS reinforced with coconut husk fiber composite had increased tensile strength of 6.5 MPa and flexural strength of 13.8 MPa as well as modulus. Udaya Kumar and co-workers investigated the mechanical properties of reinforced vinyl ester when modified with coconut waste products which showed the composites to exhibit better hardness and tensile properties [26]. Short fibers and powdered coconut husk and shell wastes are known to be used to fabricate composite boards, particleboards, and insulating panels. Particleboards prepared from blending coconut fiber with natural glue or urea-formaldehyde have been reported for flooring structures, wall partitioning, temporary shelters, and lightweight doors

[58]. In addition, boards made using coconut by-products have a naturally porous structure useful for soundproof walls and thermal and acoustic insulation [59]. Several experimental data show that coconut husk and shell waste produce lightweight bricks and other construction materials by incorporating their powder and fibers into clay and cement mixtures [60–62]. In other studies, coconut husk and shell waste ash acted as a partial alternative for cement in concrete and mortar blends, providing a sustainable solution in plastering, masonry, and non-load-bearing concrete constructions [63]. On the other hand, coconut by-products usages in construction materials have drawbacks such as low mechanical strength, vulnerability to degradation in moist environments, and high water absorption [13, 28, 29]. Furthermore, the nature of organic contents in coconut wastes makes them prone to burning, constraining their use in fire-resistant applications [9, 64]. Further studies are essential to address the limitations of coconut waste materials, including the development of fire-retardant coatings and construction materials. This involves redesigning chemical treatments to enhance water resistance and exploring the potential of blending coconut waste with other agro-wastes to improve performance.

3.2 Production of Bioenergy from Coconut Shell and Husk

Coconut husks and shells are further known to deliver significant bio-waste resources used for energy production through direct combustion and pelletization [65]. Direct combustion is used in industrial heating, electricity generation, and traditional cooking methods including burning dried husks and shells in oxygen to yield heat energy [66]. Moisture content and combustion technology is used to determine the efficiency of bioenergy from agro-wastes [65]. In the pelletization method, coconut shell and husk wastes are compacted to make dense pellets of higher energy density [67]. By doing so increases handling and combustion efficiency, and pellets of higher calorific values and lower moisture content are produced [10]. Additionally, coconut waste-based briquettes are gaining popularity as an alternative fuel source due to their sustainability, affordability, and reduced environmental impact. Due to its availability and high energy content, coconut waste is also used in the production of electricity by gasification, anaerobic digestion, and combustion, providing sustainable energy options [65]. A study by Yirijor and Bere demonstrated the potential of coconut shell bio-briquettes as a viable and sustainable energy source [68]. Their findings showed that bio-briquettes produced using various ratios of corn and cassava as binders had mechanical and energy performance properties. In another study by Hadi Santoso, the fabrication of briquettes using coconut shell and tapioca flour in a ratio of 5:1 was successfully carried out with a fuel efficiency of 37.14% [69].

3.2.1 Biogas Production

Biogas production from agro-waste biomass such as coconut wastes has been reported in use [70]. Biogas is primarily composed of carbon dioxide (CO_2) and methane (CH_4), as well as a digestate that is rich in nutrients. Biogas obtained from agro-wastes like coconut waste usually consists of energetically usable methane and carbon dioxide [71]. The process of obtaining biogas follows four stages, namely hydrolysis, acidogenesis, acetogenesis, and methanogenesis; in each stage, particular microorganisms play a role in biogas production [72].

In hydrolysis, hydrolytic bacteria break down large organic molecules like carbohydrates, proteins, and fats into monomers like sugars, amino acids, and fatty acids, a critical step since polymers such as cellulose in coconut husks are large for direct microbial digestion [70]. Acidogenesis follows, where acidogenic bacteria transform these hydrolyzed products into volatile fatty acids, alcohols, hydrogen, and CO_2, causing a drop in pH that must be achieved to sustain microbial activity [72]. During acetogenesis, acetogenic bacteria convert intermediates from acidogenesis into acetic acid, hydrogen, and CO_2, whereas hydrogen buildup is balanced by hydrogenotrophic methanogens [71].

The high lignin content of coconut husks and shells makes anaerobic digestion difficult, but by adjusting key parameters like pH, temperature, and retention time, pretreatment and co-digestion with nitrogen-rich materials can increase efficiency. Anaerobic digestion is a sustainable way to manage waste and produce renewable energy, turning coconut waste into useful biogas and organic fertilizer biogas by producing methane through processes involving hydrogen and CO_2 (hydrogenotrophic methanogenesis) and acetic acid (acetoclastic methanogenesis), respectively [73].

Several variables that affect microbial activity during anaerobic digestion enhance the production of biogas from coconut husk and shell by-products. The study by Bichang et al., [74] showed that husks with a favorable composition still require pre-chemical modification to improve biodegradability, while the chemical composition and high lignin content of coconut shells (~45%) create a barrier around cellulose and hemicellulose, making them resistant to digestion and lowering biogas output. Low nitrogen concentrations in coconut wastes cause an imbalance in the carbon-to-nitrogen ratio of about 50:1, which can inhibit microbial activity, except when nitrogen-rich materials are added [75]. Another important factor is particle size and surface area; smaller particles increase the surface area available to microorganisms, accelerating hydrolysis and boosting gas generation [76]. Thermophilic settings (50–60 °C) speed up the process at the expense of increased energy input, while mesophilic conditions (30–40 °C) provide stable methane release. To obtain the best gas yield, the retention period, which is typically 20–40 days for coconut trash, must take into consideration its complicated structure [77]. Maintaining the pH within 6.5–7.5 is important, as acid accumulation during acidogenesis can obstruct methanogenic activity [78]. Proper mixing and agitation provide equal biomass distribution while preventing the formation of scum and silt, but excessive mixing may disrupt microbial colonies [79]. The yield of biogas is increased by co-digestion with other biomass, such as manure and food waste, which balance nutrients and promote microbial development. For instance, the carbon-to-nitrogen ratio and microbial efficiency are improved when coconut husks and cow manure are digested together [80].

3.2.2 Bio-Oil and Biochar

Coconut shells and husk wastes are valuable biomass resources with significant importance for environmental management and bioenergy. Coconut waste biomass is transformed into black carbon biochar, which is a highly porous material with anti-decomposition properties and a high degree of aromatization. Gasification, pyrolysis, and self-sustained carbonization are used to produce coconut husk and shell biochar [1]. Pyrolysis and gasification procedures convert coconut wastes into biochar and bio-oil, which each have specific applications in energy, agriculture, and environmental remediation. Anaerobic digestion can produce biogas [75, 81]. Pyrolysis specifically degrades biomass at high temperatures in an aerobic condition to produce bio-oil, biochar, and some gases. Fast cooling of the pyrolysis vapors gives the bio-oil product. Crude bio-oil is dark brown and composed of a very complex mixture of oxygenated hydrocarbons with some amount of water [82]. Gasification converts biomass into a small amount of biochar and synthetic gas [1]. Coconut husks and shells' lignocellulosic content can be effectively broken down into bio-oil, biochar, and syngas through pyrolysis, a thermal breakdown process carried out in the absence of oxygen [82]. A self-sustained pyrolysis process involves carbonization, where part of the biomass is combusted to generate the heat required for the transformation [1]. The three steps of the process are drying, devolatilization, and char generation. Several methods, including slow, fast, microwave-assisted, intermediate, and catalytic pyrolysis, can be used to optimize the process. Every technique has its benefits and drawbacks, accommodating various applications and output levels [82]. Fast pyrolysis concentrates on releasing bio-oil for chemical feedstocks and renewable fuel, but slow pyrolysis optimizes the production of biochar, making it valuable for soil amendments and carbon sequestration. By balancing the production of biochar and bio-oil, intermediate pyrolysis ensures adaptability for a variety of applications. Rapid thermal breakdown with great energy efficiency is provided by microwave-assisted pyrolysis, while the addition of catalysts in catalytic pyrolysis enhances the quality of bio-oil [83, 84]. The dark brown, viscous liquid known as bio-oil is made from the leftovers of coconut husks and shells and is abundant in organic substances like acids, phenols, and ketones. It acts as a chemical feedstock, soil amendment, water treatment agent, and renewable fuel. It is also essential to the manufacture of biofuel in biorefineries. In contrast, biochar is a porous, carbon-rich solid that has a high capacity for adsorption and a high reactivity of its functional groups [84]. The global coconut industry uses coconut husk and shell biochar as potential assets for their industrial development and environmental remediation [1]. Biochar is used for water purification, a soil structure improvement to enhance fertility and water retention; it is also burned to produce energy [38, 85]. Biochar is also used as a low-cost cement substitute, in biodiesel production, capacitive deionization, and in the treatment of gas emissions and heavy metal contaminants [1]. Bio-oil can be used to substitute fuel oil or diesel in static applications including furnaces, boilers, turbines, and engines for electricity generation [82]. These developments reveal the utilization of coconut by-products in sustainable energy generation, environmental protection, and agricultural efficiency for a circular economy (Table 13.1).

Table 13.1 Various applications of coconut husk and shell products

Application	Product	Description	Reference
Textiles	Coir mats and ropes	Utilized in household dếcor and geotextiles	Ajien et al. [1]
Electronics	Carbon electrodes	Applied in supercapacitor and energy storage equipment	Koli et al. [52]
Bio-composites	Coconut fiber-reinforced composites	Useful for automotive, construction, and packaging	Srivastava [32]
Agriculture	Biochar and mulch	Improve soil fertility and moisture content It is a good fabric to control soil erosion	Werther et al. [65], Das Himadri [28]
Adsorption	Activated carbon	Applied in water purification and air filtration	Zhang et al. [38]
Food and beverages	Coconut shell power	Utilized as filler in food products and as an ingredient in traditional medicine	Ajien et al. [1]
Oil spill cleanup	Sorbent material	Used for marine oil spill recovery	Yamamoto et al. [44]
Energy production	Biomass fuel and briquettes	Applied as a renewable energy source for cooking and industrial applications	Menon et al. [66]
Building materials	Rice husk fiber and coir	Used for the production of panels, bricks, window frames, and decking Used to make sacking, twine, doormats, and items such as bags and ropes	Mohammed et al. [64], Das Himadri [28]

4 Technical Limitations in Processing and Conversion Technologies

The processing and conversion of coconut wastes are limited by several technical challenges that may limit their application on a large scale. Variability in the composition of coconut by-products, such as husks, shells, and coir, makes it more difficult to standardize processing conditions and compromises the uniformity of the final product [1]. Large energy inputs are needed for processes like pyrolysis and gasification, which limits their viability in places with little access to reasonably priced energy sources [67]. Processing time and expenses are increased by extensive pre-treatment procedures that include drying and chemical modifications [86]. Because of the slowed microbial decomposition caused by the high lignin level, biological methods such as composting are less efficient [13]. Additionally, to avoid secondary environmental problems, the thermochemical process by-products such as char and tar must be properly managed [13]. Many available technologies are still at the pilot stage, and the high infrastructure costs pose significant barriers to automation and large-scale implementation [67]. Widespread adoption is further hampered by limited access to modern technologies, particularly in rural areas [1]. Also, hazardous emissions from thermochemical processes could cause problems for the environment and regulatory issues [86]. Finally, the economic attractiveness of these technologies is diminished by high initial investment costs and uncertain market prospects for valorized products. Innovative research is needed to address these issues in order to improve sustainability, lower costs, and optimize processing.

5 Conclusion and Future Perspectives

Traditionally regarded as agro-waste, coconut husks and shells are becoming useful resources for solving environmental issues in the context of the circular economy. The increased need for sustainable solutions, renewable energy projects, and effective waste management techniques are all in line with their use. The use of biomass energy is anticipated to increase with the support of renewable energy policies and hybrid systems that mix solar and wind energy with coconut biomass. Technological developments like gasification and biorefinery methods will also improve the process of turning coconut trash into energy and valuable products like activated carbon, composites, and bioplastics. Community-based waste management programs combined with circular economy ideas offer a way to minimize environmental impact while optimizing resource efficiency. Coconut husks and shells are very promising for cleaning up marine oil spills due to their high lignocellulosic content and porous structure. This makes them the perfect environmentally friendly oil sorbents. To increase their mechanical strength, reusability, and capacity to absorb oil, future research is expected to improve their hydrophobic and oleophilic qualities through chemical and physical alterations, such as silane coatings. These developments will establish materials derived from coconuts as competitive options for mitigating oil spills. Creating biodegradable composites by mixing polymers and coconut byproducts is another exciting approach. These composites lessen long-term environmental effects by decreasing contamination from oil spills and spontaneously breaking down after usage. To enhance oil quality, hybrid sorbents that combine elements from coconuts with other agricultural wastes including kapok fiber and sugarcane bagasse are being investigated. Moreover, biochar made from coconut shells, because of its high porosity and adsorption capability, works well at cleaning up oil spills in aquatic environments. In a similar vein, coconut shell-activated carbon functions as a high-performance sorbent with uses in both industrial and marine environments. The potential of these waste materials for environmental remediation can be expanded beyond oil spill control by engineering them for multipurpose uses, like the removal of contaminants like dyes and heavy metals.

The future of materials derived from coconut waste depends on innovative research and sustainable development. Improving the design and efficiency of these materials, as well as expanding their applications, could make them essential tools for reducing marine oil spill pollution. Collaborative efforts to enhance performance and promote circular economy initiatives involving coconut by-products would ensure significant contributions to both environmental restoration and economic prosperity.

References

1. Ajien A, Idris J, Sofwan N, Husen R, Seli H (2023) Coconut shell and husk. Waste Manag Res 41(1):37–51. https://doi.org/10.1177/0734242X221127167
2. Kumar BM, Kunhamu TK (2022) Nature-based solutions nature-based solutions in agriculture: A review of the coconut (*Cocos nucifera* L) -based farming systems in Kerala, the Land of Coconut Trees. Nature-Based Solutions 2(August 2021):100012. https://doi.org/10.1016/j.nbsj.2022.100012
3. Lanka S (2016) Coconut. Academic Press, pp 201–216. https://doi.org/10.1016/B978-0-12-801309-0/00009-4
4. Van Eck J, Puchta H (2019) Food and agriculture. In: Chemistry and industry (London), vol 83. Wiley. https://doi.org/10.1002/cind.832_8.x
5. Go AW, Conag AT, Igdon RMB, Toledo AS, Malila JS (2019) Potentials of agricultural and agro-industrial crop residues for the displacement of fossil fuels: a Philippine context. Energ Strat Rev 23(December 2018):100–113. https://doi.org/10.1016/j.esr.2018.12.010
6. Khalife E, Sabouri M, Kaveh M, Szymanek M (2024) Recent advances in the application of agricultural waste in construction. Appl Sci (Switzerland) 14(6):2355. https://doi.org/10.3390/app14062355
7. Parichanon P, Matan N, Limbo S, D'Incecco P, Matana N (2021) Natural cellulosic material characteristics: a possibility to develop antimicrobial active fiber-based packaging. Bioresources 16(3):5450–5466. https://doi.org/10.15376/biores.16.3.5450-5466
8. Pode R (2016) Potential applications of rice husk ash waste from rice husk biomass power plant. Renew Sust Energ Rev 53:1468–1485. https://doi.org/10.1016/j.rser.2015.09.051

9. John MJ, Thomas S (2008) Biofibres and biocomposites. Carbohydr Polym 71(3):343–364. https://doi.org/10.1016/j.carbpol.2007.05.040
10. Chaowana P, Hnoocham W, Chaiprapat S, Yimlamai P, Chitbanyong K, Wanitpinyo K, Chaisan T, Paopun Y, Pisutpiched S, Khantayanuwong S, Puangsin B (2024) Utilization of hemp stalk as a potential resource for bioenergy. Mater Sci Energy Technol 7:19–28. https://doi.org/10.1016/j.mset.2023.07.001
11. Mohd Nurazzi N, Khalina A, Sapuan SM, Dayang Laila AHAM, Rahmah M, Hanafee Z (2017) A review: Fibres, polymer matrices and composites. Pertanika J Sci Technol 25(4):1085–1102
12. Ramamoorthy SK, Skrifvars M, Persson A (2015) A review of natural fibers used in biocomposites: plant, animal and regenerated cellulose fibers. Polym Rev 55(1):107–162. https://doi.org/10.1080/15583724.2014.971124
13. Xu C, Nasrollahzadeh M, Selva M, Issaabadi Z, Luque R (2019) Waste-to-wealth: biowaste valorization into valuable bio(nano)materials. Chem Soc Rev 48(18):4791–4822. https://doi.org/10.1039/c8cs00543e
14. Aguilar-Rosero J, Urbina-López ME, Rodríguez-González BE, León-Villegas SX, Luna-Cruz IE, Cárdenas-Chávez DL (2022) Development and characterization of bioadsorbents derived from different agricultural wastes for water reclamation: a review. Appl Sci (Switzerland) 12(5):2740. https://doi.org/10.3390/app12052740
15. Mwaikambo LY, Ansell MP (2002) Chemical modification of hemp, sisal, jute, and kapok fibers by alkalization. J Appl Polym Sci 84(12):2222–2234. https://doi.org/10.1002/app.10460
16. Sain M, Panthapulakkal S (2006) Bioprocess preparation of wheat straw fibers and their characterization. Ind Crop Prod 23(1):1–8. https://doi.org/10.1016/j.indcrop.2005.01.006
17. Adeniyi AG, Onifade DV, Ighalo JO, Adeoye AS (2019) A review of coir fiber reinforced polymer composites. Compos Eng 176(July):107305. https://doi.org/10.1016/j.compositesb.2019.107305
18. Kamali M, Sweygers N, Al-Salem S, Appels L, Aminabhavi TM, Dewil R (2022) Biochar for soil applications-sustainability aspects, challenges and future prospects. Chem Eng J 428(March 2021):131189. https://doi.org/10.1016/j.cej.2021.131189
19. Shen Y (2017) Rice husk silica derived nanomaterials for sustainable applications. Renew Sust Energ Rev 80(November 2016):453–466. https://doi.org/10.1016/j.rser.2017.05.115
20. Visco A, Scolaro C, Facchin M, Brahimi S, Belhamdi H, Gatto V, Beghetto V (2022) Agri-food wastes for bioplastics: European prospective on possible applications in their second life for a circular economy. Polymers 14(13):2752. https://doi.org/10.3390/polym14132752
21. Huang J, Khan MT, Perecin D, Coelho ST, Zhang M (2020) Sugarcane for bioethanol production: potential of bagasse in Chinese perspective. Renew Sust Energ Rev 133(March):110296. https://doi.org/10.1016/j.rser.2020.110296
22. Rainey TJ, Covey G (2016) Pulp and paper production from sugarcane bagasse. In: Sugarcane-based biofuels and bioproducts. Wiley, New York, pp 259–280
23. Abdullah NA, Sainorudin MH, Rani MSA, Mohammad M, Kadir NHA, Asim N (2021) Structure and thermal properties of microcrystalline cellulose extracted from coconut husk fiber. Polimery/Polymers 66(3):187–192. https://doi.org/10.14314/POLIMERY.2021.3.4
24. Hoang PH, Dat HT, Cuong TD, Dien LQ (2022) Pretreatment of coir lignocellulose for preparation of a porous coir-polyurethane composite with high oil adsorption capacity. RSC Adv 12(24):14976–14985. https://doi.org/10.1039/d2ra01349e
25. Mahmud MA, Anannya FR (2021) Sugarcane bagasse—a source of cellulosic fiber for diverse applications. Heliyon 7(8):e07771. https://doi.org/10.1016/j.heliyon.2021.e07771
26. Udaya Kumar PA, Ramalingaiah Suresha B, Rajini N, Satyanarayana KG (2018) Effect of treated coir fiber/coconut shell powder and aramid fiber on mechanical properties of vinyl ester. Polym Compos 39(12):4542–4550. https://doi.org/10.1002/pc.24561
27. Leng L, Xiong Q, Yang L, Li H, Zhou Y, Zhang W, Jiang S, Li H, Huang H (2021) An overview on engineering the surface area and porosity of biochar. Sci Total Environ 763:144204. https://doi.org/10.1016/j.scitotenv.2020.144204

28. Das Himadri KD (2015) Recent development of fiber reinforced composite materials recent development of fiber reinforced. North East institute of Science and Technology, Assam
29. Ganesan K, Kailasanathan C, Rajini N, Ismail SO, Ayrilmis N, Mohammad F, Al-Lohedan HA, Tawfeek AM, Issa ZA, Aldhayan DM (2021) Assessment on hybrid jute/coir fibers reinforced polyester composite with hybrid fillers under different environmental conditions. Constr Build Mater 301(April):124117. https://doi.org/10.1016/j.conbuildmat.2021.124117
30. Sundarababu J, Anandan SS, Griskevicius P (2020) Evaluation of mechanical properties of biodegradable coconut shell/rice husk powder polymer composites for light weight applications. Mater Today: Proc 39:1241–1247. https://doi.org/10.1016/j.matpr.2020.04.095
31. Hasan KMF, Horváth PG, Bak M, Alpár T (2021a) A state-of-the-art review on coir fiber-reinforced biocomposites. RSC Adv 11(18):10548–10571. https://doi.org/10.1039/d1ra00231g
32. Srivastava S (2024) Mechanical and water absorption characterization of rice husk and coconut coir reinforced biochar composites. Int J Polym Anal Charact 29(6):398–409. https://doi.org/10.1080/1023666X.2024.2375254
33. Kocaman S, Karaman M, Gursoy M, Ahmetli G (2017) Chemical and plasma surface modification of lignocellulose coconut waste for the preparation of advanced biobased composite materials. Carbohydr Polym 159:48–57. https://doi.org/10.1016/j.carbpol.2016.12.016
34. Heidarinejad Z, Dehghani MH, Heidari M, Javedan G, Ali I, Sillanpää M (2020) Methods for preparation and activation of activated carbon: a review. Environ Chem Lett 18(2):393–415. https://doi.org/10.1007/s10311-019-00955-0
35. Nawab MS, Ali T, Qureshi MZ, Zaid O, Ben Kahla N, Sun Y, Anwar N, Ajwad A (2023) A study on improving the performance of cement-based mortar with silica fume, metakaolin, and coconut fibers. Case Stud Constr Mater 19(June):e02480. https://doi.org/10.1016/j.cscm.2023.e02480
36. Guredam LL, Ossia CV (2024) Advancements in using organic waste materials for green composites in ceiling board applications: a mini review. World Sci News 196(July):114–133
37. Ichim M, Stelea L, Filip I, Lisa G, Muresan EI (2022) Thermal and mechanical characterization of coir fibre–reinforced polypropylene biocomposites. Crystals 12(9):1–16. https://doi.org/10.3390/cryst12091249
38. Zhang L, Tu LY, Liang Y, Chen Q, Li ZS, Li CH, Wang ZH, Li W (2018) Coconut-based activated carbon fibers for efficient adsorption of various organic dyes. RSC Adv 8(74):42280–42291. https://doi.org/10.1039/c8ra08990f
39. Amuda OS, Giwa AA, Bello IA (2007) Removal of heavy metal from industrial wastewater using modified activated coconut shell carbon. Biochem Eng J 36(2):174–181. https://doi.org/10.1016/j.bej.2007.02.013
40. Mikhalovsky SV, Nikolaev VG (2006) Activated carbons as medical adsorbents. Interface Sci Technol 7(C):529–561. https://doi.org/10.1016/S1573-4285(06)80020-7
41. Nguyen CH, Tran HN, Fu CC, Lu YT, Juang RS (2020) Roles of adsorption and photocatalysis in removing organic pollutants from water by activated carbon–supported titania composites: kinetic aspects. J Taiwan Inst Chem Eng 109:51–61. https://doi.org/10.1016/j.jtice.2020.02.019
42. Arena N, Lee J, & Clift R (2016) Life Cycle Assessment of activated carbon production from coconut shells. Journal of Cleaner Production, 125, 68–77. https://doi.org/10.1016/j.jclepro.2016.03.073
43. Gallego E, Roca FJ, Perales JF, & Guardino X (2013) Experimental evaluation of VOC removal efficiency of a coconut shell activated carbon filter for indoor air quality enhancement. Building and Environment, 67, 14–25. https://doi.org/10.1016/j.buildenv.2013.05.003
44. Yamamoto S, Ito T, Sato M, Goto S, Kazama JJ, Gejyo F, & Narita I (2019) Adsorption of Protein-Bound Uremic Toxins Using Activated Carbon through Direct Hemoperfusion in vitro. Blood Purification, 48(3), 215–222. https://doi.org/10.1159/000500014
45. Zieliński B, Miądlicki P, & Przepiórski J (2022) Development of activated carbon for removal of pesticides from water: case study. Scientific Reports, 12(1), 1–14. https://doi.org/10.1038/s41598-022-25247-6

46. Mudoga HL, Yucel H, & Kincal NS (2008) Decolorization of sugar syrups using commercial and sugar beet pulp based activated carbons. Bioresource Technology, 99(9), 3528–3533. https://doi.org/10.1016/j.biortech.2007.07.058
47. Fleming CA, Mezei A, Bourricaudy E, Canizares M, & Ashbury M (2011) Factors influencing the rate of gold cyanide leaching and adsorption on activated carbon, and their impact on the design of CIL and CIP circuits. Minerals Engineering, 24(6), 484–494. https://doi.org/10.1016/j.mineng.2011.03.021
48. Zhang J, Tang X, Hong Y, Chen G, Chen Y, Zhang L, Gao W, Zhou Y, & Sun B (2023) Carbon-based single-atom catalysts in advanced oxidation reactions for water remediation: From materials to reaction pathways. Eco-Environment and Health, 2(2), 47–60. https://doi.org/10.1016/j.eehl.2023.04.002
49. Chavan Professor P (2017) Preparation of activated carbon from coconut shell. *International journal of recent research in science*. Eng Technol 3(4):1–603. www.ijrrset.com
50. Ince NH, Apikyan IG (2000) Combination of activated carbon adsorption with light-enhanced chemical oxidation via hydrogen peroxide. Water Res 34(17):4169–4176. https://doi.org/10.1016/S0043-1354(00)00194-9
51. Satya Sai PM et al (1997) Production-of-activated-carbon from coconut shell char in a fluidized bed reactor. Ind Eng Chem Res 36(9):3625–3630. https://doi.org/10.1021/ie970190v
52. Koli A, Pattanshetti A, Mane-Gavade S, Dhabbe R, Kamble R, Garadkar K, Sabale S (2024) Agro-waste management through sustainable production of activated carbon for CO2 capture, dye and heavy metal ion remediation. Waste Manag Bull 2(1):97–121. https://doi.org/10.1016/j.wmb.2023.12.010
53. Rambabu N, Azargohar R, Dalai AK, Adjaye J (2013) Evaluation and comparison of enrichment efficiency of physical/chemical activations and functionalized activated carbons derived from fluid petroleum coke for environmental applications. Fuel Process Technol 106(2013):501–510. https://doi.org/10.1016/j.fuproc.2012.09.019
54. Gu H, Bergman R, Anderson N, Alanya-Rosenbaum S (2018) Life cycle assessment of activated carbon from woody biomass. Wood Fiber Sci 50(3):229–243. https://doi.org/10.22382/wfs-2018-024
55. Wasis B, Ghaida SB, Winata B (2019) Application of coconut shell charcoal and NPK fertilizer toward Acacia mangium growth on the soil of ex-limestone mining in Bogor, Indonesia. Arch Agric Environ Sci 4(1):75–82. https://doi.org/10.26832/24566632.2019.0401012
56. Marlinae L, Biyatmoko D, Husaini H, Irawan C, Arifin S, Saidi AR, Fithria A, Zubaidah T, Ratodi M, Mourad KA, Khairiyati L, Waskito A, Lasari HHD, Khoiroh I (2024) Coconut shells, water hyacinth, and rice in improving the quality of dug well water in flood areas. Jurnal Kesehatan Lingkungan 16(3):255–265. https://doi.org/10.20473/jkl.v16i3.2024.255-265
57. Jumaidin R, Gazari AS, Kamaruddin ZH, Kamaruddin ZH, Zakaria NH, Kudus SIA, Yob MS, Munir FA, Keshavarz M (2024) Mechanical properties of thermoplastic cassava starch/ coconut fibre composites: effect of fibre size. Pertanika J Sci Technol 32(S2):91–113. https://doi.org/10.47836/PJST.32.S2.07
58. Hasan KMF, Horváth PG, Kóczán Z, Alpár T (2021b) Thermo-mechanical properties of pretreated coir fiber and fibrous chips reinforced multilayered composites. Sci Rep 11(1):1–13. https://doi.org/10.1038/s41598-021-83140-0
59. Cosentino L, Fernandes J, Mateus R (2023) A review of natural bio-based insulation materials. Energies 16(12):4676. https://doi.org/10.3390/en16124676
60. Gagan, Navya, Anish, Thanmai RG (2023) Experimental investigation on brick with the partial replacement of coconut fibre. Int J Res Appl Sci Eng Tech 11(6):2773–2778. https://doi.org/10.22214/ijraset.2023.54142
61. Priyadarshini V, Felixkala T, Depaa RAB, Hemamalinie A, Francis Xavier J, Surendra Babu K, Jeya Arthi AJ (2020) Experimental investigation on properties of coir pith and its influence as partial replacement of fine aggregate in concrete. Mater Today Proc 45(February):6903–6906. https://doi.org/10.1016/j.matpr.2020.12.1122

62. Kanna V, Dhanalakshmi G (2023) Experimental investigation on brick with the partial replacement of coconut fibre. Int J Res Appl Sci Eng Technol 11(6):2773–2778. https://doi.org/10.22214/ijraset.2023.54142
63. Pereira CL, Savastano H, Payá J, Santos SF, Borrachero MV, Monzó J, Soriano L (2013) Use of highly reactive rice husk ash in the production of cement matrix reinforced with green coconut fiber. Ind Crop Prod 49:88–96. https://doi.org/10.1016/j.indcrop.2013.04.038
64. Mohammed L, Ansari MNM, Pua G, Jawaid M, Islam MS (2015) A review on natural fiber reinforced polymer composite and its applications. Int J Polymer Sci 2015:1. https://doi.org/10.1155/2015/243947
65. Werther J, Saenger M, Hartge EU, Ogada T, Siagi Z (2000) Combustion of agricultural residues. Prog Energy Combust Sci 26(1):1–27. https://doi.org/10.1016/S0360-1285(99)00005-2
66. Menon SD, Sampath K, Kaarthik SS (2021) Feasibility studies of coconut shells biomass for downdraft gasification. Mater Today Proc 44:3133–3137. https://doi.org/10.1016/j.matpr.2021.02.813
67. Ghosh SK, Samanta S, Hirani H, Vieira da Silva CR (2022) Effective waste management and circular economy. In: Effective waste management and circular economy. CRC Press, Boca Raton. https://doi.org/10.1201/9781003231608
68. Yirijor J, Bere AAT (2024) Production and characterization of coconut shell charcoal-based bio-briquettes as an alternative energy source for rural communities. Heliyon 10(16):1–13. https://doi.org/10.1016/j.heliyon.2024.e35717
69. Santoso H (2022) Study of making biomass briquettes from coconut shell and tapioca adhesive. BEST J Appl Electri Sci Technol 4(2):71–74. https://doi.org/10.36456/best.vol4.no2.5797
70. Pattanaik L, Pattnaik F, Saxena DK, Naik SN (2019) Biofuels from agricultural wastes. In: Second and third generation of feedstocks: the evolution of biofuels. Elsevier Inc. https://doi.org/10.1016/B978-0-12-815162-4.00005-7
71. Zabranska J, Pokorna D (2018) Bioconversion of carbon dioxide to methane using hydrogen and hydrogenotrophic methanogens. Biotechnol Adv 36(3):707–720. https://doi.org/10.1016/j.biotechadv.2017.12.003
72. Dahiya S, Sarkar O, Swamy YV, Venkata Mohan S (2015) Acidogenic fermentation of food waste for volatile fatty acid production with co-generation of biohydrogen. Bioresour Technol 182:103–113. https://doi.org/10.1016/j.biortech.2015.01.007
73. Amin FR, Khalid H, El-Mashad HM, Chen C, Liu G, Zhang R (2021) Functions of bacteria and archaea participating in the bioconversion of organic waste for methane production. Sci Total Environ 763:143007. https://doi.org/10.1016/j.scitotenv.2020.143007
74. Bichang DO, Oladele IO, Alabi OO, Aramide FO, Oluseye O, Borisade SG, Githinji DN, Ojemaye MO (2024) Comparative property investigation of raw and treated coconut shell biomass for potential polymer composite application. Heliyon 10(23):e40704. https://doi.org/10.1016/j.heliyon.2024.e40704
75. Raposo F, Fernández-Cegrí V, de la Rubia MA, Borja R, Béline F, Cavinato C, Demirer G, Fernández B, Fernández-Polanco M, Frigon JC, Ganesh R, Kaparaju P, Koubova J, Méndez R, Menin G, Peene A, Scherer P, Torrijos M, Uellendahl H, de Wilde V (2011) Biochemical methane potential (BMP) of solid organic substrates: evaluation of anaerobic biodegradability using data from an international interlaboratory study. J Chem Technol Biotechnol 86(8):1088–1098. https://doi.org/10.1002/jctb.2622
76. Aldin S (2010) The effect of particle size on hydrolysis and modeling of anaerobic digestion 60. http://ir.lib.uwo.ca/etd/60/
77. Gaby JC, Zamanzadeh M, Horn SJ (2017) The effect of temperature and retention time on methane production and microbial community composition in staged anaerobic digesters fed with food waste. Biotechnol Biofuels 10(1):1–13. https://doi.org/10.1186/s13068-017-0989-4
78. Zhang J, Zhang Y, Quan X, Chen S (2015) Enhancement of anaerobic acidogenesis by integrating an electrochemical system into an acidogenic reactor: effect of hydraulic retention times (HRT) and role of bacteria and acidophilic methanogenic archaea. Bioresour Technol 179:43–49. https://doi.org/10.1016/j.biortech.2014.11.102

79. Holden MN, Wolfe LM, Ogejo AJ, Cummins JE (2020) Introduction to biosystem engineering. Virginia Tech Publishing 2(4):1–232
80. Mshandete AM, Parawira W (2009) Biogas technology research in selected sub-saharan African countries—a review. Afr J Biotechnol 8(2):116–125
81. Ioannidou O, Zabaniotou A (2007) Agricultural residues as precursors for activated carbon production-a review. Renew Sust Energ Rev 11(9):1966–2005. https://doi.org/10.1016/j.rser.2006.03.013
82. Bridgwater AV (2012) Review of fast pyrolysis of biomass and product upgrading. Biomass Bioenergy 38:68–94. https://doi.org/10.1016/j.biombioe.2011.01.048
83. Abnisa F, Wan Daud WMA (2014) A review on co-pyrolysis of biomass: an optional technique to obtain a high-grade pyrolysis oil. Energy Convers Manag 87:71–85. https://doi.org/10.1016/j.enconman.2014.07.007
84. Azargohar R, Dalai AK (2006) Biochar as a precursor of activated carbon. Appl Biochem Biotechnol 131(1–3):762–773. https://doi.org/10.1385/ABAB:131:1:762
85. Demirbas A (2009) Progress and recent trends in biodiesel fuels. Energy Convers Manag 50(1):14–34. https://doi.org/10.1016/j.enconman.2008.09.001
86. Mohan D, Singh KP, Singh VK (2008) Wastewater treatment using low cost activated carbons derived from agricultural byproducts-a case study. J Hazard Mater 152(3):1045–1053. https://doi.org/10.1016/j.jhazmat.2007.07.079

Chapter 14
Management of Palm oil Mill Processing Wastes and Effluents: Biotechnological Advances

Zahra Beig-Mohammadi, Sanam Darvish, Mohammad Younesi, Elina Tavakoli, and Sahar Montaser

Abbreviations

BOD	Biological Oxygen Demand
COD	Chemical Oxygen Demand
CPO	Crude Palm Oil
EFB	Empty Fruit Bunches
FA	Fatty Acid
FFA	Free Fatty Acid
MFF	Mesocarp Fruit Fiber
OP	Oil Palm
OPF	Oil Palm Fronds
OPT	Oil Palm Trunk
PKC	Palm Kernel Cake
PKO	Palm Kernel Oil
PO	Palm Oil
POME	Palm Oil Mill Effluent
TAG	Triacylglycerol

Z. Beig-Mohammadi (✉) · S. Darvish · M. Younesi · E. Tavakoli · S. Montaser
Department of Food Science and Technology, North Tehran Branch, Islamic Azad University, Tehran, Iran
e-mail: z.beigmohammadi@iau-tnb.ac.ir

C. Trois, R. C. Ray (eds.), *Plantation Crop Wastes: Valorization for Economic Sustainability*, Waste as a Resource,
https://doi.org/10.1007/978-3-031-96805-1_14

1 Introduction

Oil palm (OP) is recognized as one of the fastest-growing and most critical vegetable oils in tropical regions, particularly in countries such as Indonesia and Malaysia, where its production has significantly contributed to economic growth by utilizing valuable natural resources. The processing of OP involves several key stages within the palm oil mill, including sterilization, bunch stripping, and fresh fruit bunch (FFB) handling [1]. OPs have a lifespan exceeding 200 years; however, their economically productive phase is limited to approximately 20–25 years. Peak yields, which account for 45–56% of the FFB, are typically achieved between 5 and 10 years of growth. The mesocarp is the primary target during oil extraction, while kernels contribute 40–50% to the total yield. On average, OP cultivation produces approximately 17 tons per hectare annually, with 5.8 tons of FFB required to generate 1 ton of crude palm oil (CPO) [2]. Effective management and utilization of by-products from PO processing are essential. Notably, palm oil mill effluent (POME), empty fruit bunches (EFB), and palm kernel cake (PKC) require proper handling to mitigate environmental risks. POME, a complex wastewater with high organic content, threatens aquatic ecosystems if discharged untreated, leading to oxygen depletion and harm to aquatic life [3]. POME production correlates directly with CPO output, with mills generating 2.5 tons of POME per ton of CPO [4]. Characterized by high BOD, COD, and acidic pH, POME contains 95–96% water, 4–5% solids, and 0.6–0.7% residual oil. Its colloidal composition imparts a viscous, brownish appearance and unpleasant odor, with methane emissions of 28 m^3 per ton of POME posing significant greenhouse gas risks [5]. Despite nonchemical extraction methods, POME's high organic content necessitates rigorous treatment before environmental discharge. Additionally, the decomposition of organic matter in POME contributes to greenhouse gas emissions, exacerbating climate change [6]. Traditional disposal methods for EFB and PKC, such as open dumping and landfilling, are unsustainable and can lead to environmental contamination and greenhouse gas release. These practices degrade the environment and forfeit opportunities to recover valuable resources from these by-products.

This chapter provides an overview of POME through bioprocessing technologies. It aims to identify innovative biotechnological approaches for sustainably managing POME and converting it into valuable products. By reviewing current strategies, the chapter further examines the challenges of scaling these technologies and evaluates their economic and environmental implications.

2 Geographical Origin, Production of OP

The OP genus *Elaeis* (family Arecaceae) consists of two primary species: *Elaeis guineensis*, originating from West Africa, and *Elaeis oleifera*, native to South America [7]. *E. guineensis*, the commercially dominant species, has a long history of culinary and medicinal use, and it is the cornerstone of global vegetable oil production [8]. Historical records indicate its introduction to Brazil and other tropical regions by Portuguese explorers in the fifteenth century. Subsequent propagation intensified in the nineteenth century through Dutch colonial initiatives in Indonesia, leading to the development of the high-yielding "Deli Dura" variety [9]. This cultivar became foundational to Southeast Asia's OP industry after its dissemination to Singapore and Malaysia in the late nineteenth century, establishing the region's dominance in global palm oil (PO) production [10]. Today, Indonesia and Malaysia collectively account for over 85% of global OP output, leveraging their tropical climates and established cultivation infrastructure [8]. Conversely, *E. oleifera*, native to South America, is distinguished by its high oleic acid content and lower saturated fatty acid profile. While not commercially cultivated on the same scale as *E. guineensis*, it serves as a critical genetic resource for interspecific hybridization programs aimed at enhancing oil quality, disease resistance, and environmental adaptability in commercial varieties [9]. These hybridization efforts reflect the growing emphasis on sustainable agricultural practices to address challenges such as climate resilience and biodiversity conservation. This synthesis underscores the historical and geographical factors that shaped the global OP industry, highlighting the interplay between colonial agricultural expansion, genetic resource utilization, and modern agroeconomic priorities.

Palm oil (PO), derived from *E. guineensis*, is the most economical and extensively utilized vegetable oil globally (Fig. 14.1). Cultivated across 19 million hectares in tropical regions, production has surged 35-fold since the 1970s (FAO, 2021). Tropical environments, characterized by substantial annual rainfall (1780–2280 mm), average temperatures of 24–30 °C, and elevated humidity, foster large-scale cultivation in Indonesia, Malaysia, Thailand, and parts of Africa. Southeast Asian countries, particularly Indonesia and Malaysia, dominate global production, accounting for approximately 56.7% and 27.36% of total output, respectively. This growth is driven by rising global demand for PO derivatives, though it generates significant

Fig. 14.1 Oil palm tree (**a**), oil palm fruit (**b**)

waste, including POME, a by-product constituting 90% of processed FFB [11]. PO is renowned for its high oil yield, and its global PO production is projected to reach 240 million tons by 2050, driven by rising demand [12].

The oil palm fruit bunch is the primary source of two distinct oils: CPO, derived from the mesocarp (fruit pulp), and PKO (Palm Kernel Oil), derived from the seed's endosperm. The mesocarp contains approximately 50–60% oil by weight, which corresponds to 20–22% of the total bunch weight. Conversely, the fruit kernels contain 48–52% oil, representing 2–3% of the bunch weight [13]. CPO is a major contributor to global vegetable oil consumption, representing over half of the market [12]. CPO and PKO are critical in global trade, and CPO, known as red palm oil due to its high carotenoid content, is also a significant source of vitamin E (600–1000 ppm), coenzyme Q10 (18–25 mg/kg), and sterols (325–365 mg/kg). Approximately 90% of palm oil is used in the edible food industry, with the remaining 10% used in soap and oleochemical manufacturing [12].

3 Nutritional Values and Uses of Oil Palm

Variations in fatty acid composition within *E. guineensis* oils have been observed across different geographical regions and varieties. For example, studies have demonstrated regional differences in the fatty acid profiles of palm oils from locations such as the Congo, Cameroon, and Nigeria [14]. Additionally, *E. oleifera* exhibits a higher unsaturated fatty acid composition. PO is a versatile ingredient within both the food and nutraceutical sectors, characterized by its balanced composition of saturated and monounsaturated fatty acids, alongside valuable fat-soluble minor components, including vitamin E (tocopherols and tocotrienols), carotenoids, and phytosterols [15]. PO's unique fatty acid (FA) and triacylglycerol (TAG) profile renders it versatile for numerous food applications, being the only vegetable oil with an almost 50–50 composition of saturated and unsaturated FAs (Table 14.1). CPO is employed in cooking, frying, and as a vitamin source. Fractionation of CPO yields palm olein (liquid) and palm stearin (solid), each with distinct physical and chemical properties (Fig. 14.2). These fractions are essential components in various food and industrial products, including shortenings, ice cream, cosmetics, lubricants, and biodiesel. Palm stearin offers solid fat functionality without hydrogenation, reducing trans-fat intake [16]. Notably, recent advancements have facilitated the recovery of water-soluble antioxidants, rich in phenolic acids and flavonoids, from POME. This unique, near-equal ratio of saturated and unsaturated fatty acids renders PO a viable substitute for trans-fat-containing solid fats across diverse food applications. Furthermore, the availability of natural fractions, such as carotenoid-rich red palm oil, allows for precise tailoring to meet specific dietary requirements.

The high tocotrienol content within PO's vitamin E composition has been the subject of extensive research, revealing potential health benefits encompassing antioxidant, cholesterol-lowering, anti-cancer, and anti-atherosclerotic properties. Additionally, the water-soluble phenolic-flavonoid complex, another valuable

Table 14.1 Components and functional properties of OP

	Component	Description	Functions	Health Effects
Major components	TGs	Glycerol esterified with three fatty acids		>95% of palm oil; define physical characteristics (melting point, crystallization)
	FAs	Primarily palmitic, oleic, linoleic	≈ 1:1 ratio of saturated to unsaturated	Balanced fatty acid profile; potential benefits for cholesterol management
Minor components	Vitamin E	Tocopherols (30%) and tocotrienols (70%)	Fat-soluble antioxidants	Antioxidant, cholesterol-lowering, anti-cancer, anti-atherosclerotic
	Carotenoids	α- and β-carotene (major)	Precursors of vitamin A; contribute to the red color	Provitamin A activity
	Phenolic-flavonoid complex	Water-soluble antioxidants	Potent antioxidant activity	Anti-cancer, skin health, anti-aging

OP oil palm, *TGs* triacylglycerols, *FAs* fatty acids

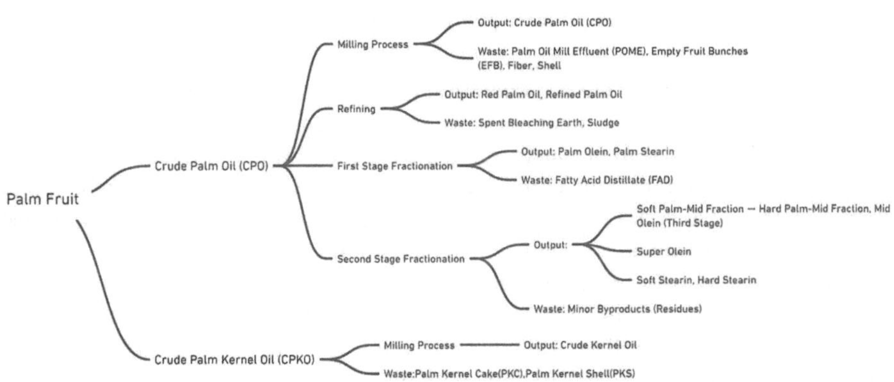

Fig. 14.2 Oil palm fruit processing

constituent, demonstrates potent antioxidant activity and potential efficacy against various cancers. Ongoing studies are investigating its utilization in nutraceuticals and cosmetics, focusing particularly on its potential dermatological anti-aging effects. PCO is primarily used in food applications, while PKO is utilized in non-food industries like oleochemical (Table 14.2). The economic importance of the oil palm is underscored by its status as the oldest and fastest-growing fruit oil crop globally [13]. However, sustainable expansion faces challenges, necessitating further research into genetic hybrids and underutilized palm species to optimize yield and nutritional quality [10]. Interesterification of CPO also expands its food

Table 14.2 Food and non-food applications of PO components and waste

Fraction/ component	Food applications	Non-food applications
PO	Cooking oil, frying oil, margarine, shortening, confectionery fats	Soaps, detergents, cosmetics, lubricants, biodiesel, candles, oleochemicals
Palm olein	Frying and cooking oil, blending with other oils	
Palm stearin	Shortenings, confectionery fats, margarines	Animal feed, soap making, candles
Red palm oil	Source of provitamin A, dietary supplements	
Tocotrienols	Dietary supplements, functional foods	Cosmetics, pharmaceuticals
Phenolic-flavonoid complex	Functional foods, beverages	Cosmetics, nutraceuticals, skin care products
Palm oil waste		
POME		Biogas production, fertilizer, animal feed, production of water-soluble antioxidants (phenolic-flavonoid complex)
EFBs		Mulch, fertilizer, fiberboard production, biofuel, animal feed

PO palm oil, *POME* palm oil mill effluent, *EFBs* empty fruit bunches

applications, allowing for the incorporation of essential polyunsaturated fatty acids and enhanced antioxidant properties. Customized blends of CPO and its fractions with other oils are used in products ranging from margarine to infant formulas. PO is a prime frying oil due to its unique fatty acid composition and high smoke point (approximately 230 °C). Frying, an ancient cooking method, involves heating food to the middle, creating a crust and a characteristic flavor [17]. The hot oil acts as a heat and mass transfer medium, with oil absorption and moisture vaporization occurring during the process. Chemical and physical changes, such as starch gelatinization, protein denaturation, and crust formation, are integral to frying. Fried food quality is directly related to oil quality, with oil degradation affecting texture, taste, and flavor. PO's stability during frying is enhanced by the synergistic activity of β-carotene and tocotrienol. Recent trends indicate a shift from animal fats and hydrogenated vegetable oils to naturally stable oils like PO and its blends, maximizing the oils' benefits.

PO's stability during frying is enhanced by the synergistic activity of β-carotene and tocotrienol. Recent trends indicate a shift from animal fats and hydrogenated vegetable oils to naturally stable oils like PO and its blends, maximizing the oils' benefits. Blending offers functional, nutritional, and technical advantages, such as tailoring oils for specific frying applications. Studies have highlighted the suitability and stability of palm oil and its by-products in frying, comparable to high oleic oils and research on the physicochemical, nutritional, and functional properties of palm oil, particularly in frying, and explores the suitability and sustainability of blending CPO with other unsaturated vegetable oils as a healthier alternative to partial hydrogenation [18].

4 Palm Oil (PO) Processing and By-Product Generation

PO extraction employs a range of methods, categorized by scale and complexity, from traditional techniques to large-scale industrial processes. To preserve oil quality, freshly harvested fruit bunches must undergo processing at an oil mill within 24 h (Fig. 14.2). Key unit operations include fruit sterilization, loosening, digestion, extraction, and clarification. Extraction methods primarily involve mechanical pressing or solvent extraction, with mechanical screw presses achieving 75–90% efficiency. CPO contains beneficial compounds (e.g., triacylglycerols, vitamins E, carotenoids) and undesirable impurities (e.g., free fatty acids, phospholipids). Extraction processes are classified as "wet" or "dry." The "wet" method uses hot water or steam to leach oil, while the "dry" method utilizes hydraulic or screw presses, with the latter preferred for continuous systems. Post-extraction, CPO is purified via centrifugation and drying, with vacuum drying recommended to minimize degradation. Solvent extraction, including supercritical carbon dioxide, is also used, especially for PKO. Refining, essential for removing impurities, employs alkali or physical methods. Alkali refining, though effective, can lead to losses and environmental pollution, favoring physical refining based on volatility differences. Quality is assessed by free fatty acid (FFA) and moisture content, with high-quality oils used in the food industry and lower-quality oils in non-edible applications. High FFA CPO may be physically refined, but temperature control is crucial to preserve carotenes. CPO can be refined chemically or physically, with physical refining omitting the acid neutralization step. Fractionation, a reversible process, separates CPO into palm olein (liquid) and palm stearin (solid) based on crystallization properties, enhancing product versatility. These fractions, like CPO, can be refined, bleached, and deodorized (RBD). Palm olein and stearin find diverse applications in food and other industries, with palm olein's oleic acid content offering potential as a substitute for other vegetable oils [4, 19].

The PO industry generates substantial residual biomass, offering potential for biofuel production. Despite the extracted oil representing only 10% of total biomass, the remaining 90% constitutes waste. With the industry's rapid expansion, waste management complexities are escalating. Each hectare of oil palm plantation yields 50–70 tons of biomass residues, including oil palm trunk (OPT), oil palm fronds (OPF), EFB, mesocarp fruit fiber (MFF), PKS, and POME, all of which, excluding POME, are characterized by high fiber content. OPF is consistently available throughout the year due to routine pruning, while OPT becomes available during replanting cycles. Current practice involves leaving dead trees to decompose naturally for soil conservation, erosion control, and nutrient recycling. However, this poses risks of insect infestations and hinders replanting. Alternatively, burning

OPT for fertilizer, while reducing insect risks, contributes to air pollution, a significant concern in Indonesia. Open burning, prevalent in Indonesian plantations, causes severe regional air pollution, prompting legal actions. Nonetheless, a sustainable, environmentally friendly OPT and OPF utilization is crucial. OPF can also be processed into livestock feed, supporting integrated crop-livestock systems. EFB, MFF, and PKS serve as boiler fuel, with ash returned to plantations as fertilizer. EFB requires drying due to high moisture content, making MFF and PKS more desirable for direct combustion. EFB is also being explored for bioplastic production, offering a biodegradable alternative to fossil-based plastics. MFFs are used as fillers in thermoplastics and thermoset composites, and oil palm biomass can be processed into absorbents for toxic gas and heavy metal removal, leveraging its high silica and mineral content [3].

Current mill boilers are inefficient; co-generation plants could enhance energy production and reduce emissions. However, grid integration limitations hinder their economic viability. POME, the sole liquid waste, results from sterilization, oil clarification, and separation processes, with 5–7.5 tons of water used per ton of oil, generating significant POME volumes. Despite being nontoxic, untreated POME, acidic and oil-laden, poses risks to aquatic ecosystems due to oxygen depletion. In addition to processing waste, used cooking oil from palm oil consumption presents disposal challenges. Direct discharge into water systems is environmentally harmful. Converting used oil into biodiesel offers an environmentally friendly and carbon-neutral alternative, as carbon emissions from biodiesel are offset by the carbon absorbed during oil palm photosynthesis.

5 Bio-Valorization of POME and Other PO By-Products in the Circular Economy

Despite its economic contributions, the PO industry generates environmental challenges due to the production of various by-products during oil extraction. These include OPT, OPF, EFB, PKS, MFF, and notably, POME. The uncontrolled release of these materials can lead to significant environmental pollution. Traditionally, OP biomass has been relegated to waste streams or primarily utilized as biofertilizers, with the traditional practice in OP cultivation involving returning biomass to the soil for natural fertilization and nutrient recycling. However, there is a discernable trend toward the utilization of this resource for wood-based products and bioenergy, reflecting a shift toward more sustainable practices. Within PO processing facilities, established energy recovery methods involve routinely combusting EFB, PKS, and MFF for steam and electricity generation. POME is also increasingly harnessed for biogas and power production, expanding its role in renewable energy generation. Beyond these established applications, OP biomass is now the subject of intense research interest for its conversion into bio-based chemicals and materials. This focus stems from the realization that these resources, once considered waste, can be transformed into high-value commodities. The valorization of OP biomass through integrated thermochemical and biological processes not only addresses critical environmental concerns but also generates novel revenue streams, contributing to a sustainable bioeconomy. Specifically, EFB, produced in substantial volumes, is commonly processed via incineration for energy recovery or composting. While mulching with EFB enhances soil moisture and nutrient retention, logistical constraints limit its widespread adoption [20]. These emerging valorization strategies present significant opportunities for conversion into diverse value-added products, including pellets, advanced biofuels, bio-derived chemicals, and innovative bio-based materials, thereby maximizing resource efficiency and promoting a circular economy. Fibers and shells are utilized as cost-effective boiler fuel and in the production of value-added materials, including activated carbon and concrete aggregates, demonstrating their versatility [21–23].

5.1 Current Waste Management Practices and Limitations

The global palm oil industry, vital to the economies of Malaysia, Indonesia, and other tropical regions, faces mounting environmental challenges due to its waste generation.

POM generates substantial waste streams, including 0.65–0.7 m^3 of POME per ton of FFB, EFB, which constitute 22% of FFB, and PKS, representing 5.5% of FFB [24]. Conventional practices, such as anaerobic digestion and open ponding, are criticized for inefficiency, methane emissions, and land intensiveness. This section examines biotechnological innovations to transform these wastes into valuable resources, aligning with circular bioeconomy principles that emphasize resource efficiency and sustainability [24, 25]. Traditional waste management methods employed by these mills include:

- **Anaerobic Digestion:** This process reduces COD by 70–85%, but necessitates a prolonged hydraulic retention time of 30–100 days and results in the emission of methane, a potent greenhouse gas [26].
- **Mulching:** EFB is reapplied to plantations to enhance soil fertility; however, issues such as uneven decomposition and high transportation costs limit the scalability of this practice [24].
- **Incineration:** Biomass combustion for energy recovery is a common practice, but it significantly contributes to air pollution and carbon dioxide emissions [27].

POME, characterized by high BOD (20,000–28,500 mg/L) and COD (45,500–65,000 mg/L) and high concentrations of organic acids, carbohydrates, minerals, and proteins, represents a potent nutrient source for biomass growth, posing significant risks to aquatic ecosystems if untreated [24, 26]. However, the untreated discharge of POME causes environmental risks that necessitate robust treatment strategies. To mitigate these concerns, both aerobic and anaerobic treatment methodologies are employed. Aerobic digestion utilizes oxygen to facilitate microbial activity and typically requires shorter retention times. However, anaerobic digestion is more widely adopted due to its superior production efficiency, reduced energy requirements, and versatility in processing diverse carbohydrate-rich wastes, including waste molasses, dairy wastewater, sewage sludge, and POME [28]. The anaerobic digestion process comprises three primary stages: hydrolysis, acidogenesis, and methanogenesis. During hydrolysis, fermentative bacteria decompose complex organic compounds into simpler monomers, such as amino acids, monosaccharides, and fatty acids. These monomers are converted during acidogenesis into hydrogen, carbon dioxide, and acetate, which serve as precursors for methane generation in the methanogenesis stage. While anaerobic digestion is effective for treating high-strength wastewater—due to its capacity to reduce COD, minimize surplus sludge production, and generate energy—it may not consistently achieve complete organic matter stabilization, particularly with highly concentrated wastewater. Consequently, the effluent from anaerobic digestion is often subjected to further treatment in an aerobic pond to ensure compliance with environmental discharge standards. These conventional waste management practices often fail to meet contemporary environmental standards, thereby highlighting the critical need for the implementation of advanced biotechnological solutions.

5.2 *Biotechnological Innovations in Waste Valorization*

Microalgae cultivation in POME presents a promising approach, offering the dual benefits of effective effluent treatment and valuable biomass production. Species such as *Chlorella sorokiniana* and *Scenedesmus dimorphus* have demonstrated significant nutrient removal capabilities, achieving 93–100% nitrogen removal and 55–96% phosphorus reduction, while simultaneously yielding lipid-rich biomass suitable for biodiesel production [26, 29]. The efficiency of these systems can be further enhanced through co-cultivation with bacteria. For instance, a co-culture of *Chlorella sorokiniana* and *Pseudomonas* sp. achieved a 94% reduction in COD and a 99% reduction in total nitrogen (TN) when treating 30% diluted POME [30]. Pilot-scale systems have validated the industrial potential of this technology, demonstrating biomass yields of 1.07 g/L with a 51% COD removal rate [26]. Moreover, microalgae contribute to carbon sequestration by utilizing CO_2, thereby mitigating emissions from anaerobic digestion. Microalgae employ diverse metabolic pathways, including autotrophic, heterotrophic, and mixotrophic, to assimilate nutrients. Mixotrophic growth, which combines photosynthesis and organic carbon uptake, is particularly advantageous in high-COD environments such as POME. *C. sorokiniana*, when cultivated in POME, has shown substantial COD removal and biomass production [30]. Microalgal lipids can be transesterified into biodiesel, with species like *Chlorella protothecoides* exhibiting high lipid yields. However, the economic viability of this process is challenged by high production costs. Co-location of microalgae systems with POM offers a potential solution by leveraging waste heat and CO_2 to reduce operational expenses [29]. The nutrient-rich microalgal biomass can also be utilized to enhance soil fertility and serve as animal feed. Vermicomposting *Chlorella*-treated POME sludge has been shown to increase soil carbon content. Additionally, *Arthrospira* and *Chlorella* have demonstrated improvements in livestock growth and disease resistance. Furthermore, microalgae synthesize high-value compounds, such as antioxidants and polyunsaturated fatty acids (PUFAs). *Haematococcus pluvialis* produces astaxanthin, a valuable compound in nutraceutical industries. Integrating POME treatment with photobioreactors offers the potential to optimize pigment yields while simultaneously mitigating wastewater pollution. Cultivation systems for POME treatment include open raceway ponds and closed photobioreactors (PBRs). While raceway ponds are cost-effective, they are susceptible to contamination risks and light penetration limitations. PBRs offer improved biomass consistency but require significant capital investment. Hybrid systems combining ponds and PBRs may provide an optimal balance between cost and efficiency. Challenges in POME treatment include high ammonium levels, which can inhibit microalgal growth. Dilution or ammonia stripping can reduce toxicity. Strain selection is also crucial, with species like *Chlorella kessleri* demonstrating tolerance to high nitrogen concentrations [31]. Traditional anaerobic–aerobic pond systems, commonly employed for palm oil mill effluent (POME) treatment, are characterized by prolonged hydraulic retention times and substantial methane emissions, contributing to greenhouse gas release. In contrast, microalgae-based systems offer a dual advantage: efficient nutrient removal and the production of commercially valuable biomass. Microalgae have demonstrated notable efficacy in the remediation of both municipal and industrial wastewater. Specifically, species such as *Chlorella vulgaris* and *Scenedesmus obliquus* have been shown to achieve high removal rates of nitrogen and phosphorus from municipal wastewater. Similarly, *Chlorella* sp. has demonstrated significant ammonium removal in dairy wastewater, and *Spirulina platensis* has effectively reduced COD in piggery effluent. These studies underscore the adaptability of microalgae to diverse wastewater matrices, thereby indicating their potential for effective POME treatment. Co-digesting POME with agricultural residues like cattle manure boosts biogas yield. Solar-assisted bioreactors optimize methane production, achieving 64.13% methane content and generating 1.9 MW of electricity in a 60-ton/hectare mill [32]. Innovations such as membrane-covered lagoons reduce methane emissions by 80%, supporting carbon neutrality goals [33].

5.3 Waste-to-Energy and Bio-Based Products

The PO industry's waste streams present significant opportunities for the production of both energy and bio-based products. Hydrothermal pretreatment of EFB facilitates a 97% glucose recovery rate, enabling efficient bioethanol production [34]. Furthermore, polyhydroxyalkanoates (PHAs), synthesized from volatile fatty acids (VFAs) derived from POME, offer a biodegradable alternative to conventional plastics, with production costs demonstrably below \$2/kg [35].

Microalgal biomass, cultivated through POME treatment, contains a significant lipid fraction, ranging from 20% to 50% of its dry weight, making it suitable for transesterification into biodiesel [26]. In addition, EFB-derived activated carbon exhibits high adsorption capacity for pollutants, thus valorizing this waste stream into valuable materials for water treatment applications [21]. Integrated biorefinery models represent a holistic approach to waste management, wherein diverse waste streams are consolidated and transformed into a spectrum of co-products. For example, empty fruit bunches of EFB and PKS can be pyrolyzed to yield bio-oil, achieving a 60% conversion rate, or processed into nanocellulose for composite materials [24]. Similarly, POME sludge, a by-product of anaerobic digestion, can be composted to produce fertilizers, while the volatile fatty acids (VFAs) derived from the same process are channeled into the production of polyhydroxyalkanoates (PHAs) [35]. POME serves as a versatile substrate for the production of a range of bioproducts through fermentation. Under controlled pH conditions, POME fermentation can yield a spectrum of organic acids, predominantly acetic and formic acids. These acids function as intermediates for further biochemical processes, such as polyhydroxyalkanoate (PHA) production, and possess direct industrial applications [36]. Aerobic fermentation systems have also been utilized for the production of citric and itaconic acids, although with variable yields [37–39]. The production of various enzymes and diverse bioactive compounds from wastes and effluents generated during oil and fat processing, as well as the application of advanced technologies—such as nanotechnology—in their utilization, has attracted considerable attention [40, 41]. Leveraging the high oil content of POME, lipase production has been achieved using both bacterial and fungal strains [42]. Anaerobic fermentation of POME has also been explored for hydrogen generation. Utilizing indigenous microflora from POME sludge and isolated strains from alternative sources like cow dung has yielded significant hydrogen production without concurrent methane generation, thereby reducing greenhouse gas emissions [43]. The implementation of sustainable practices in PO waste valorization faces several key challenges and barriers. First, technological limitations hinder the efficiency of microalgae cultivation systems. Open pond designs are susceptible to contamination risks, while closed photobioreactors, though offering greater control, present significant capital expenditure requirements [26]. Second, economic hurdles impede the scalability of biogas plants and other valorization technologies, primarily due to inconsistent subsidy support and the high pretreatment costs associated with lignocellulosic biomass conversion [24]. Third, policy gaps, such as weaknesses in the enforcement of effluent discharge standards and the limited availability of incentives for circular economy adoption, further hinder progress [24]. Despite these challenges, significant opportunities exist for advancing sustainable practices. Real-time monitoring of bioreactors, optimization of nutrient ratios and process efficiency, and genetic engineering to develop extremophile microalgae strains tolerant to high-temperature POME hold the potential to revolutionize wastewater treatment. Furthermore, expanded policy frameworks promoting sustainable practices are essential to incentivize zero-discharge systems [24]. Biotechnological advancements are instrumental in transforming palm oil waste into valuable resources, thereby achieving a balance between environmental sustainability and economic objectives. Systems utilizing microalgae, co-digestion strategies, and integrated biorefinery models serve as prime examples of the application of circular bioeconomy principles. Crucially, multidisciplinary collaboration among researchers, industries, and policymakers is essential for overcoming existing barriers and facilitating the global scaling of these innovative approaches [25].

6 Techno-Economic Analyses and Industrial Prospects

Contemporary techno-economic analyses highlight the substantial potential of the valorization of PO by-products through the implementation of sophisticated bioprocessing methodologies. Further investigations demonstrate that integrated bioprocessing of POME enables the efficient transformation of these by-products into high-value products, including biofuels and organic acids, thereby promoting the tenets of a circular bioeconomy. The increasing global demand for palm oil has driven significant expansion of plantations, particularly in tropical nations where palm oil production constitutes a pivotal economic activity. While this growth has yielded economic advantages, it has also precipitated substantial environmental challenges, encompassing deforestation, biodiversity loss, greenhouse gas emissions, and social disruption. A fundamental imperative for the palm oil industry is to balance economic prosperity with environmental and social accountability. This requires the attainment of that balance. The effective management and utilization of by-products generated from PO processing are central to addressing this imperative. Notably, POME, EFB, and PKC necessitate appropriate handling to mitigate potential environmental hazards. POME, a complex wastewater stream characterized by high organic content, poses a significant risk to aquatic ecosystems if discharged untreated, resulting in oxygen depletion and adverse impacts on aquatic biota [2]. Furthermore, the decomposition of organic matter within POME contributes to greenhouse gas emissions, exacerbating climate change. Discharging POME without treatment and utilizing unsustainable disposal methods for EFB and PKC not only degrade environmental quality but also forfeit opportunities to recover valuable resources from these by-products. Collectively, these findings provide a robust foundation for the industrial-scale implementation of applications designed to convert POME into sustainable and economically viable products.

7 Real-Time Applications

Contemporary investigations reveal a discernible trajectory wherein the inherent industrial utility of date palm by-products is being effectively harnessed for the creation of market-ready technological solutions. This progression, from raw materials to commercially viable products, is substantiated by the expanding development of novel, real-time applications, coupled with a corresponding increase in patent acquisition. These advancements illustrate a significant emphasis on the practical translation of scientific discovery into tangible, implementable technologies within the date palm industry. While PPF and PKS are commonly used for onsite energy generation through combustion, alternative thermochemical processes, such as gasification and pyrolysis, present opportunities for enhanced efficiency. Specifically, the pyrolysis of EFB conducted at 500 °C yields both bio-oil, at approximately 40% of the initial biomass, and biochar [44]. Moreover, the transesterification of CPO with methanol results in the production of biodiesel, achieving a purity level of approximately 90%, alongside glycerol, which constitutes 9–10% of the output. However, the employment of homogeneous catalysts, such as sodium hydroxide (NaOH), necessitates subsequent, often financially demanding, purification procedures [45]. Preliminary investigations, in the form of pilot initiatives focusing on biodiesel and bio-pellet production, have successfully demonstrated the technical feasibility of these processes. Nonetheless, the transition to large-scale implementation is contingent upon the establishment of a robust policy framework that provides adequate support. Furthermore, crude glycerol, a co-product generated during biodiesel synthesis, can be refined to meet the stringent quality requirements of the food and pharmaceutical industries. This purification process typically involves the application of vacuum distillation in conjunction with ion-exchange resins [46]. In Costa Rican POM, the utilization of PKS as a primary energy source is prevalent. However, implementing anaerobic digestion for POME presents a potential avenue for methane production, which can be harnessed for power generation [47]. While pilot-scale initiatives focused on biodiesel and bio-pellet production have successfully demonstrated technical feasibility, the transition to widespread commercialization necessitates the establishment of a supportive policy framework to facilitate scalability.

8 Socioeconomic Impacts, Climate Change Mitigation, and Well-being

The valorization of POME confers extensive socioeconomic and environmental benefits that are pivotal in regions where PO cultivation is a major agricultural activity. The industry faces environmental and economic challenges, including climate change impacts, deforestation concerns, and market volatility [35]. POME treatment predominantly involves anaerobic digestion, with biogas captured for electricity generation. However, a significant number of mills lack operational biogas plants, highlighting a substantial gap in current practices and indicating underutilization [32]. Zero-discharge systems, which repurpose treated POME for irrigation, offer a more sustainable alternative, minimizing environmental impact. Waste-to-energy technologies, including biomass combustion and POME-derived biogas, have the potential to contribute significantly to electricity supply [48].

Microalgae-based systems offer the potential for POME remediation and biofuel production, demonstrating high nitrogen removal rates and biomass yields [30]. Advanced thermal conversion methods, such as pyrolysis and torrefaction, offer further avenues for bio-product development from EFB, though moisture content remains a challenge [49]. Resource efficiency can be enhanced through the recycling of EFB and POME sludge into compost, reducing fertilizer costs [50].

9 Conclusion and Future Perspectives

Recent advancements in the valorization of POME and OP by-products, particularly through the development of innovative bioprocessing methodologies, have unveiled a spectrum of applications yielding substantial socioeconomic, environmental, and industrial advantages, especially in regions where PO cultivation is a foundational agricultural activity. POME has been successfully transformed into valuable commodities, including biofuels and biomaterials. The integration of these by-products within bioprocessing frameworks addresses critical waste management challenges and fosters a circular bioeconomy, thereby promoting ecological sustainability and economic advancement. Technological innovations are increasingly driving the adoption of POME across diverse sectors. Biofuels represent a promising avenue for achieving sustainable energy self-sufficiency, particularly in areas reliant on fossil fuel imports. However, the upscaling of production to meet demand and ensuring cost-competitiveness with traditional fuels remain significant hurdles. Furthermore, various bio-composite materials derived from OP have demonstrated mechanical properties comparable to conventional plastics, offering potential applications in both food and non-food industries. Biotechnological processing of POME has yielded valuable biochemicals, such as enzymes and organic acids, opening new avenues for industrial utilization. The valorization of POME generates substantial socioeconomic benefits, particularly for rural communities in tropical regions where OP cultivation is central to the local economy. The establishment of bio-refineries and processing facilities can create new employment opportunities and stimulate the growth of small-scale industries, contributing to improved livelihoods and poverty reduction.

Looking forward, the full realization of OP by-product valorization potential necessitates concerted research and development initiatives. While significant progress has been made in the direct and conversion-based utilization of oil palm biomass, the transition to full-scale commercialization of advanced technologies hinges on further enhancements in process efficiency, cost reduction, and integration strategies. Future investigations should prioritize the refinement of pretreatment methodologies, optimization of bioconversion yields, and development of scalable biorefinery models to maximize the potential of oil palm biomass as a cornerstone of a sustainable bioeconomy. The exploration of novel applications in health and nutrition, such as the development of functional food ingredients from OP by-products and POME, represents a promising area for future research. Governments can play a pivotal role by providing incentives for bio-refinery development, funding research programs, and establishing quality standards for biofuels. In conclusion, the ongoing valorization of POME holds significant promise for advancing a sustainable circular bioeconomy. However, achieving this vision requires sustained research efforts, technological innovation, supportive policy frameworks, and a commitment to comprehensive sustainability assessments.

References

1. Geng A (2014) Upgrading of oil palm biomass to value-added products. In: Hakeem K, Jawaid M, Rashid U (eds) Biomass and bioenergy. Springer, Cham. https://doi.org/10.1007/978-3-319-07578-5_10
2. Ahmad AL, Ismail S, Bhatia S (2003) Water recycling from palm oil mill effluent (POME) using membrane technology. Desalination 157(1–3):87–95. https://doi.org/10.1016/S0011-9164(03)00387-1
3. Sulaiman F, Abdullah N (2011) Optimization conditions for maximizing pyrolysis liquids of oil palm empty fruit bunches. Energy 36:2352–2359. https://doi.org/10.1016/j.energy.2010.12.067
4. Mba OI, Dumont MJ, Ngadi M (2015) Palm oil: processing, characterization and utilization in the food industry–a review. Food Biosci 10:26–41. https://doi.org/10.1016/j.fbio.2015.01.003
5. Mora-Villalobos JA, Aguilar F, Carballo-Arce AF, Vega-Baudrit JR, Trimino-Vazquez H, Villegas-Peñaranda LR et al (2021) Tropical agroindustrial biowaste revalorization through integrative biorefineries—review part I: coffee and palm oil by-products. Biomass Convers Biorefinery 1-19. https://doi.org/10.1007/s13399-021-01442-9
6. Anyaoha KE, Sakrabani R, Patchigolla K, Mouazen AM (2018) Evaluating oil palm fresh fruit bunch processing in Nigeria. Waste Manag Res 36(3):236–246
7. Paterson RRM, Lima N (2018) Climate change affecting oil palm agronomy, and oil palm cultivation increasing climate change, requires amelioration. Ecol Evol 8(1):452–461. https://doi.org/10.1002/ece3.3610
8. Absalome MA, Massara CC, Alexandre AA, Gervais K, Chantal GGA, Ferdinand D et al (2020) Biochemical properties, nutritional values, health benefits and sustainability of palm oil. Biochimie 178:81–95. https://doi.org/10.1016/j.biochi.2020.09.019
9. Cornelius JA (1966) Some technical aspects influencing the quality of palm kernels. J Sci Food Agric 17(2):57–61. https://doi.org/10.1016/0079-6832(77)90005-2
10. Sehgal S, Sharma V (2021) Palm/palm kernel (*Elaeis guineensis*). Oilseeds: Health attributes and food applications, pp. 145–161. https://doi.org/10.1007/978-981-15-4194-0_6
11. Awalludin MF, Sulaiman O, Hashim R, Nadhari WNAW (2015) An overview of the oil palm industry in Malaysia and its waste utilization through thermochemical conversion, specifically via liquefaction. Renew Sust Energ Rev 50:1469–1484. https://doi.org/10.1016/j.rser.2015.05.085
12. Gonzalez-Diaz A, Velandia CA, García-Núñez JA, Ramírez MC, Villegas PM (2024) Study of the changes in molecular composition of high–oleic palm oil (Coari× La Mé) before and after physical refining. Food Biosci 1(57):103479. https://doi.org/10.1016/j.fbio.2023.103479
13. Choudhary M, Grover K (2019) Palm (Elaeis guineensis jacq.) oil. In: Fruit oils: chemistry and functionality. Springer, Cham, pp 789–802. https://doi.org/10.1007/978-3-030-12473-1_42
14. Akpanabiatu MI, Ekpa OD, Mauro A, Rizzo R (2001) Nutrient composition of Nigerian palm kernel from the dura and tenera varieties of the oil palm (Elaeis guineensis). Food Chem 72(2):173–177. https://doi.org/10.1016/S0308-8146(00)00215-6

15. Choo YM, Yap SC, Ooi CK (1996) Recovered oil from palm-pressed fiber: a good source of natural carotenoids, vitamin E, and sterols. J Am Oil Chem Soc 73(5):599–602. https://doi.org/10.1007/BF02518114
16. Kellens M, Gibon V, Hendrix M, De Greyt W (2007) Palm oil fractionation. Eur J Lipid Sci Technol 109(4):336–349. https://doi.org/10.1002/ejlt.200600309
17. Agarwal MK, Agarwal ML, Athar M, Gupta S (2004) Tocotrienol-rich fraction of palm oil activates p53, modulates Bax/Bcl2 ratio and induces apoptosis independent of cell cycle association. Cell Cycle 3(2):200–199. https://doi.org/10.4161/cc.3.2.637
18. Begum H, Alam AF, Awang AH (2019) Sustainability of Malaysian oil palm: a critical review. Int J Environ Sustain Dev 18(4):409–429. https://doi.org/10.1504/IJESD.2019.103467
19. Kurnia JC, Jangam SV, Akhtar S, Sasmito AP, Mujumdar AS (2016) Prolysis advances in biofuel from oil palm and palm oil processing waste: a review. Biofuel Res J 9:332–346. https://doi.org/10.18331/BRJ2016.3.1.3
20. Menon NR, Ab Rahman Z, Bakar NA (2020) Empty fruit bunches evaluation: mulch in plantation vs. fuel for electricity generation. Palm Oil Eng Bull 67:15–20
21. Ho S, Khan MMH (2020) Short review on the use of oil palm shell in concrete and activated carbon. World J Nano Sci Eng 10(1):1–13. https://doi.org/10.4236/wjnse.2020.101001
22. Khoshtinat K, Beig-Mohammadi Z (2024) Plants extract and essential oil as natural preservatives in foods: one-decade editorial experiences. Appl Food Biotechnol 11(1):e13. https://journals.sbmu.ac.ir/afb/article/view/44605
23. Hashemi F, Beig-Mohammadi Z, Hashemi H (2024) Study the functional nanophytosome based on tannic acid and vitamin D3. Future Foods 1(10):100391. https://doi.org/10.1016/j.fufo.2024.100391
24. Cheah WY, Siti-Dina RP, Leng STK, Er AC, Show PL (2023) Circular bioeconomy in the palm oil industry: current practices and future perspectives. Environ Technol Innov 30:103050. https://doi.org/10.1016/j.eti.2023.103050
25. Darkwah DO, Ong-Abdullah M (2021) Sustainability of the oil palm industry. In: Elaeis guineensis. Intech Open. https://doi.org/10.5772/intechopen.100156
26. Kamarudin KF, Tao DG, Yaakob Z, Takriff MS, Rahaman MS, Salihon J (2015) A review on wastewater treatment and microalgal by-product production with a prospect of palm oil mill effluent (POME) utilization for algae. Der Pharma Chemica 7(7):73–89
27. Zamri MF, Milano J, Shamsuddin AH, Roslan ME, Salleh SF, Rahman AA et al (2022) An overview of palm oil biomass for power generation sector decarbonization in Malaysia: Progress, challenges, and prospects. Wiley interdisciplinary reviews. Energy Environ 11:6. https://doi.org/10.1002/wene.437
28. Soo PL, Bashir MJ, Wong LP (2022) Recent advancements in the treatment of palm oil mill effluent (POME) using anaerobic biofilm reactors: challenges and future perspectives. J Environ Manag 320:115750. https://doi.org/10.1016/j.jenvman.2022.115750
29. Ahmad A, Bhat AH, Buang A et al (2019) Biotechnological application of microalgae for integrated palm oil mill effluent (POME) remediation: a review. Int J Environ Sci Technol 16:1763–1788. https://doi.org/10.1007/s13762-018-2118-8
30. Cheah WY, Sankaran R, Show PL, Ibrahim TTNB, Chew KW, Culaba A, Chang J-S (2020) Pretreatment methods for lignocellulosic biofuels production: current advances, challenges, and future prospects. Biofuel Res J 7(1):1115–1127. https://doi.org/10.18331/BRJ2020.7.1.4
31. Zhu C, Cai Y, Gertz ER, La Frano MR, Burnett DJ, Burri BJ (2015) Red palm oil–supplemented and biofortified cassava gari increase the carotenoid and retinyl palmitate concentrations of triacylglycerol-rich plasma in women. Nutr Res 35(11):965–974. https://doi.org/10.1016/j.nutres.2015.08.003
32. Khalid ZB, Siddique MNI, Nasrullah M, Singh L, Wahid ZBA, Ahmad MF (2019) Application of solar-assisted bioreactor for co-digestion of POME with cattle manure for enhanced biogas production. Environ Technol Innov 16:100446. https://doi.org/10.1016/j.eti.2019.100446
33. Hosseini SE, Abdul Wahid M (2015) Pollutant in palm oil production process. J Air Waste Manag Assoc 65(7):773–781. https://doi.org/10.1080/10962247.2013.873092

34. Cardona E, Llano B, Peñuela M, Peña J, Rios LA (2018) Liquid-hot-water pretreatment of palm-oil residues for ethanol production: an economic approach to the selection of the processing conditions. Energy 160:441–451. https://doi.org/10.1016/j.energy.2018.07.045
35. Norrahim MNF, Farid MAA, Lawal AA, Yasim-Anuar TAT, Samsudin MH, Zulkifli AA (2022) Emerging technologies for value-added use of oil palm biomass. Environ Sci Adv 1:259–275. https://doi.org/10.1039/D2VA00029F
36. Hassan MA, Yee LN, Yee PL, Ariffin H, Raha AR, Shirai Y, Sudesh K (2013) Sustainable production of polyhydroxyalkanoates from renewable oil-palm biomass. Biomass Bioenergy 50:1–9. https://doi.org/10.1016/j.biombioe.2012.10.014
37. Alam MZ, Jamal P, Nadzir MM (2008) Bioconversion of palm oil mill effluent for citric acid production: statistical optimization of fermentation media and time by central composite design. World J Microbiol Biotechnol 24(8):1177–1185. https://doi.org/10.1007/s11274-007-9590-5
38. Alama MZ, Md Bari N, Muyibi SA, Jamal P, Al-Mamuna A (2010) Solid state bioconversion of oil palm empty fruit bunches for production of citric acid by wild strains of *Aspergillus niger*. Food Biotechnol 24(1):19–36. https://doi.org/10.1080/08905430903320883
39. Bari N, Alam Z, Muyibi SA, Jamal P, Al-Mamun A (2010) Statistical optimization of process parameters for the production of citric acid from oil palm empty fruit bunches. Afr J Biotechnol 9(21):554–563
40. Beig-Mohammadi Z, Hamidi-Esfahani Z, Sahari MA, Khosravi-Darani K (2018) Purification and characterization of extracellular phospholipase A1 from Trichoderma atroviride sp. ZB-ZH292. Biocatal Agric Biotechnol 1(13):176–181. https://doi.org/10.1016/j.bcab.2017.11.009
41. Mohammadi ZB, Khoshtinat K, Ghasemi S, Ahmadi Z (2022) Nanotechnology and food grade enzymes. In: Novel food grade enzymes: applications in food processing and preservation industries. Springer, Singapore, pp 455–487. https://doi.org/10.1007/978-981-19-1288-7_17
42. Ibrahim MF, Aziz SA, Razak MNA, Phang LY, Hassan MA (2012) Oil palm empty fruit bunch as an alternative substrate for acetone-butanol-ethanol production by *Clostridium butyricum* EB6. Appl Biochem Biotechnol 166(4):1615–1625. https://doi.org/10.1007/s12010-012-9538-6
43. O-Thong S, Boe K, Angelidaki I (2012) Thermophilic anaerobic co-digestion of oil palm empty fruit bunches with palm oil mill effluent for efficient biogas production. Appl Energy 93:648–654. https://doi.org/10.1016/j.apenergy.2011.12.092
44. Lawal AA, Hassan MA, Farid MAA, Yasim-Anuar TAT, Yusoff MZM, Zakaria MR, Shirai Y (2020) Production of biochar from oil palm frond by steam pyrolysis for removal of residual contaminants in palm oil mill effluent final discharge. J Clean Prod 265:121643. https://doi.org/10.1016/j.jclepro.2020.121643
45. Isahak WNRW, Ismail M, Yarmo MA, Jahim JM (2010) Purification of crude glycerol from transesterification RBD palm oil over homogeneous and heterogeneous catalysts for the biolubricant preparation. J Appl Sci 10(22):2590–2595. https://doi.org/10.3923/jas.2010.2590.2595
46. Wan Isahak WNR, Che Ramli ZA, Ismail M, Jahim JM, Yarmo MA (2015) Recovery and purification of crude glycerol from vegetable oil transesterification. Sep Purif Rev 44(3):250–267. https://doi.org/10.1080/15422119.2013.851696
47. Poh PE, Chong MF (2009) Development of anaerobic digestion methods for palm oil mill effluent (POME) treatment. Bioresour Technol 100(1):1–9. https://doi.org/10.1016/j.biortech.2008.06.022
48. Szulczyk KR, Cheema MA, Cullen R, Khan AR (2020) Bioelectricity in Malaysia: economic feasibility, environmental and deforestation implications. Aust J Agric Resour Econ 64(2):294–321. https://doi.org/10.1111/1467-8489.12345

49. Kaniapan S, Hassan S, Ya H, Patma Nesan K, Azeem M (2021) The utilisation of palm oil and oil palm residues and the related challenges as a sustainable alternative in biofuel, bioenergy, and transportation sector: a review. Sustain For 13(6):3110. https://doi.org/10.3390/su13063110
50. Rupani PF, Embrandiri A, Ibrahim MH, Ghole V, Lee CT, Abbaspour M (2018) Effects of different vermicompost extracts of palm oil mill effluent and palm-pressed fiber mixture on seed germination of mung bean and its relative toxicity. Environ Sci Pollut Res 25(36):35805–35810. https://doi.org/10.1007/s11356-018-1875-8

Chapter 15
Valorization of Olive Oil Industry Solid Wastes and Effluents into Biofuels and Value-Added Bioproducts

Vibhuti Sharma, Rohit Kumar Singh, Gaytri Mahajan, and Reena Gupta

Abbreviations

BOD	Biological oxygen demand
COD	Chemical oxygen demand
GAG	Greenhouse gases
IP	Impact pathway
LCA	Life cycle analysis
LDL	Low-density lipoprotein
MUFA	Monounsaturated fatty acids
OMSW	Olive mill solid waste
OMWW	Olive mill wastewater
PHAs	Polyhydroxyalkanoates
PUFA	Polyunsaturated fatty acids
RS	Reference scale
SDGs	Sustainable development goals
S-LCA	Social Life Cycle Assessment

1 Introduction

The olive oil industry is the cornerstone of the Mediterranean, accounting for a substantial share of global edible oil production. This industry not only meets nutritional and gastronomic needs but also has important socio-economic significance. However, this industry generates significant amounts of solid and liquid waste, posing environmental challenges due to their high organic content, phenolic compounds, and chemical oxygen demand (COD) [1]. These residues, if untreated, can

V. Sharma · R. K. Singh · G. Mahajan · R. Gupta (✉)
Department of Biotechnology, Himachal Pradesh University, Summerhill, Shimla, India

C. Trois, R. C. Ray (eds.), *Plantation Crop Wastes: Valorization for Economic Sustainability*, Waste as a Resource,
https://doi.org/10.1007/978-3-031-96805-1_15

lead to soil and water contamination, making sustainable waste management critical. Recent studies emphasize the need to transition from traditional linear production models to circular economy frameworks. Such frameworks aim at reintroducing trash into the value chain using advanced biotechnological techniques such as composting, anaerobic digestion, and biochemical extraction. This paradigm corresponds with global sustainability goals by lowering the environmental footprint of the olive oil industry and providing economically viable bioproducts, such as biofuels, bioplastics, and nutraceuticals [2]. Additionally, adding waste-to-value systems can dramatically reduce the environmental concerns associated with untreated disposal, such as groundwater contamination and methane emissions, thereby accelerating the transition of the industry to a low-carbon future [3]. The potential of these by-products is derived from their diverse chemical compositions. Olive pomace, for example, is high in lignocellulosic material, phenolic compounds, and residual oils, making it an ideal feedstock for bioenergy production and bioplastics. Similarly, olive mill wastewater (OMWW) contains high concentrations of organic acids, polyphenols, and sugars, which can be used in biotechnological procedures to make bioethanol, biosurfactants, and antioxidants. These novel applications not only enhance the olive oil manufacturing chain but also contribute to global initiatives to minimize waste and improve renewable energy sources.

This chapter delves into the valorization of olive oil industry wastes, focusing on biofuel production, bio-product development, and their socio-economic and environmental implications.

2 Geographical Origin, Distribution, Production, and Productivity of Olives and Olive Oil

2.1 Geographical Origin

The domestication of the olive tree is believed to have originated in the eastern Mediterranean, specifically in regions corresponding to modern-day Syria and Turkey. Cultivation practices began approximately 6000–8000 years ago, leading to the development of varieties of olives known today [4]. Archaeological evidence indicates that olive oil production dates back to the Pottery Neolithic to Chalcolithic transition, around 7600–7000 years before the present, in the southern Levant [5].

2.2 Distribution

From its origin, olive cultivation spread westward across the Mediterranean Basin. By the fifth millennium BCE (Before Common Era), it had become widespread in the southern Levant and the Aegean, significantly influencing the agricultural landscapes of these regions [5]. Olive farming spread to the New World after the American continents were discovered, eventually reaching places like Peru, Argentina, Chile, Uruguay, and Mexico [6]. Today, olive trees are cultivated in various parts of the world, including Mediterranean countries, parts of Africa, the Middle East, and America [7].

2.3 Production

The Mediterranean region remains the leading producer of olive oil globally. Spain is the largest producer, consistently accounting for about 45% of olive oil production in the world [8]. In the 2023/24 season, the olive oil production in Spain was approximately 766.4 thousand metric tons [9]. Other significant producers include Italy, Greece, Tunisia, Turkey, and Morocco. The European Union collectively produces around two million tons of olive oil annually, representing 68% of global production [10].

2.4 Productivity

Olive oil production is subject to fluctuations due to factors such as climate conditions, agricultural practices, and geopolitical events. For instance, global production decreased from 3.39 million tons in 2021–22 to 2.28 million tons in 2023–24, influenced by adverse weather conditions and other challenges [11]. Advancements in cultivation techniques, such as the development of super-intensive farms, have led to increased productivity in some regions, although traditional farms continue to play a crucial role in maintaining the quality and heritage of olive oil production [12]. While traditional producers like Spain, Italy, and Greece continue to dominate the olive oil market, other countries, including Turkey and Tunisia, have made significant contributions in recent years. Figure 15.1 shows the global distribution of olive oil production across different countries from 2013/14 to 2022/23 [12].

The Indian venture into olive cultivation is still in its nascent stages, with production volumes remaining low. The future of the industry will depend on addressing the challenges related to climate change, sustainable farming practices, and meeting the evolving demands of global consumers.

3 Composition and Nutritional Values of Olive

Olive trees (*Olea europaea* L.) are perennial crops, well-adapted to the Mediterranean climate. Their tenacity allows them to flourish in arid and semi-arid environments with little need for water or nutrients. Olive cultivation predominantly involves

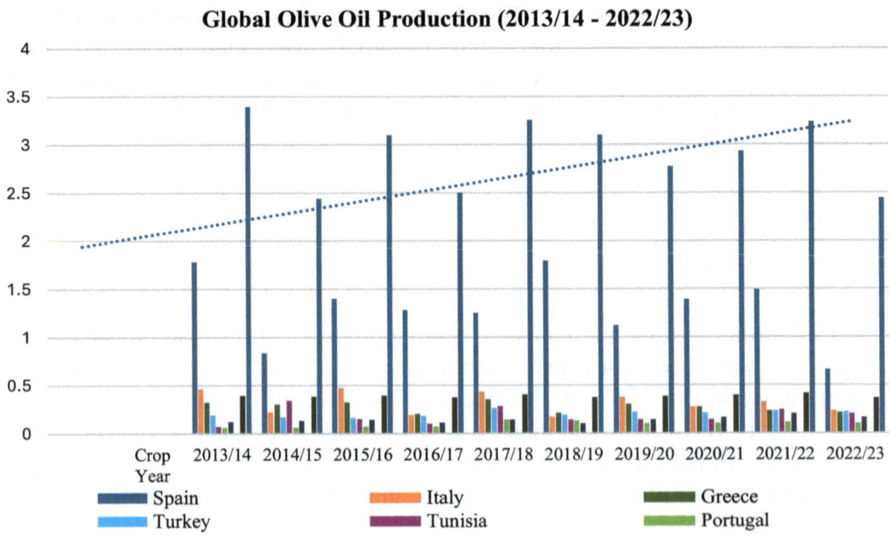

Fig. 15.1 Global olive oil production in different countries from 2013/14 to 2022/23

traditional methods; however, mechanization is becoming prevalent to enhance productivity [13].

3.1 Proximate Composition

Olives are renowned for their high monounsaturated fatty acid content, primarily oleic acid, contributing to medicinal sectors like maintaining cardiovascular health. Beyond dietary and medicinal applications, olive oil and its by-products also serve in cosmetics, pharmaceuticals, and biodiesel production. Olive pomace, the primary solid waste, contains lignocellulosic biomass, residual oils, and phenolic compounds. Conversely, effluents are rich in organic acids, sugars, and bioactive molecules [14]. These compositions offer immense potential for bio-valorization. The summary of the proximate composition and nutritional value of olives is shown in Table 15.1 [15, 16].

4 Processing of Olives for Oil Production

4.1 Olive Oil Extraction Methods

Olive oil production relies on two main technologies: The traditional methods involve manual pressing, where olives are crushed into a paste, which is then pressed to extract oil, and the modern techniques include centrifugation, which separates oil

Table 15.1 The summary of the proximate composition and nutritional value of olive

Component	Concentration (approximate)	Functionality
Total fat	98–99%	Primary energy source—Supports cardiovascular health (MUFA)
Monounsaturated fatty acids (MUFA)	55–83%	Reduce low-density lipoprotein (LDL) cholesterol—Improve heart health
Polyunsaturated fatty acids (PUFA)	3–21%	Essential fatty acids—Support brain function and cell membrane integrity
Saturated fatty acids	7.5–20%	Provide stability to olive oil—Used in moderate amounts
Vitamin E	12–43 mg/100 g	Antioxidant—Protects cells from oxidative damage
Phenolic compounds	50–800 mg/kg	Antioxidant and anti-inflammatory properties
Squalene	200–750 mg/100 g	Promotes skin health and protects against oxidative stress

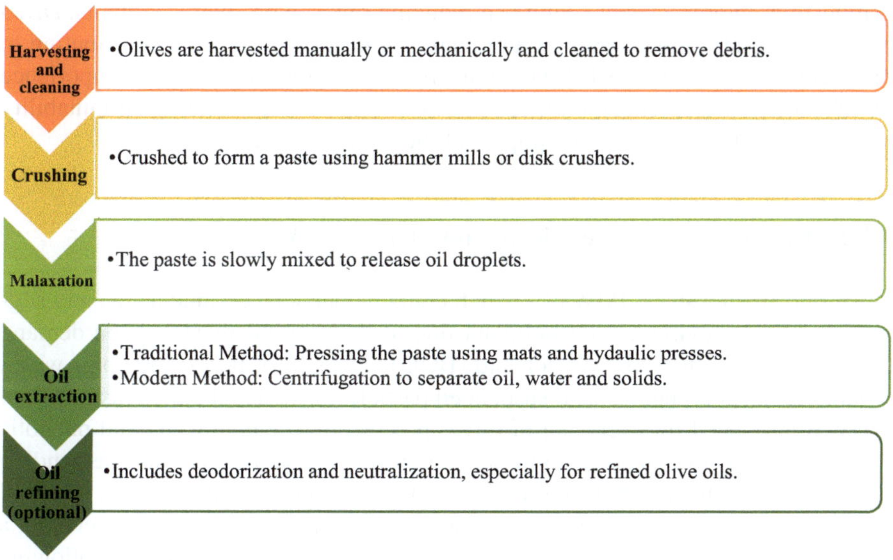

Fig. 15.2 The steps involved in olive oil production

from water and solids through high-speed spinning, offering higher efficiency and better oil quality [17]. The steps involved in olive oil production are shown in Fig. 15.2.

Despite technological advancements, the process of olive oil production produces significant waste. Traditional pressing generates large amounts of solid residues (pomace), while centrifugation creates higher quantities of liquid effluent due to water-intensive processes [18].

4.2 Residues and Effluents Generation

The main by-products produced during the processing of olive oil for food and beverages include:

4.2.1 Solid Waste (Olive Pomace)

Olive pomace, the solid waste generated during olive oil production, constitutes about 35% of the olive fruit weight and is composed of lignin, cellulose, hemicellulose, phenolic compounds, residual oil, skins, pulp, and seeds. For every ton of olive oil produced, approximately 2.5–3 tons of pomace are generated, making it a significant by-product of the industry. If left untreated, olive pomace can ferment and release greenhouse gases like methane, posing environmental concerns. However, it offers substantial potential for valorization [2]. Detoxified olive pomace can serve as livestock feed, providing a cost-effective and sustainable alternative to conventional feed. Additionally, it can be processed into organic fertilizers through composting, improving soil health and reducing reliance on chemical inputs. Dried olive pomace also holds value as a biomass for energy generation, contributing to renewable energy solutions [3]. These diverse applications highlight the importance of effective management and utilization of olive pomace in fostering sustainability and reducing the environmental impact of the olive oil industry.

4.2.2 Liquid Waste (Olive Mill Wastewater—OMWW)

Olive mill wastewater (OMWW), which contains phenols, organic acids, and sugars, is a major environmental pollutant due to its high chemical oxygen demand (COD) and biological oxygen demand (BOD) values [19]. It is made up of water, organic acids, polyphenols, and suspended particles, with about 1.5 cubic meters of OMWW produced for every ton of olives processed. The high COD and phenolic content of OMWW make it extremely harmful to aquatic habitats, demanding sophisticated management techniques to reduce its environmental impact. OMWW can be converted into biofuel by anaerobic digestion, which produces methane, a renewable energy source. OMWW can also be used as irrigation water following proper treatment, which lowers the need for freshwater in agricultural applications. The extraction of phenolic compounds, which are bioactive chemicals with usage in the nutraceutical and cosmetics industries, is one of its most beneficial applications.

The complicated nature of waste products from the olive oil industry, such as OMWW and solid olive pomace, requires creative management strategies to mitigate their effects on the environment. Sustainable solutions are offered by cutting-edge procedures, including anaerobic digestion for the production of biofuels, composting for organic fertilizers, and the extraction of bioactive phenolic compounds for cosmetics and nutraceuticals. Due to their health-promoting and antioxidant qualities, these chemicals are becoming more and more sought-after, opening up new markets for by-products of olive oil [19]. Utilizing OMWW properly increases the economic value and lessens its environmental impact, which is consistent with the circular economy views [2].

4.3 Circular Economy Integration in the Olive Oil Industry

Implementing circular business models in the olive oil industry is a transformative approach aimed at minimizing waste, optimizing resource use, and generating economic value. This paradigm shift aligns with the principles of a circular economy, which emphasizes closing the loop in production and consumption systems. By valorizing waste streams, the olive oil industry can significantly reduce its environmental footprint while creating new economic opportunities. Figure 15.3 shows the integration of a circular economy into olive oil production.

4.3.1 Key Principles of Circular Economy in the Olive Oil Industry

Waste as a Resource

- The fundamental principle is to treat by-products and waste not as disposal challenges but as raw materials for other processes.
- Solid residues (e.g., olive pomace) and liquid effluents (e.g., OMWW) can be repurposed into energy sources or high-value bioproducts.

Closed-Loop Systems

- Closed-loop systems ensure that all production outputs are reintegrated into the systems or utilized in other industries.
- For example, olive mill wastewater can be processed to extract polyphenols for nutraceuticals, while residual biomass can serve as bioenergy feedstock [20].

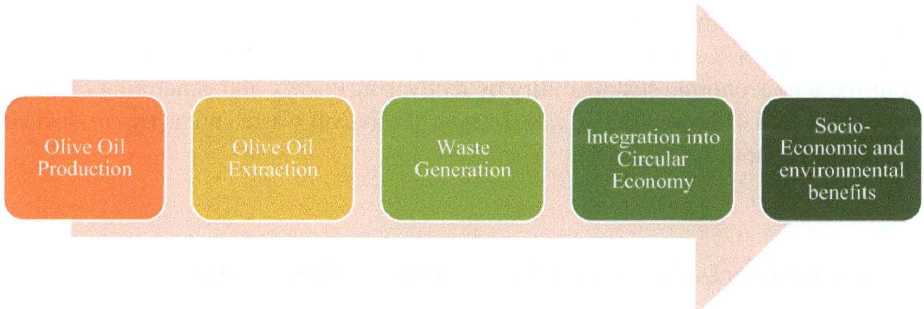

Fig. 15.3 Integration of a circular economy into olive oil production

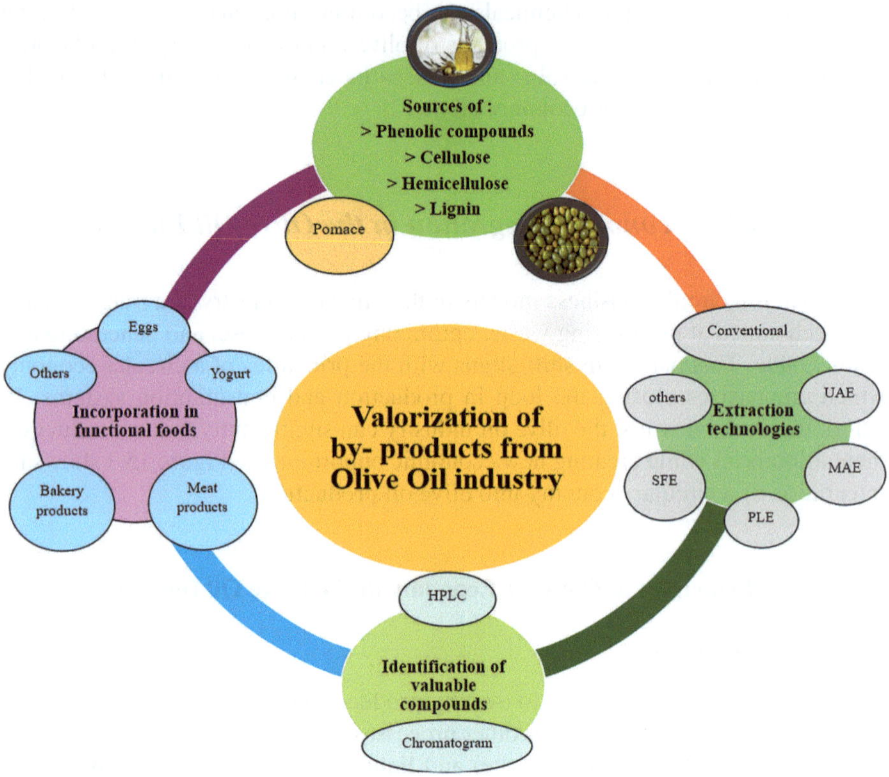

Fig. 15.4 Bio-valorization of olive oil waste and by-products in the circular economy

4.3.2 Bio-Valorization of Olive Oil Waste and By-Products in the Circular Economy

Bio-valorization integrates biological processes with waste management to convert by-products into value-added materials, aligning with the principles of a circular economy. In the olive oil industry, the high organic and phenolic content of waste streams, such as olive pomace and OMWW, presents challenges and opportunities. Bio-valorization technologies not only mitigate environmental impacts but also contribute to economic sustainability by recovering energy and generating bioproducts. Figure 15.4 shows the bio-valorization of olive oil waste and its by-products in a circular economy.

5 Bioproducts Produced by the Olive Oil Industry

The olive oil industry produces a variety of bioproducts with important economic and environmental value. Among these, olive pomace, a waste high in fibers, polyphenols, and residual oil, can be used to produce biofuels, feed animals, and

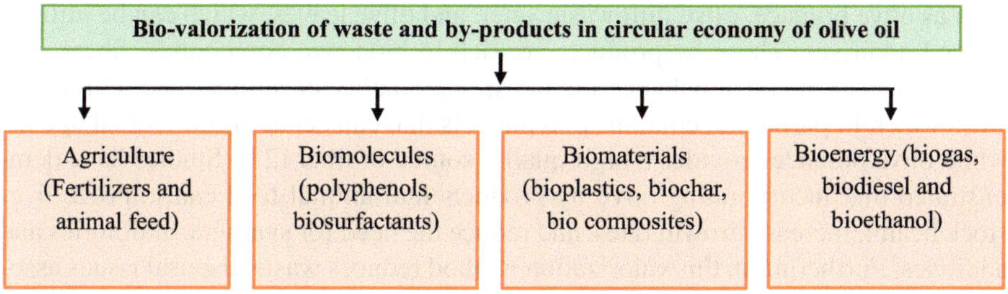

Fig. 15.5 Bioproducts produced from the olive oil industry

compost. Another prominent by-product is olive mill effluent, which contains bioactive chemicals with antibacterial and antioxidant characteristics, making it useful in medicinal, cosmetic, and agricultural applications. Furthermore, isolated polyphenols from olive residues are used in food preservation and as nutraceuticals because of their health benefits. Bioremediation and circular economy initiatives are advancing the sustainability of olive oil production by converting waste into value-added goods. The different types of bioproducts from the olive oil industry waste are shown in Fig. 15.5.

5.1 *Organic Fertilizers*

Olive pomace, the solid residue obtained from olive oil extraction, contains residual oil, water, and substantial amounts of lignocellulosic materials. OMWW is characterized by a high organic load, phenolic compounds, and acidity, making direct disposal environmentally hazardous [21]. Olive pomace is composted to produce nutrient-rich organic fertilizers. Research indicates that composting olive mill pomace enhances soil fertility and promotes sustainable agriculture by recycling essential nutrients. Long-term application of olive mill pomace compost has been shown to improve soil physical, chemical, and biological properties, thereby increasing soil quality and productivity [22].

5.2 *Feed Additives*

Olive oil mill by-products, including pomace and OMWW, contain various bioactive compounds that can be used as animal feed additives. These include phenolic compounds, fiber, and antioxidants that can enhance the nutritional value of animal feed. The incorporation of these by-products into animal diets can improve the developmental efficiency, immunity, and overall health of livestock. Detoxified olive pomace and its extracts are used as animal feed to enhance livestock health due to their antioxidant properties. Olive oil production yields several by-products, such as olive pomace, olive mill wastewater, and olive leaves, which can be utilized as feed additives. These by-products are rich in bioactive compounds, fibers, and antioxidants, which can enhance the nutritional quality of animal feed. Olive mill wastewater, in particular, contains polyphenols that can act as antimicrobial agents, while olive pomace provides a high-quality source of fiber [23]. Studies have demonstrated that incorporating olive by-products into animal feed can improve livestock health, increase growth rates, and reduce the need for synthetic antibiotics and additives. Furthermore, this valorization method reduces waste disposal issues associated with olive oil production [24].

5.3 Biomolecules

Olive wastes are rich sources of bioactive compounds, such as polyphenols, flavonoids, and antioxidants. These molecules are extracted for pharmaceutical and nutraceutical applications, offering therapeutic benefits. OMWW is rich in polyphenols, which have antioxidant properties. Innovative processes have been developed to extract these compounds for use in functional foods and nutraceuticals [25]. The biomolecules derived from olive oil industry wastes are shown in Table 15.2.

5.3.1 Enzymes

The bioconversion of olive oil waste into enzymes is an environmentally acceptable waste management method that uses agro-industrial leftovers as low-cost substrates for microbial fermentation. Olive oil waste contains organic substances such as phenolics, lipids, and polysaccharides, which can promote the growth of enzyme-producing microbes such as bacteria and fungi. Various hydrolytic and oxidative enzymes, including lipases, laccases, and peroxidases, can be produced from this waste [30]. These enzymes have numerous industrial applications, such as bioremediation, biodetergents, food processing, and biofuel production. Lipases, for example, are commonly used to hydrolyze fats and oils, whereas laccases and peroxidases are useful for breaking down phenolic pollutants. Industries that use olive oil waste for enzyme manufacturing can minimize their reliance on synthetic enzyme sources, lower production costs, and contribute to a sustainable circular economy [31].

5.3.2 Organic Acids

Microbial fermentation of olive oil waste into organic acids is an effective method of bio-valorization that converts agricultural leftovers into valuable biochemicals. Because of its high carbohydrate, lipid, and phenolic contents, olive oil waste is a good substrate for microbial metabolism, allowing the generation of organic acids such as lactic, citric, succinic, and acetic acids. These organic acids are widely used in food preservation, pharmaceuticals, and bio-based sectors, serving as acidulants,

Table 15.2 The biomolecules derived from olive oil industry wastes

Biomolecules	Source	Production process	Applications	References
Enzymes (lipase, cellulase)	Olive pomace, OMWW	Fermentation using microbial strains like *Aspergillus* and *Penicillium*	Used in food processing, detergents, and pharmaceuticals	[25]
Organic acids (citric, lactic)	OMWW	Microbial fermentation by *Aspergillus niger* or *Lactobacillus* species	Food preservatives, acidulants in beverages, and chemical intermediates	[26]
Microbial polysaccharides (xanthan, gellan gum)	OMWW, Olive Pomace	Fermentation using *Xanthomonas campestris* and *Sphingomonas elodea*	Stabilizers and thickeners in food and cosmetics, biopolymers for industrial use	[20]
Polyphenols (hydroxytyrosol, oleuropein)	OMWW	Extraction using membrane filtration or enzymatic hydrolysis	Antioxidants in nutraceuticals, functional foods, and cosmetics	[20]
Biosurfactants (rhamnolipids)	OMWW, Olive mill waste	Microbial fermentation using *Pseudomonas* species	Biodegradable cleaning agents, emulsifiers in pharmaceuticals, and enhanced oil recovery agents	[27]
Biopolymers (polyhydroxyalkanoates)	OMWW	Fermentation using *Ralstonia eutropha* or other bacterial strains	Biodegradable plastics for packaging, medical devices, and agricultural films	[25]
Squalene	Olive pomace, residual oil	Extraction through supercritical CO_2	Skin-care products, antioxidants, and as a precursor for cholesterol synthesis in pharmaceuticals	[28]
Lignin derivatives	Olive pomace	Lignin recovery during biomass processing	Biodegradable composites, adhesives, and bio-based carbon materials for energy storage	[29]

preservatives, and precursors for biodegradable plastics and biofuels [32]. Citric acid, for example, is widely used as a food additive and cleaning agent, whereas lactic acid is an important component in the manufacture of biopolymers. The conversion of olive oil waste into bio-based chemicals not only reduces environmental pollution but also promotes the development of sustainable industrial processes, reducing dependency on chemicals generated from fossil fuels [33].

5.3.3 Microbial Polysaccharides

The generation of microbial polysaccharides from olive oil waste represents a viable method for transforming agricultural leftovers into high-value biopolymers with a wide range of industrial uses. Olive oil waste contains vital nutrients for polysaccharide-producing microbes like bacteria and fungi, resulting in the manufacture of exopolysaccharides such as xanthan, alginate, and levan [34]. These biopolymers have unique functional features, such as thickening, emulsifying, and biofilm-forming capabilities, making them very attractive in the food, pharmaceutical, and environmental industries. Xanthan gum, for example, is commonly used as a food stabilizer, whereas alginate has applications in wound healing and medicinal delivery [35]. The microbial transformation of olive oil waste into useful biomaterials not only increases the economic worth of agricultural outputs but also promotes sustainability by minimizing waste accumulation and encouraging the development of bio-based alternatives to synthetic polymers [36].

5.3.4 Polyphenols

Polyphenols derived from olive oil waste provide valuable bio-valorization potential by converting agricultural leftovers into bioactive substances with important health and industrial benefits. Olive oil waste, including pomace and wastewater, contains polyphenols such as hydroxytyrosol, tyrosol, oleuropein, and flavonoids, all of which have strong antioxidant, anti-inflammatory, and antibacterial effects [37]. These bioactive compounds have been extensively researched for their potential use in nutraceuticals, functional foods, cosmetics, and pharmaceuticals due to their capacity to battle oxidative stress, prevent chronic diseases, and improve skin health. Furthermore, polyphenols generated from olive oil waste can be used in food preservation to increase shelf life by avoiding lipid oxidation and bacterial contamination. Their inclusion in biodegradable packaging materials emphasizes their importance in sustainable product development [38]. The recovery of polyphenols from olive oil waste not only lowers environmental pollution caused by the disposal of phenolic-rich leftovers but also helps to promote the circular economy by producing high-value natural antioxidants and bioactive components for a variety of sectors [39].

5.3.5 Biosurfactants

Biosurfactants are surface-active compounds produced by microorganisms that have potential applications in various industries, particularly in the bio-valorization of waste and by-products within a circular economy framework. Their ability to lower surface and interfacial tension makes them useful for environmental remediation, waste treatment, and enhancing the recovery of valuable materials from waste streams [40]. Effluents rich in organic compounds are used to produce biosurfactants, reducing environmental toxicity and serving industries like cosmetics and detergents. Biosurfactants have been employed in the recovery of biofuels from waste oils, oils from food processing, and agricultural by-products. Additionally, they play a role in the biodegradation of pollutants, promoting the breakdown of hydrocarbons, heavy metals, and pesticides in contaminated environments, and facilitating waste treatment processes [41].

5.4 Biomaterials (Biocomposites and Biopolymers)

Olive residues are integrated into bioplastics to produce eco-friendly alternatives. Research highlights their potential in the packaging and construction industries. Bioactive substances that can be recovered from the solid wastes of the olive oil sector, such as polyphenols, are renowned for their strong antioxidant and antibacterial activities, with potential applications in the nutraceutical, cosmetic, and food industries [42]. Studies have revealed the potential of converting olive mill waste into biodegradable polymers, contributing to sustainable materials development [20]. Composting olive pomace results in organic fertilizers that enhance soil fertility and structure, promoting sustainable agricultural practices.

5.4.1 Biochar

The bio-valorization of olive oil waste into biochar is an environmentally benign technique for increasing the sustainability of the olive oil industry. Olive mill waste, including pomace and wastewater, is high in organic matter and can be thermochemically transformed into biochar by pyrolysis. This carbon-rich substance has a variety of applications, including soil amendment, wastewater treatment, and carbon sequestration. In agriculture, biochar boosts soil fertility, increases water retention, and promotes microbial activity. Furthermore, its high adsorption capability makes it useful for removing heavy metals and organic pollutants from wastewater. Producing biochar from olive oil waste not only minimizes environmental pollution but also adds economic value by developing long-term waste management solutions.

5.5 Biofuels

Olive oil waste products are also valuable resources for biofuel production, particularly bioethanol and biogas. Olive pomace, which is typically discarded, can be fermented to produce bioethanol [43]. Additionally, anaerobic digestion of OMWW and pomace can generate biogas, providing a renewable energy source. This process contributes to waste-to-energy initiatives, supporting the circular economy by turning waste materials into valuable energy products and reducing dependence on fossil fuels.

Biofuels from olive oil by-products also offer lower carbon emissions compared to traditional fuels, promoting a more sustainable energy sector. The adoption of biofuel production technologies in olive oil-producing regions can create local economic opportunities and contribute to energy self-sufficiency [44].

5.5.1 Biodiesel

The residual olive oil recovered from pomace is turned into biodiesel through transesterification, creating fatty acid methyl esters with qualities similar to those of conventional diesel. The environmental impacts of biodiesel generation from olive pomace oil (B20 and B100) and petroleum diesel were compared using life cycle analysis [45]. In addition, energy and economic evaluations of biodiesel generation from pomace oil were conducted. They concluded that among the investigated categories, biodiesel made from pomace oil performed significantly better than conventional fuels in terms of climate change and resource harm. The economic analysis demonstrated the economic potential of producing biodiesel from pomace oil [46]. Olive oil by-products, particularly olive pomace, contain residual oils that can be extracted and used as feedstock for biodiesel production [47].

By transforming these by-products into biodiesel, the olive oil business can cut waste disposal expenses while also creating a sustainable energy source. Also, this technique can aid in mitigating climate change by lowering greenhouse gas emissions when compared to conventional fossil fuels [48].

5.5.2 Biogas

Anaerobic digestion of solid and liquid wastes produces biogas, a renewable energy source for electricity and heating. Anaerobic digestion of olive mill wastes can produce biogas, a renewable energy source. This process not only generates energy but also reduces the pollutant load of waste. Moreover, novel approaches are being investigated to improve the efficiency of biogas generation. Researchers have explored the co-digestion of processed olive waste under anaerobic circumstances, producing sustainable models that increase biogas generation. Such solutions reduce the environmental effects of olive oil production while simultaneously providing a useful source of renewable energy [49].

5.5.3 Bioethanol

Bioethanol, an alcohol produced from the fermentation of sugars, is another renewable energy source that can be derived from olive oil waste. OMWW, a by-product of olive oil extraction, contains organic compounds such as sugars and phenolic compounds that can be fermented to produce bioethanol [50]. The valorization of

OMWW not only offers an alternative to conventional waste disposal methods but also enhances the economic viability of the olive oil industry by adding value to waste streams. Lignocellulosic components of olive pomace are hydrolyzed and fermented to produce bioethanol. Furthermore, Dantas Palmeira et al. [51] reviewed the potential of fermenting olive pomace to make bioethanol, emphasizing its significance in valorizing agricultural waste and contributing to sustainable energy solutions.

6 Techno-Economic Analyses and Industrial Prospects

Waste valorization in the olive oil industry is economically viable, with significant potential for profit generation. A techno-economic analysis of biofuel production from olive pomace reveals an internal rate of return of 18–25% under optimized conditions.

The techno-economic feasibility of valorizing olive oil industry waste hinges on the efficiency of conversion technologies, scalability, and market potential. Pilot studies have demonstrated that integrating bioprocessing technologies, such as anaerobic digestion for biogas production and enzymatic hydrolysis for bioethanol production, can yield significant returns on investment [52].

6.1 Technological Approaches

Anaerobic digestion is a key technology for converting olive mill waste into biogas, which is primarily composed of methane and carbon dioxide. This process involves the decomposition of organic matter under oxygen-free conditions using microorganisms, enabling energy recovery, reducing environmental pollution, and stabilizing high COD waste. One kilogram of solid waste can produce approximately 0.5 m^3 of biogas, offering significant energy savings in olive oil-producing regions [53]. Similarly, the production of polyhydroxyalkanoates (PHAs) from OMWW provides an eco-friendly alternative to petrochemical plastics. Polyhydroxyalkanoates, which are biodegradable polymers, have applications in food packaging, coatings, and medical prosthetics. The growing demand for sustainable plastics enhances their market potential [54].

Composting is another viable approach, converting solid olive mill waste into organic fertilizers that improve soil health and reduce reliance on chemical inputs. Compost sales offer an additional revenue stream while promoting agricultural sustainability [53]. Additionally, biohydrogen production via microbial processes during OMWW treatment presents a clean energy solution, particularly for decentralized energy systems. While the initial production costs are high, biohydrogen holds promise as a renewable energy source with long-term economic viability [54].

Techno-economic studies underscore the feasibility of these valorization methods. For instance, a biogas power plant in Corfu, Greece, capable of processing 70,000 tons of waste annually, requires an initial investment of approximately €19.6 million. However, under a subsidy model and through the sale of by-products like soil improvers, the payback period could be reduced to 3 years. The plant's projected annual energy production of 25,000 MWh could meet 14.13% of Corfu's energy needs [53]. Furthermore, biorefineries using OMWW align with circular bioeconomy principles by producing biopolymers, biohydrogen, and biogas. Studies in Greece demonstrated that such facilities reduce environmental impacts while generating diverse revenue streams [54]. The industrial prospects for olive mill waste valorization are promising. The rising global demand for renewable energy and biodegradable materials positions this sector for high growth, with the bioplastics market expanding at a compound annual growth rate of 20%, driven by legislative pressures and changing consumer preferences [54]. Government subsidies, carbon credit mechanisms, and EU policies supporting waste-to-energy projects further enhance the economic feasibility of these initiatives [53]. However, challenges such as high initial capital costs, regulatory hurdles, and variability in waste composition persist. Despite these barriers, advancements in bioreactor technologies and value-added product development present opportunities for improving the economic and environmental sustainability of olive mill waste valorization [53].

Challenges in scaling include logistical constraints in waste collection, high initial capital costs, and the need for advanced infrastructure. However, government incentives and carbon credits make these ventures attractive to investors. Industries adopting circular economy models gain competitive advantages while reducing their ecological footprints and improving yield efficiency [55].

7 Real-Time Applications, Patents, and Innovations

7.1 Real-Time Applications of Olive Oil Waste

Valorized products from olive oil waste have found applications across industries, including renewable energy, agriculture, and materials science. Recent innovations have led to the development of bioplastics and advanced composites derived from olive pomace, which are being patented and commercialized globally [18]. The

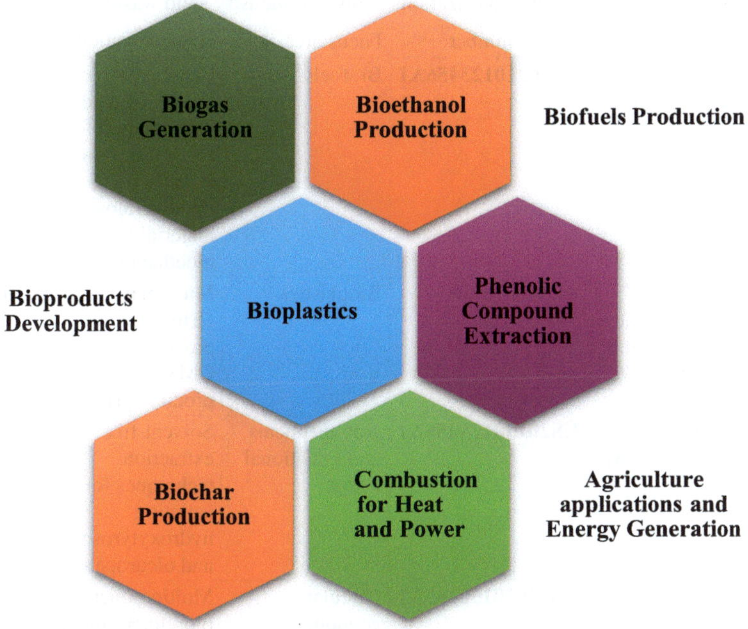

Fig. 15.6 The real-time applications of olive oil wastes in different sectors

real-time applications of olive oil wastes in different sectors are mentioned in Fig. 15.6.

7.2 Patents and Innovations

The patents, such as enzymatic processes for bioethanol production and microbial fermentation techniques for biopolymers, underscore the technological progress in this domain. Collaborative projects between academia and industry are further accelerating the commercialization of these innovations [56]. Table 15.3 shows the patents related to the valorization of the olive oil industry's solid waste and effluents.

Table 15.3 Patents related to the valorization of olive oil industry solid waste and effluents

Patent title	Patent number	Focus area	Key innovations	Reference
Process for Producing Bioethanol from olive pomace	**US20210123456A1**	Biofuels	Enzymatic hydrolysis and fermentation for converting olive pomace lignocellulosic material to bioethanol	[57]
Production of polyhydroxyalkanoates (PHAs) from olive mill wastewater (OMWW)	**EP3456789B1**	Bioplastics	Microbial fermentation process utilizing OMWW to produce PHAs	[54]
Extraction of polyphenols from olive mill waste	**US20190123456A1**	Nutraceuticals and Functional Foods	Solvent-free extraction techniques for isolating hydroxytyrosol and oleuropein	Abbattista et al. [57]
Integrated biorefinery for olive waste utilization	**WO2020156783A1**	Circular economy	Multiproduct biorefinery model for producing biofuels, biochar, and biofertilizers from olive waste	Mouzakitis and Adamides [54]
Methods for producing lightweight aggregates from olive pomace	**EP2233445A1**	Construction materials	Incorporation of olive pomace into lightweight aggregates for sustainable building materials	[57]
Process for producing biodiesel from olive waste oil	**US20170234567A1**	Biofuels	Conversion of high-free fatty acid olive waste oil to biodiesel through enzymatic transesterification	[58]
Method for the recovery of bioactive compounds from olive leaves	**JP2020134567A**	Pharmaceuticals and Cosmetics	Use of mass spectrometry-guided extraction techniques for isolating antioxidants from olive leaves	[57]
Utilization of olive mill wastewater in fertilizer production	**EP3141592B1**	Agriculture	Integration of OMWW into fertilizer formulations with high nutrient value	[54]

8 Socio-Economic and Environmental Impacts on Olive Oil Production

Using circular economy models for assessing olive oil waste boosts local economies and generates job possibilities. The development of biorefineries and related sectors supports rural communities, particularly in the Mediterranean region [59]. In terms of climate change mitigation, converting waste into biofuels reduces greenhouse gas emissions by displacing fossil fuels. The production of biochar from olive pomace contributes to carbon sequestration, enhancing soil health and agricultural productivity.

8.1 Socio-Economic Impacts

The olive oil industry plays a pivotal role in the economies of Mediterranean countries and beyond. It provides livelihoods for millions of individuals across farming, production, and supply chains. However, the socio-economic impacts of olive oil production are complex and often underrepresented in the academic literature, especially when compared to its environmental and economic dimensions. For example, while regions like Spain, Italy, and Greece dominate global production, North Africa and the Middle East, which have a rich history of olive cultivation, remain understudied in terms of socio-economic outcomes [60].

Labor practices, community engagement, and cultural preservation are central to the social sustainability of an industry. Employment opportunities generated by the olive oil sector are substantial, particularly in rural areas. Yet, the working conditions, wage levels, and overall labor standards in many regions warrant further investigation to ensure fair treatment and economic security for workers. According to a study, achieving social sustainability in this sector requires addressing issues like worker safety, equitable pay, and community well-being [61]. Despite its potential, the socio-economic dimension often takes a backseat in sustainability assessments, as highlighted by Wei et al. [62]. This limited perspective could hinder comprehensive progress, particularly in regions where olive oil production serves as a cultural and economic backbone. For example, in the Guilan region of Iran, olive oil production has historical and cultural significance, yet challenges like inadequate labor practices and insufficient community development remain prevalent [63]. Addressing these socio-economic gaps is essential for a more inclusive approach to sustainability.

8.2 Impact of Climate Change on Olive Oil Production

Olive oil production is not only impacted by climate change but also contributes to it through greenhouse gas (GHG) emissions, particularly during the agricultural phase. Fertilization, irrigation, and phytosanitary treatments are key contributors, with the agricultural stage accounting for nearly 94% of the total GHG emissions in some studies [63]. Using chemical fertilizers alone accounts for over half of these emissions, underscoring the need for environmentally friendly alternatives.

Implementing life cycle assessment (LCA) methodologies has provided valuable insights into mitigating these environmental impacts. Studies by Khounani et al. [64] emphasize transitioning from linear olive production models to bioeconomy-centric agro-biorefinery systems. This shift reduces the environmental effects per ton of olive oil and enables the production of a wider variety of value-added bio-products. For example, bioethanol production from olive waste cake in Iran has been shown to significantly reduce waste and contribute to renewable energy goals [65].

Another critical area for climate change mitigation is the adoption of closed-loop supply chains. Seydanlou et al. [66] propose this approach as a feasible strategy for integrating economic, environmental, and social dimensions of sustainability in the olive oil industry. Such systems recycle and repurpose waste, minimizing environmental degradation while enhancing resource efficiency. Additionally, the cultivation of drought-resistant olive varieties and the optimization of irrigation systems are essential strategies for adapting to climate variability.

9 Wellness and Sustainability in Society

Social sustainability in olive oil production has often been overshadowed by environmental and economic considerations, but its importance cannot be overstated. Human well-being, as defined by the United Nations Sustainable Development Goals, is deeply intertwined with factors like job security, equitable labor practices, and community health [61]. In the olive oil sector, these elements manifest through indicators such as worker safety, societal altruism, and cultural preservation [67].

One approach to measuring social sustainability is through the Social Life Cycle Assessment (S-LCA). Two primary methods—the reference scale (RS) approach and the impact pathway (IP) approach—have emerged to address social impacts. The RS method evaluates the behavior of producers and actors across life cycles, while the IP approach focuses on the final social consequences of production activities. However, integrating S-LCA with environmental LCA remains challenging due to the scarcity of social data and the complexity of selecting appropriate indicators [68].

In regions like Iran, social sustainability faces additional challenges, including underdeveloped infrastructure and a lack of emphasis on cultural heritage in the production cycle. Addressing these issues requires targeted policies that balance economic growth with community well-being. For instance, promoting educational programs and professional training in olive oil production can enhance both worker competency and product quality, as demonstrated by studies in Andalusia [69].

Additionally, incorporating socio-economic issues into sustainability frameworks can have major consequences for the olive oil business. A holistic strategy that considers environmental, economic, and social factors can result in a more resilient and fair system. By addressing labor practices, encouraging community engagement, and safeguarding cultural heritage, the sector can achieve long-term sustainability while improving the quality of life for everyone involved in the manufacturing process.

10 Future Perspectives

Future perspectives on waste valorization in the olive oil industry are promising, but the sector must address key technological, economic, and social challenges to fully realize its potential. A multifaceted approach that incorporates advanced waste processing techniques, policy support, and industry collaboration will be essential for overcoming these barriers. For instance, the development of cost-effective and efficient conversion technologies for bioethanol and biodiesel production can enhance the economic feasibility of waste valorization projects [70]. Additionally, the optimization of waste-to-energy systems, such as anaerobic digestion for biogas production, could provide a sustainable solution for energy generation in olive oil-producing regions. From an economic perspective, government incentives and carbon credit schemes could encourage investment in waste valorization technologies. Such financial support would reduce the initial capital burden and make biofuel production more attractive to investors. The creation of biorefineries and other valorization facilities could also generate employment opportunities, particularly in rural areas where olive oil production is a key economic activity. These facilities could act as hubs for local economies, offering training and skill development for workers while promoting sustainable agricultural practices.

Social sustainability in the olive oil industry must also be prioritized to ensure that the benefits of waste valorization extend to the workers and communities involved in olive cultivation and oil production. Labor conditions, worker safety, and equitable pay are critical aspects that need to be addressed to ensure fair treatment and well-being. Additionally, fostering community engagement and preserving cultural heritage in olive oil-producing regions will contribute to the long-term sustainability of the sector [71]. Collaborative efforts between government, industry, and academia can help establish best practices in social sustainability, ensuring that all stakeholders benefit from the transition to a circular economy. As the olive oil industry continues to grow, climate change mitigation will become an increasingly important focus. The implementation of LCA methodologies will play a key role in evaluating the environmental impact of olive oil production and waste valorization processes. By adopting a bioeconomy-centric approach, the industry can reduce its environmental footprint while creating new opportunities for renewable energy production and sustainable materials. The cultivation of drought-resistant olive varieties and the optimization of irrigation systems will also be crucial in adapting to the challenges posed by climate change.

The future of the olive oil industry lies in its ability to integrate circular economy principles, innovate waste valorization technologies, and ensure social and environmental sustainability. By harnessing the potential of olive oil by-products, the industry can reduce waste, generate renewable energy, and create new bioproducts, contributing to a more sustainable and resilient economy [71]. The adoption of circular economy models will not only benefit the olive oil sector but also promote the broader goals of environmental protection, economic growth, and social well-being. As such, a collaborative and multidisciplinary approach will be essential for realizing the full potential of the olive oil industry in the context of a circular economy.

11 Conclusion

The olive oil business faces substantial waste management issues, yet these wastes, including olive pomace and OMWW, have enormous potential for conversion into biofuels and bioproducts. Adopting waste-to-value systems is consistent with circular economy ideas, as it converts trash into renewable energy sources like biodiesel, bioethanol, and biogas, as well as bioproducts like organic fertilizers, biosurfactants, and biopolymers. This method decreases environmental effects, prevents resource depletion, and creates new market potential in pharmaceuticals, cosmetics, and nutraceuticals by extracting important components, such as polyphenols. While barriers to wider adoption include significant initial investment, logistical obstacles, and technological complications, the integration of sustainable practices and innovation can propel the industry toward economic and environmental sustainability.

References

1. Niaounakis M (2020) Olive mill waste: recent advances for sustainable management. Waste Manag 113:1–4
2. Roig A, Cayuela ML, Sánchez-Monedero MA (2021) An overview of the olive oil industry by-products. Bioresour Technol 99:5261–5279
3. Khdair A, Abu-Rumman G (2020) Sustainable environmental management and valorization options for olive mill byproducts in the Middle East and North Africa (MENA) region. PRO 8:671. https://doi.org/10.3390/pr8060671
4. Oliva A (2021) The history of the olive tree. https://artemoliva.com/the-history-of-the-olive-tree/
5. Barazani O, Dag A, Dunseth Z (2023) The history of olive cultivation in the southern Levant. Front Plant Sci 14:1131557. https://doi.org/10.3389/fpls.2023.1131557
6. Land Arch Concepts (2021) Olive distribution in the New World and beyond. https://landarchconcepts.wordpress.com/olive-distribution-in-the-new-world-and-beyond/
7. Green PS (2002) A revision of *Olea* L. (Oleaceae) and related genera in Africa. Kew Bull 57(1):91–140
8. North American Olive Oil Association. (2015). Olive oil production by country. Retrieved from https://www.aboutoliveoil.org/olive-oil-production-by-country
9. Best Diplomats (2024) Top 10 countries by olive oil production in 2024. Retrieved from https://bestdiplomats.org/countries-by-olive-oil-production/

10. European Commission (2022) Olive oil and table olives—Agri-food data portal. Retrieved from https://agridata.ec.europa.eu/extensions/DataPortal/olive-oil.html
11. The Times (2024) Why is olive oil so expensive? The liquid gold that took over the world. Retrieved from https://www.thetimes.co.uk/article/why-olive-oil-so-expensive-q5snc9jjk
12. Financial Times. (2024). Olive oil heartlands battle over the industry's future. Retrieved from https://www.ft.com/content/dddaa80e-7ef3-41b3-b9a7-c5b56770ee04
13. Martins S, Pereira S, Dinis L-T, Brito C (2024) Enhancing olive cultivation resilience: sustainable long-term and short-term adaptation strategies to alleviate climate change impacts. Horticulturae 10:1066. https://doi.org/10.3390/horticulturae10101066
14. Hamdi M, Frikha F, Ksibi M (2022) Olive oil wastewater: valorization technologies and future perspectives. Environ Technol Rev 11:75–90
15. Servili M, Montedoro G (2002) Contribution of phenolic compounds to virgin olive oil quality. Eur J Lipid Sci Technol 104:602–613
16. Tsimidou M, Papoti V, Karakostas KX (2020) Phenolic compounds in olive oil: role in quality and health. J Food Sci 85:3456–3464
17. Inglese P, Famiani F, Galvano F, Servili M (2017) Advances in olive oil extraction. J Agric Sci 9:78–85
18. Zhang J, Zhang W, Cheng HM, Silva SRP (2020) Critical review of recent progress of flexible perovskite solar cells. Mater Today 39:66–88
19. Azbar N, Bayram A, Filibeli A, Muezzinoglu A (2020) A review of olive mill wastewater management. Bioresour Technol 136:80–88
20. Galanakis CM (2012) Recovery of high added-value components from food wastes: conventional, emerging technologies and commercialized applications. Trends Food Sci Technol 26:68–87
21. Morillo JA, Antizar-Ladislao B, Monteoliva-Sánchez M, Ramos-Cormenzana A, Russell NJ (2009) Bioremediation and biovalorisation of olive-mill wastes. Appl Microbiol Biotechnol 82:25–39
22. García-Ruiz, Ochoa M, Hinojosa M, Gómez-Muñoz (2012) Improved soil quality after 16 years of olive. Agron Sustain Dev 32:803–810
23. Kalogeropoulos N, Sergent M, Kanellopoulos N (2020) The use of olive oil mill wastewater and other olive by-products in animal feed. Waste Manag Res 38:601–610
24. Kailis SG, Harris M (2020) Olive by-products as feed additives: nutritional and health benefits. Olives and olive oil in health and disease prevention, 1: 45–59
25. Galanakis CM, Tornberg E, Gekas V (2010) Clarification of high-added value products from olive mill wastewater. J Food Eng 99:190–197
26. Bampidis VA, Robinson PH (2006) Citrus by-products as ruminant feeds: a review. Anim Feed Sci Technol 128:175–217
27. Billaud F, Gornay J, Coniglio L (2007) Pyrolysis of secondary raw material from used frying oils. https://doi.org/10.48550/arXiv.0709.4340
28. Otero P, Garcia-Oliveira P, Carpena M, Barral-Martinez M, Chamorro F, Echave J et al (2021) Applications of by-products from the olive oil processing: revalorization strategies based on target molecules and green extraction technologies. Trends Food Sci Technol 116:1084–1104
29. La Cara F, Ionata E, Del Monaco G, Marcolongo L, Gonçalves MR, Marques IP (2012) Olive mill wastewater anaerobically digested: phenolic compounds with antiradical activity. Chem Eng Trans 27. https://doi.org/10.3303/CET1227055
30. Sharma V, Tsai M-L, Nargotra P, Chen C-W, Kuo C-H, Sun P-P, Dong C-D (2022) Agro-industrial food waste as a low-cost substrate for sustainable production of industrial enzymes: a critical review. Catalysts 12:1373. https://doi.org/10.3390/catal12111373
31. Di Giacomo G, Romano P (2022) Evolution of the olive oil industry along the entire production chain and related waste management. Energies 15(2):465
32. ElMekawy A, Diels L, Bertin L, De Wever H, Pant D (2014) Potential biovalorization techniques for olive mill biorefinery wastewater. Biofuels Bioprod Biorefin 8:283–293

33. Gautam K, Vishvakarma R, Sharma P, Singh A, Gaur VK, Varjani S, Srivastava JK (2022) Production of biopolymers from food waste: constraints and perspectives. Bioresour Technol 361:127650. https://doi.org/10.1016/j.biortech.2022.127650
34. Mouro C, Gomes AP, Gouveia IC (2024) Microbial exopolysaccharides: structure, diversity, applications, and future frontiers in sustainable functional materials. Polysaccharides 5:241–287
35. Petri DF (2015) Xanthan gum: a versatile biopolymer for biomedical and technological applications. J Appl Polym Sci 132:803–810
36. Demichelis F, Lenzuni M, Converti A, Del Borghi A, Freyria FS, Gagliano E et al (2025) Agro-food waste conversion into valuable products in the Italian scenario: current practices and innovative approaches. J Environ Chem Eng 115458. https://doi.org/10.1016/j.jece.2025.115458
37. Ronca CL, Marques SS, Ritieni A, Giménez-Martínez R, Barreiros L, Segundo MA (2024) Olive oil waste as a source of functional food ingredients: assessing polyphenolic content and antioxidant activity in olive leaves. Food Secur 13:189. https://doi.org/10.3390/foods13020189
38. Selim S, Albqmi M, Al-Sanea MM, Alnusaire TS, Almuhayawi MS, AbdElgawad H, El-Saadony MT (2022) Valorizing the usage of olive leaves, bioactive compounds, biological activities, and food applications: a comprehensive review. Front Nutr 9:1008349. https://doi.org/10.3389/fnut.2022.1008349
39. Muíño I, Díaz MT, Apeleo E, Pérez-Santaescolástica C, Rivas-Cañedo A, Pérez C et al (2017) Valorisation of an extract from olive oil waste as a natural antioxidant for reducing meat waste resulting from oxidative processes. J Clean Prod 140:924–932
40. Soberón-Chávez G, Maier RM (2015) Biosurfactants: from microorganisms to industrial applications. Springer
41. Wei Y, Zhan Y (2018) A review of biosurfactant production and its applications in waste valorization and environmental remediation. Environ Sci Pollut Res 25:15071–15087
42. Fotiadou R, Fragkaki I, Pettas K, Stamatis H (2024) Valorization of olive pomace using ultrasound-assisted extraction for application in active packaging films. Int J Mol Sci 25:6541. https://doi.org/10.3390/ijms25126541
43. Fernandes MC, Torrado I, Carvalheiro F, Dores V, Guerra V, Lourenço PM, Duarte LC (2016) Bioethanol production from extracted olive pomace: dilute acid hydrolysis. Bioethanol 2. https://doi.org/10.1515/bioeth-2016-0007
44. Moreno JA, Soto M, García ML (2019) Biofuels from olive oil production by-products: technologies, economic potential, and environmental impact. Renew Sust Energ Rev 99:400–413
45. Rajaeifar MA, Akram A, Ghobadian B, Rafiee S, Heijungs R, Tabatabaei M (2016) Environmental impact assessment of olive pomace oil biodiesel production and consumption: a comparative lifecycle assessment. Energy 106:87–102
46. Alvarez Serafini MS, Tonetto GM (2019) Production of fatty acid methyl esters from an olive oil industry waste. Braz J Chem Eng 36:285–297
47. Che F, Sarantopoulos I, Tsoutsos T, Gekas V (2012) Exploring a promising feedstock for biodiesel production in Mediterranean countries: a study on free fatty acid esterification of olive pomace oil. Biomass Bioenergy 36:427–431
48. Amin S (2016) Biofuels: biodiesel production from olive oil waste. Energy Environ 27:42–54
49. Uddin MA, Siddiki SYA, Ahmed SF, Rony ZI, Chowdhury MAK, Mofijur M (2021) Estimation of sustainable bioenergy production from olive mill solid waste. Energies 14:7654. https://doi.org/10.3390/en14227654
50. Amor SR, Gharbi S, Souilem S (2017) Bioethanol production from olive mill wastewater: opportunities and challenges. J Clean Prod 156:99–110
51. Dantas Palmeira J, Araújo D, Mota C, Alves RCPP, Oliveira MB, Ferreira HMN (2023) Fermentation as a strategy to valorize olive pomace, a by-product of the olive oil industry. Fermentation 9:442. https://doi.org/10.3390/fermentation9050442
52. Abdalla M, Ahmed W, Khan S (2020) Economic feasibility of biofuel production from olive oil wastes. Renew Energy 157:1346–1356

53. Tasionas A, Dounavis AS (2019) Techno-economic analysis of the olive oil mills waste valorisation for energy production: a case study of Corfu. Environ Res Eng Manag 75:18–29
54. Mouzakitis Y, Adamides ED (2022) Techno-economic assessment of an olive mill wastewater biorefinery in the context of circular bioeconomy. Eng 3:488–503
55. Li S, Xu YJ, Ni M (2021) Changes in sediment, nutrients and major ions in the world largest reservoir: effects of damming and reservoir operation. J Clean Prod 318:128601. https://doi.org/10.1016/j.jclepro.2021.128601
56. Ryberg MW, Bjerre TK, Nielsen PH, Hauschild M (2021) Absolute environmental sustainability assessment of a Danish utility company relative to the planetary boundaries. J Ind Ecol 25:765–777
57. Abbattista R, Ventura G, Calvano CD, Cataldi TRI, Losito I (2021) Bioactive compounds in waste by-products from olive oil production: applications and structural characterization by mass spectrometry techniques. Food Secur 10:1236. https://doi.org/10.3390/foods10061236
58. Hansen C (2022) Environmental impact of olive oil processing. *EcoMENA*. Retrieved from https://www.ecomena.org/olive-oil-wastes/
59. Tsendsuren C, Yadav PL, Han SH, Mun S (2021) The effect of corporate environmental responsibility and religiosity on corporate cash holding decisions and profitability: evidence from the United States' policies for sustainable development. Sustain Dev 29:987–1000
60. Espadas-Aldana G, Vialle C, Belaud JP et al (2019) Analysis and trends for life cycle assessment of olive oil production. Sustain Prod Consump 19:216–230
61. Kühnen M, Hahn R (2017) Indicators in social life cycle assessment: a review of frameworks, theories, and empirical experience. J Ind Ecol 21:1547–1565
62. Wei J, Cui J, Xu Y et al (2022) Social life cycle assessment of major staple grain crops in China. Agriculture 12:535. https://doi.org/10.3390/agriculture12040535
63. Rajaeifar MA, Akram A, Ghobadian B, Rafiee S, Heidari MD (2014) Energy-economic life cycle assessment (LCA) and greenhouse gas emissions analysis of olive oil production in Iran. Energy 66:139–149
64. Khounani Z, Hosseinzadeh-Bandbafha H, Moustakas K, Talebi AF, Goli SAH, Rajaeifar MA, Lam SS (2021) Environmental life cycle assessment of different biorefinery platforms valorizing olive wastes to biofuel, phosphate salts, natural antioxidant, and an oxygenated fuel additive (triacetin). J Clean Prod 278:123916. https://doi.org/10.1016/j.jclepro.2020.123916
65. Salikandi M, Salehi S, Ashoori S, Soleimani M (2021) Techno-economic study of bioethanol production from olive wastes cake (OWC) in Iran. Clean Techn Environ Policy 23:1851–1862
66. Seydanlou P, Jolai F, Tavakkoli-Moghaddam R, Fathollahi-Fard AM (2022) A multi-objective optimization framework for a sustainable closed-loop supply chain network in the olive industry: hybrid meta-heuristic algorithms. Expert Syst Appl 203:117566. https://doi.org/10.1016/j.eswa.2022.117566
67. Mota B, Gomes MI, Carvalho A, Barbosa-Povoa AP (2015) Towards supply chain sustainability: economic, environmental and social design and planning. J Clean Prod 105:14–27
68. Busset G, Belaud JP, Montréjaud-Vignoles M, Sablayrolles C (2014) Integration of social LCA into sustainability assessment: limitations and challenges. Int J Life Cycle Assess 19:1506–1515
69. Vicario-Modroño V, Gallardo-Cobos R, Sánchez-Zamora P (2023) Sustainability evaluation of olive oil mills in Andalusia (Spain): a study based on composite indicators. Environ Dev Sustain 25:6363–6392
70. Donner M, Radić I, Erraach Y, El Hadad-Gauthier F (2022) Implementation of circular business models for olive oil waste and by-product valorization. Resources 11(7):68. https://doi.org/10.3390/resources11070068
71. Angeloni G, Spadi A, Corti F, Calcaprina M, Carpi G, Maioli F, Parenti A, Masella P (2024) Advancing circular economy in olive oil production: comparing maturation Systems for Vermicompost Creation from olive pomace. Biomass 4(4):1178–1190

Chapter 16
Valorization of Olive Mill Wastewater: A Review

Manal Haj-Zaroubi and Hassan Azaizeh

Abbreviations

AAAs	Alkyl aromatic alcohols
AD	Anaerobic digestion
BOD	Biological oxygen demand
COD	Chemical oxygen demand
CW	Critical wavelength
DPPH	2,2-Diphenyl-1-picrylhydrazyl
EC	Electrical conductivity
EC50	Half maximal effective concentration
EU	European Union
GAE	Gallic acid
HPLC-PAD	High-performance liquid chromatography with photodiode array detection
IR	Infrared
MBC	Minimum bactericidal concentration
MED	Minimal erythema dose
MF	Microfiltration
MFC	Minimum fungicidal concentration
MIC	Minimum inhibitory concentrations

M. Haj-Zaroubi (✉)
The Institute of Applied Research, The Galilee Society, Shefa-Amr, Israel

Department of Natural Resources and Environmental Management, Faculty of Management, University of Haifa, Haifa, Israel
e-mail: manalhz@gal-soc.org

H. Azaizeh
Department of Environmental Sciences, Biotechnology, and Water Sciences, Tel Hai College, Upper Galilee, Israel
e-mail: hassn@telhai.ac.il

C. Trois, R. C. Ray (eds.), *Plantation Crop Wastes: Valorization for Economic Sustainability*, Waste as a Resource,
https://doi.org/10.1007/978-3-031-96805-1_16

NF	Nanofiltration
OMSW	Olive mill solid waste
OMWW	Olive mill wastewater
ORAC	Oxygen radical absorbance capacity
PUFAs	Polyunsaturated fatty acids
RO	Reverse osmosis
ROS	Reactive oxygen species
SPF	Sun protection factor
UASB	Using up-flow anaerobic sludge blanket
UF	Ultrafiltration
UV	Ultraviolet
WWTPs	Wastewater treatment plants

1 Introduction

Olive cultivation covers approximately 10.8 million hectares worldwide, with over 98% of this area concentrated in the Mediterranean region, which has historically dominated global olive oil production and consumption. Annually, more than three million tons of olive oil is produced globally, with more than two million tons originating from the European Union (EU), where the EU is considered the leading producer, exporter, and consumer of olive oil worldwide [1, 2]. Almost 99% of the EU's olive oil is produced in Spain, Italy, Greece, and Portugal [1, 3]. In recent decades, both production and consumption have grown substantially, where olive oil, one of the most ancient food products, has played an important role in the culture and economy of Mediterranean populations. Spain leads global production, contributing approximately 826,000 tons annually (over 52% of global production), followed by Italy (accounting for 33% of the European Union's production), Greece, and Portugal. Collectively, these countries produce up to 2.5×10^6 tons of olive oil annually [4, 5].

The olive oil industry is one of the most economically significant agro-food sectors in the Mediterranean and Middle Eastern countries. However, olive oil production generates substantial quantities of unexploited agronomic residues, primarily in the form of olive mill wastewater (OMWW) and olive mill solid waste (OMSW). These by-products pose severe environmental challenges, particularly in regions where olive cultivation and processing are intensive. Annually, as much as 30 million m^3 of OMWW is generated during the brief olive harvesting and oil extraction season, which typically spans 2–3 months [6, 7].

The olive fruit comprises pulp (75–85% by weight), stones (13–23% by weight), and seeds (2–3% by weight). Phenolic compounds, such as hydroxytyrosol and tyrosol, present in olive oil, contribute to its bitter taste, astringency, and moderate resistance to biodegradation. Currently, three primary methods are employed for olive oil extraction: (a) the traditional press process, (b) the two-phase decanter process, and (c) the three-phase decanter process (Figs. 16.1 and 16.2) [8]. The two-phase and three-phase systems are the most widely used today.

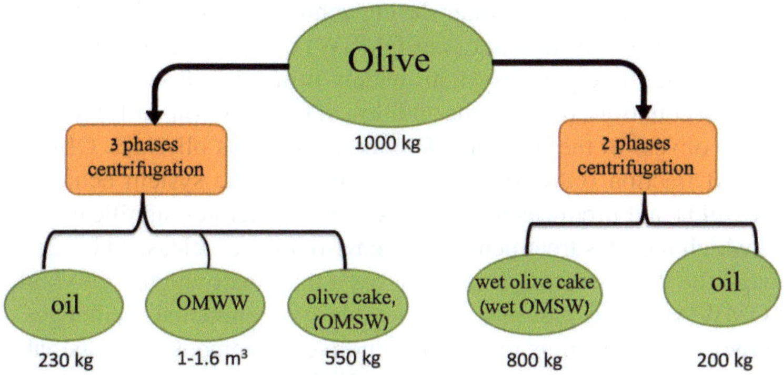

Fig. 16.1 Olive oil extraction systems

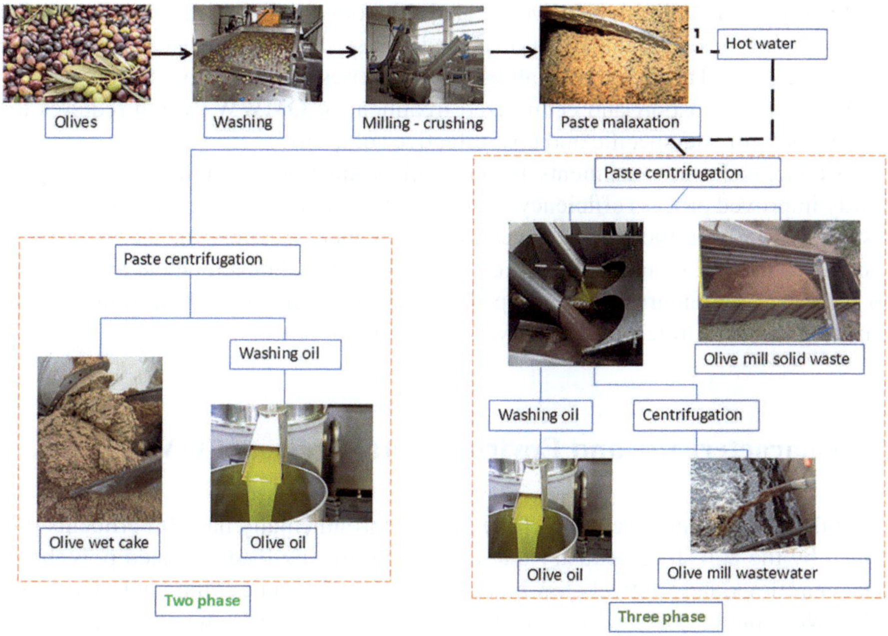

Fig. 16.2 Olive oil extraction process using the two-phase and three-phase decanter processes, based on [8]

In the traditional press process, olives are washed, crushed, and malaxed, with a small amount of water added to facilitate oil separation. The resulting paste is pressed to extract the oil and liquid waste, which comprises olive juice, added water, and residual oil. The oil is subsequently separated from the water via vertical centrifugation or decanting [9]. This method is primarily used in small-scale mills, while larger mills have adopted two-phase and three-phase systems. The two-phase system, predominantly used in Spain, relies on the natural moisture content of olives (except for a small addition of water during centrifugation) and produces olive oil alongside a semi-solid by-product (wet OMSW) comprising both liquid and solid residues [10]. The three-phase system, widely utilized in Italy, Greece, and other Mediterranean countries, requires the addition of hot water (up to 1.3 m^3 per 1000 kg of processed olives) to the decanter. This process yields olive oil, OMWW, and a residual solid fraction (OMSW). While the three-phase system achieves slightly higher oil yields and produces less solid waste, it generates significant volumes of OMWW, which requires treatment before environmental release [11, 12].

Although the three-phase system achieves a marginally higher oil yield, its substantial OMWW output presents considerable environmental challenges. In contrast, the two-phase system is regarded as more environmentally sustainable due to its reduced water consumption and lower OMWW production. It is well known that approximately 98% of the polyphenols and natural antioxidants extracted from olives during the oil extraction process are retained in the OMWW and OMSW, while only 2% remain in the olive oil [13]. The olive milling season is short, with tens of thousands of small mills dispersed across the Mediterranean and Middle Eastern regions. This geographical distribution poses logistical and economic challenges for the transport and centralized treatment of OMWW, which requires the urgent need for site-specific and cost-effective solutions.

In conclusion, advancements in olive oil production technology have significantly improved process efficiency. However, challenges in waste management persist, particularly in the Mediterranean region, which accounts for the majority of global production. Continued research and technological development are essential to minimize the environmental impact of OMWW and other by-products while ensuring the economic sustainability of the olive oil industry.

2 Characteristics and Environmental Impact of OMWW

The OMWW is considered one of the most environmentally harmful effluents produced by the agro-food industry due to its high organic load, elevated polyphenol concentrations, and relatively low biodegradability of organic compounds. Untreated OMWW exhibits extreme biological oxygen demand (BOD) and chemical oxygen demand (COD), reaching levels as high as 100 and 200 g/L, respectively. This effluent is acidic (pH $\leq$5), exhibits high electrical conductivity (EC: 7–11 dS/m), and contains significant ionic content, predominantly potassium. It is further characterized by its high phenolic content (3–15 g/L), strong odor, and distinct colorations [7, 8]. Due to its high resistance to biodegradation, OMWW exhibits potential toxicity to both plants and microorganisms. These specific properties require proper treatment of OMWW prior to its release into the environment, as it cannot be directly discharged into environmental compartments such as soil, water bodies, or municipal wastewater treatment plants (WWTPs). Untreated land disposal poses significant risks of soil and water contamination, threatening water resource quality in arid Mediterranean regions and representing a major environmental concern. Additionally, in European countries, the cost of OMWW land disposal accounts for approximately 5–10% of the olive oil industry's turnover, reaching as high as

15–20% in certain areas [14]. The direct application of OMWW to agricultural soils has detrimental effects, primarily due to its high mineral salt content, low pH, and significant concentrations of polyphenols and tannins. OMWW poses a severe environmental threat, particularly in Mediterranean regions, due to its phytotoxic effects on plants, soils, microbial communities, and aquatic ecosystems [15]. The main phenolic compounds present in OMWW include hydroxytyrosol, tyrosol, verbascoside, and cinnamic acids, such as caffeic, gallic, vanillic, and syringic acids (Fig. 16.3), which are responsible for phytotoxicity [16]. Olive mills commonly dispose of OMWW in settling ponds, which are often inadequately sized and prone to overloading, leading to spillage into adjacent valleys [17, 18]. Although OMWW

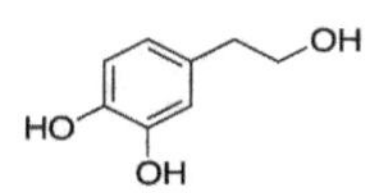

Hydroxytyrosol (HT)

1

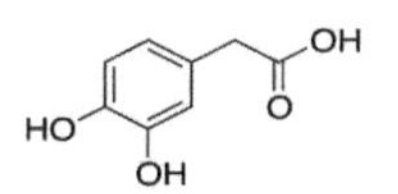

3,4-dihydroxyphenylacetic acid

2

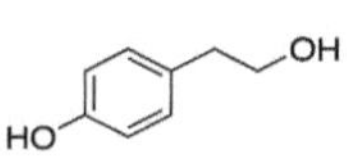

Tyrosol

3

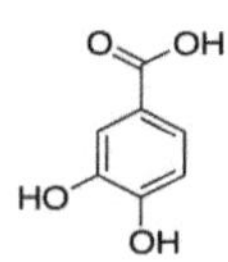

Protocatechuic acid

4

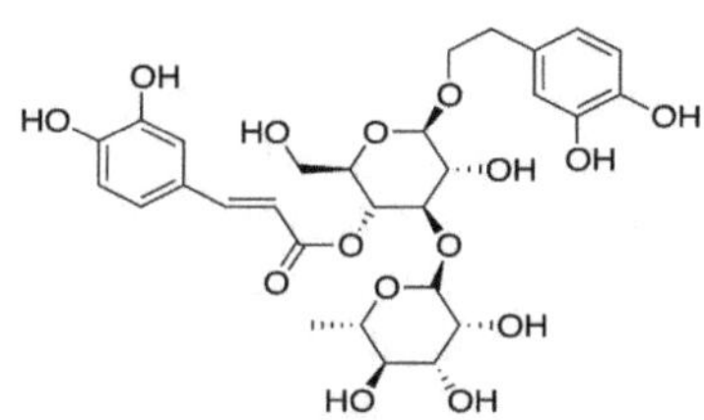

Verbascoside

5

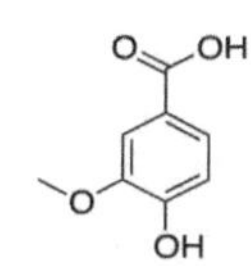

Vanilic acid

6

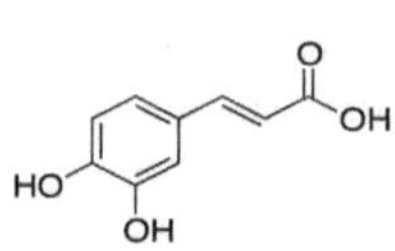

Caffeic acid

7

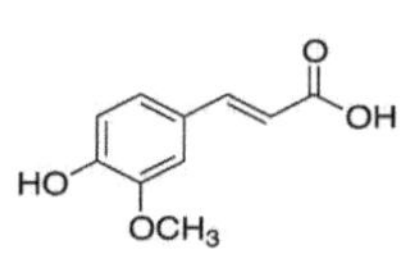

Ferulic acid

8

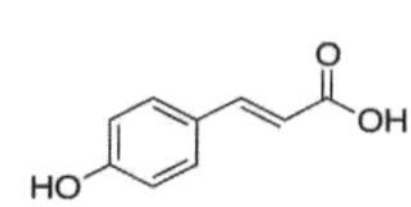

p-coumaric acid

9

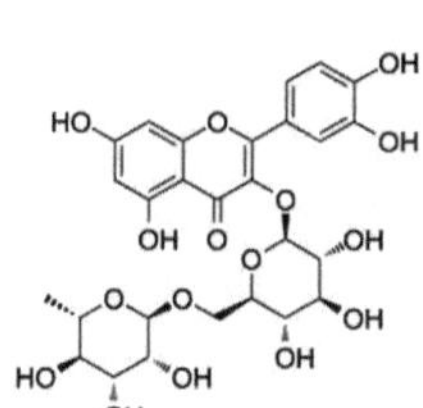

Rutin

10

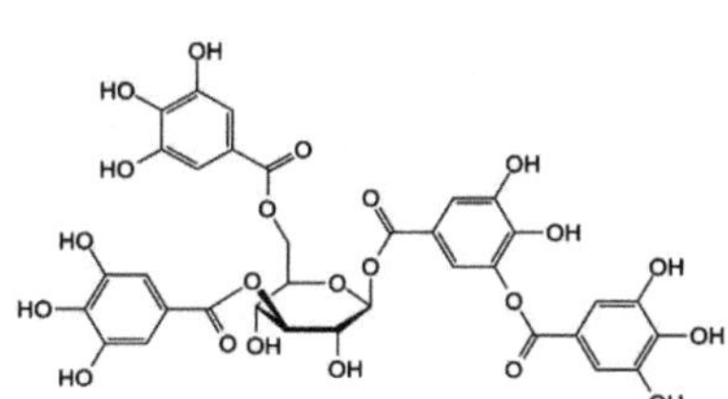

Tannins

11

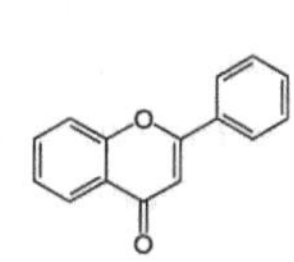

Flavonoid

12

Fig. 16.3 Chemical structures of selected phenolic compounds present in OMWW based on [21]

is produced seasonally and in moderate quantities relative to other agricultural wastes, its environmental impact is disproportionately significant. The pollution potential of 1 m^3 of OMWW has been estimated to be equivalent to that of 200 m^3 of domestic sewage [19]. The unregulated discharge of OMWW into soils significantly reduces water retention and infiltration rates, increases soil hydrophobicity, and causes pronounced phytotoxic and antimicrobial effects. Additionally, it alters soil properties by affecting acidity, salinity, nitrogen immobilization, and nutrient dynamics. Other adverse effects include the accumulation of organic acids, lipids, naturally occurring phenols, and disruptions to microbial activity [19]. OMWW also contributes to odor pollution and can be detrimental to plant health if applied at high concentrations.

The absence of proper treatment facilities at individual mills has necessitated extensive research into methods for minimizing, preventing, and treating OMWW to mitigate its environmental impact [20].

3 Various Methods for Treating OMWW

The COD and BOD values of OMWW can reach as high as 220 and 100 g/L, respectively. These elevated levels of COD and BOD reflect the high concentrations of organic pollutants in the OMWW, underscoring its significant environmental impact. Furthermore, OMWW is recognized as one of the most challenging industrial effluents due to its substantial organic load, which includes large quantities of (phyto)toxic compounds, such as phenolic substances [22]. Unregulated disposal of OMWW leads to severe environmental consequences, including contamination of groundwater, surface water bodies, coastal areas, and marine ecosystems [23]. The environmental burden is further amplified by the vast quantities of OMWW generated by olive oil production in Mediterranean countries, which contributes to a pressing pollution problem affecting the soil-air-water continuum and adversely impacting organisms, including plants, animals, aquatic species, and microorganisms. Therefore, effective treatment strategies are essential to mitigate the detrimental effects of OMWW before it is discharged into the environment or transferred to WWTPs [24, 25]. Numerous studies have assessed the efficacy of various treatment approaches to develop environmentally sustainable and economically feasible solutions for OMWW management [24]. As a result, a variety of technologies have been proposed and developed to mitigate the environmental impact of OMWW. These methods are broadly categorized into physical, chemical, and biological processes, with some employing integrated approaches that combine multiple treatment techniques to achieve optimal pollutant reduction. The goal is to ensure that the OMWW meets the acceptable standards for environmental discharge or further treatment in WWTPs [24].

To address the environmental challenges posed by OMWW, several physicochemical, biological, and hybrid processes have been explored. Among the promising technologies for OMWW treatment is membrane filtration, which also facilitates the concentration of its phenolic compounds [26]. Research on membrane-based processes for OMWW treatment has predominantly focused on pressure-driven techniques, such as microfiltration (MF), ultrafiltration (UF), reverse osmosis (RO), and nanofiltration (NF). Recently, emerging membrane technologies, like membrane distillation, have gained attention [26]. An innovative approach reported by Bottino et al. [27] integrates an MF stage followed by RO for OMWW treatment. To further concentrate the RO permeate and its polyphenolic content, vacuum membrane distillation using hydrophobic hollow fiber membranes was employed.

Notably, over the past few decades, significant attention has been directed toward the recovery and fractionation of polyphenolic compounds from OMWW as part of circular economy initiatives. This valorization strategy aims to repurpose OMWW by-products, adding economic and environmental value to the treatment process [28–30].

Anaerobic digestion (AD) of waste materials has garnered significant attention due to its ability to convert organic compounds into methane, an economically valuable biogas that can be used for heat and electricity production. The high organic load of OMWW makes AD an attractive treatment option. However, achieving high AD efficiency for OMWW has been challenging due to several factors: (a) its extremely high COD and low BOD [31, 32], (b) high acidity, (c) low nitrogen content, and (d) the presence of compounds such as phenols, lipids, and long-chain fatty acids [8, 31, 32], many of which are inhibitory to the sludge bacteria (primarily methanogens) responsible for the biodegradation process [31–33].

To address these limitations, anaerobic co-digestion has been proposed as a potential solution, wherein OMWW is combined with another type of organic waste—in this case, swine waste—in a single bioreactor for biogas production [34]. Using an up-flow anaerobic sludge blanket (UASB) continuous reactors over 170 days of operation, it was demonstrated that seeding the reactor with different types of sludge enhanced the biodegradation potential. The highest COD removal efficiency (85–95%) and biogas production (0.55 L/g COD) were achieved using a mixture containing 33% OMWW and 67% swine waste. Under these conditions, the initial organic load of 28,000 mg/L COD was reduced to 1500–3500 mg/L [34]. These findings underscore the potential of UASB reactors as a reliable and promising technology for OMWW treatment and biogas production, despite the requirement for long retention times.

However, it is important to note that no single treatment method is universally applicable to all types of OMWW, as their chemical and physical properties vary significantly depending on the region of origin. While OMWW is currently classified as waste, it holds the potential to become a valuable resource for energy production and high-value-added natural products. Based on the reviewed studies, it can be concluded that no ideal or economically feasible solution for OMWW treatment exists at present. As such, multiple environmentally friendly and economically viable strategies should be explored to extract valuable products from OMWW and address their environmental impact.

4 Valorization of OMWW: From Pollutant to High-Value Products

OMWW has long been regarded as a hazardous waste due to its high COD and negative environmental impacts, which also pose an economic burden on the olive oil industry. However, this perception is evolving, as recent research highlights its potential as a low-cost starting material rich in bioactive compounds, particularly phenolic compounds. These phenolics can be extracted and utilized as natural antioxidants in the food and pharmaceutical industries, thereby transforming OMWW into a valuable resource [35–40].

The production of biologically active compounds from OMWW offers a promising alternative for adding value to this problematic waste. Recognizing the antioxidant activity of OMWW [35, 41, 42] and the role of oxidative stress in various diseases, OMWW has been identified as a potential source of antioxidants derived from olive processing by-products. In this context, the EU initiated a project focused on the extraction of phenolic compounds from OMWW for use as antioxidant products [38–43].

There is increasing evidence that free radical-mediated processes are implicated in the pathogenesis of numerous human diseases. Free radicals attack biomolecules such as lipids, proteins, DNA, and cellular membranes, contributing to oxidative damage in biological systems and the degradation of food, animal feed, and cosmetics. Reactive oxygen species (ROS), which are by-products of normal aerobic respiration, can inflict significant damage to biological molecules. Diets rich in fruits and vegetables, known for their abundant antioxidant content, have been shown to counteract these effects [44]. Key antioxidants in foods include vitamins C and E, polyphenols, and carotenoids, which are believed to contribute substantially to antioxidant activity [38–40, 44–46].

Phenolic compounds, in particular, are thought to account for a significant portion of the antioxidant capacity in plants and OMWW [8, 47]. The growing scientific interest in OMWW as a natural source of bioactive compounds is largely driven by the biological activities associated with its phenolic fraction, positioning it as a promising candidate for innovative applications in health and industry.

4.1 Antioxidant Activity

Olive oil, derived from the olive tree (*Olea europaea* L.), represents one of the oldest domesticated plant species and ranks as the second most significant fruit crop cultivated globally [48]. Renowned for its integral role in the Mediterranean diet, olive oil is well known for its nutritional richness and numerous health benefits, establishing it as a cornerstone of this dietary tradition.

Polyphenolic compounds are well-documented for their extensive health-promoting properties, including antioxidant, cardioprotective, anticancer, anti-inflammatory, and antimicrobial activities [49]. These bioactive compounds are widely recognized for their relevance in pharmaceutical and nutraceutical applications [50]. Olive-derived polyphenols, such as hydroxytyrosol, tyrosol, oleuropein, and verbascoside, have been the focus of extensive pharmacological research. These compounds exhibit a broad spectrum of biological activities, encompassing antioxidant, anti-inflammatory, cardiovascular, immunomodulatory, gastrointestinal, respiratory, autonomic, central nervous system, antimicrobial, anticancer, and chemopreventive effects [51]. Their health-promoting attributes, particularly within the context of the Mediterranean diet, have been well-characterized, with particular attention focused on their molecular mechanisms of action [52]. Furthermore, their potential in the prevention and management of chronic diseases, including cardiovascular disorders, diabetes, neurodegenerative diseases, and cancer, has been extensively highlighted [53]. Polyphenols and tocopherols, present in both dietary sources and plants, provide protective effects against oxidative stress. The underlying mechanisms of these compounds, particularly in preventing cardiovascular diseases and cancer, are believed to involve direct or indirect antioxidant activities [41].

OMWW, a by-product of olive oil production, contains significant concentrations of phenolic compounds with potent antioxidant activity, making them valuable for industrial applications as food additives and pharmaceutical agent. Using fractionation techniques, five distinct OMWW fractions were isolated, and their antioxidant potential was evaluated through three methodologies: oxygen radical absorbance capacity (ORAC), 2,2-diphenyl-1-picrylhydrazyl (DPPH) assay, and β-carotene bleaching assay [35]. The total phenolic content of OMWW ranged from 115 to 170 mg/L, and high-performance liquid chromatography with photodiode array detection (HPLC-PAD) identified the major phenolic constituents as hydroxytyrosol, tyrosol, caffeic acid, vanillic acid, verbascoside, oleuropein, ferulic acid, and *p*-coumaric acid. The antioxidant capacities of the five OMWW fractions varied depending on the assay employed. In the DPPH assay, the fraction containing alkyl aromatic alcohols (AAAs) demonstrated the highest antioxidant activity, with an EC50 value of 20 mg/L, while pure hydroxytyrosol exhibited an EC50 of 2 mg/L. Similarly, the ORAC assay revealed that the AAA fraction and semi-hydrolyzed total phenol fraction exhibited significantly greater antioxidant activity compared to the positive control Trolox at a concentration of 20 mg/L [35].

Fractions obtained from Italian OMWW have been tested for their antioxidant properties (DPPH assay) and polyphenol content (Folin-Ciocalteu assay) and characterized via infrared (IR) spectroscopy. These analyses revealed distinct antioxidant activities and varying total polyphenol contents, which correlated with differences in qualitative composition, as highlighted by the IR data [54]. The tested fractions demonstrated significant antioxidant activity (10–80%) and total phenolic content ranging from 1.5 to 15 g GAE/L. Notably, antioxidant properties are primarily attributed to the total phenol content and specific phenolic compounds, such as hydroxytyrosol [54]. These phenolic and non-phenolic compounds, along with hydroxytyrosol, represent key factors for producing distinct antioxidant fractions from OMWW for diverse applications. Investigating the synergistic effects between phenolic compounds derived from different olive cultivars remains an important area for further research.

Although polyphenols exhibit significant potential as therapeutic agents, their clinical applications are hindered by limitations such as poor stability, low bioavailability, and limited membrane permeability [55]. The challenge of low oral bioavailability remains a critical barrier, necessitating the re-evaluation of polyphenol bioactivity based on its metabolites prior to clinical use [56]. To address these issues, various studies have investigated nano-delivery systems and encapsulation technologies aimed at enhancing the bioavailability of polyphenols [57–59]. Furthermore, synergistic interactions among different polyphenols have been proposed as a means to improve their oral bioavailability [35, 60].

4.2 Antimicrobial Activities

Antibiotics derived from biotechnological processes are the primary treatment for bacterial infections; however, their overuse has led to significant issues, including the emergence of antibiotic-resistant strains, the persistence of drug residues, and associated environmental concerns. Thus, identifying alternative antimicrobial agents is imperative. Polyphenolic compounds recovered from OMWW demonstrate antimicrobial activity, with potential applications in agriculture to combat phytopathogens and as novel therapeutic agents against antibiotic-resistant human pathogens [61, 62]. Specific compounds, such as hydroxytyrosol [35, 63], oleuropein [64], and other phenolic acids (e.g., 4-hydroxybenzoic acid, vanillic acid, and p-coumaric acid) [35, 65], have demonstrated antibacterial activity. For example, hydroxytyrosol has shown in vitro activity against both Gram-positive and Gram-negative bacteria responsible for respiratory and gastrointestinal infections [66]. Additionally, OMWW application to soil has been shown to suppress soil-borne diseases caused by *Rhizoctonia solani* and *Fusarium solani* [67].

The extraction of bioactive compounds, particularly tyrosol, hydroxytyrosol, aromatic acids, and conjugated aromatic acids, from OMWW is of great interest due to its antioxidant and antimicrobial properties [35, 68, 69]. Bacterial infections in aquaculture, caused by various pathogens, pose significant threats to fish health and productivity, leading to substantial economic losses [70, 71]. Furthermore, contamination of fish and shellfish by food-poisoning bacterial strains poses serious health risks, contributing to foodborne diseases and infection outbreaks [72, 73]. The widespread use of antibiotics in aquaculture has resulted in antimicrobial resistance [74], environmental damage [75], and the presence of drug residues in fish fillets, thereby increasing health risks to humans [76, 77].

Research has shown that OMWW extracts exhibit antimicrobial effects against a range of bacterial isolates and yeasts, such as *Candida albicans* [78, 79]. For instance, OMWW extracts demonstrated potent inhibitory effects against *Pseudomonas fluorescens* isolated from mozzarella cheese [80]. Additionally, OMWW extracts have been shown to enhance the nutritional quality of fish fillets, support antioxidant defense mechanisms, and improve growth performance and immune responses in fish [81, 82].

A specific OMWW fraction, termed "AntiSolvent," predominantly consisting of hydroxytyrosol (10.3%), verbascoside (7.4%), and tyrosol (2.6%), has demonstrated antimicrobial activity. In vitro evaluations of antisolvent/gallic acid combinations revealed strong antimicrobial potential against Gram-positive (*Streptococcus pyogenes*, *Staphylococcus aureus*) and Gram-negative (*Escherichia coli*, *Klebsiella pneumoniae*) bacteria, with minimal inhibitory concentrations (MIC) of 50–100 μg/mL leading to complete growth inhibition [83]. These findings highlight the potential of OMWW-derived fractions, augmented with natural phenolic compounds, as bioactive mixtures for controlling pathogenic bacteria.

Furthermore, an ethanolic extract of OMWW exhibited significant antimicrobial activity against 38 bacterial strains, including fish-associated pathogens and non-pathogenic isolates, as well as against the yeast *Candida albicans*. Disk diffusion, minimum inhibitory concentration (MIC), and minimum bactericidal/fungicidal concentration (MBC/MFC) assays have revealed strong inhibitory effects against *Shewanella baltica* and moderate effects on *Plesiomonas shigelloides* and *Vibrio anguillarum* [84]. MIC and MBC values for *S. baltica*, *Vibrio gigantis* strain C24, and *V. anguillarum* strain SY-L24 were determined to be 500 μg/mL. These results suggest that OMWW extracts can serve as eco-friendly additives for disease prevention and microbial control in aquaculture while also improving food safety by reducing microbial contamination [84]. The varying efficacies of OMWW extracts across studies emphasize the need for further exploration to optimize their use as antimicrobial and antioxidant agents.

5 Impact of OMWW Polyphenols on Animal Health

In the Mediterranean region, where dairy sheep and goats are prevalent, the inclusion of natural products in the diet of small ruminants has been a long-standing tradition. Currently, the utilization of locally available industrial by-products, such as grape, olive, and tomato residues, represents a promising strategy to mitigate competition with edible human foods. Additionally, these co-feeds contribute to reducing the ecological and water footprints associated with conventional crop cultivation. The presence of bioactive compounds, particularly polyphenols, adds significant value to these by-products. Numerous studies have demonstrated the positive effects of incorporating these materials into the diets of small dairy ruminants, particularly on ruminal metabolism, animal health, and the quality of the derived products [85–87].

Given the widespread occurrence of oxidative stress and inflammation in livestock, polyphenols are increasingly recognized as promising feed additives in the nutrition of small dairy ruminants. The use of natural materials rich in phenolic compounds, such as OMWW, as sources of bioactive ingredients has garnered significant attention in the animal production sector [86, 88]. Research has shown that moderate concentrations of dietary polyphenols can improve animal performance and health by enhancing dietary protein utilization and improving feed quality [85, 87].

Dietary polyphenols have also been shown to modulate the rumen biohydrogenation of polyunsaturated fatty acids (PUFAs) [89], thereby improving the lipid profile of dairy products. This includes increasing the concentrations of beneficial fatty acids, such as PUFAs, vaccenic acid, and rumenic acid, while also enhancing the oxidative stability of these products [90]. Notably, Branciari et al. [91] demonstrated that supplementing the diets of dairy sheep with OMWW significantly increases the levels of phenolic compounds, such as tyrosol, hydroxytyrosol, verbascoside, and pinoresinol, in milk and cheese. This enrichment enhances the nutraceutical value of the products. While the inclusion of polyphenols derived from OMWW does not alter the composition of cheese, these compounds exert potent antioxidant effects, thereby improving the nutraceutical properties of cheese. This approach provides an innovative means to transform OMWW, a significant environmental pollutant, into a high-value feed ingredient [91].

The health benefits of polyphenols in livestock have been well-documented. These include reductions in the intestinal parasitic load in sheep [92], improvements in cell-mediated immune responses, decreased incidence of inflammatory diseases [93], and enhanced antioxidant status in treated animals [94]. Furthermore, the addition of spray-dried phenolic extracts from OMWW to the diets of lactating ewes has been shown to enhance the nutritional value of milk by increasing the levels of beneficial fatty acids, such as rumenic and vaccenic acids. These polyphenols may also reduce the somatic cell count in milk, an important marker of udder health, thus warranting further investigation [95].

The production of innovative feed enriched with natural phenols derived from OMWW presents a sustainable strategy to improve the nutritional quality of milk while valorizing the by-products of the olive oil industry. This approach aligns with the principles of the circular economy by reducing the environmental impact of OMWW and promoting the reuse of natural compounds. Overall, the inclusion of polyphenols in the nutrition of small ruminants offers a cost-effective and sustainable solution to enhance dairy product quality, support farmers' livelihoods, and reduce agricultural waste.

6 Potential Sunscreen Activity of OMWW

The harmful effects of solar radiation on the skin are caused predominantly by the UV region of the electromagnetic spectrum, which can be subdivided into three parts: UVA, from 320 to 400 nm; UVB, from 290 to 320 nm; and UVC, from 200 to 290 nm. The atmosphere filters outUVC radiation before it reaches Earth; however, the UVB (B for burning) radiation is not completely filtered out by the ozone layer and is responsible for sunburn damage. UVA (A for aging) radiation reaches the deeper layers of the epidermis and provokes the aging process and other serious problems of the skin. UV radiation has been considered a causative factor of skin cancer [96]. The efficacy of a sunscreen is usually expressed in terms of its sun protection factor (SPF), which is defined as the ratio of the UVB energy from 290 to 320 nm required to produce a minimal erythema dose (MED) on protected skin to that of the unprotected skin. The higher the SPF, the more effective the product is in preventing sunburns (the SPF value must exceed 15, indicating the level of protection against UVB radiation). Nevertheless, it is necessary to standardize the methods used to determine the SPF of these products. In scientific research, the in vitro method is followed, in which the absorption characteristics (SPF) of sunscreen agents are determined based on spectrophotometric analysis of dilute solutions. Mansur et al. [97] developed a very simple equation that substitutes the in vitro method, utilizing UV spectrophotometry:

$$\mathrm{SPF} = \mathrm{CF} \times \sum_{290}^{320} \mathrm{EE}(\lambda) \times I(\lambda) \times \mathrm{Abs}(\lambda), \tag{16.1}$$

where EE (λ) is the erythemal effect spectrum, I (λ) is the solar intensity spectrum, Abs (λ) is the absorbance of the sunscreen product, and CF is the correction factor (=10). The EE × I values are constants [97, 98].

Methods used to evaluate only UVB SPF values have proven to be inadequate because the harmful consequences of UVA exposure have become known [99]. Therefore, increasing concerns regarding the effects of UVA radiation have led to the development of broad-spectrum sunscreens. The European Cosmetic, Toiletry, and Perfumery Association group has developed in vitro measures of UVA wavelength or the critical wavelength (CW). This method is based on the use of a spectrophotometer with integration boundaries from 290 nm to 400 nm. The wavelength at which the summed absorbance reaches 90% of the total absorbance is defined as the critical wavelength (Eq. 16.2):

$$\int_{290}^{\lambda_{\mathrm{CW}}} \mathrm{Abs}(\lambda) \times d(\lambda) = 0.9 \int_{290}^{400} \mathrm{Abs}(\lambda) \times d(\lambda), \tag{16.2}$$

where the wavelength increment $d(\lambda)$ is taken to be 1 nm.

The sunscreen products are then classified as broad spectrum when the CW is higher than 370 nm. The determination of SPF and CW values (using Eq.16.1 and Eq.16.2) of the active fractions is further confirmed using special equipment with a bio-membrane.

The organic components of sunscreens are generally responsible for shielding the skin from UVB radiation, while the inorganic components offer protection from UVA radiation, which constitutes over 90% of the UV spectrum. Examples of synthetic ingredients commonly found in conventional sunscreens, including formulations utilizing nanotechnology, include

- **Organic compounds:** Benzophenones, parabens, retinyl palmitate, and PABA (para-aminobenzoic acid).
- **Inorganic compounds:** Titanium dioxide (TiO_2) and zinc oxide (ZnO).

Many of these substances are known to be toxic to the skin, with the potential to penetrate the epidermal layer, enter the bloodstream, and cause adverse effects such as allergies, toxicity, and other health concerns [100, 101]. Over time, some of these synthetic ingredients were removed from the list of sunscreen ingredients approved by the FDA due to the above-mentioned side effects. The growing public awareness of these risks and the increasing demand for reduced reliance on synthetic chemical sunscreens present researchers with a unique opportunity to develop natural products for UV protection.

One claimed natural sunscreen product is the so-called Supergoop, which is made with natural extracts of citrus, basil, sunflower, rosemary, and bois de rose. These substances, which are rich in antioxidant extracts, are excellent but expensive and need wide areas to grow enough plants. Therefore, there is a need for developing a natural, cost-effective, and environmentally friendly sunscreen product. One of the promising avenues for the quest for finding such a product is utilizing phenols recovered from OMWW as UV boosters in synthetic sunscreens [102].

Phenolic compounds are abundant in OMWW, which represents a rich source of bioactive compounds and natural phenols such as hydroxytyrosol and others [35, 37, 103]. This composition underscores its potential for utilization in various commercial products. When treated correctly, OMWW has substantial potential as an inexpensive and widely available starting material for the extraction of polyphenols.

Recent studies have focused on developing organic alternatives for the preparation of natural protective prototypes derived from OMWW. Morgane et al. [104], in their review, highlighted the potential of OMWW as a source of polyphenols like hydroxytyrosol for the treatment of skin disorders, as well as the use of OMWW phenolics as sun protection factors. Galanakis [2, 102] demonstrated that phenolic compounds from OMWW, along with other natural antioxidants, can function as UV filters (Fig. 16.4) and showed the synergistic effect of OMWW phenolics when added to commercial sunscreens.

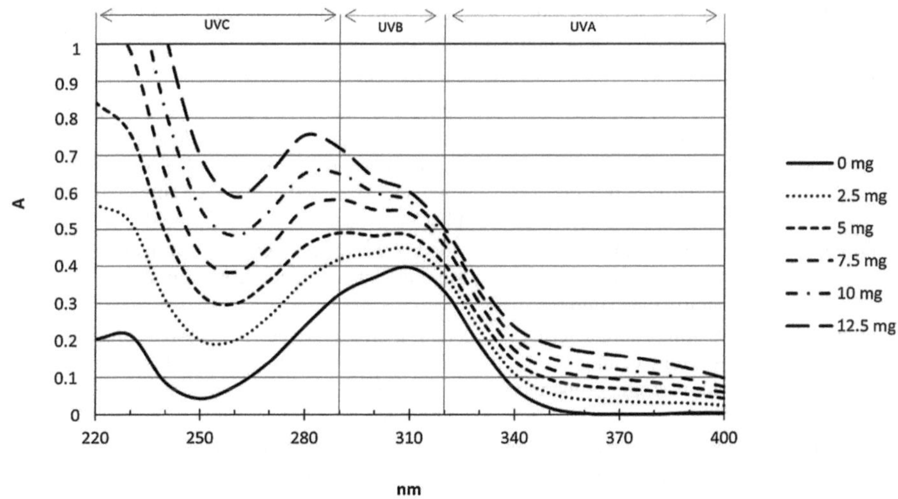

Fig. 16.4 Spectra of 1 mg octylmethoxy cinnamate (OMC)/L solutions boosted with olive phenols in different concentrations (0–12.5 mg/L) [102]. With permission under License No.5963660095483 from Elsevier

Extracting polyphenols from OMWW significantly reduces COD and total phenols, thereby producing effluents with a lower organic load. Preliminary studies conducted by our group (unpublished data) revealed that fractionation [35] and resin-based recovery techniques [105] successfully isolated polyphenolic compounds from OMWW with promising SPF (>30 nm) and CW (> 370 nm) values. These polyphenols demonstrate skin-protective activity against solar radiation, positioning them as viable green natural products for cosmetic applications. Consequently, their practical use in the cosmetic industry is ready to be pursued.

OMWW phenolics provide promising potential for the development of commercial, natural, and active sunscreens. OMWW typically contains polyphenols and residual lignocellulose. The extraction of polyphenols through specialized resins or other techniques, combined with the recovery and filtration of lignocellulose for use as fuel or compost, simplifies the chemical composition of the remaining waste. This process reduces the polluting properties of OMWW, enabling the liquid effluent to be safely disposed of in receiving bodies such as olive orchards, general crops, or local WWTPs.

The value-added products derived from olive mill solid waste (OMSW) and olive mill wastewater (OMWW) and their applications are summarized in Table 16.1.

Table 16.1 Summary of value-added products derived from olive mill solid waste (OMSW) and olive mill wastewater (OMWW) and their applications

Valued product/ benefit	Description	Applications	Reference
Phenolic compounds	Contains hydroxytyrosol, tyrosol, oleuropein, and other bioactive phenols	Used in health supplements, cosmetics, food additives, and pharmaceuticals	Azaizeh et al. [35]; Obied et al. [37]; Rahmanian et al. [103]
Antioxidants	High polyphenol content makes OMWW a natural antioxidant source	Food preservatives, dietary supplements, nutraceuticals	Azaizeh et al. [35]; Russo et al. [54]
Antimicrobial agents	Exhibits antibacterial, antifungal, and antiparasitic properties	Natural preservatives, agricultural pathogen control, and alternative antibiotics	Abu-Lafi et al. [78]; Tafesh et al. [83]; Sar and Akbas [79]
Animal feed additive	Polyphenols improve gut health, immunity, and milk quality in livestock	Used in animal nutrition to enhance dairy product quality	Branciari et al. [91]; Cifuni et al. [95]
Biogas production	Anaerobic digestion of OMWW generates methane-rich biogas	Renewable energy source for heat and electricity	Azaizeh and Jadoun [34]; Andreozzi et al. [31]
Compost and soil amendment	Organic matter and nutrients improve soil quality	Sustainable agriculture, fertilizer substitute	Sierra et al. [106]; Mekki et al. [15]
Sunscreen ingredient	OMWW-derived phenolics provide UV protection	Used in natural sunscreens as UV absorbers	Galanakis et al. [102]; Morgane et al. [104]
Bio-based chemicals	Recovery of valuable bioactive compounds through membrane filtration	Functional ingredients in pharmaceuticals and cosmetics	Cifuentes-Cabezas et al. [28]; Tapia-Quirós et al. [30]
Biochar from OMSW	Pyrolysis of solid waste produces carbon-rich biochar	Soil enhancement, carbon sequestration	Roig et al. [10]
Biofuels	Extraction and conversion of lipid residues from OMSW	Production of biodiesel and other biofuels	Azaizeh et al. [8]

7 Technologies for Polyphenol Recovery from OMWW

The recovery of polyphenols from OMWW is critical for advancing environmental sustainability and optimizing resource valorization, with several advanced methodologies having been developed for this purpose:

1. *Liquid–Liquid Extraction*, which employs acidification of OMWW to a target pH and the application of organic solvents, such as ethyl acetate, to efficiently extract dietary phenolic compounds like hydroxytyrosol and tyrosol, although solvent recovery and environmental impact present significant challenges [107];

2. *Membrane Filtration Techniques*, including MF, UF, NF, and RO, are often used in sequential configurations to enhance separation efficiency while simultaneously minimizing wastewater volumes [108];
3. *Adsorption Techniques*, which utilize advanced adsorbent materials, such as XAD resins made of styrene-divinylbenzene or acrylate-divinylbenzene with varying surface and porosity characteristics, to selectively capture phenolic compounds, thereby yielding extracts with antioxidant, antibacterial, and sunscreen properties [105]; and
4. *Green Extraction Techniques*, such as aqueous extraction, which avoid the use of organic solvents and demonstrate the potential to recover phenolic compounds with substantial antioxidant activities [109], with the selection of an appropriate method governed by considerations of purity, environmental impact, and cost-effectiveness.

The various available technologies will play a crucial role in advancing future research on the extraction of polyphenols and other bioactive compounds from OMWW. These efforts aim to mitigate the environmental contamination associated with the discharge of such effluents while fostering the reuse of natural compounds in alignment with the principles of the circular economy.

8 Conclusions and Future Perspectives

OMWW has historically been viewed as a hazardous waste due to its high COD and negative environmental impacts, which also impose economic burdens on the olive oil industry. However, this perspective is evolving as OMWW is increasingly recognized as a valuable feedstock rich in bioactive compounds, particularly phenolic compounds such as tyrosol, hydroxytyrosol, aromatic acids, and conjugated aromatic acids. These compounds hold significant potential for application as natural antioxidants and antimicrobials in the food, pharmaceutical, and cosmetic industries.

The growing interest in the valorization of OMWW stems from its potential to transform a problematic by-product into a source of high-value materials (Table 16.1). Advances in extraction technologies, such as resin adsorption, membrane filtration, and green extraction methods, have demonstrated promising methods for recovering phenolic compounds efficiently and sustainably. These compounds not only exhibit potent antioxidant and antimicrobial activities but also have potential applications in animal nutrition as feed additives, enhancing the quality of milk and dairy products while supporting sustainable agricultural practices.

Furthermore, OMWW valorization aligns with the principles of the circular economy, promoting the reuse of natural resources while reducing the environmental footprint of olive oil production. The integration of polyphenols into small ruminant diets, for instance, represents an innovative solution to reduce feed costs for farmers and improve the nutritional value of livestock-derived products. However, further research is required to optimize the use of these compounds, particularly through in vivo studies, to establish their efficacy and safety across different species and applications. In addition to polyphenol recovery, the co-utilization of lignocellulosic residues from OMWW for biofuel production or composting offers a complementary approach to reduce the pollutant load of OMWW and facilitate its safe disposal. This integrated valorization strategy minimizes waste while creating opportunities for the development of novel, eco-friendly products.

Future research should focus on improving the extraction efficiency, reducing costs, and scaling up the recovery of bioactive compounds from OMWW. Emphasis should also be placed on developing standardized methods for evaluating the bioavailability, stability, and safety of these compounds in various applications. Moreover, the potential of OMWW-derived phenolic compounds in emerging fields, such as natural UV protection agents in cosmetics, warrants further exploration.

In conclusion, the valorization of OMWW offers a multifaceted approach to addressing environmental challenges, supporting sustainable agricultural practices, and driving innovation in various industries. By transforming OMWW from an environmental pollutant into a valuable resource, this paradigm shift underscores the importance of integrating science, sustainability, and economic viability into waste management strategies.

References

1. Attard K, Lia F (2023) The antioxidant and bioactive potential of olive mill waste. Open Access. https://doi.org/10.5772/intechopen.1004127
2. Galanakis CM (2017) Chapter 2—olive mill waste recent advances for sustainable management. In: Olive mill waste, recent advances for sustainable management, 1st edn. Elsevier, Greece
3. Alkhalidi A, Halaweh G, Khawaja, and M.K. (2023) Recommendations for olive mills waste treatment in hot and dry climate. J Saudi Soc Agric Sci 22(6):361–373
4. FaoStat. (2020) Food and Agriculture Organization of the United Nations. Crops. Available online: http://www.fao.org/faostat/en/?#data/QC (Accessed 1 Jan 2021)
5. Khdair A, Abu-Rumman G (2020) Sustainable environmental management and valorization options for olive mill byproducts in the Middle East and North Africa (MENA) region. PRO 8:671
6. Cabrera F, Lopez R, Martinez-Bordiu A, de Lome ED et al (1996) Land treatment of olive oil mill wastewater. Int Biodeterior Biodegrad 38:215–225
7. El Mekawy A, Diels L, Bertin L, De Wever H et al (2014) Potential biovalorization techniques for olive mill biorefinery wastewater. Biofuels Bioprod Biorefin 8:283–293
8. Azaizeh H, Abu Tayeh H, Gerchman Y (2020) Valorization of olive oil industry solid waste and production of ethanol and high value-added biomolecules. In: Biovalorisation of wastes to renewable chemicals and biofuels. Elsevier Press, pp 27–40
9. Foti P, Romeo FV, Russo N, Pino A et al (2021) Olive mill wastewater as renewable raw materials to generate high added-value ingredients for agro-food industries. Appl Sci 11:7511
10. Roig A, Cayuela ML, Sanchez-Monedero M (2006) An overview on olive mill wastes and their valorisation methods. Waste Manag 26:960–969
11. Paraskeva CA, Papadakis VG, Kanellopoulou DG, Koutsoukos PG et al (2007a) Membrane filtration of olive mill wastewater and exploitation of its fractions. Water Environ Res 79:421–429

12. Paraskeva CA, Papadakis VG, Tsarouchi E, Kanellopoulou DG et al (2007b) Membrane processing for olive mill wastewater fractionation. Desalination 213:218–229
13. Di Nunzio M, Picone G, Pasini F, Chiarello E et al (2019) Olive oil by-product as functional ingredient in bakery products. Influence of processing and evaluation of biological effects. Food Res Int 131:108940
14. Vavouraki AI, Zakoura MV, Dareioti MA, Kornaro M (2019) Biodegradation of polyphenolic compounds from olive mill wastewaters (OMW) during two-stage anaerobic co-digestion of agro-industrial mixtures. Waste Biomass Valori:1–9
15. Mekki A, Dhouib A, Sayadi S (2007) Polyphenols dynamics and phytotoxicity in a soil amended by olive mill wastewaters. J Environ Manag 84(2):134–140
16. Abbattista R, Ventura G, Calvano CD, Cataldi TRI et al (2021) Bioactive compounds in waste by-products from olive oil production: applications and structural characterization by mass spectrometry techniques. Food Secur 10:1236
17. Khdair IA, Abu-Rumman G (2017) Evaluation of the environmental pollution from olive mills wastewater. Fresenius Environ Bull 26:2537–2540
18. Khdair IA, Abu-Rumman G (2019) Pollution estimation from olive wastewater in Jordan. Heliyon 5:e02386
19. Paraskeva P, Diamadopoulos E (2006) Technologies for olive mill wastewater (OMW) treatment: a review. J Chem Technol Biotechnol 81:1475–1485
20. Galanakis CM (2012) Recovery of high added-value components from food wastes: conventional, emerging technologies and commercialized applications. Trends Food Sci Technol 26:68–87
21. Asli S, Diab M, Haj-Zaroubi M (2023) Valuable materials from olive mill wastewater. in: wastewater from olive oil production: Environmental impacts, treatment, and valorization. Springer Water, p 406
22. Fiorentino A, Gentili A, Isidori M, Monaco P et al (2003) Environmental effects caused by olive mill wastewaters: toxicity comparison of low-molecular-weight phenol components. J Agric Food Chem 51:1005–1009
23. Rupani PF, Singh RP, Ibrahim MH, Esa N (2010) Review of current palm oil mill effluent (POME) treatment methods: vermicomposting as a sustainable practice. World Appl Sci J 11:70–81
24. Al-Hmoud L, Al-Saida B, Sandouqa A (2020) Olive mill wastewater treatment: a recent review. Jordanian J Eng Chem Ind 3(3):91–106. https://orcid.org/0000-0002-5560-1060
25. Tsagaraki E, Lazarides HN, Petrotos KB (2007) Olive mill wastewater treatment.In. In: Oreopoulou V, Russ W (eds) Utilization of by-products and treatment of waste in the food industry. Springer, Boston, MA
26. Carnevale M, Gnisci E, Hilal J, Criscuoli A (2016) Direct contact and vacuum membrane distillation application for the olive mill wastewater treatment. Sep Purif Technol 16:121–127
27. Bottino A, Capannelli G, Comite A, Costa C et al (2020) Treatment of olive mill wastewater through integrated pressure-driven membrane processes. Membranes 10:334
28. Cifuentes-Cabezas M, Carbonell-Alcaina C, Vincent-Vela MC, Mendoza-Roca JA et al (2021) Comparison of different ultrafiltration membranes as first step for the recovery of phenolic compounds from olive-oil washing wastewater. Process Saf Environ Prot 149:724–734
29. Sygouni V, Pantziaros AG, Iakovides IC, Sfetsa E et al (2019) Treatment of two-phase olive mill wastewater and recovery of phenolic compounds using membrane technology. Membranes 9:27
30. Tapia-Quirós P, Montenegro-Landívar MF, Reig M, Vecino X et al (2022) Recovery of polyphenols from Agri-food by-products: the olive oil and winery industries cases. Food Secur 11:362
31. Andreozzi R, Longo G, Majone M, Modeti G (1998) Integrated treatment of olive oil mill effluents (OME): study of ozonation coupled with anaerobic digestion. Water Res 32(8):2357–2364

32. Ubay G, Ozturk I (1997) Anaerobic treatment of olive mill effluents. Water Sci Technol 36(2–3):287–294
33. Hanaki K, Matsuo T, Nagase M (1981) Mechanism of inhibition caused by long chain fatty acids in anaerobic digestion process. Biotechnol Bioeng 23(7):1591–1610
34. Azaizeh H, Jadoun J (2010) Codigestion of olive mill wastewater and swine manure using UASB bioreactors for biogas production. J Water Resour Protect 2(4):314–321
35. Azaizeh H, Halahleh F, Najami N, Brunner D et al (2012) Antioxidant activity of phenolic fractions in olive mill wastewater. Food Chem 134:2226–2234
36. Cardinali A, Cicco N, Linsalata V, Minervini F et al (2010) Biological activity of high molecular weight phenolics from olive mill wastewater. J Agric Food Chem 58:8585–8590
37. Obied HK, Allen MS, Bedgood DR, Prenzler PD et al (2005) Bioactivity and analysis of biophenols recovered from olive mill waste. J Agric Food Chem 53:823–837
38. Obied HK, Bedgood D, Mailer R, Prenzler PD et al (2008a) Impact of cultivar, harvesting time, and seasonal variation on the content of biophenols in olive mill waste. J Agric Food Chem 56:8851–8858
39. Obied HK, Prenzler PD, Robards K (2008b) Potent antioxidant biophenols from olive mill waste. Food Chem 111:171–178
40. Obied HK, Prenzler PD, Konczak I, Rehman AU et al (2009) Chemistry and bioactivity of olive biophenols in some antioxidant and antiproliferative in vitro bioassays. Chem Res Toxicol 22:227–234
41. Visioli F, Poli A, Galli C (2002) Antioxidant and other biological activities of phenols from olives and olive oil. Med Care Res Rev 22:65–75
42. Visioli F, Wolfram R, Richard D, Abdullah MICB et al (2009) Olive phenolics increase glutathione levels in healthy volunteers. J Agric Food Chem 57:1793–1796
43. Mulinacci N, Romani A, Galardi C, Pinelli P et al (2001) Polyphenolic content in olive oil wastewaters and related olive samples. J Agric Food Chem 49:3509–3514
44. Ness AR, Powles JW (1997) Fruit and vegetables, and cardiovascular disease: a review. Int J Epidemiol 26:1–13
45. Cooper DA, Eldridge AL, Pet JC (1999) Dietary carotenoids and certain cancers, heart diseases, and age-related macular degeneration: a review of recent research. Nutr Rev 57:201–214
46. Esterbauer H, Dieber-Rotheneder M, Striegl G, Waeg G (1991) Role of vitamin E in preventing the oxidation of low-density lipoprotein. Am J Clin Nutr 53:314s–321s
47. Duthie G, Crozier A (2000) Plant-derived phenolic antioxidants. Curr Opin Lipidol 11:43–47
48. Tsagaraki E, Lazarides HN, Petrotos K (2006) Olive mill wastewater treatment. In: Utilization of by-products and treatment of waste in the food industry. Springer, Iceland, pp 133–157
49. Fraga CG, Croft KD, Kennedy DO, Tandomas-Barberan FA (2019) The effects of polyphenols and other bioactives on human health. Food Funct 10:514–528
50. Smiljković M, Dias MI, Stojković D, Barros L et al (2018) Characterization of phenolic compounds in tincture of edible Nepeta nuda: development of antimicrobial mouthwash. Food Funct 9:5417–5425
51. Obied HK, Prenzler PD, Omar SH, Ismael R et al (2012) Pharmacology of olive biophenols. Adv Mol Toxicol 6:195–242
52. Crespo MC, Tome-Carneiro J, Davalos A, Visioli F (2018) Pharma-nutritional properties of olive oil phenols. Transfer of new findings to human nutrition. Food Secur 7(6):90
53. Bucciantini M, Leri M, Nardiello P, Casamenti F et al (2021) Olive polyphenols: antioxidant and anti-inflammatory properties. Antioxidants 10(7):1044
54. Russo E, Spallarossa A, Comite A, Pagliero M et al (2022) Valorization and potential antimicrobial use of olive mill wastewater (OMW) from Italian olive oil production. Antioxidants 11:903
55. Oliver S, Vittorio O, Cirillo GBC (2016) Enhancing the therapeutic effects of polyphenols with macromolecules. Polym Chem 7:1529–1544

56. Teng H, Chen L (2019) Polyphenols and bioavailability: an update. Crit Rev Food Sci Nutr 59(13):2040–2051
57. Aguirre A, Borneo R (2019) Improving bioavailability of polyphenols using nano delivery systems based on food polymers. Polyphenols Plants:59–65
58. Hu B, Liu X, Zhang C, Zeng X (2017) Food macromolecule-based nanodelivery systems for enhancing the bioavailability of polyphenols. J Food Drug Anal 25(1):3–15
59. Vivekanandhan DK, Verma P, Singh S (2016) Emerging technologies for improving bioavailability of polyphenols. Curr Nutr Food Sci 12(1):12–22
60. Scheepens A, Tan K, Paxton JW (2009) Improving the oral bioavailability of beneficial polyphenols through designed synergies. Genes Nutr 5:75–87
61. Jovanović AA, Djordjević VB, Petrović PM, Pljevljakušić DS et al (2021) The influence of different extraction conditions on polyphenol content, antioxidant and antimicrobial activities of wild thyme. J Appl Res Med Aromatic Plants 25:100328
62. Sun S, Huang S, Shi Y, Shao Y et al (2021) Extraction, isolation, characterization, and antimicrobial activities of non-extractable polyphenols from pomegranate peel. Food Chem 351:129232
63. Capasso R, Evidente A, Schivo L, Orrù G et al (1995) Antibacterial polyphenols from olive oil mill waste waters. J Appl Bacteriol 79(4):393–398
64. Sousa A, Ferreira ICFR, Calhelha R (2006) Phenolics and antimicrobial activity of traditional stoned table olives 'alcaparra'. Bioorg Med Chem 14(24):8533–8538
65. Soler-Rivas C, Espín JC, Wichers HJ (2000) Oleuropein and related compounds. J Sci Food Agric 80(7):1013–1023
66. Ibarra A, Sniderman Z (2008). Olive polyphenol concentrate. US Patent Application 20080014322 A1
67. Yangui T, Rhouma A, Triki MA, Gargouri K et al (2008) Control of damping-off caused by *Rhizoctonia solani* and *Fusarium solani* using olive mill waste water and some of its indigenous bacterial strains. Crop Prot 27(2):189–197
68. Guinda A (2006) Use of solid residue from the olive industry. Grasas Aceites 57(1):107–115
69. Khoufi S, Aloui F, Sayadi S (2008) Extraction of antioxidants from olive mill wastewater and electro-coagulation of exhausted fraction to reduce its toxicity on anaerobic digestion. J Hazard Mater 151(2–3):531–539
70. Abdel-Latif HMR, Citarasu T, Turgay E, Yilmaz E et al (2024) Control of yersiniosis in rainbow trout, *Oncorhynchus mykiss*: innovative non-antibiotic feed-based strategies. Ann Anim Sci. https://doi.org/10.2478/aoas-2024-0087
71. El-Son MAM, Nofal MI, Abdel-Latif HMR (2021) Co-infection of *Aeromonas hydrophila* and *Vibrio parahaemolyticus* isolated from diseased farmed striped mullet (*Mugil cephalus*) in Manzala, Egypt—case report. Aquaculture 530:735738
72. Ghaly AE, Dave D, Budge S, Brooks M (2010) Fish spoilage mechanisms and preservation techniques. Am J Appl Sci 7:859
73. Gram L, Dalgaard P (2020) Fish spoilage bacteria—problems and solutions. Curr Opin Biotechnol 13:262–266
74. Santos L, Ramos F (2018) Antimicrobial resistance in aquaculture: current knowledge and alternatives to tackle the problem. Int J Antimicrob Agents 52:135–143
75. Ahmad AL, Chin JY, Mohd Harun MHZ, Low SC (2022) Environmental impacts and imperative technologies towards sustainable treatment of aquaculture wastewater: a review. J Water Process Eng 46:102553
76. He X, Deng M, Wang Q, Yang et al (2016) Residues and health risk assessment of quinolones and sulfonamides in cultured fish from Pearl River Delta, China. Aquaculture 458:38–46
77. Liu X, Steele JC, Meng X-Z (2017) Usage, residue, and human health risk of antibiotics in Chinese aquaculture: a review. Environ Pollut 223:161–169
78. Abu-Lafi S, Al-Natsheh MS, Yaghmoor R, Al-Rimawi F (2017) Enrichment of phenolic compounds from olive mill wastewater and in vitro evaluation of their antimicrobial activities. Evid Based Complement Alternat Med 3706915:9. https://doi.org/10.1155/2017/3706915

79. Sar T, Akbas MY (2023) Antimicrobial activities of olive oil mill wastewater extracts against selected microorganisms. Sustain For 15:8179
80. Roila R, Branciari R, Ranucci D, Ortenzi R et al (2016) Antimicrobial activity of olive mill wastewater extract against *Pseudomonas fluorescens* isolated from mozzarella cheese. Ital J Food Saf 5:5760
81. Hazreen-Nita MK, Abdul Kari Z, Mat K, Rusli et al (2022) Olive oil by-products in aquafeeds: opportunities and challenges. Aquacult Rep 22:100998
82. Sicuro B, Badino P, Daprà F, Gai F et al (2010) Physiological effects of natural olive oil antioxidants utilization in rainbow trout (*Onchorynchus mykiss*) feeding. Aquac Int 18:415–431
83. Tafesh A, Najami N, Jadoun J, Halahleh F et al (2011) Synergistic antibacterial effects of polyphenolic compounds from olive mill wastewater. J Evid-Based Complement Alternat Med:1–9
84. Yilmaz DK, Kesbic FI, Celik ES, Odabasi DA et al (2024) Evaluation of the antimicrobial effects of olive mill wastewater extract against food spoiling/poisoning, fish-pathogenic and non-pathogenic microorganisms. Microorganisms 12:2216
85. Castrica M, Rebucci R, Giromini C, Tretola M et al (2019) Total phenolic content and antioxidant capacity of Agri-food waste and by-products. Ital J Anim Sci 18:336–341
86. Correddu F, Lunesu MF, Buffa G, Atzori AS et al (2020) Can agro-industrial by-products rich in polyphenols be advantageously used in the feeding and nutrition of dairy small ruminants? Animals 10:131
87. Frutos P, Hervás G, Giráldez FJ, Mantecón AR (2004) Review. Tannins and ruminant nutrition. Span J Agric Res 2:191–202
88. Zhao J, Kagami M, Yano K, Kawasaki K (2023) Evaluation of the effect of incorporating olive mill wastewater on nutrients, quality, and bacterial flora in fermented total mixed ration. Fermentation 9:665
89. Vasta V, Daghio M, Cappucci A, Buccioni A et al (2019) Invited review: plant polyphenols and rumen microbiota responsible for fatty acid biohydrogenation, fiber digestion, and methane emission: experimental evidence and methodological approaches. J Dairy Sci 102:3781–3804
90. Toral PG, Monahan FJ, Hervás G, Frutos P et al (2018) Review: modulating ruminal lipid metabolism to improve the fatty acid composition of meat and milk. Challenges and opportunities. Animal 12:s272–s281
91. Branciari R, Galarini R, Miraglia D, Ranucci D, at al. (2020) Dietary supplementation with olive mill wastewater in dairy sheep: evaluation of cheese characteristics and presence of bioactive molecules. Animals 10:1941
92. Athanasiadou S, Kyriazakis I, Jackson F, Coop RL (2001) Direct anthelmintic effects of condensed tannins towards different gastrointestinal nematodes of sheep: *in vitro* and *in vivo* studies. Vet Parasitol 99:205–219
93. Pathak AK (2013) Potential of using condensed tannins to control gastrointestinal nematodes and improve small ruminant performance. Int J Mol Vet Res 3:36–50
94. Gessner DK, Ringseis R, Eder K (2017) Potential of plant polyphenols to combat oxidative stress and inflammatory processes in farm animals. J Anim Physiol Anim Nutr 101:605–628
95. Cifuni GF, Claps S, Morone G, Sepe L et al (2023) Valorization of olive mill byproducts: recovery of biophenol compounds and application in animal feed. Plan Theory 12:3062
96. Stierner U, Rosdahl I, Augustsson A, Kãgedal B (1989) UVB irradiation induces melanocyte increase in both exposed and shielded human skin. J Invest Dermatol 92(4):561–564
97. Mansur JS, Breder MNR, Mansur MCA, Azulay RD (1986) Determinação do fator de proteção solar por espectrofotometria. An Bras Dermatol 61:121–124
98. Sayre RM, Agin PP, Levee GJ, Marlowe E (1979) Comparison of *in vivo* and *in vitro* testing of sunscreening formulas. Photochem Photobiol 29:559–566
99. Hao Y, Georgios S, Claude S, Nikiforos A (2004) A chemiluminescence study of UVA-induced oxidative stress in human skin *in vivo*. J Invest Dermatol 122(4):1020–1029
100. Kyu-Bong K, Young Woo K, Seong KWL, Tae Hyun R et al (2017) Risk assessment of zinc oxide, a cosmetic ingredient used as a UV filter of sunscreens. J Toxic Environ Health, Part B 20(3):155–182

101. Suzanna AL, William W, Brittany L, Elizabeth Q et al (2013) Photochemical behavior of nanoscale TiO2 and ZnO sunscreen ingredients. J Photochem Photobiol A Chem 263:24–33
102. Galanakis CM, Tsatalas P, Galanakis I (2018) Implementation of phenols recovered from olive mill wastewater as UV booster in cosmetics. Ind Crop Prod 111:30–37
103. Rahmanian N, Jafari SM, Galanakis CM (2014) Recovery and removal of phenolic compounds from olive mill wastewater. J Am Oil Chem Soc 91:1–18
104. Morgane C, Mary TK, Florence R, Michel L et al (2021) Potential of olive oil mill wastewater as a source of polyphenols for the treatment of skin disorders: a review. J Agric Food Chem 69(26):7268–7284
105. Apostolis A, Prokopios M, Alexios-Leandros S, Emmanuel M et al (2007) A new process for the management of olive oil mill waste water and recovery of natural antioxidants. J Agric Food Chem 55:2671–2676
106. Sierra J, Marti E, Garau AM, Cruanas R (2007) Effects of the agronomic use of olive oil mill wastewater: field experiment. Sci Total Environ 378:90–94
107. Cuffaro D, Bertolini A, Bertini S, Ricci C et al (2023) Olive mill wastewater as source of polyphenols with nutraceutical properties. Nutrients 15(17):3746
108. Cassano A (2016) Recovery of polyphenols from olive mill wastewaters by membrane operations. In: Drioli E, Giorno L (eds) Encyclopedia of membranes. Springer, Berlin, Heidelberg, pp 163–187
109. Hernández-Fernández A, Garrido Y, Iniesta-López E, Pérez de los Ríos A et al (2023) Recovering polyphenols in aqueous solutions from olive mill wastewater and olive leaf for biological applications. PRO 11:2668

Part VI
Industrial Crops

Chapter 17
Bamboo Crop Waste Management: A Review of Methods, Mechanisms, and Prospects

Anju Singhwane, Harsh Bajpai, Medha Mili, M. C. Shekhar Nayak, A. K. Srivastava, and Sarika Verma

Abbreviations

AD	Anaerobic digestion
PHAs	Polyhydroxyalkanoates
PLA	Polylactic acid

1 Introduction

Bamboo, a fast-growing and highly versatile plant, plays a significant role in various industries, including construction, furniture, textiles, and paper production. Its rapid growth, carbon sequestration ability, and renewable nature make it an eco-friendly alternative to conventional materials [1]. Bamboo waste, including sawdust, shavings, leaves, and culm residues, can be effectively managed and repurposed for various applications. These wastes can be utilized for biochar production, bio-composites, paper manufacturing, biomass energy, and organic fertilizers, reducing

A. Singhwane
Council of Scientific and Industrial Research - Advanced Materials and Processes Research Institute, Bhopal, MP, India

H. Bajpai
AcSIR-Advanced Materials and Processes Research Institute (AMPRI), Bhopal, MP, India

M. Mili · M. C. S. Nayak · A. K. Srivastava · S. Verma (✉)
Council of Scientific and Industrial Research - Advanced Materials and Processes Research Institute, Bhopal, MP, India

AcSIR-Advanced Materials and Processes Research Institute (AMPRI), Bhopal, MP, India
e-mail: sarika.verma@ampri.res.in

C. Trois, R. C. Ray (eds.), *Plantation Crop Wastes: Valorization for Economic Sustainability*, Waste as a Resource,
https://doi.org/10.1007/978-3-031-96805-1_17

environmental impact and promoting circular economy principles [2, 3]. Proper management of bamboo waste minimizes landfill accumulation and contributes to sustainable material innovation and resource conservation. However, the extensive utilization of bamboo generates substantial waste, posing environmental and disposal challenges. Effective bamboo crop waste management is crucial for mitigating these issues and harnessing the potential of bamboo by-products for sustainable applications [4]. Composting facilitates the biodegradation of bamboo residues into nutrient-rich fertilizer, although it requires substantial time and space [5]. The process involves microbial decomposition under controlled conditions, releasing essential nutrients back into the soil and improving the soil structure. Conversely, Mulching repurposes shredded bamboo waste to enhance soil moisture retention, suppress weeds, and promote soil health, making it a practical recycling approach [6]. Mulch layers also help regulate soil temperature and minimize erosion, contributing to soil conservation [7].

Chemical waste management techniques, particularly pyrolysis, efficiently decompose bamboo waste under oxygen-limited conditions [8], thermo-chemically. This process yields biochar, bio-oil, and syngas, with biochar proving beneficial for carbon sequestration and soil fertility enhancement [9]. Biochar, a porous carbon-rich material, enhances soil aeration, retains essential nutrients, and reduces greenhouse gas emissions when incorporated into agricultural systems [10]. Bio-oil can serve as a potential biofuel source, and syngas can be utilized for electricity generation or chemical synthesis. The controlled pyrolysis conditions [11], including temperature, heating rate, and residence time, significantly influence the yield and composition of these products [12, 13].

Biochemical methods such as anaerobic digestion (AD) have emerged as a promising solution for converting bamboo waste into renewable energy in biogas, alongside digestate production as an organic fertilizer [14]. The AD process involves the microbial breakdown of organic matter in oxygen-deprived environments, producing methane-rich biogas, which can be harnessed for electricity or heat generation. The nutrient-rich by-product digestate can enhance soil fertility, promoting sustainable agricultural practices [15]. Factors such as pH, temperature, and microbial community composition are crucial in optimizing AD efficiency.

Emerging biotechnological advancements present further opportunities for the value-added conversion of bamboo waste. Enzymatic hydrolysis and microbial fermentation leverage the high cellulose content of bamboo residues to produce bioethanol and bioplastics, aligning with global efforts toward sustainable material development. Enzymatic hydrolysis breaks down complex polysaccharides into fermentable sugars, which are then converted into bioethanol through microbial fermentation [16]. Additionally, bamboo-derived lignocellulosic biomass can be processed into biodegradable plastics, offering an eco-friendly alternative to petroleum-based polymers. Physical methods, such as composting, mulching, and pyrolysis, are examined for their roles in converting bamboo waste into organic matter and biochar. Composting and mulching are discussed for their environmental benefits, while pyrolysis is highlighted for its potential to produce valuable by-products like biochar, which can improve soil quality and sequester carbon. Chemical processes, including hydrolysis and pyrochemical treatments, are explored for their ability to extract valuable compounds from bamboo waste, offering opportunities for high-value applications. Biological approaches, such as microbial degradation [17] and enzymatic treatments, are emphasized for their efficiency in breaking down bamboo waste into biofuels and bioproducts, thus presenting a sustainable avenue for energy and resource recovery.

This chapter comprehensively reviews various bamboo waste management strategies, incorporating insights from recent research. It begins by investigating the sources and characteristics of bamboo waste, categorizing it into distinct types based on its origin and form. Further, it explores various physical, chemical, and biological methods for managing bamboo waste. It also stresses the need for policy support, technological advancements, and interdisciplinary collaborations to enhance the efficiency of bamboo waste management systems and drive sustainable industrial applications.

2 Sources and Characteristics of Bamboo Waste

Bamboo waste is generated during cultivation, processing, and product manufacturing, contributing to environmental challenges if not managed efficiently. This waste originates from sawmills, furniture production, pulp and paper industries, and construction activities. Proper utilization of bamboo waste can lead to the development of value-added products, supporting sustainability and circular economy principles [18].

2.1 Types of Bamboo Waste

Bamboo waste can be classified into various types based on its source and form [19]:

- Sawdust: Fine particulate waste from cutting and milling processes.
- Chips: Small, irregular fragments generated during mechanical processing.
- Leaves: Biomass residue from bamboo plantations and harvesting.
- Fibers: Extracted during processing for textiles, paper, and composite applications.
- Nodes and Internodes: Solid structural waste from bamboo poles used in construction and furniture.
- Bio-residues: Leftovers from pulping, boiling, or chemical treatment processes.

2.2 Chemical Composition of Bamboo Waste

Bamboo waste is rich in lignocellulosic components [20], making it a potential raw material for bio-based applications (Fig. 17.1). The key constituents include:

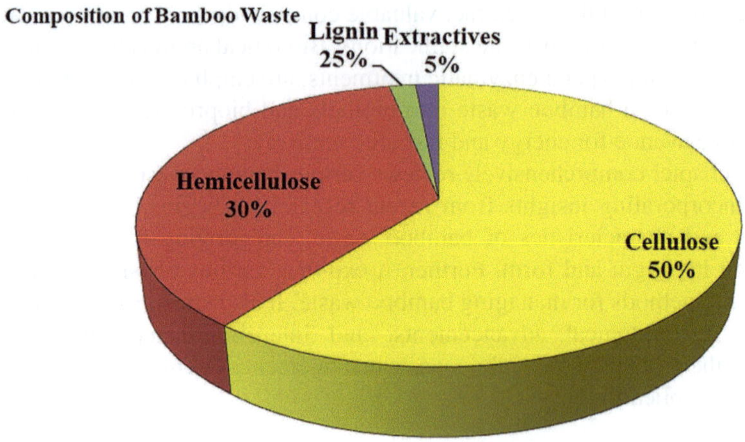

Fig. 17.1 Illustrating the chemical composition of bamboo waste, highlighting the relative proportions of cellulose, hemicellulose, lignin, and extractives

- Cellulose (40–50%): It provides structural strength and is a precursor of bioethanol and biodegradable materials.
- Hemicellulose (20–30%): A polysaccharide that enhances flexibility and plays a role in enzymatic hydrolysis.
- Lignin (20–25%): A complex aromatic polymer contributing to rigidity; it can be used in biochar and energy applications.
- Extractives (2–5%): These include tannins, resins, and essential oils, which influence bamboo's durability and decay resistance.

3 Methods of Bamboo Waste Management

Bamboo waste management can be approached through physical, chemical, and biochemical methods (Fig. 17.2), each with distinct advantages and applications. Physical methods include mechanical processes, such as shredding, grinding, or heating, to reduce the size or alter the structure of bamboo waste. These methods are often employed for the recycling and reuse of composite materials, construction, or energy generation. Chemical methods involve using solvents or acids to break down bamboo into its components, facilitating the extraction of valuable materials like cellulose, lignin, or sugars for various applications, including biofuels or chemical production. Biochemical methods leverage microbial or enzymatic processes to decompose bamboo waste, converting it into valuable products like bioethanol, biogas, or compost. These methods are environmentally friendly and sustainable and can lead to the creation of bio-based products.

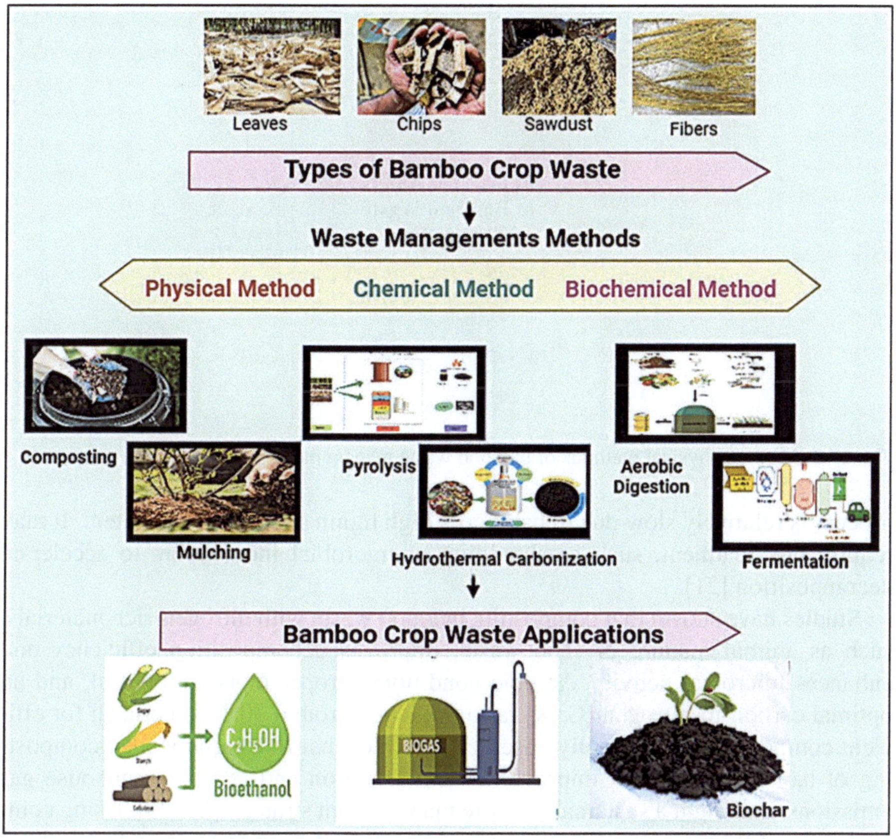

Fig. 17.2 Illustrating the waste management methods of different types of bamboo crop waste

3.1 *Physical Methods of Bamboo Waste Management*

Physical methods for managing bamboo waste involve mechanical and natural degradation processes that facilitate recycling and resource recovery. Composting is a widely used technique that converts bamboo residues into nutrient-rich organic matter, thereby promoting soil health and sustainable agriculture. Figure 17.3 outlines these methods and their respective pathways for bamboo waste processing.

3.1.1 Composting

Composting is a biological process in which microorganisms break down organic bamboo waste under controlled conditions to produce compost, which is a nutrient-rich soil amendment. Bamboo composting enhances soil fertility, improves water retention, and reduces the need for chemical fertilizers. However, the composting

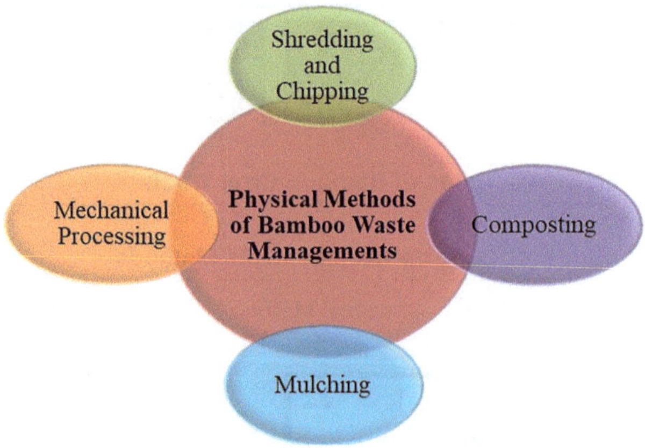

Fig. 17.3 Various physical methods of bamboo waste management

process is relatively slow due to bamboo's high lignin and cellulose content. It may require pre-treatment, such as shredding or microbial inoculation, to accelerate decomposition [21].

Studies have shown that composting bamboo waste with nitrogen-rich materials, such as animal manure or food waste, improves decomposition efficiency and enhances microbial activity. Aerobic conditions, proper moisture control, and an optimal carbon-to-nitrogen (C: N) ratio (typically around 30:1) are crucial for efficient composting. Additionally, research indicates that biochar-amended composting of bamboo waste can improve nutrient retention and reduce greenhouse gas emissions, making it a sustainable waste management strategy [22, 23]. Using compost derived from bamboo waste in agriculture has demonstrated benefits, such as increased crop yield and enhanced soil microbial diversity.

3.1.2 Mulching

Mulching is a physical method of bamboo waste management in which shredded bamboo residues, such as leaves, chips, and sawdust, are applied to soil surfaces. This practice improves soil moisture retention, suppresses weeds, and enhances soil fertility by gradually decomposing and releasing nutrients [24]. Studies have shown that bamboo mulch can regulate soil temperature and reduce erosion, making it beneficial for sustainable agriculture. Additionally, bamboo bio-mulch has been explored as an eco-friendly alternative to synthetic mulching films [25].

3.1.3 Mechanical Processing

Mechanical processing involves transforming bamboo waste into usable forms, such as pellets, briquettes, fibers, and particle boards. This method includes shredding, grinding, and compacting bamboo residues into high-value products. Bamboo sawdust and chips are commonly used for manufacturing engineered wood panels and bio-composite materials [26]. Moreover, bamboo briquettes and pellets serve as sustainable biomass fuel, providing an alternative to conventional fossil fuels [27].

3.2 Chemical Methods of Bamboo Waste Management

Chemical methods involve the transformation of bamboo waste through controlled chemical reactions to produce valuable by-products such as biofuels, biochar, and functional biomaterials. These methods enhance resource efficiency and contribute to sustainable waste valorization.

3.2.1 Pyrolysis

Pyrolysis is a thermochemical decomposition process that converts bamboo waste into biochar, bio-oil, and syngas in oxygen-limited environments (Fig. 17.4). Biochar enhances soil fertility and sequesters carbon, while bio-oil and syngas are alternative energy sources. The efficiency of pyrolysis depends on process parameters such as temperature, heating rate, and residence time [28]. Studies have shown that bamboo-derived biochar improves soil microbial activity and water retention, making it a promising agricultural amendment [29].

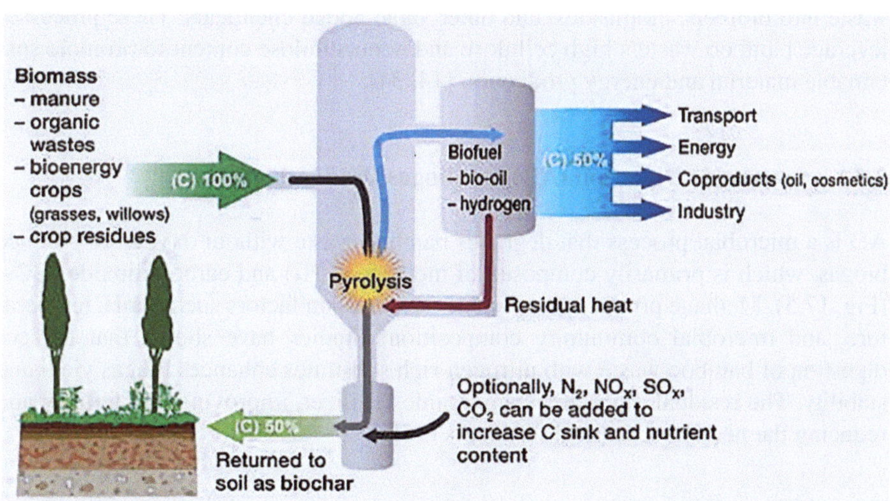

Fig. 17.4 Schematic representation of the pyrolysis process for bamboo waste management, illustrating the thermal decomposition of bamboo in an oxygen-limited environment to produce biochar, bio-oil, and syngas (Adapted with permission from [30]; License no 5961821257041)

3.2.2 Hydrothermal Carbonization

Hydrothermal carbonization occurs when bamboo waste is treated in hot, pressurized water to produce hydrochar, a carbon-rich solid material. Hydrochars have applications in soil conditioning, solid fuels, and bio-based adsorbents for wastewater treatment. This energy-efficient process effectively converts high-moisture bamboo biomass into stable carbon products. Additionally, hydrochar has been explored for pollutant removal due to its high porosity and adsorption capacity [31, 32].

3.2.3 Alkaline and Acid Treatment for Biomaterial Processing

Alkaline and acid treatments modify the structure of bamboo fibers by breaking down lignin and hemicellulose, thereby enhancing their usability as biomaterials. Alkaline treatment using sodium hydroxide (NaOH) improves the fiber strength and enhances the compatibility of bamboo fibers with polymer composites. Acid hydrolysis is used to extract nanocellulose from bamboo waste, which has applications in bioplastics, coatings, and biomedical materials [33].

3.3 Biochemical Methods of Bamboo Waste Management

Biochemical methods utilize microbial and enzymatic processes to convert bamboo waste into biofuels, bioplastics, and other value-added chemicals. These processes leverage bamboo waste's high cellulose and hemicellulose content to promote sustainable material and energy production [14, 34].

3.3.1 Anaerobic Digestion (AD) for Biogas Production

AD is a microbial process that degrades bamboo waste without oxygen to produce biogas, which is primarily composed of methane (CH_4) and carbon dioxide (CO_2) (Fig. 17.5). Methane production efficiency depends on factors such as pH, temperature, and microbial community composition. Studies have shown that the co-digestion of bamboo waste with nitrogen-rich substrates enhances biogas yield and stability. The residual digestate is an organic fertilizer, improving soil fertility and reducing the need for chemical fertilizers [35].

3.3.2 Fermentation for Bioethanol and Bioplastics Production

Fermentation processes utilize microbial pathways to convert bamboo-derived sugars into bioethanol and bioplastics. Enzymatic hydrolysis breaks down cellulose into fermentable sugars metabolized by microorganisms, such as *Saccharomyces*

Fig. 17.5 Proposed anaerobic digestion systems for biogas production (adapted with permission under CC BY 4.0 from [36])

cerevisiae and *Zymomonas mobilis,* to produce ethanol [37]. Bioethanol from bamboo waste can be a sustainable fuel alternative, reducing the dependence on fossil fuels. Additionally, microbial fermentation can synthesize biopolymers like polyhydroxyalkanoates (PHAs), offering biodegradable plastic alternatives.

3.3.3 Enzymatic Hydrolysis for Value-Added Chemicals

Enzymatic hydrolysis employs cellulase and hemicellulase enzymes to degrade bamboo lignocellulose into simple sugars and other biochemicals. These sugars can be further processed into platform chemicals, such as furfural, lactic acid, and levulinic acid, which have applications in bioplastics, pharmaceuticals, and green solvents. Optimizing enzyme concentration, reaction time, and temperature improves hydrolysis efficiency and product yield [38].

4 Advanced and Emerging Technologies

Recent advancements in material science and processing technologies have transformed bamboo waste into high-value materials such as nanomaterials, biodegradable composites, and bio-based filaments. These innovations contribute to sustainable product development in diverse industries, including packaging, construction, and automotive manufacturing.

4.1 Nanomaterials and Bamboo Waste

Bamboo waste is a valuable source of nanocellulose, which can be extracted through mechanical, chemical, or enzymatic processes. Nanocellulose-based biodegradable films and coatings have applications in food packaging, biomedical devices, and sustainable textiles due to their high strength, flexibility, and biodegradability. Additionally, bamboo-derived carbon nanofibers and aerogels are being explored for energy storage, water purification, and thermal insulation due to their lightweight and porous structures [22, 23, 39].

4.2 Bamboo Waste in 3D Printing and Additive Manufacturing

Bamboo waste can be processed into bio-based filaments suitable for 3D printing applications Fig. 17.6. These filaments, often blended with biopolymers like polylactic acid (PLA), offer an eco-friendly alternative to petroleum-based materials. Bamboo-based filaments have been used in the production of sustainable furniture, customized eco-friendly packaging, and biodegradable consumer products [40, 41].

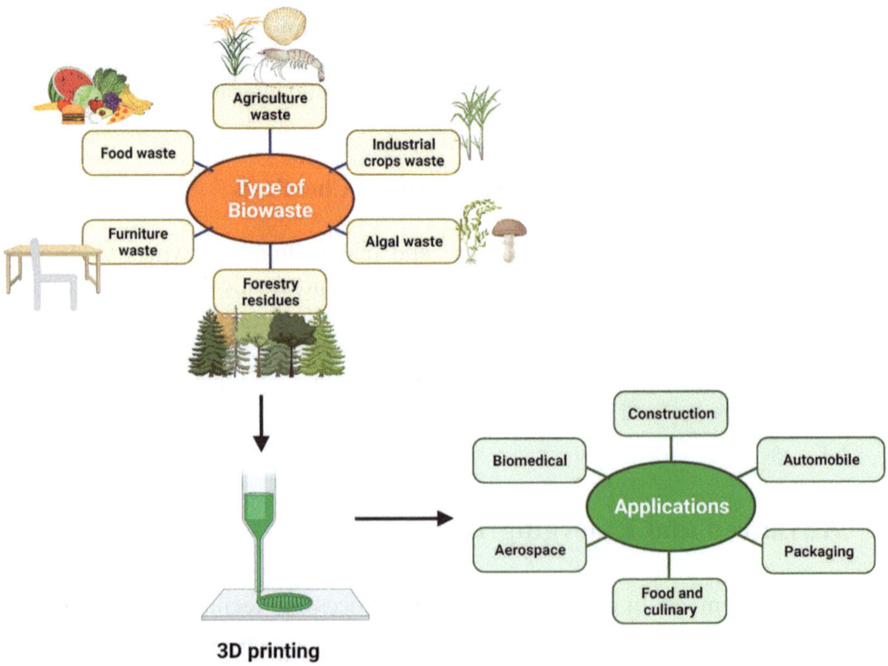

Fig. 17.6 Bamboo crop waste utilization in 3D printing, converting it into bio-based filaments for sustainable applications (adapted with permission under CC BY 4.0 from [42])

4.3 *Bamboo Waste as a Source for Biodegradable Composites*

Bamboo fibers are integrated into bioplastics to create reinforced composites with improved mechanical strength and environmental sustainability. These composites have potential applications in the automotive industry for lightweight, high-strength interior components, in the construction sector for durable panels, and in packaging as an alternative to synthetic plastics [43, 44]. Incorporating bamboo fibers into biodegradable polymer matrices enhances their durability while maintaining biodegradability, aligning with circular economy principles.

Table 17.1 summarizes Sects. 3 and 4 and provides an overview of the different bamboo waste-derived products and their respective applications.

This table highlights how bamboo waste-derived products contribute to reducing the environmental impact and provide opportunities for innovation in sectors that are key to sustainable development. Bamboo can be harnessed by integrating these materials into production processes to promote a circular economy.

5 Environmental and Economic Impact of Bamboo Waste Management

Effective bamboo waste management addresses environmental concerns and provides economic opportunities. By adopting sustainable practices, industries can minimize their carbon footprints while unlocking new revenue streams through value-added bamboo-derived products.

5.1 *Carbon Footprint Reduction Through Sustainable Waste Utilization*

The conversion of bamboo waste into biochar, bioplastics, biofuels, and other renewable materials contributes to carbon sequestration and greenhouse gas reduction. For instance, biochar application in soil improves carbon storage, reducing CO_2 emissions [45]. Similarly, anaerobic digestion for biogas production reduces methane emissions from decomposing organic waste, enhancing overall sustainability [19].

5.2 *Economic Benefits and Market Potential of Bamboo-Derived Products*

Bamboo waste valorization presents economic advantages by creating new market opportunities. The demand for sustainable materials, such as biodegradable composites and bio-based textiles, is growing rapidly. Bamboo fiber-reinforced

Table 17.1 Value-added products derived from various types of bamboo waste and their applications

Bamboo waste type	Value-added products	Applications	References
Bamboo residues (leaves, chips, sawdust, stems)	Compost	Soil amendment improves soil fertility and water retention	Zhong et al. [21]; Zhang et al. [22, 23]
Bamboo mulch	Mulching material	Weed suppression, soil moisture retention, and erosion control	Xu et al. [24]; Zhang et al. [25]
Sawdust, chips, bamboo fibers	Engineered wood panels, bio-composites	Construction, furniture, and interior structures	Li et al. [26]
Bamboo waste biomass	Briquettes, pellets	Renewable biofuel, an alternative to fossil fuels	Muche and Gebremedhen [27]
Bamboo biomass	Biochar (via pyrolysis)	Soil amendment, carbon sequestration, and alternative energy sources	Nam et al. [28]
Bamboo biomass	Hydrochar (via hydrothermal carbonization)	Soil conditioning, solid fuels, wastewater treatment adsorbent	Deng et al. [31]; Schneider et al. [32]
Bamboo fibers	Nanocellulose (via acid/alkaline treatment)	Bioplastics, coatings, biomedical applications	Lin et al. [33]
Bamboo waste biomass	Biogas (via anaerobic digestion)	Renewable energy, methane production	Huang et al. [35]
Bamboo waste sugars	Bioethanol (via fermentation)	Sustainable biofuel alternative	Rovera [37]
Bamboo biomass	Bioplastics (via fermentation)	Biodegradable plastic alternatives	Shen et al. [14]; Ngabala and Emmanuel [34]
Bamboo lignocellulose	Value-added chemicals (via enzymatic hydrolysis)	Bioplastics, pharmaceuticals, green solvents	Soni et al. [38]
Bamboo fibers	Nanocellulose, carbon nanofibers	Packaging, textiles, biomedical, energy storage	Zhang et al. [22, 23]; Iroegbu and Ray [39]
Bamboo powder/fibers	3D printing filaments	Sustainable furniture, eco-friendly packaging	Moraes [40]; Wong et al. [41]
Bamboo fibers	Biodegradable composites	Automotive, construction, and packaging industries	Sholokhova et al. [43]; Wang et al. [44]

bioplastics and nanocellulose-based films have commercial potential in the packaging, automotive, and construction industries [46]. Establishing bamboo waste recycling enterprises can generate employment and support rural economies.

5.3 *Comparison of Different Waste Management Strategies in Terms of Sustainability*

Among the various waste management techniques, composting and mulching are cost-effective but require time and land. Chemical processes like pyrolysis and hydrothermal carbonization provide high-value products but necessitate advanced infrastructure. Biochemical approaches, such as enzymatic hydrolysis and fermentation, offer promising bioethanol and bioplastics production pathways but require optimized processing conditions. A comparative analysis highlights the need for integrated approaches to maximize sustainability benefits [47].

6 Conclusion

Bamboo waste management plays a pivotal role in sustainable resource utilization, addressing environmental and economic challenges associated with bamboo processing. This chapter provides an in-depth examination of various waste management strategies, encompassing physical, chemical, and biochemical methods, alongside emerging technologies that aim to enhance the value of bamboo residues. Traditional approaches, such as composting, mulching, and pyrolysis, remain integral in converting bamboo waste into valuable products. Composting and mulching processes offer ecological benefits by transforming bamboo residues into organic fertilizers, which can enrich soil health and improve agricultural productivity. Pyrolysis, on the other hand, thermally decomposes bamboo waste in an oxygen-limited environment, producing biochar, which improves soil fertility and sequesters carbon, mitigating greenhouse gas emissions. These techniques are foundational to creating renewable energy sources, reducing waste, and contributing to sustainable agricultural practices. Enzymatic hydrolysis, microbial fermentation, and nanomaterial extraction provide innovative pathways for developing high-value products, such as bioethanol, bioplastics, and biodegradable composites. Nanomaterial extraction from bamboo waste opens up new avenues for developing advanced materials used in electronics, medicine, and environmental applications. Integrating these diverse waste management strategies contributes significantly to reducing the carbon footprint of bamboo processing, promoting the principles of the circular economy, and enhancing resource efficiency. By transforming bamboo waste into high-value products, these strategies help create closed-loop systems in which waste is minimized and resources are continuously reused. Sustainable bamboo waste valorization holds substantial economic potential, opening new markets for bio-based materials, promoting rural development, supporting green industries, and contributing to sustainable economic growth. However, successfully implementing these strategies requires technological advancements, policy support, and interdisciplinary collaboration. The optimization of processing techniques, along with the scaling up of commercial applications, will be essential for unlocking the full potential of bamboo waste management. Ongoing research and innovation will be critical in expanding the applications of bamboo waste across diverse sectors such as packaging, construction, and energy production.

Acknowledgments The authors thank the Director, CSIR-AMPRI, Bhopal, for providing the necessary institutional facilities and encouragement.

Funding The authors declare that the funding for this research is granted by CSIR-AMPRI, Bhopal (Grant No. OLP-303) for scientific research support.

Conflicts of Interest The authors declare no conflicts of interest.

References

1. Verma S, Hashmi SAR, Mili M, Hada V, Prashant N, Naik A et al (2021) Extraction and applications of lignin from bamboo: a critical review. Eur J Wood Wood Prod 79(6):1341–1357
2. Chaturvedi K, Singhwane A, Dhangar M, Mili M, Gorhae N, Naik A et al (2024) Bamboo for producing charcoal and biochar for versatile applications. Biomass Convers Biorefinery 14(14):15159–15185
3. Verma S, Mili M, Hada V, Kalita P, Hashmi SAR, Rathore SKS, Srivastava AK (2022) Bamboo leaves: sustainable feedstock in pharmaceuticals, food, and agricultural sector. Environ Technol Rev 11(1):266–281
4. Mili M, Singhwane A, Hada V, Naik A, Nair P, Srivastava AK, Verma S (2023) Advances in bamboo composites for structural applications: a review. In Book: Bamboo - Recent Development and Application. IntechOpen, pp.1–18. https://doi.org/10.5772/intechopen 110489
5. Atanda J (2015) Environmental impacts of bamboo as a substitute constructional material in Nigeria. Case Stud Constr Mater 3:33–39
6. Sawarkar AD, Shrimankar DD, Manekar SC, Kumar M, Garlapati PK, Singh L (2023) Bamboo as a sustainable crop for land restoration in India: challenges and opportunities. Environ Dev Sustain 27(1):1–33
7. Ben-Zhi Z, Mao-Yi F, Jin-Zhong X, Xiao-Sheng Y, Zheng-Cai L (2005) Ecological functions of bamboo forest: research and application. J For Res 16:143–147
8. Koul B, Yakoob M, Shah MP (2022) Agricultural waste management strategies for environmental sustainability. Environ Res 206:112285
9. Balat M (2011) Production of bioethanol from lignocellulosic materials via the biochemical pathway: a review. Energy Convers Manag 52(2):858–875
10. Demirbas A (2011) Waste management, waste resource facilities and waste conversion processes. Energy Convers Manag 52(2):1280–1287
11. Bhatt PK, Singh RP (2018) The effects of toxic agricultural wastes on the environment and their management. Cosmos. : An International Journal of Art and Higher Education 7(1):1–7

12. Ding Z, Awasthi SK, Kumar M, Kumar V, Dregulo AM, Yadav V et al (2023) A thermochemical and biotechnological approaches for bamboo waste recycling and conversion to value added product: towards a zero-waste biorefinery and circular bioeconomy. Fuel 333:126469
13. Tagade A, Kirti N, Sawarkar AN (2021) Pyrolysis of agricultural crop residues: an overview of researches by Indian scientific community. Bioresour Technol Rep 15:100761
14. Shen S, Nges IA, Yun J, Liu J (2014) Pre-treatments for enhanced biochemical methane potential of bamboo waste. Chem Eng J 240:253–259
15. Esfahani MR, Taylor A, Serwinowski N, Parkerson ZJ, Confer MP, Kammakakam I et al (2020) Sustainable novel bamboo-based membranes for water treatment fabricated by regeneration of bamboo waste fibers. ACS Sustain Chem Eng 8(10):4225–4235
16. Zhang X, Zhang P, Yuan X, Li Y, Han L (2020) Effect of pyrolysis temperature and correlation analysis on the yield and physicochemical properties of crop residue biochar. Bioresour Technol 296:122318
17. Durán-Valle CJ, López-Coca IM (2024) Biochemical and thermochemical conversion technologies for agriculture waste transformation. In: Transforming agriculture residues for sustainable development: from waste to wealth. Springer Nature Switzerland, Cham, pp 47–84
18. Debnath B, Haldar D, Purkait MK (2021) A critical review on the techniques used for the synthesis and applications of crystalline cellulose derived from agricultural wastes and forest residues. Carbohydr Polym 273:118537
19. Wijitkosum S, Sriburi T, Krutnoi L (2024) Taking advantage of disposal bamboo chopsticks to produce biochar for greenhouse crop cultivation. Emerg Sci J 8(3):917–932
20. Yeh CH, Yang TC (2020) Utilization of waste bamboo fibers in thermoplastic composites: influence of the chemical composition and thermal decomposition behavior. Polymers 12(3):636
21. Zhong Z, Bian F, Zhang X (2018) Testing composted bamboo residues with and without added effective microorganisms as a renewable alternative to peat in horticultural production. Ind Crop Prod 112:602–607
22. Zhang X, Zhong Z, Bian F, Yang C (2019a) Effects of composted bamboo residue amendments on soil microbial communities in an intensively managed bamboo (Phyllostachys praecox) plantation. Appl Soil Ecol 136:178–183
23. Zhang Y, Xu W, Wang X, Ni S, Rosqvist E, Smått JH et al (2019b) From biomass to nanomaterials: a green procedure for preparation of holistic bamboo multifunctional nanocomposites based on formic acid rapid fractionation. ACS Sustain Chem Eng 7(7):6592–6600
24. Xu Y, Li Q, Man L (2021) Bamboo-derived carboxymethyl cellulose for liquid film as renewable and biodegradable agriculture mulching. Int J Biol Macromol 192:611–617
25. Zhang Y, Jiang P, Li Y, Wu J, Xu K, Hill S, Wang H (2013) Chemistry of decomposing mulching materials and the effect on soil carbon dynamics under a Phyllostachys praecox bamboo stand. J Soils Sediments 13:24–33
26. Li X, Ji S, Li T, Liu Z, Hao X, Chen Z et al (2023) The physical, mechanical and fire performance of bamboo scrimber processed with thermal-treated bamboo bundles. Ind Crop Prod 205:117549
27. Muche BA, Gebremedhen HS (2024) Experimental investigation on the mechanical and physical properties of particleboard from residual materials with lowland bamboo chip composition. Adv Mater Sci Eng 2024(1):8396809
28. Nam NH, Linh VN, Dung LD, Ha VTT (2020) Physico-chemical characterization of forest and agricultural residues for energy conversion processes. Vietnam J Chem 58(6):735–741
29. Yang H, Huan B, Chen Y, Gao Y, Li J, Chen H (2016) Biomass-based pyrolytic polygeneration system for bamboo industry waste: evolution of the char structure and the pyrolysis mechanism. Energy Fuels 30(8):6430–6439
30. Lehmann J (2007) Bio-energy in the black. Front Ecol Environ 5(7):381–387
31. Deng F, Jiang H, Xie Z, Chen Y, Zhou P, Liu X, Li D (2022) Nutrient recovery from biogas slurry *via* hydrothermal carbonization with different agricultural and forestry residue. Ind Crop Prod 189:115891

32. Schneider D, Escala M, Supawittayayothin K, Tippayawong N (2011) Characterization of biochar from hydrothermal carbonization of bamboo. Int J Energy Environ 2(4):647–652
33. Lin W, Yang J, Zheng Y, Huang C, Yong Q (2021) Understanding the effects of different residual lignin fractions in acid-pretreated bamboo residues on its enzymatic digestibility. Biotechnol Biofuels 14(1):143
34. Ngabala FJ, Emmanuel JK (2024) Potential substrates for biogas production through anaerobic digestion-an alternative energy source. Heliyon 10:23
35. Huang C, Li R, Zheng Y, Lan K (2024) Life cycle assessment of bioproducts from bamboo residues with biphasic phenoxyethanol-acid based biorefinery technology. Chem Eng J 489:151181
36. Black MJ, Roy A, Twinomunuji E, Kemausuor F, Oduro R, Leach M, Sadhukhan J, Murphy RJ (2021) Bottled biogas—an opportunity for clean cooking in Ghana and Uganda. Energies 14:3856. https://doi.org/10.3390/en14133856
37. Rovera C (2022) Reinforced plastics and bioplastics using added value additives extracted from lignocellulosic agro-waste feedstocks. Doctoral Thesis. Department of Food, Environmental and Nutritional Sciences (DeFENS), The University of Milano. Italy.
38. Soni SK, Parkash O, Manhas R, Tewari R, Soni R (2019) Value added products from lignocellulosic agricultural residues: an overview. Int J Food Ferment Technol 9(2):101–115
39. Iroegbu AOC, Ray SS (2021) Bamboos: from bioresource to sustainable materials and chemicals. Sustain For 13(21):12200
40. Moraes MJBD (2024) Valorization of bamboo leaf ash in biocomposites for fibercement and 3D printing, Doctoral dissertation, Universidade de SÃ£ o Paulo, Brazil
41. Wong J, Aşut S, Brancart S (2024) 3D printing with bamboo: an early-stage exploration towards its use in the built environment. Sustain For 16(11):4619
42. Rahman AM, Rahman TT, Pei Z, Ufodike CO, Lee J, Elwany A (2023) Additive manufacturing using agriculturally derived biowastes: a systematic literature review. Bioengineering 10(7):845. https://doi.org/10.3390/bioengineering10070845
43. Sholokhova A, Varžinskas V, Rutkaitė R (2024) Valorization of agro-waste in bio-based and biodegradable polymer composites: a comprehensive review with emphasis on Europe perspective. Waste Biomass Valori:1–35
44. Wang R, Huang Q, Hong G, Liu Y, Liu C, Li J et al (2024) Eco-friendly versatile shielding revolution: tannin tailored bamboo waste composite with wave-absorbing, flame retardancy, and antibacterial abilities. Int J Biol Macromol 277:134162
45. Pang B, Zhou T, Cao XF, Zhao BC, Sun Z, Liu X et al (2022) Performance and environmental implication assessments of green bio-composite from rice straw and bamboo. J Clean Prod 375:134037
46. Kaur PJ, Yadav P, Gupta M, Khandegar V, Jain A (2022) Bamboo as a source for value added products: paving way to global circular economy. Bioresources 17:3
47. Gan J, Chen M, Semple K, Liu X, Dai C, Tu Q (2022) Life cycle assessment of bamboo products: review and harmonization. Sci Total Environ 849:157937

Chapter 18
Valorization of Rubber Wood Waste and Crumb Rubber Factory as a Source of Bioenergy and Bioproducts

Vibhuti Sharma, Sasane Rohan Gautam, Gaytri Mahajan, and Reena Gupta

Abbreviations

CO	Carbon monoxide
CR	Crumb rubber
ELTs	End-of-life tires
EU	European Union
LiP	Lignin peroxidase
L-MA	L-malic acid
MP	Microplastics
NOx	Nitrogen oxides
NR	Natural rubber
PAH	Polycyclic aromatic hydrocarbons
PHA	Polyhydroxyalkanoate
PLA	Polylactic acid
PM	Particulate matter
PMA	Poly(β-L-malic acid)
RSO	Rubber seed oil
RW	Rubber wood
RWH	Rubber wood enzymatic hydrolysates
RWS	Rubber wood sawdust
SOx	Sulfur oxides
SS	Sewage sludge
SSA	Sub-Saharan African
TDF	Tire-derived fuel
VOC	Volatile organic compounds
WPC	Wood plastic composites

V. Sharma · S. R. Gautam · G. Mahajan · R. Gupta (✉)
Department of Biotechnology, Himachal Pradesh University, Summerhill, Shimla, India

C. Trois, R. C. Ray (eds.), *Plantation Crop Wastes: Valorization for Economic Sustainability*, Waste as a Resource,
https://doi.org/10.1007/978-3-031-96805-1_18

1 Introduction

The rubber tree, *Hevea brasiliensis* L., is a perennial plant that grows quickly and belongs to the Euphorbiaceae family [1]. Rubber is a fast-growing tropical tree crop, primarily cultivated for its latex. In 1496, Columbus became the first to report on latex [2]. The tree begins to produce latex at 6 years of age. Originally used to remove charcoal or graphite stains from paper, latex is now utilized for various purposes, including biodiesel production and the cosmetics industry. The tree bole is an alternative to more common timber-bearing trees and provides natural latex [1]. Rubber is a well-established economic crop that generates income for several nations [2]. The concentrated rubber latex industry, which produces concentrated latex as a raw material for various downstream markets, such as rubber gloves, medical supplies, boots, and balloons, has seen a significant increase in demand in both domestic and international markets [3]. After the rubber plantations were industrially cultivated, they began to achieve significant success. The rubber tree is often referred to as "white gold" due to the pale color of its latex. Rubber plantations are expanding rapidly due to government incentives, investor interest, rising rubber sheet prices, and increased rubber consumption [2].

India is the world's third-largest user (8.70%) and fourth-largest producer (8.90%) of natural rubber. The southern Indian states have made significant contributions to traditional rubber cultivation: Kerala accounts for over 90% of the country's total rubber production, while Tamil Nadu contributes 2.89%. According to government estimates, growing rubber in the northeastern states will improve smallholders' socioeconomic circumstances, and the National Bureau of Soil Survey found about 450,000 hectares of land suitable for rubber cultivation [4]. The composition of natural rubber wood is 2.63% lignin, 20.13% cellulose, and 65.04% hemicellulose. The moisture content of fresh rubber wood is about 60–80%. The chemical component of natural rubber (cis-1,4-polyisoprene) is a polymer of isoprene (C_5H_8). *Parthenium argentatum* Gray, commonly known as the guayule shrub, along with *Ficus carica, Ficus racemosa, Ficus benghalensis, Euphorbia lactiflua, Ficus elastica* (Indian rubber tree), *Hevea brasiliensis*, and over 2500 other latex-producing plant species, serve as sources of natural rubber (NR), a vital raw material for the polymer industry. The history of natural rubber (NR) production highlights ongoing efforts to identify alternative sources in both temperate and tropical regions. However, only a few crop species capable of producing NR generate substantial amounts of high-molecular-weight rubber [5]. Because fossil fuels have detrimental effects on the environment, including climate change, global warming, and human health, renewable energy sources are essential for developing sustainable energy systems in the future. As a result, many nations are seeking and developing alternative and renewable energy sources to address these issues [6]. Following various structural changes in rubber plantations, small rubber landholders now have a significant impact. Rubber plantation growth depends on the price of rubber sheets. Increased investment is needed in the rubber plantation sector [7]. Conversely, almost 1.5 billion tires are discarded each year at the end of their life cycle. Most waste tires and rubber goods, other than tires, such as belts and hoses, go through accumulation or disposal processes. While a significant portion of waste tires is disposed of, a smaller portion is recycled into retaining wall construction, boat safety guards, and weather-resistant products [8]. This chapter explores rubber wood waste production, its use in bioenergy and bioproducts, and the challenges of its non-biodegradability.

2 Global Rubber Production

Rubber trees are the primary source of natural rubber and latex. The rubber plantations in Southeast Asian countries are well-known. The top three countries in terms of rubber production are Malaysia, Indonesia, and Thailand. In India, large estates initially controlled the rubber plantations. Southern Thailand hosts a large natural rubber manufacturing sector, which accounted for approximately 37% of global natural rubber production in 2019 [9]. Rubber plantations in Southeast Asia are well-known, with Malaysia, Indonesia, and Thailand being the top three producers. Initially, large estates controlled rubber plantations in India, but after various structural changes, small rubber landholders now play a significant role. An overview of global rubber production is shown in Fig. 18.1. The growth of rubber plantations depends on the cost of rubber sheets, and more investment is needed in the sector [7]. Natural rubber is significant in both area and production, with smallholder farmers accounting for over 80% of rubber output [4].

The global amount of rubber product waste exceeds 80 million tons and grows by at least 10% each year. The danger class includes waste from worn tires, which takes 12–140 years to decompose in the environment. The composition of rubber tires is shown in Fig. 18.2. Currently, used tires can be recycled in various ways [10]. Research into different waste tire disposal options has been prompted by the negative environmental effects of tire waste accumulation in several countries.

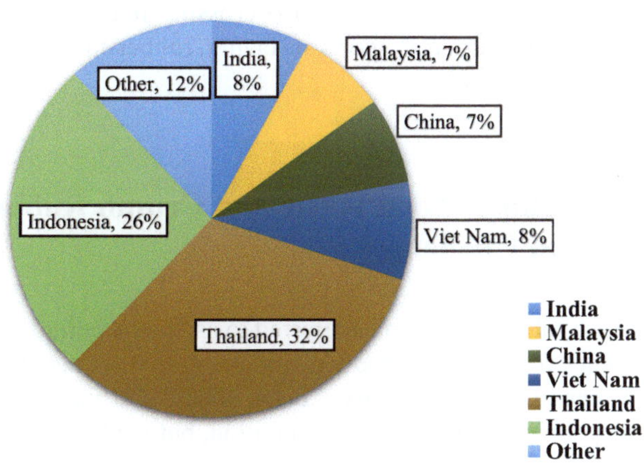

Fig. 18.1 Overview of global rubber production

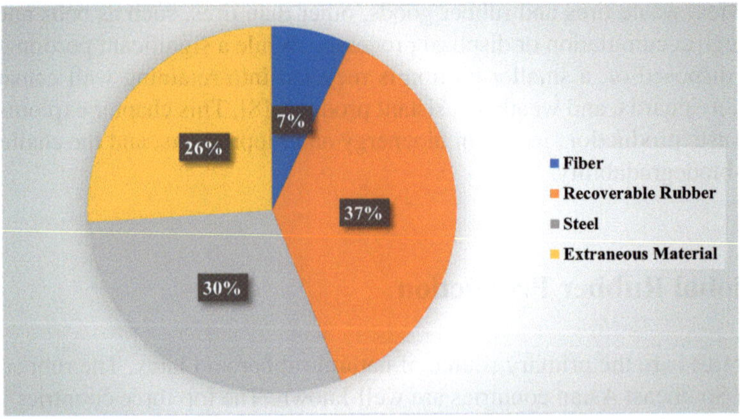

Fig. 18.2 The composition of rubber tires

Rubber waste is a serious global issue. In 2005, more than 10 billion tires were disposed of worldwide, and over one billion tires were removed from service annually. By 2030, up to 1.2 billion tires may be disposed of. In 2011, it was estimated that there were four billion discarded tires in landfills and reserves globally [11]. The high carbohydrate content of rubber wood makes it a potentially suitable substrate for biogas production. Rubber seeds are an intriguing feedstock for bioenergy due to their abundance, residual nature, and annual availability, especially when treated as waste. Additionally, as an inedible oil source, they do not compete with food production. The kernel contains 19–23% crude protein and 40–50% non-edible vegetable oil. Defatted kernels or meals can be used as animal feed for fish, pigs, ruminants, and poultry due to their high protein content and amino acid profile [12]. While rubber comprises the majority of tires, they also contain metals, textiles, fillers, and other chemical additives, posing challenges for tire recycling systems. The four R's—Reduce, Reuse, Recycle, and Recover—are fundamental options to address waste tire issues. Waste tires can be recycled into new products for various applications, including railroad crossing mats, gaskets, doormats, wheel chocks, playground surfaces, sports surfacing, rubber-modified asphalts, and civil engineering applications [8].

3 Rubber Wood Waste and Crumb Rubber: Characteristics and Challenges

Trash rubber, sometimes referred to as "black pollution," is a type of solid trash that comes from rubber products that are no longer needed for their intended use, such as car tires, rubber gloves, belts, shoes, and industrial byproducts. Rubber waste recycling has always drawn attention, but neither the general public nor researchers have paid enough attention to studies on long-term pollutant release and new eco-toxicological mechanisms. The volume and variety of waste rubber have increased significantly every year, which has a significant negative impact on the environment. In particular, the strong and pungent smells and flammability of storage sites are extremely upsetting for the surrounding population. It can be very difficult to put out fires caused by waste rubber burning, accidents, or major disasters, and the toxins produced can easily cause respiratory problems in people [13].

3.1 Composition and Characteristics of Rubber Wood Waste

An important feedstock for biomass power plants in Southern Thailand is rubber wood waste (RW), an industrial waste made from an agricultural product. The burning process produces two significant carcinogens: fine particulate matter with a particle size of 2.5 μm and polycyclic aromatic hydrocarbons (PAHs). Additionally, the burning process releases contaminated woody biomass ash that contains heavy metals, which has raised concerns for the environment. This procedure is widely accepted due to its low cost and cleanliness. Holocellulose, a polysaccharide, accounts for 54–74% of the weight of rubber wood waste, whereas lignin accounts for approximately 13–22.4% [14].

3.1.1 Physical Properties of Rubber Wood

Because of its high lignin content, which can improve the mechanical qualities of the composites, recycled rubber wood shows promise as a material for wood-plastic composites. Recycled polypropylene can be used as the matrix material to enhance the qualities of recycled rubber wood composites. Recycled rubber wood composites exhibit impact resistance and thermal stability, which can be improved by adding recycled polypropylene [15]. Investigating how moisture affects wood plastic composites (WPC) composed of recycled rubber wood and polypropylene may lead to several innovative solutions to difficulties, such as increasing the durability of WPC, creating sustainable materials, comprehending how WPC react to moisture, and improving product performance for a range of uses [16].

3.1.2 Sources and Types of Rubber Waste Generated

The rubber wood industry depletes timber supplies in addition to causing environmental deterioration through wood waste. Although rubber wood is frequently regarded as an eco-friendly substitute, the manufacture of this material can lead to several environmental problems, such as the creation of solid waste and pollution of the air and water. Sanding, sawing, and timber machining all produce wood dust and particles, which are released into the air during wood processing. Dust emissions are also a result of traffic zones and woodyards. Combustion products, such as carbon monoxide (CO), sulfur oxides (SOx), nitrogen oxides (NOx), particulate matter (PM), and volatile organic compounds (VOCs), are released by industrial boilers when wood and bark are burned. The fuel used determines the precise kinds and amounts of emissions. VOC emissions are also caused by kiln drying and the application of varnishes, lacquers, and solvents [17].

3.2 Crumb Rubber: Production and Waste Streams

Crumb rubber can be produced from scrap tires in a wide range of particle sizes and quality levels. In 2001, the United States produced over 281 million scrap tires. Approximately 75% of these tires were repurposed in a secondary market of some kind. Tire-derived fuel (TDF), which likely accounts for around 33% of all scrap tires produced, has the largest market for reuse, mostly for use as auxiliary fuel in cement kilns. About 15% of trash tires come from civil engineering uses, where they are shredded for things like highway embankments and landfill leachate collection. Crumb rubber, sometimes referred to as ground rubber, was the third most significant reuse application [18]. The mechanically produced crumb rubber has a very rough surface with grains no bigger than 0.3 mm. Rubber that has been devulcanized is cooled to below its glass-transition temperature by cryogenic grinding in liquid nitrogen (−80 °C). After that, the frozen, brittle rubber is delivered to hammer mills, where it is ground into uniformly sized particles (75 μm) with sharp edges and smooth surfaces. Although it can be obtained, pure rubber powder has a significant humidity content (about 12–15 weight percent). The primary drawback of cryogenic grinding is the high expense of liquid nitrogen needed to cool the rubber waste, which renders it less cost-effective than traditional grinding at room temperature [19].

3.2.1 Manufacturing Process of Crumb Rubber

Granulated crumb rubber (CR), which is made from used tires, has gained attention as a bituminous mixed ingredient. The material composition of the tires is shown in Table 18.1. CR in asphalt pavements is frequently regarded as a long-term method to cut waste. As a substitute polymer, CR enhances hot mix asphalt's performance. Numerous investigations have revealed an overview of how CR affects the stiffness, fatigue resistance, and rutting levels of road pavements [20]. Several studies have been carried out in recent years to improve the material cycle, where rubber, which is produced as a by-product of the crushing and sieving of scrap tires, becomes the resource required in other production processes for the construction of road pavement layers. This is due to the growing popularity of crumb rubber, a recycled material derived from End-of-Life Tires (ELTs). There are various uses for crumb rubber

Table 18.1 The material composition of tires

Material (wt.%)	Passenger car tires	Truck tires	Off-road tires
Natural rubber	14–22	27–30	14–22
Synthetic rubber	23–27	14–15	23–27
Carbon black	28	20–28	22
Steel	13–15	14–25	12
Zinc oxide	1	2	2
Fabric, fillers, accelerators, antiozonants	14–17	10–17	10
Others	8.5	5	7
Average weight of a new tire (kg)	8.5–11	54–65	27–74
Average weight of a waste tire (kg)	7–9	45–56	21–64

(CR) in the creation of asphalt mixtures. The goal is to enhance asphalt mixture performance and identify a substitute for tire disposal [21].

3.3 Rubber Wood and Crumb Rubber: Impact on the Environment

By 2030, it is anticipated that there will be five billion waste tires disposed of. The majority of end-of-life tires were disposed of as rubbish, with only one billion being recycled. Wasted tires are recognized to be difficult for the environment to biodegrade. The three existing tire disposal techniques—stockpiling, landfilling, and burning—all have detrimental effects on the environment and health of the general public. One of the waste tire-derived products, crumb rubber (CR), can be used as modifiers and fillers in the manufacturing of concrete and asphalt, potentially lowering the demand for virgin materials, raising the market value of rubber products, and recycling waste tires with the least amount of adverse environmental effects [22].

4 Valorization Pathways for Rubber Wood Waste and Crumb Rubber

A promising alternative is the valorization of discarded tires, which involves turning them into useful goods or sources of energy [23]. However, there are obstacles to optimizing the valorization process because of the intricacy of waste tires and several paths for valorization. To address these issues, simulation and optimization methods are being employed more and more. In the context of valorization processes, simulation models are useful for investigating mechanisms and predicting behavior [24]. Numerous studies have looked into thermochemical processes like gasification, pyrolysis, and hydrothermal liquefaction as potential avenues for waste tire valorization [25]. Different methods of valorization of waste tires or crumb rubber are shown in Fig. 18.3. There are different pathways for the valorization of rubber waste, such as grinding and shredding. Grinding waste tires are ground into tiny particles to create crumb rubber, which, depending on the size of the particles, can be used in a variety of ways. Shredding-waste rubber wood can be broken down into smaller bits that can be used to create composite materials [25].

By using waste as aggregates in concrete, it might be possible to produce environmentally friendly concrete, which would encourage decarbonization in the building industry, the valorization of environmental waste, and a decrease in the use of non-renewable resources. It is especially possible to incorporate trash into concrete through methods like modular building using sustainable cellular concrete blocks and building using 3D printing. For these reasons, the use of different kinds of waste in concrete has been investigated by several researchers [26]. Rubber wood and rubber crumb factory waste can be valorized into various bioproducts and bioenergy. A summary of bioproducts and bioenergy from rubber wood waste and rubber crumb factory waste is shown in Table 18.2.

The three main methods of waste tire management are landfilling, incineration, and recycling, and each has itsvarious benefits and drawbacks. Recycling used tires is becoming a more sustainable and environmentally friendly method. On the other hand, landfilling and combustion pollute the environment. A previous study focuses on tire recycling to investigate the sustainable valorization of discarded tires [27]. For the years 2014–2023, the terms "waste tires," "recycling," and "valorization" were used in a search on the Web of Science and Scopus, demonstrating the increasing interest of academics in the use of scrap tires as a resource throughout time. Keywords were then taken from the statistical literature, and the results showed that pyrolysis, reused rubber, building materials, and other processes are the most

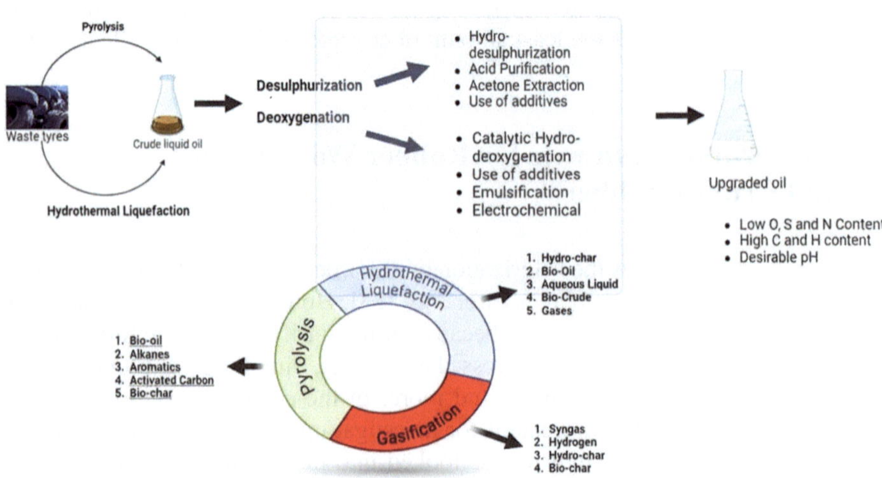

Fig. 18.3 Different methods of valorization of waste tire or crumb rubber

Table 18.2 A summary of bioproducts and bioenergy from rubber wood waste and rubber crumb factory waste

Waste type	Bioproducts	Bioenergy	Processes involved
Rubber wood waste	Wood composites	Biomass combustion for heat and electricity	Pyrolysis, gasification, and direct combustion
	Activated carbon (adsorbents)	Syngas for energy production	Torrefaction
	Biochar (carbon sequestration, soil amendment)	Bio-oil for alternative fuel	Hydrothermal carbonization
Rubber crumb factory waste	Rubber-derived oil for industrial applications	Biodiesel from rubber seed oil	Transesterification and fermentation
	Rubber-derived bioplastics	Biogas (methane)	Anaerobic digestion
	Carbon black substitute for rubber products	Solid biofuels	Palletization, torrefaction

common ways to generate resources from discarded tires. The aforementioned investigation indicates that there are a variety of methods for using waste tires as resources, each with pros and cons of its own. As a result, it is difficult to identify the most sustainable technique plainly and impartially [28].

The utilization of waste materials in long-lasting products, such as concrete, and the possibility of recycling or additional valorization after disposal are crucial environmental factors because they can lower the carbon footprint and resource consumption connected with construction operations. Given the limited demand for these resources and the tendency for lower costs as compared to other waste products with established or well-defined valorization pathways, an inquiry into these pathways may prove to be a cost-effective strategy [26]. Co-gasification accounts for the majority of plastic gasification research. The properties of the gas generated between 800 and 900 °C during the gasification and pyrolysis of plastic samples that also contained rubber were also analyzed. Significant hydrogen production was demonstrated by the gasification results at any temperature. Gasification demonstrated a seven-fold increase in hydrogen generation above 900 °C and a five-fold increase over the pyrolysis process at 800 °C. Rubber gasification and forest biomass comparisons yielded comparable findings. Synthesis gas production from rubber was higher than that from forest biomass. These findings imply that the co-gasification of rubber additives and biomass-based plastics has a significant impact while operating at high temperatures [29].

4.1 Bioenergy Production

Crumb rubber is a widely used bitumen additive, primarily valued for its affordability. As a by-product of recycling old tires, rubber crumbs are essentially a waste product. Bitumen qualities like elasticity, wear resistance, and fatigue strength are all improved by worn rubber. Normally, bitumen does not create a homogenous system with crumb rubber because it is insoluble. The dispersion of rubber particles onto the surface may negatively impact the pavement's ability to maintain its quality while in use. The stabilization of rubber crumb by adding special substances or dispersing it to the sizes at which the system of rubber bitumen is subject to much weaker stratification while maintaining the original elasticity properties of the rubber particles is one of the solutions that researchers have suggested to address this issue. To partially dissolve the rubber in the bitumen, another, more energy-intensive method is to destroy the vulcanizate over the entire volume. When rubber dissolves between 50% and 60%, the optimum results are obtained [30]. Many individuals are interested in using pyrolysis oil made from scrap rubber as an alternative fuel. It contains a variety of substances, including diesel, petrol, and kerosene. In addition, any concoction developed to enhance the quality and efficiency of petrol is frequently referred to as a petrol additive [31]. Wood biomass ranks among the most abundant biomass materials and is widely utilized for the production of renewable bioenergy, high-value chemicals, and materials. Southern Thailand is the primary region for rubber cultivation, encompassing the majority of the regions in the nation that grow rubber. This region has a planting area of approximately 7–8 million rai, producing around 3–4 million tons of rubber annually [32]. The steps for energy production from waste rubber wood are shown in Fig. 18.4.

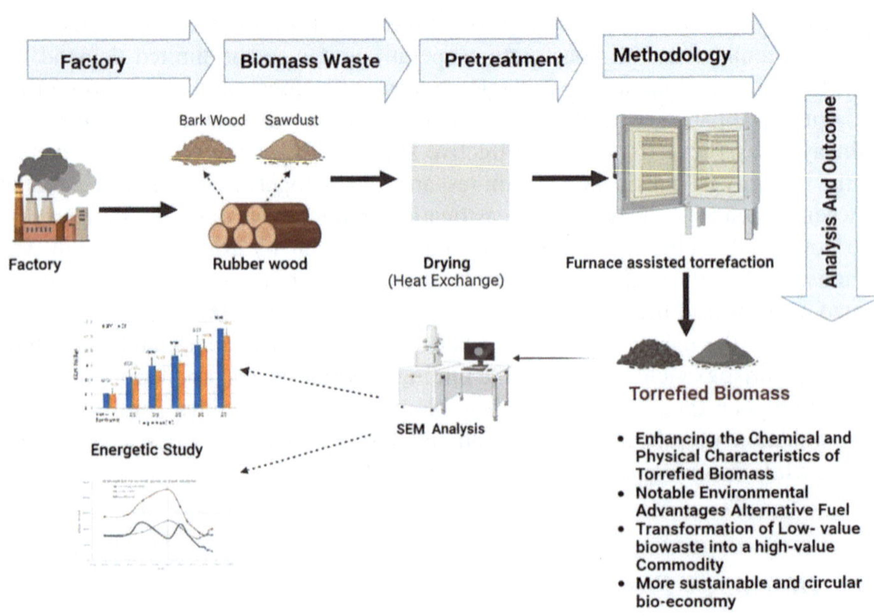

Fig. 18.4 Steps for the energy production from waste rubber wood

4.2 Biodiesel Production

Rubber farmers who intercrop with arable crops such as cassava, pineapple, yam, pepper, bitter leaf, maize, groundnut, and rice for sustenance, soil enrichment, and weed control before beginning natural rubber production may benefit from an additional source of income by collecting rubber seeds for biodiesel production [33]. In addition to its good lubrication, biodegradability, non-toxicity, and eco-friendliness when used in diesel engines, biodiesel has been identified as a good supplement and a possible substitute for fossil diesel due to its many characteristics. Food consumption and the manufacture of biodiesel from edible vegetable oils are in competition, which drives up the price of both. According to studies, rubber seed has 35–45% oil by weight, making it a more competitive option for biodiesel production than other non-edible oil-bearing plants. A viable alternative for the sustainable development of Sub-Saharan African (SSA) nations that rely significantly on fossil diesel is biodiesel made from non-edible rubber seed oil (RSO). The cost of biodiesel will be further decreased by using the plentiful plantain (*Musa paradisiaca*) peels, which are regarded as waste in SSA nations, as a heterogeneous base catalyst in the synthesis of RSO biodiesel. In most tropical SSA countries, rubber trees (*Hevea brasiliensis*) are produced domestically in enormous plantations for the express purpose of producing latex. The seeds of these plants fall to the ground as waste, either to germinate or to be gathered and thrown away every year. The aggregate capacity of the SSA countries, however, is predicted to be 717,750 hectares of rubber trees, or 251 million trees, which can produce roughly 107,662.5 tons of rubber seed, 17,947.339 tons of rubber seed oil, and 16,691.025 tons of biodiesel annually. Natural rubber tree planting will continue to support the manufacturing of rubber products from latex and seed oils for biodiesel production, which will boost local economies in SSA nations and provide extra financial benefits to plantation farmers [34].

4.3 Biogas Production

A natural biological process known as anaerobic digestion or biomethanation occurs when bacteria decompose organic materials under conditions with little to no oxygen.Among the many renewable technologies, biomethanation is a commonly used, commercially established method for processing biomass, turning it into a stable compost, and producing methane, which is used to make biofuels. The biomethanation method is used in this study to treat agricultural waste and wastewater produced during the production of natural rubber sheets. For organic materials to be digested efficiently, the C/N ratio should be between 25 and 30. The production of H_2S gas, which impeded the two methanogenic activities, caused the method to fail when it was tried in Vietnam. The high concentration of sulfates from sulfuric acid, which is utilized as a coagulant in the rubber manufacturing process, resulted in the formation of H_2S gas. In a previous study, sulfates were not added to the wastewater since formic acid is the coagulant utilized in the production of rubber sheets [14].

4.4 Production of Bioproducts

The application and disposal of biomass is another important subject from the perspective of the ecologically friendly concept that supports the sustainable development of energy, much like the state of waste tires. The qualities of biomass include a vast supply, a broad spread, and a variety of sources. Using rapid pyrolysis or thermochemical liquefaction, biomass can be converted into bio-oil, bio-char, and other byproducts. Many people use these bioproducts as clean, sustainable green energy. In the study of road materials, bio-oil was typically mixed with asphalt to change or replace it and create bio-asphalt with better qualities. The impact of bio-oil on asphalt modification was also slightly different because of the variations in the preparation method and source of bio-oil. When combined with the body of current literature, bio-oil greatly enhanced asphalt's fatigue performance [35].

4.5 Production of High-Value Chemicals Such as Proteins

Although it is not a top priority at the moment, using non-latex fractions from rubber trees, such as wood or rubber seed oil, could become significant in the future. One of the primary components of biomass and, in certain situations, its most valuable component, protein can improve the overall profitability of rubber plants. Rubber tree latex contains well-identified proteins, although some are noted for their potential to cause allergies. The proteins found in rubber tree seeds and leaves receive little attention. Rubber seed has been recognized as a possible source of protein, especially following the extraction of oil for the creation of biodiesel [36].

4.6 Production of Enzymes Such as Laccases and Lignin Peroxidase

Laccases catalyze the four-electron reduction of molecular oxygen to water that occurs when aromatic amines and many phenolic compounds undergo oxidation [37]. The active site of lignin peroxidase contains iron ions, while the active site of laccase (LaC) contains copper ions. These enzymes are used in dye decolorization, the food business, the pulp industry, and bioremediation. They are also highly specific to different substrates, effective catalysts, and environmentally friendly [38]. *Trichoderma homatum* strain NGL1 was used in solid-state fermentation to bio-convert rubber tree sawdust. Laccase and lignin peroxidase production were enhanced with the application of the response surface approach [39].

Lignin peroxidase (LiP) is an oxidative enzyme that degrades lignin into smaller aromatic molecules. LiP can degrade lignocellulosic components found in rubber waste, allowing rubber wood waste to be converted into biofuels and biochemicals. This enzymatic technique increases the value of rubber waste by allowing for efficient lignin depolymerization, leading to sustainable bioproducts [39].

4.7 Production of Polymer Composites and Bioplastics

The effect of treated ground tire rubber (treatment by various natural oils, $KMnO_4$, or H_2O_2) on polyurethane/ground tire rubber composite foams showed that suitable modification of ground tire rubber is a promising approach for tailoring the structure and performance of polyurethane composite foams [40].

Good biocompatibility, complete degradability, water solubility, non-immunogenicity, and ease of modification are characteristics of poly(β-L-malic acid) (PMA), a polymeric substance that contains L-malic acid (L-MA) as the only monomer and is naturally formed in living organisms. PMA is now widely utilized in chemicals, food additives, cosmetics, biopharmaceuticals, and biodegradable bioplastics. Co-fermentation can increase target product concentration and substrate utilization. It is far less expensive to use a lignocellulose-based substrate for co-fermentation, like rubber wood. The cost of sugar and PMA products can be decreased by 61% and 13%, respectively, by using *Aureobasidium pullulans* to ferment PMA in a co-substrate mixture of waste xylose mother liquor (which contains xylose, arabinose, and glucose) and gluconate (which contains sodium gluconate and glucose). The fermentation of rubber wood hydrolysates to produce polyhydroxyalkanoate (PHA) revealed that xylose is a synergistic carbon source that aided in the screening of the predominant PHA strain. To obtain an inexpensive fermentation medium for the synthesis of PMA by employing rubber wood enzymatic hydrolysates (RWH) as substrates and *A. pullulans* as the production strain. To further increase the microbial growth rate and PMA production, a novel biological method was created that involved co-fermenting RWH with an external carbon source (glucose and xylose). Following an investigation of the usage of xylose and glucose during fermentation, the PMA products were isolated and described [41].

4.8 Production of Toys Using Rubber Wood Sawdust

These days, the best toys for kids are thought to be entertaining and instructive. Children can enjoy themselves while fostering their natural growth and development with the aid of these toys [42]. Among all the materials used to make toys for kids, wooden toys have grown in popularity recently because of their straightforward design, eco-friendliness, and superior hygienic qualities over plastic and metal toys. Bio-composite materials, such as wood-polymer composites, are new materials suitable for making toys for kids. These materials are biodegradable and biocompatible composites that can be made from waste wood, natural rubber (NR), and polylactic acid (PLA), among other renewable resources. One of the commercially available bio-based thermoplastic polymers with outstanding qualities, environmental friendliness, and biodegradability is PLA. As a second-phase polymer, natural rubber has been utilized to toughen fragile PLA to further improve its properties and reduce its limitations. By acting as stress concentrators, NR particles boost the ability of the brittle polymer to absorb fracture energy, leading to a material with higher toughness. PLA toughness can be effectively increased by combining PLA and NR as polymer blends.

5 Technologies for Valorization

Valorization of rubber waste entails a variety of technologies targeted at transforming waste materials into valuable goods while lowering environmental effects and boosting sustainability. Key approaches include thermochemical processes (pyrolysis, gasification, and torrefaction) for bioenergy production, biotechnological methods (microbial and enzymatic degradation) for bioproduct synthesis, and mechanical and chemical recycling for raw material recovery [25]. These methods enable the conversion of rubber waste into biofuels, bioplastics, carbon black replacements, and other high-value industrial goods, thus contributing to circular economy practices.

5.1 Mechanical and Physical Processes

Discarded rubber products like tires, belts, shoes, and hoses are the main sources of waste rubber. Reclamation, retreading, landfilling, incineration, pyrolysis, ultrasonic devulcanization, use in road pavement through asphalt changes, and repurposing as a filler in polymers and rubber compounds are some of the processes for recycling rubber. The largest source of rubber waste comes from tires that are nearing the end of their useful lives. Rubbers, both natural and synthetic, as well as plasticizers and additives, are commonly utilized [43]. Using a variety of external energy sources, physical reclamation processes involve rupturing the three-dimensional network of cross-linked rubber. Rubber from waste tires and rubber wood is broken down by networks, resulting in low molecular weight fragments that are then mixed with virgin rubber. Products from physical processes can be used directly for valuable purposes like making concrete, or they can be used as feedstock for other chemical and biological processes [44]. Cryo-mechanical methods for extracting rubber from used tires have been documented since the middle of the 1960s. Nowadays, cyano-mechanical methods are employed as a pretreatment to produce fine rubber powder before the devulcanization process. Liquid nitrogen is used to immerse tiny bits of crumb rubber until they reach their glass transition temperature. They are taken to a ball mill, where they are ground into particles with a mesh size of 30 to 100. In addition to producing finer, more valuable powders, this method enables the alteration of physical properties like tensile strength, elongation modulus, and elongation at break [45].

5.2 *Thermo-Mechanical Processes*

Thermo-mechanical methods are usually the most popular among modern rubber reclamation technologies. These methods use friction between the crumb rubbers, mechanical shearing, and a high-temperature increase of about 200 °C brought on by an external heat source. Devulcanization and the formation of shorter polymer chains are caused by the reorganization of polysulfide cross-links and the dissolution of carbon-carbon bonds. Under the effect of sulfur dioxide, carbon disulfide, and hydrogen sulfide, the polysulfide cross-links are removed. Solvents like hexane, supercritical fluids, subcritical water, or oils can be added before or during the grinding process to assist in speeding it up, dissolve the small chains, and create inflated rubber [46].

5.3 *Grinding Processes*

Cryo-mechanical methods for extracting rubber from used tires have been documented since the middle of the 1960s. To produce fine, uniformly sized rubber powder, cyano-mechanical procedures are currently employed as a pretreatment before the devulcanization process. Liquid nitrogen is used to immerse small bits of crumb rubber until they reach their glass transition temperature. They are taken to a ball mill, where they are ground into particles with a mesh size of 30–100. In addition to producing finer, more valuable powders, this method enables the alteration of physical properties like tensile strength, elongation modulus, and elongation at break. Scrap tire pieces are ground mechanically using a grinder with a jagged edge in a technique known as dry ambient grinding. The size of rubber particles is reduced from 10 to 30 mesh. Although this method is known as "ambient grinding," the name is not true because the heat produced during the grinding process might cause the polymer chain to break, particularly in tires with high modulus or older tires. The reclaimed rubber may have a direct connection to the virgin rubber matrix because pendant groups are created during the grinding process [45].

5.4 Chemical Processes

Industrial companies use chemical processes the most out of all the rubber reclamation techniques. Numerous organic and inorganic compounds are used to recover rubber. Sulfides and mercaptans are frequently used in these processes [47]. However, because they are more economical and environmentally benign, non-sulfured compounds are currently being researched. Moreover, pyrolysis is regarded as a chemical reclamation process because of the chemical breakdown that occurs even in the absence of chemicals [48].

5.5 Pyrolysis

Pyrolysis is preferred for converting biomass or organic wastes into valuable bio-oil, biochar, and pyrolysis gas. At high temperatures (400–900 °C) in an inert environment, the high molecular chains undergo pyrolysis, which results in the breakdown of stable low-molecular products and solid residue. As the most significant governing process varies, the system temperature influences processes like cracking, dehydrogenation, cyclization, and aromatization. Furthermore, it was discovered that additional process variables and parameters such as particle size, heating rate, volatile residence time, and pyrolysis time had a substantial impact [49]. The flammable components of pyrolysis gas, such as CO, H, and CH_4, allow it to be used to generate heat and electricity. According to a survey of the literature on the processes of pyrolysis and co-pyrolysis of woody and non-woody biomasses as well as sewage sludge, there is no previous investigation of the pyrolysis and co-pyrolysis of rubber wood sawdust with sewage sludge from wastewater treatment of seafood processing. In an agitated pyrolysis bed reactor, the pyrolysis of tires or crumb rubber in a fixed bed reactor in a nitrogen atmosphere at temperatures ranging from 300 to 720 °C at a heating rate of 5 °C/min showed that higher temperatures resulted in higher oil yields, with yields of 3% and 54.8% at 300 °C and 720 °C, respectively [49]. The yield and properties of the liquid product and biochar produced by pyrolyzing and co-pyrolyzing rubber wood sawdust (RWS), sewage sludge (SS), and RWS depend upon the combination with SS in different ratios [50].

5.6 Gasification and Combustion

The combustion temperature, excess oxygen in the reactor, and material particle size all have a major impact on the gaseous emissions produced during the burning of used tires. An intensive initial phase of volatile combustion was followed by a less intense phase of combustion with simultaneous burning of volatiles and coals, according to thermogravimetric research. Burning times were also shown to be significantly shorter than those of coal. The most effective waste management method is fluidized bed testing. It is a more costly technology, though, and requires a lot of material preparation before processing. It also has high running and maintenance requirements [51]. According to several studies, NOx emissions from burning old tires with charcoal are three to four times lower than those from burning tires with charcoal. Concerning SO_2 emissions, the same has been noted. Tire burn concentrations of CO and PAH were higher than those of coal, although both materials had comparable amounts of indirect PAH products. However, a number of studies suggest that co-burning used tires with coal is a low-emission method of recovering energy from this trash [29].

Gasification is a sub-stoichiometric thermochemical process that produces fuel gas from carbon-based feedstock by using a controlled supply of gasifying agents [52]. Compared to pyrolysis, the gasification process is more complex and includes both homogeneous and heterogeneous reaction mechanisms. Since chemical reactions take place across the surface of the material during a heterogeneous reaction, variables like surface area, surface accessibility, carbon active sites, inorganic matter concentration, and gasification agent composition all have a significant impact on conversion rates. Tires break down during the first depolymerization step, producing solid char and heavy and light hydrocarbons. Waste tires are one of the many feedstocks that can be gasified to produce high-quality syngas. Nevertheless, standard gasification methods have not been effectively designed to lower the significant amounts of energy needed to break down the feed material [53].

5.7 Biochemical Processes

Due to their polymeric composition and refractory character, particularly in a microbial environment, there are not many studies in the literature on the direct biological conversion (anaerobic digestion and fermentation) of waste tires. Nonetheless, a number of researchers have documented the use of carbon produced from waste tires to enhance the anaerobic digestion process's efficiency. Pyrolytic carbon made from discarded tires has better conductivity and produces 50.3% more volatile fatty acids. Using pyrolysis, carbon generated from discarded tires also resulted in a 104.4% increase in methane output [25]. When rubber products are used and disposed of at the end of their useful lives, they fragment and degrade very slowly (by UV and infrared rays, microorganisms, bacteria and germs, enzymes, etc.), which causes microplastics (MPs) to accumulate. The most advantageous biodegradation occurs when the degradant is carbon-based. Biodegradation occurs when larger molecules are broken down into smaller components by the secretion of enzymes by microorganisms. But when synthetic rubbers with non-biodegradable additives and persistent nanofillers are used, biodegradation becomes more complicated, and bacteria are less able to eat them [54].

It has been investigated to enhance the interaction between asphalt and crumb rubber by the use of UV, plasma, microwaves, bio-oils, and chemical grafting. Apart from physical and chemical approaches, rubber has also been etched or desulfurized using microbiological methods. In order to produce energy, microorganisms such as *Acidithiobacillus* sp. and *Sphingomonas* sp. can combine the oxidation of sulfur with the reduction of oxygen. Microbial desulfurization using the Neospagnol T-20 method was 22% more successful than a chemical treatment in a study employing *Acidithiobacillus* to treat crumb rubber. According to reports, *Sphingomonas* sp. caused the S–C and S–S bonds of desulfurized rubber to scission. In a study using *Alicyclobacillus* sp., a notable degree of desulfurization (62.5%) was stated [55].

6 Emerging Technologies for Valorization

New emerging technologies for rubber waste valorization include hydrothermal liquefaction, advanced pyrolysis, and nano-engineering, which turn waste into high-value bio-oils and syngas, and biocatalytic processes that use microbes and enzymes that have been engineered to break down rubber polymers into reusable monomers for green chemicals and bioplastics [54].

6.1 Liquefaction Technology

Liquefaction is the thermochemical conversion of an organic solid into a petroleum-like liquid. By breaking down carbon-based materials at high temperatures and pressures, a process known as "direct liquefaction," a crude liquid product is created that can be further distilled to create liquid fuels. The fundamental process involves dissolving carbonaceous material in water or a solvent at relatively low temperatures (250–350 °C) and high pressures (5–20 MPa), followed by hydrocracking (adding hydrogen over a catalyst). There are a number of various production techniques [56]. Rubbers undergo liquefaction due to the commencement of a radical reaction and the breakdown of dual bonds in the central polymer by monomer radicals (also called allylic radicals) produced by β-scission reactions. These highly volatile radicals undergo cyclization and alkylation production to produce a variety of functional groups in alkenes and alkyls. Aromatic hydrocarbons are produced by the dissolution and dehydrogenation reaction mechanism of cyclic compounds from aliphatic compounds at high temperatures. Other functional groups, including alcohols, ketones, and aldehydes, are also produced during the hydrolysis reaction [53].

6.2 Nano-Engineered Materials from Rubber Waste

The issue is waste from various sources, such as consumers or industries. Waste can contain basic structures or value-added materials that can be used to create nanoparticles. For example, waste from rubber tires, large batteries, wastewater, bio-solids, etc., contains carbon, copper, zinc, and lead, which could be used to create nanoparticles [57]. Due to its low strength, crumbed rubber concrete was first created for low-profile non-structural components, including furniture, acoustic wall panels, pedestrian pathways, and low-stress pavement/floors. According to reports, many people have experienced major problems while utilizing aggregate crumb rubbers since they increase resilience but diminish strength. As a result, some structural applications are now restricted. Recent studies have demonstrated that smaller crumb rubber particles can function as micro-fillers, preserving mechanical strength while boosting material resilience in areas like hysteresis and damping [58].

7 Environmental and Economic Benefits

Recycling used tires into crumb rubbers (CR) for concrete has emerged as a viable option for maintaining ecological balance while also adding commercial value [59]. In a study, the potential uses of crumb rubber in rural and industrial environments are examined. Finding the availability and commercial potential of proteins found in various rubber tree components was the aim of this investigation [36]. Tire waste can be treated via pyrolysis, combustion, and gasification to produce fuel and recover energy. By-products like carbon black, metals, and nylon strands can be recovered in addition to the tires' partial embodied energy. Pyrolysis was found to be thermodynamically more efficient than gasification based on a number of thermodynamic and economic factors; nonetheless, it results in larger emissions and significantly lower profits than gasification. Rubberized asphalt is an environmentally friendly alternative to concrete since Portland concrete pavement generates approximately 2.6 times more air pollutants than hot-mixed rubberized asphalt. Additionally, it has been demonstrated that using discarded tire textile fibers to produce normal bituminous conglomerate instead of the customary reinforcement cellulose used in asphalt is technically, environmentally, and financially possible. By adding waste tire fibers to bituminous conglomerates, their mechanical performance can be enhanced, extending the pavement's usable life. Bitumen made from discarded tire fibers is highly profitable and has a smaller environmental impact than incineration [19].

7.1 Reduction in Carbon Footprint

As the temperature increased, a lesser amount of carbon was present. The synthesis gas's decreased calorific value increased with temperature and declined linearly as the equivalence ratio increased. Rapid gasification was made possible by a high equivalence ratio and a temperature below 600 °C. Tar production decreases at temperatures that exceed 600 °C. It was determined that the synthesis gas has calorific powers between 4 and 9 MJ/Nm^3 with a generation of 1.8 to 3.7 Nm^3/kg when operating at 650–750 °C and equivalency ratios between 0.2 and 0.4. The creation of charcoals that can be used, for instance, in wastewater treatment procedures is another intriguing feature of gasification. Coals with surface areas ranging from 20 to 30 m^2/g were created in this experiment [29].

7.2 *Economic Feasibility*

It is essential from an economic and environmental perspective to recycle worn tires for automobiles and recover the materials and energy they contain. Numerous methods, such as recovery, devulcanization, which creates rubber regeneration, pyrolysis, and mechanical grinding to make rubber flour, can be used to recycle waste vehicle tires. Reusing rubber is a crucial way that we can preserve the environment. The reclamation process depends on the detachment of three-dimensional cross-links. After that, vulcanizing and reprocessing rubber will be easier. The loading of ground rubber powder was examined as a filler for different types of polymers [31].

8 Policy and Regulations

Many nations are thinking about establishing rules and regulations to control the procedures that propel these efforts in the present push for circular economies. For example, Cyprus and other nations have various ELT management systems. Common treatment techniques include limited pyrolysis, the production of rubber granules for artificial surfaces, and the use of these tires as alternative fuel in the cement industry [60]. Both the European Union and the majority of US states have laws prohibiting the disposal of tires that have reached the end of their useful lives in landfills. Setting up end-of-life tire management systems in these economies has been greatly aided by the policies underlying this law. In the EU, for example, there are three types of regulations: i) extended producer responsibility, ii) free market, and iii) taxes. The most popular system is extended producer responsibility. The European Union has effectively adopted the extended producer responsibility framework, which is now the cornerstone of contemporary end-of-life tire management [61].

9 Conclusion

In conclusion, the world's production of rubber wood, which is primarily centered in Asia, offers a substantial chance for sustainable development by valuing waste rubber wood and by-products from crumb rubber factories. These resources may be converted into valuable bioproducts and renewable bioenergy by tackling waste management issues and utilizing cutting-edge bioconversion technologies like pyrolysis, gasification, liquefaction technology, and anaerobic digestion. Lowering greenhouse gas emissions and encouraging the sustainable use of resources not only lessens the environmental effects related to their disposal but also advances a circular economy. Achieving environmental sustainability and energy resilience requires utilizing the potential of rubber wood waste and crumb rubber by-products, which emphasizes the need to incorporate cutting-edge waste management techniques into the procedures of the rubber wood sector.

References

1. Bhattacharjee A, Bhowmik M, Paul C, Chowdhury BD, Debnath B (2021) Rubber tree seed utilization for green energy, revenue generation and sustainable development–a comprehensive review. Ind Crop Prod 174:114186
2. Panda BK, Sarkar S (2020) Environmental impact of rubber plantation: ecological vs. economical perspectives. Asian J Microbiol Biotechnol Environ Sci 22:657–661
3. Goldthorpe CC (2015) Rubber manufacturing in Malaysia: resource-based industrialization in practice. NUS Press
4. Adikari SB, Sharma A (2018) Marketing pattern of rubber plantation in Tripura, India. Int J Curr Microbiol App Sci 7:847–853
5. Eakvanich V, Kalasee W, Dangwilailux P, Wattana W, Taweekun J (2023) A review of the parameters related to the rubber sheets production: major constituents of rubber latex, rubber particles, particle interaction of rubber latex, rubber porous structure, rubber drying kinetics and energy consumption. J Adv Res Appl Sci Eng Technol 29:159–184
6. Kongto P, Palamanit A, Chaiprapat S, Tippayawong N (2021) Enhancing the fuel properties of rubberwood biomass by moving bed torrefaction process for further applications. Renew Energy 170:703–713
7. Sarkar S (2024) Growth and challenges of rubber plantation in the Northeastern States of India. J Glob Econ 20:1
8. Kumar A, Dhanorkar R, Mohanty S, Gupta V (2024) Advances in recycling of waste vulcanized rubber products via different sustainable approaches. Mater Adv 5:7584–7600
9. Sultan SH, Palamanit A, Techato KA, Amin M, Baloch KA (2021) Physiochemical characterization and potential of synthesis gas production from rubber wood biomass by using downdraft gasifier. Mehran University Res J Eng Technol 40:1–15
10. Obukhova S, Budkina A, Korolev E, Gladkikh V (2024) Impacts of waste rubber products on the structure and properties of modified asphalt binder: part I—crumb rubber. Materials 17:4685
11. Harinder D, Srujana M, Reddy YJC, Goud DA, Priya KS (2024) Comprehensive review on utilization of construction demolished waste and crumb rubber for sustainable pavement construction. J Phys Conf Ser 2779(1):012023. IOP Publishing
12. Pizzi A, Duca D, Rossini G, Fabrizi S, Toscano G (2020) Biofuel, bioenergy and feed valorization of by-products and residues from hevea brasiliensis cultivation to enhance sustainability. Resources 9:114
13. Zhao S, Liu M, Meng X, Liu A (2024) Waste rubber–black pollution reframed as a global issue: ecological challenges and sustainability initiatives. Environ Pollut 356:124291
14. Tongbuekeaw T, Sawangkeaw R, Chaiprapat S, Charnnok B (2021) Conversion of rubber wood waste to methane by ethanol organosolv pretreatment. Biomass Convers Biorefinery 11:999–1011
15. Rahman AA, Adeboye OJ, Adebayo A, Salleh MR (2023) Behaviour and some properties of wood plastic composite made from recycled polypropylene and rubberwood. Jordan J Mech Indust Eng 17:281–287

16. Nuryawan A, Risnasari I, Rahmawaty PE, Hutauruk NO (2018) Preliminary results of wood plastics composite: an innovative eco-friendly product. In: International conference of science, technology, engineering, environmental and ramification researches (ICOSTEERR) August: 30–31
17. Jawjit S, Jawjit W, Pibul P, Wongcharee S, Suwannahong K (2024) Enhancing the sustainability of the sawn rubberwood industry through resource-efficient and cleaner production approaches. J Clean Prod 479:143913
18. Sunthonpagasit N, Duffey MR (2004) Scrap tires to crumb rubber: feasibility analysis for processing facilities. Resour Conserv Recycl 40:281–299
19. Abdullah ZT (2024) Remanufactured waste tire by-product valorization: quantitative–qualitative sustainability-based assessment. Results Eng 22:102229
20. Picado-Santos LG, Capitão SD, Neves JM (2020) Crumb rubber asphalt mixtures: a literature review. Constr Build Mater 247:118577
21. Tushar Q, Santos J, Zhang G, Bhuiyan MA, Giustozzi F (2022) Recycling waste vehicle tyres into crumb rubber and the transition to renewable energy sources: a comprehensive life cycle assessment. J Environ Manag 323:116289
22. Kantasiri T, Kasemsiri P, Pongsa U, Posi P, Chindaprasirt P (2024) Optimization of concrete containing waste crumb rubber mix design for thermal insulating applications using Taguchi method. Constr Build Mater 434:136636
23. Palos R, Gutierrez A, Vela FJ, Olazar M, Arandes JM, Bilbao J (2021) Waste refinery: the valorization of waste plastics and end-of-life tires in refinery units. A review. Energy Fuels 35:3529–3557
24. Atsonios K, Nesiadis A, Detsios N, Koutita K, Nikolopoulos N, Grammelis P (2020) Review on dynamic process modeling of gasification-based biorefineries and bio-based heat and power plants. Fuel Process Technol 197:106188
25. Rogachuk BE, Okolie JA (2023) Waste tires-based biorefinery for biofuels and value-added materials production. Chem Eng J Adv 14:100476
26. Miranda A, Muñoz R, Aedo C, Bustos F, Tuninetti V, Valenzuela M, Oñate A (2024) High-performance concrete from rubber and shell waste materials: experimental and computational analysis. Materials 17:5516
27. Yusha H, Xiaoping Y, Jingzheng R, Zhiqiang Z, Qiming Q (2024) Waste tire valorization: advanced technologies, process simulation, system optimization, and sustainability. Sci Total Environ 942, 173561
28. Formela K (2021) Sustainable development of waste tires recycling technologies–recent advances, challenges and future trends. Adv Ind Eng Polym Res 4(3):209–222
29. Carmo-Calado L, Hermoso-Orzáez MJ, Mota-Panizio R, Guilherme-Garcia B, Brito P (2020) Co-combustion of waste tires and plastic-rubber wastes with biomass technical and environmental analysis. Sustain For 12:1036
30. Zvereva AE, Ershov MA, Kozhevnikova YV, Serdyukova EY, Burov NO, Savelenko VD, Abdellatief TM (2024) Biomass valorization toward sustainable bio-oil obtained by pyrolysis of lignocellulosic raw materials for bitumen modification. Process Saf Environ Prot 188:251–267
31. Singh MV, Pan D (2024) Deformation of waste inner tube Tyre into valorized reused petrochemicals through pyrolysis process. ES Energy Environ 25:1201
32. Kasawapat J, Khamwichit A, Dechapanya W (2024) Waste-to-energy conversion of rubberwood residues for enhanced biomass fuels: process optimization and eco-efficiency evaluation. Energies 17:5444
33. Esekhade TU, Idoko SO, Abolo CU, Igberase S (2021) Sustainability of rubber agroforestry strategies in boosting smallholders' resilience to cope with the realities of new global challenges. In: E3S web of conferences, vol 305. EDP Sciences, p 02004
34. Onoji SE, Iyuke SE, Igbafe AI, Nkazi DB (2016) Rubber seed oil: a potential renewable source of biodiesel for sustainable development in sub-Saharan Africa. Energy Convers Manag 110:125–134

35. Lei Y, Wang H, Fini EH, You Z, Yang X, Gao J et al (2018) Evaluation of the effect of bio-oil on the high-temperature performance of rubber modified asphalt. Constr Build Mater 191:692–701
36. Widyarani, Coulen SCW, Sanders JPM, Bruins ME (2017) Valorisation of proteins from rubber tree. Waste Biomass Valori 8:1027–1041
37. Sharma ND (2025) Reaction mechanism and redox potential of laccase. In: Laccase and polyphenol oxidase. Academic Press, pp 111–132. https://doi.org/10.1016/B978-0-443-13301-5.00005-6
38. Khan SI, Sahinkaya M, Colak DN, Zada NS, Uzuner U, Belduz AO et al (2024) Production and characterization of novel thermostable CotA-laccase from *Bacillus altitudinis* SL7 and its application for lignin degradation. Enzym Microb Technol 172:110329
39. Arokiyaraj S, Elshikh MS, Alarjani KM, Alghmdi MA, Vijayaraghavan P (2024) Valorization of rubber tree sawdust biomass for the production of laccase-lignin peroxidase and bioconversion of sawdust. Biomass Convers Biorefinery:1–14. https://doi.org/10.1007/s13399-024-05322-w
40. Formela K, Kurańska M, Barczewski M (2022) Recent advances in the development of waste-based polymer materials: a review. Polymers 14:1050
41. Zeng D, Zhang Y, Ma X, Li J, Yin F, Li D, Bie W (2024) Biosynthesis of poly (β-L-malic acid) from rubberwood enzymatic hydrolysates in co-fermentation by *Aureobasidium pullulans*. Int J Biol Macromol 257:128605
42. Dag NC, Turkkan E, Kacar A, Dag H (2021) Children's only profession: playing with toys. North Clin Istanb 8:414–420
43. Sathiskumar C, Karthikeyan S (2019) Recycling of waste tires and its energy storage application of by-products–a review. Sustain Mater Technol 22:e00125
44. Myhre M, Saiwari S, Dierkes W, Noordermeer J (2012) Rubber recycling: chemistry, processing, and applications. Rubber Chem Technol 85:408–449
45. Matar MS (2022) Investigating the impact and mechanical properties of reclaimed rubber and sawdust composite (master's thesis). Qatar University. http://hdl.handle.net/10576/26367
46. Asaro L, Gratton M, Seghar S, Hocine NA (2018) Recycling of rubber wastes by devulcanization. Resour Conserv Recycl 133:250–262
47. Zhang T, Asaro L, Gratton M, Hocine NA (2024) An overview on waste rubber recycling by microwave devulcanization. J Environ Manag 353:120122
48. He Z, Zhao A, Liu S, Chen Y, Liu J, Zhao W, Bi D (2024) Preparation of nitrogen-containing chemicals from lignocellulosic biomass and nitrogen-rich organic solid waste by pyrolysis: characteristics, reaction mechanisms, and feedstock interactions. Chem Eng J 496:153793
49. Singh RK, Ruj B, Jana A, Mondal S, Jana B, Sadhukhan AK, Gupta P (2018) Pyrolysis of three different categories of automotive Tyre wastes: product yield analysis and characterization. J Anal Appl Pyrolysis 135:379–389
50. Ali L, Palamanit A, Techato K, Baloch KA, Jutidamrongphan W (2022) Valorization of rubberwood sawdust and sewage sludge by pyrolysis and co-pyrolysis using agitated bed reactor for producing biofuel or value-added products. Environ Sci Pollut Res 29:1338–1363
51. Mikulčić H, Klemeš JJ, Vujanović M, Urbaniec K, Duić N (2016) Reducing greenhouse gasses emissions by fostering the deployment of alternative raw materials and energy sources in the cleaner cement manufacturing process. J Clean Prod 136:119–132
52. Sansaniwal SK, Pal K, Rosen MA, Tyagi SK (2017) Recent advances in the development of biomass gasification technology: a comprehensive review. Renew Sust Energ Rev 72:363–384
53. Nkosi N, Muzenda E, Gorimbo J, Belaid M (2021) Developments in waste tyre thermochemical conversion processes: gasification, pyrolysis and liquefaction. RSC Adv 11:11844–11871
54. Yasin S, Hafeez M, Hussain M, Mahmood A, Yihu S, Qiang Z et al (2024) Cellulose nanocrystal filled synthetic rubber: sustainable production, environmental assessment and end-of-life biodegradation by polyphagous pests. Sustain Mater Technol 40:e00947
55. Kabir SF, Sundar SV, Robles A, Miranda EM, Delgado AG, Fini EH (2024) Microbially mediated rubber recycling to facilitate the valorization of scrap tires. Polymers 16:1017

56. He L, Tao M, Liu Z, Cao Z, Zhu J, Gao J, Ma Y (2023) Biomass valorization toward sustainable asphalt pavements: Progress and prospects. Waste Manag 165:159–178
57. Geetha K, Akshitha S, Kiran RS, Rao TR (2023) An overview of nanotechnology in waste management. Int J Novel Res Dev 8(10):a281–a289. (www.IJNRD.org), ISSN:2456-4184 October-2023, Available: https://ijnrd.org/papers/IJNRD2310033.pdf
58. Kaewunruen S, Meesit R, Mondal P (2017) Early-age dynamic moduli of crumbed rubber concrete for compliant railway structures. J Sustain Cem-Based Mater 6:281–292
59. Guerrero-Bustamante O, Camargo R, Dawd I, Duque J, Polo-Mendoza R, Gálvis J et al (2024) Implementation of crumb rubber (CR) in road pavements: a comprehensive literature review. Inf Dent 9:223
60. Magagula SI, Lebelo K, Motloung TM, Mokhena TC, Mochane MJ (2023) Recent advances on waste tires: bibliometric analysis, processes, and waste management approaches. Environ Sci Pollut Res 30:118213–118245
61. Martínez JD (2021) An overview of the end-of-life tires status in some Latin American countries: proposing pyrolysis for a circular economy. Renew Sust Energ Rev 144:111032

Part VII
Life Cycle Analysis

Chapter 19
Life Cycle Analysis of Plantation Crop Waste Valorization Processes: Economic Assessment and Sustainability

Denis Nsubuga and Stanslaus Mtavangu

Abbreviations

AD	Anaerobic digestion
BaU	Business-as-usual
EFB	Empty fruit bunches
GWP	Global warming potential
LCA	Life cycle analysis
LCC	Life cycle costing
LCI	Life cycle inventory
LCIA	Life cycle impact assessment
PKS	Palm kernel shell
POME	Palm oil mill effluent
PPF	Palm pressed fiber
SOC	Soil organic carbon
SCBT	Sugarcane bagasse and trash

D. Nsubuga (✉)
Department of Agricultural and Biosystems Engineering, Makerere University, Kampala, Uganda
e-mail: denis.nsubuga@mak.ac.ug

S. Mtavangu
Department of Chemistry, Dar es Salaam University College of Education, Dar es Salaam, Tanzania

C. Trois, R. C. Ray (eds.), *Plantation Crop Wastes: Valorization for Economic Sustainability*, Waste as a Resource,
https://doi.org/10.1007/978-3-031-96805-1_19

1 Introduction

The term "plantation" is broadly defined but typically refers to perennial crops cultivated in extensive monoculture systems. These crops produce a diverse array of products, including spices, condiments, beverages, oilseeds, industrial materials, and medicinal products [1]. Predominantly grown in tropical regions, these crops are highly adapted species, mostly cross-pollinated, resulting in heterozygous and heterogeneous populations [2]. Major plantation crops include coconut, areca nut, cocoa, rubber, tea, coffee, cashew, oil palm, and sugarcane, among others [3].

Globally, plantation crops play a vital role in agriculture and the economy. They contribute significantly to national income and international trade, making them an essential economic driver in many tropical countries. For instance, in Ghana, cocoa exports account for approximately 60% of the nation's earnings, while in Indonesia, cocoa generates an estimated USD 600 million annually [4]. Additionally, rising global demand for plantation commodities, such as palm oil, sugarcane, cocoa, and rubber, has fueled the rapid expansion of plantation agriculture. Between 1990 and 2005, palm oil plantation areas increased from 1.8 million to 4.2 million hectares in Malaysia and from 4.4 million to 6.1 million hectares in Indonesia [4]. However, despite their economic significance, plantations face growing scrutiny due to environmental challenges. Plantation agriculture generates large volumes of waste, including biomass residues, husks, shells, and other by-products, posing significant environmental concerns, including pollution, greenhouse gas emissions, and inefficient resource utilization [5].

The contributions of waste valorization to the circular economy are extensive and multifaceted. Firstly, it enhances resource efficiency by converting waste into secondary raw materials, reducing the demand for virgin resource extraction [6, 7]. Secondly, it facilitates energy recovery through methods such as anaerobic digestion (AD) and waste-to-energy technologies, generating renewable energy from organic waste [8]. Thirdly, it drives economic growth by fostering innovation, enabling the creation of new industries, and generating green jobs through value-added waste management activities [9, 10].

By integrating waste valorization into plantation crop management and broader economic systems, societies can address global challenges such as resource depletion and environmental pollution [11, 12]. This approach fosters innovation, protects ecosystems, and builds economic resilience, making waste valorization a cornerstone of the circular economy [13].

2 Life Cycle Analysis (LCA)

2.1 Introduction

Life cycle assessment (LCA) is a robust analytical tool used to assess the environmental impacts associated with all stages of a product, process, or service's life cycle [14]. It evaluates the entire lifecycle, from raw material extraction and manufacturing to transportation, use, and end-of-life disposal. By examining inputs like energy, water, and raw materials, as well as outputs such as emissions and waste, LCA provides a comprehensive understanding of the environmental trade-offs involved in a system [15].

The LCA methodology adheres to internationally recognized standards, including ISO 14040 and ISO 14044, and is divided into four key stages [16]. The goal and scope definition phase outlines the study's purpose, system boundaries, and functional unit—a standardized measure of the product or service. The life cycle inventory (LCI) phase involves gathering detailed data on all material and energy flows within the system. During the life cycle impact assessment (LCIA) phase, these flows are analyzed to quantify environmental impacts, such as greenhouse gas emissions, water consumption, and resource depletion [17]. Finally, the interpretation phase synthesizes the findings to offer actionable recommendations that align with the study's objectives.

LCA serves as a critical decision-making tool for identifying areas of significant environmental impact, comparing alternative solutions, and designing sustainable products and processes [18]. By providing a lifecycle perspective, LCA helps stakeholders make informed decisions that reduce environmental footprints and support sustainability goals. Integrating LCA into waste valorization studies is vital for ensuring the environmental and economic sustainability of waste management practices [19]. Waste valorization involves converting waste materials into valuable products, such as fuels, chemicals, and materials, often through complex processes that span multiple stages. LCA provides a comprehensive framework to assess the environmental impacts of these stages, ensuring that the benefits of waste valorization, such as resource recovery, do not lead to unintended negative consequences, such as increased emissions or energy use [20].

LCA plays a crucial role in identifying the most sustainable waste valorization pathways by enabling the comparison of alternative strategies [21]. For instance, it can determine whether converting waste into bioenergy or recycling it into raw materials yields a lower environmental impact. This analysis supports decision-making by highlighting the option that maximizes resource efficiency and minimizes adverse effects on the environment. Additionally, LCA facilitates alignment with circular economy principles by quantifying how waste valorization contributes to resource conservation, emissions reductions, and waste minimization [22]. Policymakers and industries benefit significantly from LCA integration, as it provides robust, science-based data to justify investments in waste valorization technologies [23]. It also guides improvements in efficiency by identifying environmental hotspots and inefficiencies within valorization systems, enabling targeted interventions to reduce negative impacts.

2.2 Flowchart for Assessing Environmental, Economic, and Social Impacts of Plantation Crops Using Life Cycle Assessment

The process of determining the environmental, cost, and social impacts of plantation crops through life cycle assessment (LCA) begins with Goal and Scope Definition (Fig. 19.1). This step involves setting the boundaries of the study, identifying the objectives, and determining the functional units that will be used to compare different processes or systems. The goal is to clearly define what the assessment aims to achieve and which stages of the crop's life cycle will be considered, such as cultivation, processing, and distribution.

Next, in the Inventory Analysis (Life Cycle Inventory—LCI) phase, data is collected on the inputs (e.g., water, labor, materials) and outputs (e.g., emissions, waste, energy use) throughout the life cycle of the plantation crops (Fig. 19.1). This stage forms the foundation for the assessment, as it provides detailed data on the environmental flows associated with the crop's production. The inventory also includes data on resource use, land management practices, and the emissions generated during various stages of the crop's life cycle.

The third phase involves conducting a comprehensive assessment across three key impact categories: environmental, cost, and social (Fig. 19.1). For the Environmental Impact, this includes analyzing factors such as greenhouse gas emissions (e.g., CO_2, CH_4, NOx), resource use (e.g., water, land, energy), and the effects on biodiversity and ecosystem services. This phase helps identify the crop's contribution to environmental degradation or benefits. The Cost Assessment evaluates both direct and indirect costs. Direct costs include inputs like labor and materials, while indirect costs encompass supply chain, transport, and logistics expenses. Understanding the financial implications is crucial for decision-making in plantation crop management. Lastly, the Social Impact Analysis looks at how the plantation crop affects local communities, including employment opportunities, health and safety concerns, and the broader socio-economic benefits or challenges faced by the communities involved in the crop's production.

Following the assessment, the Data Validation phase ensures the accuracy and reliability of the data collected (Fig. 19.1). This involves quality checks and sensitivity analysis to identify any uncertainties or gaps in the data. Ensuring the validity of the data is critical for making informed decisions.

Finally, the Interpretation of Results phase brings all the findings together. The results from the environmental, cost, and social assessments are interpreted to evaluate the trade-offs between the three impact categories (Fig. 19.1). This stage helps decision-makers understand the overall sustainability of the plantation crop and provides insights into areas that may require improvement or adjustment. The outcome is a comprehensive view of the plantation crop's lifecycle impacts, which can guide future decisions regarding sustainability practices, crop management, and policy-making.

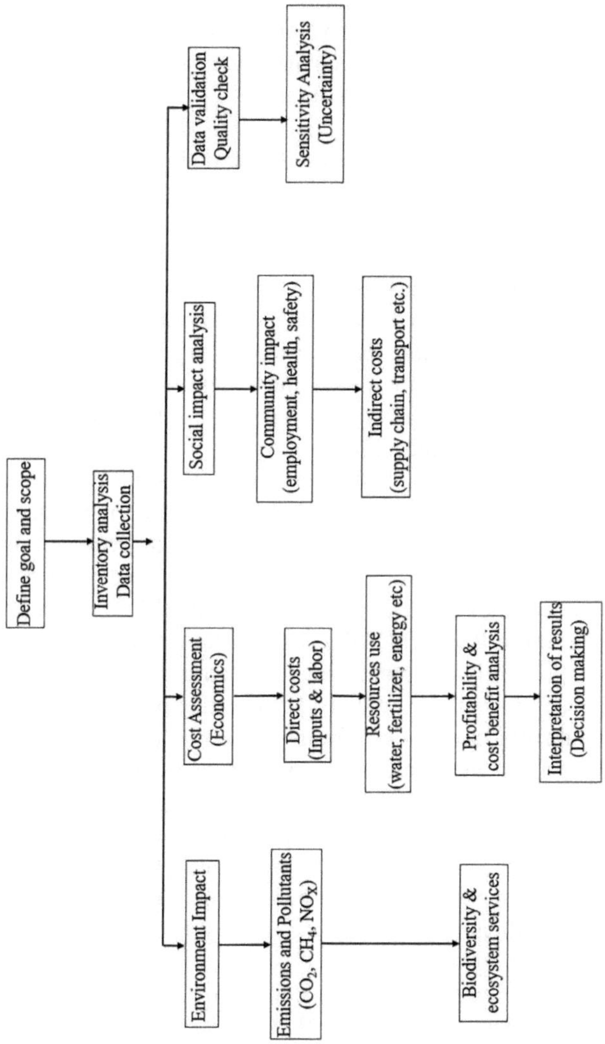

Fig. 19.1 Flowchart for assessing environmental, economic, and social impacts of plantation crops using life cycle assessment

3 Plantation Crop Waste: Sources and Properties

3.1 Sources of Plantation Crop Waste

Plantation agriculture significantly contributes to the global economy but also generates substantial volumes of waste, much of which remains underutilized or improperly managed [24]. Effective waste valorization strategies, such as bioenergy production, composting, and material recovery, can help mitigate environmental impacts while creating economic opportunities.

Coconut (*Cocos nucifera* L.) plantations produce large amounts of waste, including husks, shells, coir dust, and leaves. These materials have potential applications in the production of biochar, activated carbon, and compost, yet improper disposal often results in environmental challenges. Globally, 64,000 million coconuts are harvested each year [25], of which 55–80% are non-edible components (waste) [26]. This biomass is characterized by its fibrous and heterogeneous structure. Coconut waste can be utilized to produce handicraft items such as carpets, mattresses, and upholstery. Additionally, they can undergo chemical transformation processes that enable their use as environmentally friendly alternatives, such as bio-based polymers, biodegradable packaging materials, activated carbon for water purification, and sustainable biofuels [26].

Oil palm (*Elaeis guineensis* L.) plantations generate residues such as empty fruit bunches, palm kernel shells, and palm oil mill effluent (POME), which can be transformed into bioenergy, organic fertilizers, and animal feed. However, much of this waste is often left unused. Waste from oil palm cultivation and palm oil processing includes empty fruit bunches (EFB), palm kernel shells (PKS), palm pressed fiber (PPF), and palm oil mill effluent (POME). The global oil palm production is 72.2 million metric tons [27]. It is estimated that one metric ton of FFB produces 23% EFB, 14–15% PPF, and 6–7% PKS, the rest being crude palm oil (CPO) [28].

The annual global **sugarcane** (*Saccharum officinarum* L.) production is 1.89 billion tons [29]. Starting from sugarcane plantation to sugar processing, the total waste biomass from this crop is about 30 percent, and sugarcane bagasse and trash (SCBT) are the main waste parts. Sugarcane farming produces by-products like bagasse, molasses, and cane tops. Bagasse is widely used for bioenergy production, while other residues can be converted into bioethanol, paper, or compost. Despite these opportunities, significant portions of sugarcane waste are discarded or improperly handled.

The global **coffee** (*Coffea arabica* L.) production was ten million tons in 2023, and the roasted coffee beans constitute only 9.5% by weight of the total fruit and are used for preparing the beverage. The remaining 90.5% consists of the outer layer, which becomes a waste material [30]. The coffee waste materials are rich in organic content and suitable for composting, biochar production, and biogas generation, yet untreated disposal remains a common practice.

Cocoa (*Theobroma cacao* L.) plantations also produce significant waste, including cocoa pods, husks, and processing residues. The global cocoa production is approximately 4.2 million metric tons per year [31], of which 70–80% is waste products [32]. This waste can be converted into bioenergy, biofertilizers, or animal feed, but inefficient waste management frequently leads to underutilization. Likewise, cashew nut (*Anacardium occidentale* L.) production generates waste such as shells and testa (outer coverings). Cashew shells are rich in oil and are suitable for energy generation or chemical extraction, though they are often discarded.

3.2 Challenges in Waste Management

One of the primary barriers to efficient waste management is the decentralized and scattered nature of plantation crops. Plantations are often spread over large areas, making it difficult to systematically collect residues such as husks, shells, and pruned materials [33]. Seasonal variations in waste generation further complicate collection efforts. Additionally, residues like EFB and POME have high moisture content, which increases their weight and volume, leading to higher transportation costs and energy requirements [34]. The lack of adequate infrastructure in many plantation regions, particularly in developing countries, exacerbates these issues. Storage facilities, collection systems, and preprocessing technologies are often unavailable or inaccessible, resulting in waste accumulation and improper disposal [35].

Logistical challenges also impede efficient waste management. The bulky and heterogeneous nature of plantation crop waste make its handling and transportation difficult. Poor road networks and transportation infrastructure in remote plantation areas further complicate logistics, leading to increased transportation costs that reduce the economic viability of waste management projects [36]. These logistical hurdles often discourage stakeholders from investing in sustainable waste solutions.

4 Waste Valorization Techniques

4.1 Composting

Composting is a biological process that converts solid organic waste from plants and animals into a nutrient-rich matrix with the help of various microorganisms, including actinomycetes, bacteria, and fungi, under aerobic conditions [37]. The introduction of diverse microorganisms into solid waste facilitates its transformation into compost and produces by-products such as heat, water, and CO_2 [38]. The resulting solid and stable matrix, known as humus, can be applied as an organic fertilizer to improve soil fertility and structure.

Despite the diversity of processes and techniques, some essential aspects of composting remain consistent, such as natural, chemical, and physical properties [39].

Key considerations for successful composting include selecting appropriate input materials, adjusting their composition and structure, managing substrate degradability, moisture, energy, porosity, and air space, and ensuring proper stabilization and degradation. Further research is needed to refine and differentiate composting methods and their outcomes [40]. Composting processes progress through four stages: mesophilic, thermophilic, cooling, and compost maturation. These stages may occur simultaneously rather than sequentially [37]. The duration of each stage is influenced by factors such as the initial composition of the mixture, moisture content, air circulation, and the microbiological community present [37].

Composting significantly enhances soil health by improving its physical, chemical, and biological properties [41]. It enriches the soil with essential nutrients like nitrogen, phosphorus, and potassium, along with micronutrients that are vital for plant growth. Compost also improves soil structure, promoting better water retention, aeration, and root development [42]. Additionally, it reduces soil compaction and erosion while fostering microbial activity that aids nutrient cycling and suppresses pathogens [43]. In terms of carbon sequestration, composting plays a critical role in reducing atmospheric carbon dioxide levels. The process converts organic waste into stable humus, a long-term carbon reservoir in the soil. By increasing soil organic carbon (SOC) levels, compost enhances the soil's ability to act as a carbon sink while also preventing methane emissions from organic waste decomposition in landfills [44]. Furthermore, compost application supports regenerative agriculture by restoring degraded soils and making them more resilient to environmental changes [45].

4.2 Anaerobic Digestion (AD)

AD is a process enabling the production of biogas (methane-rich gas) via microbiological decomposition of the organic fraction of residues in the absence of oxygen [46]. In AD, organic carbon is degraded in the absence of oxygen into the most reduced state (methane) and its most oxidized state, carbon dioxide [47]. Trace gases such as hydrogen sulfide, nitrogen, ammonia, and hydrogen are also formed in the process. The by-product of AD, known as digestate, can be treated as a residue to be used as an agricultural fertilizer. Linville et al. [48] noted that the digestate is less odorous and has a lower density of pathogens and enriched nutrients compared to raw waste. Compared to aerobic treatment, the anaerobic process saves the energy needed for aeration, generates much less sludge (almost 50 times less), and also produces an energy-rich biogas. AD is a biological process performed by many classes of bacteria [49]. The main product of this process, methane, can be used as a vehicle fuel or for the cogeneration of electricity and heat and thus can lead to a reduction in greenhouse gas emissions when fossil fuels are used.

The process of AD takes place through four successive stages: hydrolysis, acidogenesis, acetogenesis, and methanogenesis. The AD process is dependent on the interactions between the diverse microorganisms that can carry out the four aforementioned stages [50]. In a single-stage batch reactor, all the waste is loaded simultaneously, and all four processes are allowed to occur in the same reactor sequentially. The compost is then emptied after a given retention period or the cessation of biogas production.

4.2.1 Suitability of AD for Different Types of Plantation Waste

AD is a versatile and sustainable process for converting plantation waste into biogas and nutrient-rich digestates. Its suitability for various types of plantation waste depends on the waste's composition, biodegradability, and the necessity for pretreatment. For example, oil palm waste, such as POME, EFB, and PKS, is highly suitable for AD [51]. POME is particularly efficient due to its high degradability, while solid residues like EFB require pretreatment to break down their lignocellulosic structure for optimal digestion [51]. Similarly, sugarcane waste, including bagasse and press mud, is moderately suitable. While press mud is readily digestible, bagasse's high lignocellulose content necessitates pretreatment to enhance its suitability for biogas production.

Coconut waste, such as husks, shells, and coir pith, poses challenges for AD due to its high lignin content [52]. Husks and shells have low digestibility without extensive pretreatment, while coir pith is moderately suitable after proper conditioning. In contrast, coffee and cocoa waste, which includes husks, shells, and pulp, is more suitable due to its carbohydrate content, though high lignin and tannin levels may require pretreatment or co-digestion to optimize biogas yields [53].

4.3 Pyrolysis

Pyrolysis of lignocellulosic biomass is the thermochemical decomposition of this biomass at high temperatures and in the absence of oxygen to produce solid, liquid, and gaseous products in different proportions depending on the operating conditions such as temperature and heating rate [54].

4.3.1 Suitability of Plantation Crop Waste for Pyrolysis

The plantation of farming waste is highly suitable for pyrolysis, a thermochemical process that converts organic materials into biochar, bio-oil, and syngas under oxygen-limited conditions. The suitability of different types of plantation waste for pyrolysis depends on their chemical composition, moisture content, and lignocellulosic structure.

Oil Palm Waste: This includes EFB, PKS, and fibers, which are rich in lignocellulose. These wastes are highly suitable for pyrolysis due to their low moisture content and high carbon content, which contribute to efficient biochar production and energy recovery. Palm kernel shells, in particular, yield biochar with high calorific value and stability.

Sugarcane Waste: Bagasse and press mud are by-products of sugarcane processing. Bagasse is particularly suitable for pyrolysis because of its high cellulose and hemicellulose content, which produces quality bio-oil and biochar. Press mud, while containing more moisture, can still be processed after drying to reduce energy losses during pyrolysis.

Coconut Waste: Coconut shells, husks, and coir pith are well-suited for pyrolysis. Coconut shells, in particular, are dense and have a high lignin content, making them ideal for producing biochar with excellent adsorption properties and useful in water treatment and soil enhancement.

Coffee and Cocoa Waste: Husks, shells, and pulp from coffee and cocoa processing are moderately suitable for pyrolysis. While these wastes are rich in carbon, their high volatile matter and tannin contents may influence the composition of the bio-oil and biochar produced. Pretreatment or blending with other feedstocks can improve their pyrolysis performance.

Tea and Rubber Plantation Waste: Pruning, leaves, and latex residues are moderately suitable for pyrolysis. While their lignocellulosic structure supports biochar production, the presence of polyphenols and other complex compounds may affect the quality of the bio-oil and syngas. Rubber residues, however, can yield valuable energy-rich products.

5 Life Cycle Analysis (LCA) Framework

LCA is a globally recognized methodology for assessing the energy consumption and environmental impacts of a product or system throughout its life cycle [55]. It can also be defined as a tool for quantifying the environmental performance of products, taking into account the complete life cycle, starting from the production of raw materials to the final disposal of the products, including material recycling if needed [56]. LCA can provide insights into a product or system in all stages of its life cycle.

The technical framework of the LCA methodology, as it is defined in ISO 14040, consists of four phases, namely goal and scope definition, inventory analysis, impact assessment, and interpretation [57]. These phases are not followed just one after the other, but they form an iterative process, which can be followed in different rounds, achieving increasing levels of detail, or which may lead to changes in the first phase because of the results of the last phase.

The first phase sets the boundaries for the analysis and defines the level of detail and the functional basis for comparison. This phase defines why the LCA is to be performed, who the target audiences are, and the product under the LCA study. To understand the system boundary, both the unit process and product system need to be defined. A unit process is the smallest portion of a product system for which data were collected when performing an LCA [58]. A product system is the collection of materially and energetically connected unit processes that perform one or more defined functions. A system boundary defines the unit processes to be included in the system to be modeled.

5.1 Application of LCA in the Plantation Crop Waste Management

LCA is a valuable tool for evaluating the environmental impacts associated with the management of plantation crop waste. By assessing the inputs, outputs, and potential environmental effects throughout the waste management process, LCA provides a comprehensive framework for identifying sustainable strategies and optimizing resource use. LCA is used to compare different waste management options such as composting, AD, pyrolysis, incineration, and landfilling [21]. It helps determine the most sustainable pathway by evaluating factors such as greenhouse gas emissions, energy consumption, and resource recovery potential. For instance, LCA can identify AD as a preferable option for biogas production from oil palm or sugarcane waste due to its lower carbon footprint compared to landfilling. LCA enables the quantification of energy recovery potential from plantation waste through processes like pyrolysis or incineration [59]. By comparing scenarios, such as bioenergy production versus direct combustion, LCA highlights the efficiency and environmental advantages of converting biomass into biochar, bio-oil, or biogas, thereby promoting circular economy practices.

The use of plantation waste for biochar production and its application in soil amendment is another area where LCA proves essential [60]. By modeling carbon sequestration benefits, nutrient recycling, and soil health improvement, LCA supports decision-making in reducing carbon emissions and enhancing soil productivity.

Plantation waste often contains lignocellulosic materials that can emit methane in anaerobic conditions when mismanaged. LCA identifies the environmental burden of improper waste disposal and guides the adoption of sustainable practices, such as composting or co-digestion, to minimize pollution and optimize resource use [61]. LCA findings inform policymakers and stakeholders about the trade-offs between different waste management strategies [62]. By providing data on life cycle emissions, energy use, and environmental impacts, LCA supports the development of regulations and incentives that promote sustainable plantation waste management.

The application of LCA in managing plantation crop waste ensures a science-based approach to identifying environmentally sound and economically viable waste management solutions. It promotes resource efficiency, reduces environmental impacts, and supports the transition to a sustainable and circular agricultural system.

5.2 Economic Analysis

The economic analysis of managing plantation crop waste involves assessing various waste management strategies' costs, benefits, and financial feasibility [63]. These strategies aim to optimize resource utilization while creating economic value. Key cost components include the collection and transportation of waste from plantations to processing facilities, which can be significant, especially in dispersed or remote areas. Additionally, the costs of processing and treatment vary depending on the chosen method. For instance, composting may have lower operational expenses, while advanced technologies like anaerobic digestion or pyrolysis require significant initial investment in equipment and infrastructure [64]. Pretreatment processes, such as drying, grinding, or chemical treatment, also add to the operational costs.

Despite these costs, plantation waste management offers substantial revenue streams. Energy production from biogas, bio-oil, or syngas through anaerobic digestion or pyrolysis provides a renewable energy source that can be sold or used on-site, reducing energy expenditures [65]. Value-added products like biochar, compost, or liquid fertilizers create opportunities for additional income. Furthermore, sustainable waste management practices that reduce greenhouse gas emissions may qualify for carbon credits, thus trading for financial gains. Reusing processed materials within plantations, such as compost, enhances soil fertility and reduces dependence on external inputs, thereby lowering overall operational costs [66].

The economic benefits of waste management extend beyond direct financial returns. Transforming waste into valuable products promotes a circular economy, increasing profitability and reducing waste disposal expenses [67]. Additionally, waste management initiatives create employment opportunities in collection, transportation, and processing, contributing to local economic development [68]. Proper waste management also prevents environmental degradation, helping plantation owners avoid fines or penalties related to pollution. However, a cost-benefit analysis highlights some challenges. High initial investments in advanced technologies may not be financially accessible to small-scale farmers without subsidies or government support [69]. Market volatility can also impact the profitability of byproducts like biochar, organic fertilizers, and renewable energy [70]. Moreover, the scale of operation significantly influences economic feasibility, as small-scale operations often face higher per-unit costs compared to large-scale enterprises.

6 Sustainability and Economic Trade-Offs

Balancing sustainability with economic trade-offs requires weighing short-term costs against long-term benefits like resource efficiency and environmental protection. Businesses and policymakers must ensure sustainable investments align with financial goals while remaining competitive. Despite initial challenges, sustainability drives innovation, enhances brand value, and strengthens economic resilience.

6.1 Balancing Environmental and Economic Goals

From an environmental perspective, plantation waste management aims to reduce greenhouse gas emissions, prevent soil and water pollution, and promote resource conservation. Techniques such as AD, composting, and pyrolysis help in converting waste into valuable products like biogas, biochar, and organic compost, which can enhance soil health and reduce reliance on chemical fertilizers [71]. These methods also contribute to carbon sequestration by capturing and storing carbon in stable forms, such as biochar or soil organic matter, thus mitigating the effects of climate change [72].

Economically, waste management must provide financial incentives to encourage adoption. Value-added products, such as renewable energy and fertilizers, can generate significant revenue streams for plantation owners [73]. However, high initial investments in technologies like pyrolysis units or biogas plants can be a barrier, particularly for small-scale farmers. Balancing costs with benefits involves identifying scalable solutions, leveraging subsidies, or accessing financing options to reduce the economic burden [74]. Additionally, efficient resource recovery reduces operational costs by substituting external inputs like synthetic fertilizers and fossil fuels with sustainable alternatives derived from waste.

To reconcile environmental and economic goals, integrated approaches are essential. Policies that support green technologies, such as carbon credit programs and tax incentives, can encourage the adoption of sustainable waste management practices [75]. Co-digestion or blending of plantation waste with other feedstocks can enhance biogas yields, improving both environmental and financial outcomes. Educating farmers about the long-term economic and environmental benefits of sustainable waste management is also crucial for its widespread acceptance [9].

7 Promoting Circular Economy

Adopting a circular economy approach to managing plantation crop waste presents a sustainable pathway for minimizing environmental impacts and maximizing resource efficiency. This model emphasizes waste valorization, resource recovery, and the creation of closed-loop systems that transform waste into valuable inputs for other processes, driving both economic and environmental benefits.

Waste Valorization and Resource Efficiency: In a circular economy, plantation crop waste is viewed as a resource rather than a liability. Waste such as oil palm residues, sugarcane bagasse, coconut husks, and coffee or cocoa shells can be transformed into value-added products [76]. These include biochar for soil amendment, biogas for renewable energy, compost for organic fertilizers, and bio-oil for industrial applications. By converting waste into usable materials, the circular economy reduces reliance on virgin resources, enhances resource efficiency, and lowers the overall environmental footprint.

Energy Recovery: Technologies such as AD and pyrolysis enable the conversion of organic waste into renewable energy forms, including biogas, syngas, and bio-oil. These energy products can be used to power farming operations, reducing dependency on fossil fuels and closing energy loops within agricultural systems [77]. Energy recovery also provides economic incentives for adopting circular practices, as communities and industries can offset energy costs or generate income by selling surplus energy.

Soil Health and Carbon Sequestration: The application of waste-derived products like compost and biochar in agriculture not only enriches soil fertility but also promotes carbon sequestration. Compost improves soil structure, water retention, and microbial activity, while biochar locks carbon into the soil, reducing greenhouse gas emissions [78]. These practices align with circular economy principles by reintegrating nutrients and organic matter into the soil, creating sustainable agricultural cycles that boost productivity while mitigating climate change.

Reduction of Waste and Pollution: Circular economy strategies reduce waste generation and pollution by diverting plantation crop residues from landfills and open burning. Processes such as co-digestion, recycling, and material recovery minimize harmful emissions and ensure that all components of the waste are utilized effectively [79]. This approach helps address environmental issues such as air and water pollution by closing waste loops, contributing to healthier ecosystems and communities.

Economic Growth and Job Creation: The circular economy model stimulates economic growth by creating opportunities in waste management, recycling, and the production of value-added products [80]. Small-scale enterprises and cooperatives can thrive by adopting circular practices, empowering local communities, and fostering entrepreneurship. Moreover, industries that embrace circular principles gain competitive advantages by reducing costs, enhancing sustainability credentials, and meeting the increasing demand for eco-friendly products.

Collaboration and Innovation: Promoting a circular economy in plantation crop waste management requires collaboration between governments, industries, academia, and local communities [80]. Policies and incentives that support sustainable practices, such as subsidies for renewable energy projects and tax breaks for waste recycling initiatives, can drive this transition. Research and innovation in waste-to-value technologies further enhance the viability and scalability of circular solutions.

8 Case Studies of LCA of Different Plantation Crop Waste

Wulandari et al. [81] conducted an LCA of empty oil palm fruit bunch waste generated from palm oil mill operations. This study, utilizing the ISO 14040 framework and the CML-Baseline IA method, aimed to evaluate the environmental impacts associated with the utilization of empty fruit bunches. The results indicated that the global warming potential (GWP) of the furnace scenario was significantly higher than that of the plantation scenario, with values of 234.719 kg CO_2 eq and 234.689 kg CO_2 eq, respectively. Additionally, the eutrophication impact was 0.951 kg PO_4 eq for both the furnace and plantation scenarios, while the acidification impact was slightly higher for the furnace at 0.525 kg SO_2 eq compared to 0.523 kg SO_2 eq for the plantation. The study concluded that reducing the input of empty fruit bunches by up to 80% and repurposing the material for co-composting or alternative technologies could significantly mitigate the dominant environmental impacts.

Kim et al., [82] evaluated the environmental impact of activated carbon production from various carbonaceous materials, including coconut shells, coal, and wood. The study focused on assessing the environmental and human health impacts of activated carbon production through carbonization and activation processes. Using this waste as feedstock in the system was found to lower fossil fuel consumption and disposal costs. The analysis employed LCA methodology, with a functional unit of one ton of waste. A boundary expansion approach was applied to evaluate this waste recycling process for activated carbon production, and an environmental credit was calculated using avoided impact analysis. The study specifically assessed greenhouse gas emissions associated with producing 1 kg of activated carbon from this waste. Results indicated that this system reduced the global warming potential by approximately 9.69 kg CO_2-eq. compared to coal-based production processes. Additionally, the system demonstrated significant environmental benefits when compared to conventional coconut shell waste disposal methods. For instance, utilizing one ton of coconut shell waste in activated carbon production reduced global warming potential by 163 kg CO_2-eq. compared to landfilling the same amount of waste. These findings highlight the advantages of using coconut shell waste as a feedstock for activated carbon production over other disposal methods.

Modeste et al. [83] conducted an LCA to evaluate the environmental and energy performance of using agricultural residues, including cocoa cortex, sugarcane bagasse, and oil palm fiber, for power generation in boilers through fuel combustion. The study focused on the production of 1 kWh of electricity from these residues in the Ivory Coast using the LCA methodology. This method assessed the environmental impacts of a product throughout its entire life cycle, from collection and transportation to the combustion of these biomasses in cogeneration unit boilers. The results revealed that the global warming potential for cocoa cortex, oil palm fiber, and sugarcane bagasse were 0.12, 0.16, and 0.19 kg CO_2 eq./kWh, respectively. In terms of depletion of non-renewable resources, the impacts were 0.007, 0.010, and 0.012 MJ/kWh for cocoa cortex, oil palm fiber, and sugarcane bagasse, respectively. For eutrophication impacts, the findings were 0.38, 0.4, and 0.91 g PO_4^{3-}/kWh for cocoa cortex, oil palm fiber, and sugarcane bagasse, respectively. The study demonstrated that these agricultural residues provide significant environmental and energy advantages compared to conventional fossil fuels such as diesel, oil, and coal. Their use for electricity production was shown to have superior environmental performance and lower environmental impacts. The overarching goal of the research was to provide credible data to policymakers, government agencies, private and public companies, national and international organizations, and the general public. By doing so, it seeks to encourage the establishment of energy production units producing both heat and electricity based on the combustion of these agricultural residues in rural and industrial settings.

Ansanelli et al. [84] conducted a sustainability assessment of coffee silverskin waste management in the metropolitan city of Naples, Italy, from a life cycle perspective. The study employed LCA and life cycle costing (LCC) methodologies to compare the environmental and economic performance of two disposal scenarios for coffee silverskin, the primary waste generated during coffee roasting. The first scenario, termed the business-as-usual (BaU) scenario, involves composting the silverskin, a common practice in Naples. The alternative scenario, however, explored the valorization of silverskin by using it as a functional ingredient in bakery products. The findings revealed that the alternative scenario is significantly more advantageous, reducing environmental impact by 96% compared to the BaU scenario, where compost replaces synthetic fertilizers. Additionally, coffee roasters in the alternative scenario experienced a substantial reduction in silverskin disposal costs, with costs dropping from €447.55 per ton in the BaU scenario to €190.09 per ton in the alternative scenario. The study also highlighted the primary contributors to environmental and economic burdens. For environmental impacts, resource use for equipment construction accounted for 37% in the BaU scenario and 62% in the alternative scenario, while electricity consumption contributed 30% and 67%, respectively. On the economic side, personnel costs were the most significant expense, representing 58% in the BaU scenario and 88% in the alternative scenario. These findings emphasize the environmental and economic benefits of valorizing coffee silverskin as a functional ingredient, offering a sustainable alternative to traditional composting practices.

9 Conclusion and Recommendations

The sustainable management of plantation crop waste represents a transformative opportunity to tackle environmental challenges, enhance economic resilience, and drive social development. This study underscores the importance of valorization techniques such as composting, anaerobic digestion, and pyrolysis, which convert plantation residues into valuable products like renewable energy, biochar, and organic fertilizers. Integrating LCA into these strategies has proven vital for evaluating both environmental and economic impacts, enabling more informed decision-making. By adopting circular economy principles, plantation waste management can optimize resource use, reduce greenhouse gas emissions, and support regenerative agricultural systems. Furthermore, the involvement of communities and the implementation of supportive policies are crucial for overcoming challenges, particularly in rural and resource-constrained regions. Together, these efforts can create a more sustainable and equitable future for plantation agriculture.

It is recommended that, to advance sustainable plantation waste management, policymakers should prioritize the development of supportive frameworks, including subsidies for equipment, tax incentives for sustainable practices, and carbon credit systems, to alleviate financial barriers for farmers and stakeholders. Additionally, investment in research and innovation is essential, particularly in developing advanced pretreatment methods, hybrid energy systems, and decentralized waste valorization technologies. Collaborative efforts between academia, industry, and governments can accelerate the scalability and efficiency of these solutions. Finally, community engagement and capacity-building initiatives are equally important, empowering local populations through education, training, and the establishment of cooperatives to ensure equitable access to the benefits of waste valorization.

References

1. Venkateswarlu B, Shanker AK, Shanker C, Maheswari M (2012) Crop stress and its management: perspectives and strategies. In: Crop stress and its management: perspectives and strategies (Vol. 9789400722). https://doi.org/10.1007/978-94-007-2220-0
2. Singh HCP, Shivashankara KS, Rao NKS (2013b) Climate-resilient horticulture: adaptation and mitigation strategies. Climate-resilient horticulture: adaptation and mitigation strategies, 1–302. https://doi.org/10.1007/978-81-322-0974-4
3. Mukesh KBH, Berwal K (2016) Plantation crops : climatic risks and adaptation strategies. Indian J Plant Physiol 21:428. https://doi.org/10.1007/s40502-016-0265-9
4. Nadaraja D, Lu C, Islam MM (2021) The sustainability assessment of plantation agriculture—a systematic review of sustainability indicators. Sustain Prod Consum 26:892–910. https://doi.org/10.1016/j.spc.2020.12.042
5. Arancon RAD, Lin CSK, Chan KM, Kwan TH, Luque R (2013) Advances on waste valorization: new horizons for a more sustainable society. Energy Sci Eng 1(2):53–71. https://doi.org/10.1002/ese3.9
6. Arun M, Barik D, Chandran SSR (2024) Exploration of material recovery framework from waste—A revolutionary move towards clean environment. Chem Eng J Adv 18(January):100589. https://doi.org/10.1016/j.ceja.2024.100589
7. Geissdoerfer M, Savaget P, Bocken NMP, Jan E (2016) The circular economy e A new sustainability paradigm ? J Clean Prod 143:757. https://doi.org/10.1016/j.jclepro.2016.12.048
8. Labatut RA, Pronto JL (2018) Sustainable waste-to-energy technologies : anaerobic digestion. In: Sustainable Food Waste-to-Energy Systems. Elsevier Inc. https://doi.org/10.1016/B978-0-12-811157-4.00004-8
9. Ahmad Y, Kumar R, Kumar A (2023) Environmental waste management strategies and vermi transformation for sustainable development. Environ Chall 13(April):100747. https://doi.org/10.1016/j.envc.2023.100747
10. Hwang H, Kweon T, Kang H, Hwang Y (2024) Resource and greenhouse gas reduction effects through recycling of platinum-containing waste. Sustain For 16(1):80. https://doi.org/10.3390/su16010080
11. Abubakar IR, Maniruzzaman KM, Dano UL, AlShihri FS, AlShammari MS, Ahmed SMS, Al-Gehlani WAG, Alrawaf TI (2022) Environmental sustainability impacts of solid waste management practices in the global south. Int J Environ Res Public Health 19(19):12717. https://doi.org/10.3390/ijerph191912717
12. Rathore D, Sevda S, Prasad S, Venkatramanan V, Chandel AK, Kataki R, Bhadra S, Channashettar V, Bora N, Singh A (2022) Bioengineering to accelerate biodiesel production for a sustainable biorefinery. Bioengineering 2022(9):618. https://doi.org/10.3390/bioengineering9110618

13. El N, Chaher H, Nassour A, Nelles M (2024) The (FWE) nexus: bridging food, food waste, water, energy, and ecosystems for circular systems and sustainable development. Trends Food Sci Technol 108:24–34. https://doi.org/10.1016/j.tifs.2020.10.005
14. Li H, Saboori A, Cao X (2016) Information synthesis and preliminary case study for life cycle assessment of reflective coatings for cool pavements. Int J Transport Sci Technol 5(1):38–46. https://doi.org/10.1016/j.ijtst.2016.06.005
15. Ciacci, Passarini (2020) Life cycle assessment (LCA) of environmental and energy systems. Energies 13(22):5892. https://doi.org/10.3390/en13225892
16. Tascione V, Simboli A, Taddeo R, Del M, Raggi A (2024) A comparative analysis of recent life cycle assessment guidelines and frameworks : methodological evidence from the packaging industry. Environ Impact Assess Rev 108(March):107590. https://doi.org/10.1016/j.eiar.2024.107590
17. Arief M, Fajarini I, Wahyuningrum S (2024) Market, and Complexity Enhancing waste management in Batang regency: Integrating circular economy principles with life cycle assessment and material flow analysis. J Open Innov: Technol Mark Complex 10(4):100437. https://doi.org/10.1016/j.joitmc.2024.100437
18. Pryshlakivsky J, Searcy C (2021) Life cycle assessment as a decision-making tool : practitioner and managerial considerations. J Clean Prod 309(April):127344. https://doi.org/10.1016/j.jclepro.2021.127344
19. Ebadi A, Ali M, Rostom M, Malmir B, Yazdani M, Suh S, Heidrich O (2022) Integrating life cycle assessment and multi criteria decision making for sustainable waste management : key issues and recommendations for future studies. Renew Sust Energ Rev 168(August):112819. https://doi.org/10.1016/j.rser.2022.112819
20. Adhikari S, Mahmud MAP, Moon E, Timms W (2024) Comprehensive life cycle assessment of garden organic waste valorisation : A case study in regional Australia. *J Clean Prod* 472(March):143496. https://doi.org/10.1016/j.jclepro.2024.143496
21. Cherubini F, Bargigli S, Ulgiati S (2009) Life cycle assessment (LCA) of waste management strategies: landfilling , sorting plant and incineration. Energy 34(12):2116–2123. https://doi.org/10.1016/j.energy.2008.08.023
22. Oyejobi DO, Firoozi AA, Blanco D (2024) Results in engineering integrating circular economy principles into concrete technology : enhancing sustainability through industrial waste utilization. Results Eng 24(12):102846. https://doi.org/10.1016/j.rineng.2024.102846
23. Ahmed J, Hamidul Q, Lorber P, Rafizul IM (2024) Integrated approach of waste analysis and life cycle assessment for the management of non-recyclable plastics in recycling shops. Clean Environ Syst 15(April):100229. https://doi.org/10.1016/j.cesys.2024.100229
24. Wan-Mohtar WAAQI, Khalid NI, Rahim MHA, Luthfi AAI, Zaini NSM, Din NAS, Mohd Zaini NA (2023) Underutilized Malaysian agro-industrial wastes as sustainable carbon sources for lactic acid production. Fermentation 9(10):905. https://doi.org/10.3390/fermentation9100905
25. Bradley WL, Conroy S (2019) Using agricultural waste to create more environmentally friendly and affordable products and help poor coconut farmers. In: The 1st international conference on automotive, manufacturing, and mechanical engineering, E3S web of conferences, 130, 01034. https://doi.org/10.1051/e3sconf/201913001034
26. Vieira F, Santana EP, Jesus M, Santos J, Pires P, Vaz-velho M, Silva DP, Ruzene DS (2024) Coconut waste : discovering sustainable approaches to advance a circular economy. Sustainability 16(7):3066. https://doi.org/10.3390/su16073066
27. Murphy DJ, Goggin K, Paterson RRM (2021) Oil palm in the 2020s and beyond : challenges and solutions. CABI Agric Biosci 1–22. https://doi.org/10.1186/s43170-021-00058-3
28. Terence E, Kiplagat J, Kimutai SK, Mecha AC (2023) Current trends in palm oil waste management : a comparative review of Cameroon and Malaysia. Heliyon 9(11):e21410. https://doi.org/10.1016/j.heliyon.2023.e21410
29. Nunes LJR, Loureiro LMEF, Sa CR, Silva HFC (2020) Applied sciences sugarcane industry waste recovery: a case study using thermochemical conversion technologies to increase sustainability. Appl Sci 10(18):6481. https://doi.org/10.3390/app10186481

30. Biswas S, Hossiney N, Krishna M (2024) Case studies in construction materials valorisation of coffee husk as replacement of sand in alkali-activated bricks. Case Stud Construct Mater 21(March):e03440. https://doi.org/10.1016/j.cscm.2024.e03440
31. Suh NN, Molua EL (2022) Cocoa production under climate variability and farm management challenges : some farmers ' perspective. J Agric Food Res 8(March):100282. https://doi.org/10.1016/j.jafr.2022.100282
32. Vásquez ZS, Carvalho DPD, Pereira GVM, Vandenberghe LPS, Oliveira PZD, Tiburcio PB, Rogez HLG, Góes A, Soccol CR (2019) Biotechnological approaches for cocoa waste management: a review. Waste Manag 90:72–83. https://doi.org/10.1016/j.wasman.2019.04.030
33. Amasuomo E, Baird J (2016) The concept of waste and waste management. J Manag Sustain 6(4):88. https://doi.org/10.5539/jms.v6n4p88
34. Suksong W, Tukanghan W, Promnuan K, Kongjan P, Reungsang A, Insam H, O-Thong S (2020) Biogas production from palm oil mill effluent and empty fruit bunches by coupled liquid and solid-state anaerobic digestion. Bioresour Technol 296:122304. https://doi.org/10.1016/j.biortech.2019.122304
35. Hajam YA, Kumar R, Kumar A (2023) Environmental waste management strategies and vermi transformation for sustainable development. Environ Chall 13(June):100747. https://doi.org/10.1016/j.envc.2023.100747
36. Kaiser N, Barstow CK (2022) Rural transportation infrastructure in low-and middle-income countries: a review of impacts, implications, and interventions. Sustainability (Switzerland) 14:4. https://doi.org/10.3390/su14042149
37. Waqas M, Hashim S, Humphries UW, Ahmad S, Noor R, Shoaib M, Naseem A, Hlaing PT, Lin HA (2023) Composting processes for agricultural waste management: a comprehensive review. PRO 11:3. https://doi.org/10.3390/pr11030731
38. Dróżdż D, Malińska K, Kacprzak M, Mrowiec M, Szczypiór A, Postawa P, Stachowiak T (2020) Potential of fish pond sediments composts as organic fertilizers. Waste Biomass Valori 11(10):5151–5163. https://doi.org/10.1007/s12649-020-01074-6
39. Manea EE, Bumbac C, Dinu LR, Bumbac M, Nicolescu CM (2024) Composting as a sustainable solution for organic solid waste management: current practices and potential improvements. Sustain For 16(15):1–25. https://doi.org/10.3390/su16156329
40. Avidov R, Saadi I, Krassnovsky A, Hanan A, Medina S, Raviv M, Chen Y, Laor Y (2017) Composting municipal biosolids in polyethylene sleeves with forced aeration: process control, air emissions, sanitary and agronomic aspects. Waste Manag 67:32–42. https://doi.org/10.1016/j.wasman.2017.05.035
41. Mazumder P, Akhil PM, Jyoti, Khwairakpam M, Mishra U, Kalamdhad AS (2021) Enhancement of soil physico-chemical properties post compost application: optimization using response surface methodology comprehending central composite design. J Environ Manag 289:112461. https://doi.org/10.1016/j.jenvman.2021.112461
42. Adugna G (2016) A review on impact of compost on soil properties, water use and crop productivity. Acad Res J Agric Sci Res:93–104. https://doi.org/10.14662/ARJASR2016.010
43. Al-Shammary AAG, Al-Shihmani LSS, Fernández-Gálvez J, Caballero-Calvo A (2024) Optimizing sustainable agriculture: A comprehensive review of agronomic practices and their impacts on soil attributes. J Environ Manag 364(February):121487. https://doi.org/10.1016/j.jenvman.2024.121487
44. Malone Z, Berhe AA, Ryals R (2023) Impacts of organic matter amendments on urban soil carbon and soil quality: A meta-analysis. J Clean Prod 419(July):138148. https://doi.org/10.1016/j.jclepro.2023.138148
45. Atoloye IA (2024) Revitalizing soil health for climate resilience in Africa: the need for high rates of carbon inputs. Farm Syst 2(3):100096. https://doi.org/10.1016/j.farsys.2024.100096
46. Morales-polo C, Cledera M, Soria BYM (2018) Reviewing the anaerobic digestion of food waste : from waste generation and anaerobic process to its perspectives. Appl Sci 8(10):1804. https://doi.org/10.3390/app8101804

47. Teghammar A (2013) Biogas production from lignocelluloses: pretreatment, substrate characterization, co-digestion and economic evaluation (unplished doctoral thesis). Chalmers University of Technology, Gothenburg, Sweden. Retrieved from https://publications.lib.chalmers.se/records/fulltext/176194/176194.pdf
48. Linville JL, Shen Y, Wu MM, Urgun-demirtas M (2015) Current state of anaerobic digestion of organic wastes in North America. Sustain Renew Fuel 2:136. https://doi.org/10.1007/s40518-015-0039-4
49. Sebola R, Tesfagiorgis H, Muzenda E (2014) Production of biogas through anaerobic digestion of various waste: review. International conference on chemical, integrated waste management and environmental engineering (ICCIWEE'2014) April 15–16, 2014 Johannesburg., October 2015
50. Meegoda JN, Li B, Patel K, Wang LB (2018) A review of the processes, parameters , and optimization of anaerobic digestion. Int J Environ Res Public Health. https://doi.org/10.3390/ijerph15102224
51. Jönsson LJ, Martín C (2016) Pretreatment of lignocellulose: formation of inhibitory byproducts and strategies for minimizing their effects. Bioresour Technol 199:103–112. https://doi.org/10.1016/j.biortech.2015.10.009
52. Adediran AA, Akinwande AA, Balogun OA, Olasoju OS, Adesina OS (2021) Experimental evaluation of bamboo fiber/particulate coconut shell hybrid PVC composite. *Sci Rep* 11(1):1–15. https://doi.org/10.1038/s41598-021-85038-3
53. Salcedo-Puerto O, Mendoza-Martinez C, Vakkilainen E (2025) Solid residues from cocoa production chain: assessment of thermochemical valorization routes. Renew Sust Energ Rev 208:115048. https://doi.org/10.1016/j.rser.2024.115048
54. Czajczyńska D, Nannou T, Anguilano L, Krzyzyńska R, Ghazal H, Spencer N, Jouhara H (2017) Potentials of pyrolysis processes in the waste management sector. Energy Procedia 123(2017):387–394. https://doi.org/10.1016/j.egypro.2017.07.275
55. Yang Y, Ni J, Zhu W, Xie G (2019) Life cycle assessment of large-scale compressed bionatural gas production in China: a case study. Energies. https://doi.org/10.3390/en12030429
56. Mark G., Michiel O., Jorrit L., Tommie P. E. M., (2016). Introduction to LCA with SimaPro (1st ed.). Retrieved from https://pre-sustainability.com/files/2014/05/SimaPro8IntroductionToLCA.pdf
57. Singh A, Olsen SI, Pant D (2013a) Life cycle assessment of renewable energy sources, 1st edn. Springer London Publisher, London. https://doi.org/10.1007/978-1-4471
58. Nsubuga D (2023) Production and life cycle assessment of jackfruit waste nutrient-enriched biochar for soil amendment and carbon sequestration [unpublished Ph.D. Thesis]. Makerere University, Kampala. http://hdl.handle.net/10570/12821
59. Dong J, Tang Y, Nzihou A, Chi Y, Weiss-Hortala E, Ni M (2018) Life cycle assessment of pyrolysis, gasification and incineration waste-to-energy technologies: theoretical analysis and case study of commercial plants. Sci Total Environ 626:744–753. https://doi.org/10.1016/j.scitotenv.2018.01.151
60. Patel MR, Panwar NL (2023) Biochar from agricultural crop residues: environmental, production, and life cycle assessment overview. Resour Conserv Recycl Adv 19(July):200173. https://doi.org/10.1016/j.rcradv.2023.200173
61. Oviedo-Ocaña ER, Abendroth C, Domínguez IC, Sánchez A, Dornack C (2023) Life cycle assessment of biowaste and green waste composting systems: A review of applications and implementation challenges. Waste Manag 171(August 2022):350–364. https://doi.org/10.1016/j.wasman.2023.09.004
62. De Laurentiis V, Caldeira C, Sala S, Tonini D (2024) Life cycle thinking for the assessment of waste and circular economy policy: status and perspectives from the EU example. Waste Manag 179(February):205–215. https://doi.org/10.1016/j.wasman.2024.02.037
63. Lohri CR, Camenzind EJ, Zurbrügg C (2014) Financial sustainability in municipal solid waste management—costs and revenues in Bahir Dar, Ethiopia. Waste Manag 34(2):542–552. https://doi.org/10.1016/j.wasman.2013.10.014

64. Yaser AZ, Lamaming J, Suali E, Rajin M, Saalah S, Kamin Z, Safie NN, Aji NAS, Wid N (2022) Composting and anaerobic digestion of food waste and sewage sludge for campus sustainability: A review. Int J Chem Eng 2022:1. https://doi.org/10.1155/2022/6455889
65. Tayibi S, Monlau F, Bargaz A, Jimenez R, Barakat A (2021) Synergy of anaerobic digestion and pyrolysis processes for sustainable waste management: a critical review and future perspectives. Renew Sust Energ Rev 152:1–64. https://doi.org/10.1016/j.rser.2021.111603
66. Omotoso AB, Omotayo AO (2024) The interplay between agriculture, greenhouse gases, and climate change in Sub-Saharan Africa. Reg Environ Chang 24(1):1–13. https://doi.org/10.1007/s10113-023-02159-3
67. Edirisinghe LGLM, de Alwis AAP, Wijayasundara M, Hemali NA (2024) Quantifying circularity factor of waste: assessing the circular economy potential of industrial zones. Clean Environ Syst 12(October 2023):100160. https://doi.org/10.1016/j.cesys.2023.100160
68. Mngomezulu S, Mbanga S, Adeniran A (2024) The factors influencing waste management for economic development—the perspective of Nelson Mandela Bay municipality residents. Front Sustain 5. https://doi.org/10.3389/frsus.2024.1469207
69. Yigezu YA, Mugera A, El-Shater T, Aw-Hassan A, Piggin C, Haddad A, Khalil Y, Loss S (2018) Enhancing adoption of agricultural technologies requiring high initial investment among smallholders. Technol Forecast Soc Chang 134(April):199–206. https://doi.org/10.1016/j.techfore.2018.06.006
70. Popp J, Lakner Z, Harangi-Rákos M, Fári M (2014) The effect of bioenergy expansion: food, energy, and environment. Renew Sust Energ Rev 32:559–578. https://doi.org/10.1016/j.rser.2014.01.056
71. Alengebawy A, Ran Y, Osman AI, Jin K, Samer M, Ai P (2024) Anaerobic digestion of agricultural waste for biogas production and sustainable bioenergy recovery: a review. Environ Chem Lett 22(6):2641–2668. https://doi.org/10.1007/s10311-024-01789-1
72. Nayak N, Mehrotra R, Mehrotra S (2022) Carbon biosequestration strategies: a review. Carbon Capture Sci Technol 4(August):100065. https://doi.org/10.1016/j.ccst.2022.100065
73. Piñeiro V, Arias J, Dürr J, Elverdin P, Ibáñez AM, Kinengyere A, Opazo CM, Owoo N, Page JR, Prager SD, Torero M (2020) A scoping review on incentives for adoption of sustainable agricultural practices and their outcomes. Nat Sustain 3(10):809–820. https://doi.org/10.1038/s41893-020-00617-y
74. Clark R, Reed J, Sunderland T (2018) Bridging funding gaps for climate and sustainable development: pitfalls, progress and potential of private finance. Land Use Policy 71(August 2017):335–346. https://doi.org/10.1016/j.landusepol.2017.12.013
75. Imran M, Jijian Z, Sharif A, Magazzino C (2024) Evolving waste management: the impact of environmental technology, taxes, and carbon emissions on incineration in EU countries. J Environ Manag 364(June):121440. https://doi.org/10.1016/j.jenvman.2024.121440
76. Vasileiadou A (2024) From organic wastes to bioenergy, biofuels, and value-added products for urban sustainability and circular economy: a review. Urban Sci 8(3):121. https://doi.org/10.3390/urbansci8030121
77. Cucina M (2023) Integrating anaerobic digestion and composting to boost energy and material recovery from organic wastes in the circular economy framework in Europe: A review. Bioresour Technol Rep 24(September):101642. https://doi.org/10.1016/j.biteb.2023.101642
78. Afshar M, Mofatteh S (2024) Biochar for a sustainable future: environmentally friendly production and diverse applications. Results Eng 23:102433. https://doi.org/10.1016/j.rineng.2024.102433
79. Prasanna Kumar DJ, Mishra RK, Chinnam S, Binnal P, Dwivedi N (2024) A comprehensive study on anaerobic digestion of organic solid waste: A review on configurations, operating parameters, techno-economic analysis and current trends. Biotechnol Notes 5(February):33–49. https://doi.org/10.1016/j.biotno.2024.02.001
80. Savini F (2023) Futures of the social metabolism: degrowth, circular economy and the value of waste. Futures 150:103180. https://doi.org/10.1016/j.futures.2023.103180

81. Wulandari A, Hartono DM, Dahlan AV (2023) Life cycle assessment analysis of empty oil palm fruit bunches waste from palm oil mill activities. E3S Web Conf 422. https://doi.org/10.1051/e3sconf/202342201001
82. Kim MH, Jeong IT, Park SB, Kim JW (2019) Analysis of environmental impact of activated carbon production from wood waste. Environ Eng Res 24(1):117–126. https://doi.org/10.4491/eer.2018.104
83. Modeste BY, Bouich A, Oladapo OS, El Messaoudi N, Soucase BM, Boko A (2023) Life cycle assessment (LCA) of agricultural residues (cocoa cortex, sugarcane bagasse and oil palm fiber) for power generation in boilers through fuel combustion. SHS Web Conf 175:01050. https://doi.org/10.1051/shsconf/202317501050
84. Ansanelli G, Fiorentino G, Chifari R, Meisterl K, Leccisi E, Zucaro A (2023) Sustainability assessment of coffee silverskin waste management in the metropolitan city of Naples (Italy): a life cycle perspective. Sustainability (Switzerland) 15(23). https://doi.org/10.3390/su152316281

Chapter 20
Policy Options, Development of Sustainability Indicators, Environmental Impacts, and Management Options for Plantation Crop Residues

Noé Aguilar-Rivera

Abbreviations

GHGs	Greenhouse gases
SCT	Sugarcane trash

1 Introduction

The sustainability of agriculture and agro-related social and economic activities in a country or region depends fundamentally on the technical, productive, environmental, and socioeconomic management of the agricultural, livestock, and forestry lands on which these activities are carried out. Plantation crops, cash crops, or monocultures are cultivated under humid tropical conditions, and the cultivation of certain plantation crops is also extended to subtropical conditions. They are consumed either directly or used for industrial purposes as sugarcane, bamboo, timber, coffee, cocoa, tea, rubber, oil palm, eucalyptus, banana, citrus, coconut, fruit trees, and others play a crucial role in the regional and global agricultural economy [1].

In the past, plantations were the way to mobilize production factors such as land, labor, and capital to grow tropical products in large operational units under centralized management. Nowadays, these crops satisfy various human needs, such as food, fuel, oil, industrial raw materials, beverages, and sugars. The growth profile and the potential for supplying large quantities of recyclable biomass from

N. Aguilar-Rivera (✉)
Facultad de Ciencias Biológicas y Agropecuarias, Universidad Veracruzana, Córdoba, Veracruz, Mexico
e-mail: naguilar@uv.mx

C. Trois, R. C. Ray (eds.), *Plantation Crop Wastes: Valorization for Economic Sustainability*, Waste as a Resource,
https://doi.org/10.1007/978-3-031-96805-1_20

plantations make adopting organic crops achievable. Nevertheless, due to their monoculture structures and management practices, they are highly vulnerable to externalities, such as climate change, affecting their resilience and economic performance. The adoption of innovative cropping practices from the first to the fifth green revolution in agriculture has led to improved yields, sustainability, and economic viability of these crops [2]. However, environmental and socioeconomic impacts have been severe, mainly due to land use change, where the availability of land for cultivation is limited, deforestation, loss of biodiversity, intensive use of agrochemicals and agricultural machinery, increased emissions of greenhouse gases (GHGs), water cycle destabilization, soil erosion, and nutrient loss, in addition to forced appropriation of land and related resources, thus displacing local people, disrupting their livelihoods, and causing land conflicts and governance concerns [3].

The adverse impact of plantation management practices on the environment is also due to underutilized and untreated agricultural residues, which are mainly disposed of in landfills, clandestine dumps, and unplanned incinerations. In this regard, it is a priority to incorporate emerging practices in such plantations through agricultural reconversion, the use of nature-based solutions, agroecological strategies, productive diversification, intercropping, and mixed cropping such as flower, spice, medicinal and aromatic, green manure and cover, tuber and fodder crops, and agroforestry systems with innovative technologies from precision agriculture to artificial intelligence to increase productivity, profitability, and sustainability of plantations as well as to optimize inputs, mainly water. Concerning the aforementioned points, Jena and Singh [4] conducted a review on the use of plantation and crop residues. These constitute unconsumed and unused parts of vegetables, fruits, and other agricultural products derived from plantations and constitute agro-residues. Among the main ones are husks and pulps, cereal husks, nuts, coffee pulp from the coffee industry, trash, and bagasse from the sugar industry. Nonetheless, plantation residues become raw materials to obtain biomolecules because these residues are rich in cellulose, hemicellulose, lignin, enzymes, and other organic molecules, and their application is mostly focused on the "waste to energy" concept [5].

2 Plantation Crop Residues

In plantations, as in the food, petrochemical, mining, coffee, soybean, corn, and other industries, as continuous production enterprises, there are several products derived or generated from the same raw material, normally classified as co-products or by-products. Consequently, process industries that generate several co-products from one raw material can be classified into three main categories:

P1. Set of Main Products. They are the main or desirable products of a company and are derived from the direct transformation of the raw material, for example, cane sugar, soybean oil, soluble coffee, coconut oil, cocoa, etc., as well as petrochemicals.

P2. Set of Secondary Products. These are productions that are not necessarily desirable but are of significant economic importance. They are derived from the transformation of the raw material, for example, livestock feed derived from soybean processing, ethanol from bagasse, compost from coffee pulp, as well as asphalt from petrochemicals.

P3. Set of By-products. They do not represent a relevant economic importance under the company's criteria; they are considered by-products or residues from the transformation of the raw material. Normally, they are considered industrial waste, for example, trash, coffee husks, and coconut fiber, which are processing residues of the target products of a plantation.

The outputs from P1 and P2 are called products and co-products because their production involves unitary separation or extraction operations that require raw materials, labor, energy, and production inputs, which implies production costs that are difficult to identify in the products of origin unless the co-products are physically and economically homogeneous. The co-products are marketed as such after the separation process, without additional conditioning processes, as are the by-products (P3).

In this regard, residues, co-products, and by-products should be economically and environmentally transformed into raw materials for new production cycles used for various applications, obtaining significant economic returns. Therefore, it is necessary to consider the search for alternative uses for these by-products and not only their effective and harmless elimination, since this possibility, in addition to avoiding environmental costs, would create new sources of wealth that would provide greater profitability for the industrial process, generating new industries of all kinds with the consequent social advantages that this would bring. However, it is evident that despite all these economic and social-environmental advantages that would result from the technical use of residues from the industrialization and harvesting of plantation crops, these are not being used or are only incipient. The treatment and/or diversified use of plantation crop residues and by-products through green technologies is generally considered a costly activity, and both public institutions and private companies have not carried out this task effectively, either due to a lack of foresight, excessive legislation, political support, diverse interests, or a lack of application of innovative economic models. All these constraints are contributing to environmental degradation in large areas of the plantation-producing regions [6].

The conclusions of Haloui et al. [7] and Eixenberger et al. [8] suggest that a change is needed in the traditional approaches of the green and industrial revolutions, betting on innovation, transformation, and the creation of new productive chains considering the quality of the raw material, the composition of the by-products, and not the survival of the traditional plantation crop, although currently sugar, cocoa, coffee, coconut, rubber, soybean, wood, etc., have been the main objective for their cultivation without considering the multiple strategies for diversification and multiplication of goods obtained from agricultural and industrial activity. Have been the main objective for their cultivation without considering the multiple strategies of diversification and multiplication of goods obtained from agricultural and industrial activity. Notwithstanding, it is not enough to have a single option for plantations, because different schemes should be considered which are possible to carry out with the available capital, selecting the alternatives that allow reaching the best economic, environmental, and social performance, which should, in turn, show the greatest flexibility so that it is possible to shift the production toward those products that have greater economic benefit in the market. Consequently, diversification projects or biorefinery of plantation crop residues should be based on scientific knowledge of the products, co-products, and by-products, since the potential for diversification derives from the volume of organic by-products generated each harvest season, physiological elements that constitute the raw material, and its chemical composition of the co-products and by-products due to these are composed of sugars, structural carbohydrates of the lignocellulosic complex, and inorganic material, which offer different possibilities of industrialization with various physical, chemical and biotechnological or hybrid routes and to establish possible productive alternatives of the sustainable future of these residues [9].

3 Challenges in Valorization and Industrialization of Plantation Residues

According to Nadaraja et al. [10], the profitability of plantations (target products and waste management) depends mainly on the analysis of indicators related to the 'Environmental' dimension of sustainability, followed by the 'Social' and 'Economic' evaluations. However, the lowest proportion belonged to the 'Governance' dimension and public policies, which weaken sustainability by not influencing decision-making in plantations and waste management in a highly vulnerable domestic and global environment, with low resilience, and further impacted by climate change. These facilities must now be restructured to contribute to bioenergy with carbon capture and storage production. However, the following factors should be considered:

(a) Practically, the entire production of food and agricultural raw materials originates from plantation crops (including inputs for fishing), meaning the supply of products (domestic and external) is fundamental to food security and food sovereignty.
(b) Agricultural products are the basis for many commercial and industrial activities.
(c) Plantation farming is a fundamental activity in rural areas, which are still populated by a highly significant proportion of the world's population.
(d) Plantations are the basis for activities beyond agriculture, such as local commerce, handicrafts, material extraction, ecotourism, environmental services, or salaried work in various occupations.

(e) Rural development derived from plantations involves the incorporation of an important economic potential for the country's competitiveness.
(f) Long-term domestic development priorities include environmental sustainability and the conservation of natural resources, in addition to reducing the impact of climate change, depletion, pollution, and deterioration of natural resources, prioritizing the preservation of aquifers, soils, biodiversity, forests, the density of marine life and inland waters, and other elements of environmental sustainability impacted by plantations.
(g) Agricultural and rural development is also very important in the strategy for improving the international insertion of developing countries. In international trade negotiations, the agrifood sector is considered highly sensitive because of the multifunctionality of agriculture, i.e., its important impact on food, employment, the conservation of natural resources, the environment—including the pace of global warming and climate change—land-use planning, ecotourism, the preservation of the rural landscape, and the conservation of cultural heritage.
(h) There is a large amount of biomass left over from harvesting, use, and transformation of biomass into plantation crops. Stems, branches, and leaves are kept as leftovers in the field in addition to seeds, shells, cobs, etc.

Therefore, possibly, the great problem for the valorization and use of plantation residues lies in the dispersion with which they are generated in the fields and the high cost of their collection. However, these factors will become less and less important as conditioning factors for their use because of the growing importance of their utilization for economic and environmental reasons. The Mexican government (https://www.gob.mx/cms/uploads/attachment/file/346978/Manejo_de_residuos_Detallado.pdf) reported the factors that constrain the valorization of these wastes. The management of agricultural waste is not carried out in a planned manner.

- Its management is oriented toward treatment and final disposal rather than prevention and use.
- Regulations, laws, mandates, and public policies are scattered or outdated.
- There is a lack of knowledge on the part of generators, managers, receivers, authorities, and the community in general about the issue, which exacerbates the problem.
- There is no separate management system for agricultural waste. In general, non-hazardous waste is disposed of as final waste.
- There is a lack of commitment on the part of waste generators to assume responsibility for waste management and handling.
- Consumers are unaware of the risks associated with improper management of agricultural waste, which is due to a culture based on the tendency to consume the cheapest product, regardless of its quality and post-consumer management.
- There are few economic and financial incentives to promote integrated management.
- There is a limited supply of authorized companies that provide agricultural waste management services.

- Most of the economic regions do not have specialized technical regulations on the topic that would allow for the classification, identification, characterization, and adequate management of agricultural wastes.
- There is a limited analytical offer for the characterization of agricultural residues.
- The lack of adoption of cleaner production alternatives by the productive sectors leads to the prevention and minimization of agricultural waste at the source.
- The supply of specialized agricultural waste transportation is limited and, in many cases, does not meet legal requirements.
- Lack of institutional coordination and low management capacity of the authorities to promote agricultural waste management.

In relation to the above, the survival of humankind globally is directly related to agriculture, and factors such as population and income growth, mainly in developing countries with high populations, urbanization, changes in diet, and increased life expectancy will drive global demand for agricultural products, mainly from plantations.

To address the challenges facing plantation agriculture and agro-commodity systems globally, it is recommended to increase investment in farming research and development to improve productivity, promote sustainable cropping practices, improve access to credit and insurance for smallholder farmers, invest in rural infrastructure, and promote trade policies that support sustainable and inclusive agriculture [11]. Mukhovi and Jacobi [12] suggest that increasing the resilience of plantation crops as socio-ecological systems with the participation of stakeholders and decision-makers will yield numerous benefits:

- Improves soil fertility.
- Yield improvement.
- Enhances agroecosystem productivity.
- Enhances ecological resilience.
- Improves food security.
- Increases the diversity of vegetation because it provides habitat for biodiversity and pollinators.
- Helps to recharge the water system.
- Contributes to yield improvement.
- Improves livelihood resilience.
- Improves crop yield.
- Improves soil health.
- Improves provision of agroecosystem services.
- Crop rotation supports soil health.
- Improves productivity.
- Enhances soil carbon.
- Livelihood capital (human, social, financial, natural, and physical capital).
- Aggregation ensures economies of scale.

- Advocacy by associations.
- Multiple credit sources, shareholding, and savings cushion against price.
- Fluctuations support the procurement of inputs and farm maintenance in event of shocks.
- Human skills and knowledge are important for efficient production.
- Proximity to roads, railways, and airports is important for the procurement of inputs and the sale of outputs.
- Water is important for irrigation.
- Land for the expansion of farms and maintaining biodiversity.

To address the above benefits, Nadaraja et al. [10] identified a set of indicators that can be used to assess the sustainability of plantation farming systems covering the four dimensions of sustainability: environmental, social, economic, and governance. Therefore, policy- and decision-makers at the regional level are advised to leverage the region's comparative and competitive advantages for the development of plantations such as oil palm, rubber, coconut, cocoa, cacao, coffee, and sago. The development of these products could contribute to the economy, both regionally and nationally. However, government policies should be guided by regional commodity-based development (Fig. 20.1).

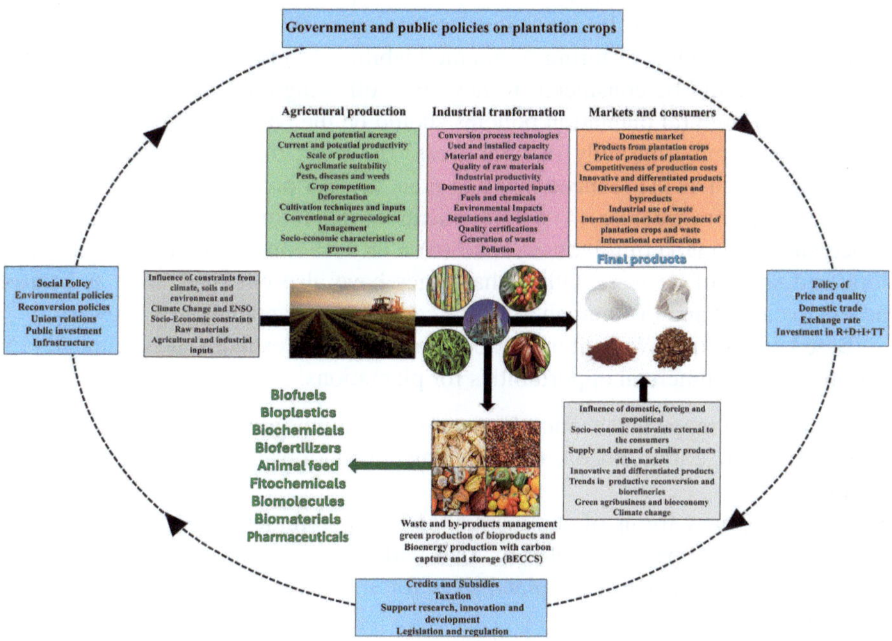

Fig. 20.1 Integration of factors related to the sustainability of plantation crops and waste

4 Public Policy on Plantation Crops and Waste

The ability of agriculture to meet current and future human needs is threatened by a wide range of constraints, such as loss of biodiversity, soil degradation through erosion, compaction, salinization, depletion and pollution of water resources, rising oil prices, climate change, soil erosion, compaction, salinization, depletion and pollution of water resources, rising production costs, the steady decline in the number of farms, and, associated with all of these, poverty and the decline in the rural population. Consequently, sustainable agriculture is a commitment to meet current and long-term food and fiber needs while improving the standard of living of farmers and society in general [13]. Consequently, long-term domestic agricultural plantation policies should be the basis for agricultural development programs, especially in rural areas. However, there are several technical constraints, such as (1) reduced planting area, (2) low productivity, (3) inadequate facilities and infrastructure, (4) poor application of planting technology, (5) a limited processing and marketing industry, and (6) a lack of appropriate attitudes, behaviors, and skills of growers. Other challenges have also emerged, such as limited access to capital and the low human resource capacity of plantation crop growers.

Gamage et al. [14] concluded that the need to improve food security and access to commodities and raw materials derived from plantations should promote environmental sustainability by innovative technologies that effectively address ecological problems while ensuring economic viability. To move toward sustainability, plantations should be considered to have the following characteristics: long-term domestic agricultural plantation policies should be the basis for agricultural development programs, especially in rural areas. However, there are several technical constraints, such as (1) reduced planting area, (2) low productivity, (3) inadequate facilities and infrastructure, (4) poor application of planting technology, (5) limited processing and marketing industry, and (6) a lack of appropriate attitudes, behaviors, and skills of growers. Other challenges have also emerged, such as limited access to capital and low human resource capacity of plantation crop growers.

Based on the improvement in the above indicators, Alexandra and Campbell [15] propose new commercial opportunities for plantations.

- Broaden the range and scope of existing plantation-based industries, e.g., coconut, oil palm, coffee, cocoa, and sugarcane plantations in agro-ecologically suitable areas.
- To develop new forms of commercial and semi-commercial exploitation with a wider range of species adapted to areas with plantations as intercrops.
- Develop new plantation cropland uses for short-rotation species, e.g., oil and fodder crops.
- Use of biomass for rural energy purposes, e.g., charcoal, compressed bales, briquettes, pellets, etc.
- Promote agro-tourism through a program designed by specialists.

To move toward sustainable development in existing plantations, it is necessary to holistically assess the importance of understanding the prospects of plantation crops in the regional, domestic, and international physical, cultural, and political contexts. The prospects for regional plantations depend on how stakeholders and decision-makers manage a few critical relationships between them [16].

5 Case Study: Sugarcane Trash

Sugarcane is one of the most efficient plants in terms of biomass production per unit area. It is a major crop in tropical and subtropical areas of the world due to its wide adaptability and resilience to various climatic and soil externalities and its high biomass production. It is one among the plants with the highest photosynthetic efficiency in the use of solar energy. Of the plant material, it can be estimated that 1/3 is leaves and tops and 2/3 is stalk, of which 1/3 would be fiber and 1/3 the juice from which sugar, ethanol, electric energy, and potentially more than 100 bioproducts are manufactured.

The environmental issues of sugarcane plantations were recently reviewed by Chaya [17], Ravish and Chaudhry [18], and Mugica-Álvarez et al. [19], concluding that in these plantations, sugarcane production patterns are associated with the prevailing production patterns at different stages of the production chain, which require the use of chemical fertilizers, pesticides, and high water consumption in the fields. Besides, the burning harvest practiced during the harvest, which includes soil erosion, decreasing its long-term productive potential and the emission of CH_4, CO, N_2O, SO_X, and NO_X and a large number of other potentially carcinogenic compounds into the atmosphere, occupational hazards, and, in the industrial sector, the great impact on water resources due to the use of large quantities in the production process and the generation of waste that is discharged into surface water flows, in addition to other emissions of gases and particles into the environment, due to the use of fossil fuels that persist in agroindustry with production patterns that date back many decades, in which value creation is also present, which does not consider the repair of the environmental damage generated by producers and also by the businessmen who participate in it.

As a result of these farming practices. Sugarcane trash (SCT) is defined as the whole of the cane tops and leaves that are left in the field. According to Sari et al. [20], in each agricultural season, about 279 million metric tons of sugarcane trash are produced globally (8–13 tons of trash per ha), which could provide material for the elaboration of various bioproducts. SCT containing one-third of the sugarcane biomass.

The reasons for pre-harvest burning, according to Dhanushkodi et al. [21], are as follows:

- The time available between crop harvesting and sowing is very narrow.

- Transportation of crop residues from the field to other areas for composting is cost-involving.
- The natural decomposition of crop residues takes a long time.
- Insufficient or unavailability of effective decomposing microorganisms in the soil.

Nowadays, the use of SCTs is very low, and their utilization is scarce. It is preferable to burn them during the sugarcane harvest because the use of sugarcane residues for productive diversification or biorefineries is not generalized, since it depends on certain factors, among which the following stand out: lack of economic stimulation, lack of processing companies and buyers, technological culture, environmental impact, and resistance to change: absence of economic stimulation, lack of processing companies and buyers, technological culture, impact on the environment, and resistance to change; to this is added its low density, which implies the need to handle large volumes for its supply, and the generalization of its use becomes more difficult [22].

Nevertheless, in the majority of sugarcane-producing countries, there is no production chain or market for sugarcane plantation waste to improve its performance, despite the existence of potential derivatives based on its composition (Table 20.1), restricting its use, in some countries, to the boilers of sugar mills, replacing the consumption of fuel oil (fossil fuel) and bagasse, contributing to environmental protection through its substitution in the generation and/or cogeneration of electric power, as well as in the production processes of ethanol or other by-products and as mulch in sugarcane fields for temperature regulation of soil and plant, roots by protection from direct radiation, reduced nutrient losses, reduced soil erosion and compactness, and enhanced physical soil conditions due to nutrient recycling and favor meso and microflora [23, 24].

Sugarcane trash contains a high proportion of cellulose, hemicellulose, lignin, and extractives that may be used to create more valuable products. Therefore, based on this chemical composition and abundant biomass generation, it is possible to establish, on a sustainable basis, numerous applications in agriculture and livestock, bioenergy generation, and novel products (Fig. 20.2).

The bioproducts obtained can have various applications (Table 20.2).

Powar et al. [35] conducted a review of production alternatives, technological routes, and potential bioproducts for the sustainable use of trash after it is used as mulch, fodder, or compost on farms.

Concerning the actual use of the trash, the findings of Powar et al. [36] concluded that farmers have a good knowledge of how to use sugarcane waste as organic

Table 20.1 Chemical composition of sugarcane harvest residue [25–28]

Component (%)	Value
Lignin	21.5–24.5
Holocellulose	70–78.7
α-Cellulose	39.2–47
Pentosan	25.6–33

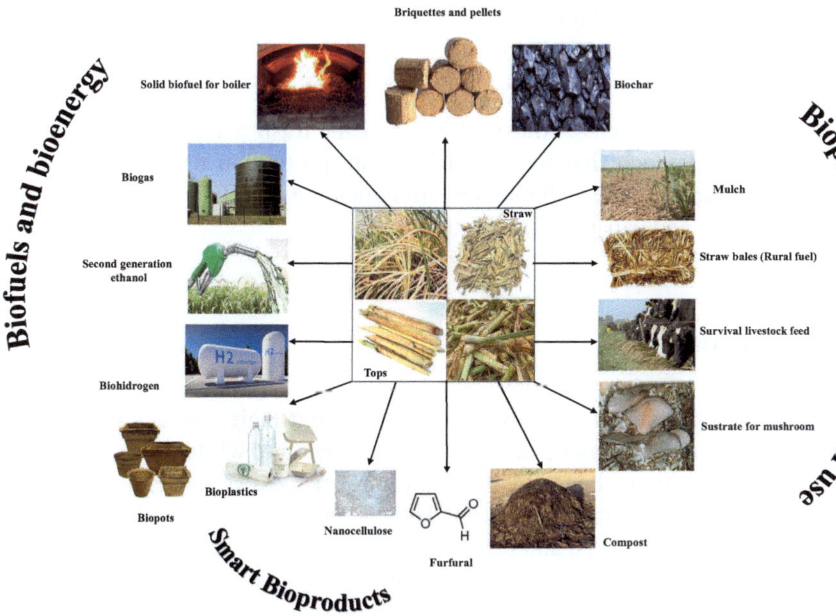

Fig. 20.2 Sustainable use of sugarcane trash

Table 20.2 Novel technological options for the conversion of trash into bioproducts

Bioproduct	References	Bioproduct	References
Nanolignin	Karimah et al. [29]	Adsorption for the removal of dyes, metals, and other contaminants.	Paulose and Kaparaju [30]
Biodegradation of sugarcane trash	Abro et al. [31]	Biorefinery-produced sugarcane straw or trash	Sari et al. [20]
Biohydrogen production	Tomasini et al. [32]	Nanocrystalline cellulose	Wibowo et al. [33]
Pulp and paper	Bhardwaj et al. [26]	Textile fibers	Costa et al. [34]

fertilizer and mulch, even though they burn sugarcane waste in the field to avoid waste load on the next crop; the risk of rodents, scorpions, and snakes; the lack of a market; the damage to weeds and their breeding sites; and the lack of technical knowledge. Through the effective implementation of policies and technology, it is possible to prevent the open burning of sugarcane waste. An overall strategy for crop residue management should aim to do the following:

- Educate farmers and create awareness.
- Promote alternative uses of crop residues.
- Government policies and incentives can promote the adoption of such alternative uses, leading to reduced burning of crop residue.

- Develop appropriate technologies.
- Strengthen policies and regulations.
- Build infrastructure and capacity.

In this regard, an effective strategy for sugarcane trash management requires a comprehensive approach that includes education, promotion of alternative uses, development of appropriate technologies, strengthening of policies and regulations, and infrastructure and capacity building.

6 Conclusions

Agro-industrial diversification of plantation crops and their residues provides an increase in regional sustainable development and income, while at the same time reducing vulnerability to change by making future business strategies more flexible and increasing resilience to externalities. The chemical and structural composition of crop residues, co-products, and plantation by-products makes them attractive for transformation through chemical or biotechnological processes into industrial processes into high value-added and energy products of interest to the market. Therefore, the success of diversification projects at the industrial level, where the composition of products, co-products, and by-products and their availability play a fundamental role, will depend on the capacity of those involved to transform the comparative advantages derived from their geographical location and the economic and technological characteristics existing in the location of the plantations into dynamic competitive advantages with a prospective plan and adopting a holistic vision, capable of being maintained over time. In addition, aspects such as the social relations generated around the projects, market and environmental planning processes, and community participation are basic in environmental management in plantations, recognizing the actors involved in environmental problems (the community, local and environmental authorities, etc.) and interacting with them to achieve common diversification objectives.

References

1. Maxwell S, Fernando A (1989) Cash crops in developing countries: the issues, the facts, the policies. World Dev 17(11):1677–1708
2. Thomas GV, Krishnakumar V (2024) Plantation crops and soil health management: an overview. In: Soil health management for plantation crops: recent advances and new paradigms. Springer, pp 1–36. https://doi.org/10.1007/978-981-97-0092-9_1
3. Diczbalis Y, Daniells J, Lambert S, Searle C (2014). Plantation crops. Horticulture: plants for people and places, Volume 1: Production horticulture, 263–299. https://doi.org/10.1007/978-94-017-8578-5_8
4. Jena S, Singh R (2022) Agricultural crop waste materials–a potential reservoir of molecules. Environ Res 206:112284. https://doi.org/10.1016/j.envres.2021.112284

5. Rani GM, Pathania D, Umapathi R, Rustagi S, Huh YS, Gupta VK et al (2023) Agro-waste to sustainable energy: a green strategy of converting agricultural waste to nano-enabled energy applications. Sci Total Environ 875:162667. https://doi.org/10.1016/j.scitotenv.2023.162667
6. Phang LY, Abd-Aziz S, Gozan M, Ibrahim MF (2024) Sustainability of chemical substitutes from agricultural and industrial by-products. In: Chemical substitutes from agricultural and industrial by-products: bioconversion, bioprocessing, and biorefining. Wiley, pp 355–373. https://doi.org/10.1002/9783527841141.ch18
7. Haloui D, Oufaska K, Oudani M, El Yassini K (2024) Bridging industry 5.0 and agriculture 5.0: historical perspectives, opportunities, and future perspectives. Sustain For 16(9):3507. https://doi.org/10.3390/su16093507
8. Eixenberger D, Carballo-Arce AF, Vega-Baudrit JR, Trimino-Vazquez H, Villegas-Peñaranda LR, Stöbener A et al (2024) Tropical agroindustrial biowaste revalorization through integrative biorefineries—review part II: pineapple, sugarcane and banana by-products in Costa Rica. Biomass Convers Biorefinery 14(4):4391–4418. https://doi.org/10.1007/s13399-022-02721-9
9. Byerlee D, Viswanathan PK (2018) Plantations and economic development in the twentieth century: the end of an era? Agricultural development in the world periphery: a global economic history approach, 89–117. https://doi.org/10.1007/978-3-319-66020-2_4
10. Nadaraja D, Lu C, Islam MM (2021) The sustainability assessment of plantation agriculture- a systematic review of sustainability indicators. Sustain Prod Consum 26:892–910. https://doi.org/10.1016/j.spc.2020.12.042
11. Polymeni S, Plastras S, Skoutas DN, Kormentzas G, Skianis C (2023) The impact of 6G-IoT technologies on the development of agriculture 5.0: a review. Electronics 12(12):2651. https://doi.org/10.3390/electronics12122651
12. Mukhovi S, Jacobi J (2022) Can monocultures be resilient? Assessment of buffer capacity in two agroindustrial cropping systems in Africa and South America. Agric Food Secur 11(1):19. https://doi.org/10.1186/s40066-022-00356-7
13. Bathaei A, Štreimikienė D (2023) A systematic review of agricultural sustainability indicators. Agriculture 13(2):241. https://doi.org/10.3390/agriculture13020241
14. Gamage A, Gangahagedara R, Subasinghe S, Gamage J, Guruge C, Senaratne S et al (2024) Advancing sustainability: the impact of emerging technologies in agriculture. Curr Plant Biol 100420. https://doi.org/10.1016/j.cpb.2024.100420
15. Alexandra J, Campbell A (2003) Plantations and sustainability science: the environmental and political settings. Aust For 66(1):12–19. https://doi.org/10.1080/00049158.2003.10674883
16. Waqas M, Hashim S, Humphries UW, Ahmad S, Noor R, Shoaib M et al (2023) Composting processes for agricultural waste management: a comprehensive review. PRO 11(3):731
17. Chaya W (2024) Reframing the wicked problem of pre-harvest burning: a case study of Thailand's sugarcane. Heliyon 10(7):e29327. https://doi.org/10.1016/j.heliyon.2024.e29327
18. Ravish P, Chaudhry S (2024) Impact of sugarcane trash burning on ambient air quality in agricultural regions of northern Haryana, India. Int J Environ Sci Technol 1-14. https://doi.org/10.1007/s13762-024-06192-5
19. Mugica-Álvarez V, Millán-Vázquez F, de Jesús Figueroa-Lara J, Valle-Hernández BL, Querol X, Hernández-Rosas F (2024) Fine particles composition and emission chemical profiles from sugarcane production for source reconciliation applying the chemical mass balance. J Mex Chem Soc 68(4):576–592. https://doi.org/10.29356/jmcs.v68i4.2280
20. Sari FP, Solihat NN, Azelee NI, Fatriasari W (2024) Biorefinery from plant biomass: a case study on sugarcane straw. In: Plant biomass derived materials: sources, extractions, and applications. Wiley, pp 243–269. https://doi.org/10.1002/9783527839032.ch10
21. Dhanushkodi V, Krishnaveni A, Nageswari R, Senthil K, Ramesh KR, Sangeetha S, Anitha R (2023) Harmful effect of sugarcane trash burning and significant benefit on utilization of sugarcane trash—a review. Transportation 13(20):2. https://doi.org/10.58321/AATCCReview.2023.11.01.03
22. Hamawand I, da Silva W, Seneweera S, Bundschuh J (2021) Value proposition of different methods for utilisation of sugarcane wastes. Energies 14(17):5483. https://doi.org/10.3390/en14175483

23. Sen M, Roy A, Rani K, Nalia A, Das T, Tigga P et al (2024) Crop residue: status, distribution, management, and agricultural sustainability. In: Waste management for sustainable and restored agricultural soil. Academic Press, pp 167–201. https://doi.org/10.1016/B978-0-443-18486-4.00017-8
24. Tandon V, Rani M, Kumar R, Lathwal M, Kalyan N, Kumar S et al (2025) The role of sugarcane trash in soil fertility and plant growth: a review. Asian J Soil Sci Plant Nutr 11(1):67–79
25. Aguilar RN (2010) Edible mushroom *Pleurotus ostreatus* production on cellulosic biomass of sugar cane. Sugar Tech 12(2):176–178. https://doi.org/10.1007/s12355-010-0034-4
26. Bhardwaj NK, Kaur D, Chaudhry S, Sharma M, Arya S (2019) Approaches for converting sugarcane trash, a promising agro residue, into pulp and paper using soda pulping and elemental chlorine-free bleaching. J Clean Prod 217:225–233. https://doi.org/10.1016/j.jclepro.2019.01.223
27. Franco HCJ, Pimenta MTB, Carvalho JLN, Magalhães PSG, Rossell CEV, Braunbeck OA et al (2013) Assessment of sugarcane trash for agronomic and energy purposes in Brazil. Sci Agric 70:305–312
28. Sukphun P, Wongarmat W, Imai T, Sittijunda S, Chaiprapat S, Reungsang A (2023) Two-stage biohydrogen and methane production from sugarcane-based sugar and ethanol industrial wastes: a comprehensive review. Bioresour Technol 386:129519. https://doi.org/10.1016/j.biortech.2023.129519
29. Karimah A, Wistara NJ, Fatriasari W, Watanabe T, Hussin MH (2024) Green preparations of nanolignin from acid-saccharification-treated sugarcane trash. Process Biochem 144:266–277. https://doi.org/10.1016/j.procbio.2024.06.009
30. Paulose P, Kaparaju P (2021) Anaerobic mono-digestion of sugarcane trash and bagasse with and without pretreatment. Ind Crops Prod 170:113712. https://doi.org/10.1016/j.indcrop.2021.113712
31. Abro SA, Ahmed N, Ali M, Tunio S (2022) Biodegradation of sugarcane trash with agro-industrial and farm wastes for value added organic byproduct. Int J Econ Environ Geol 13(2):62–66. https://doi.org/10.46660/ijeeg.v13i2.34
32. Tomasini M, de Oliveira Faber M, Ferreira-Leitão VS (2023) Sequential production of hydrogen and methane using hemicellulose hydrolysate from diluted acid pretreatment of sugarcane straw. Int J Hydrog Energy 48(27):9971–9987. https://doi.org/10.1016/j.ijhydene.2022.11.276
33. Wibowo A, Chiarasumran N, Thanapimmetha A, Saisriyoot M, Srinophakun P, Suriyachai N, Champreda V (2022) Conversion of sugarcane trash to nanocrystalline cellulose and its life cycle assessment. Catalysts 12(10):1215. https://doi.org/10.3390/catal12101215
34. Costa SMD, Aguiar A, Luz SM, Pessoa A, Costa SAD (2015) Sugarcane straw and its cellulosic fraction as raw materials for obtainment of textile fibers and other bioproducts. Polysaccharides:1–17. https://doi.org/10.1007/978-3-319-16298-0_53
35. Powar RV, Mehetre SA, Powar TR, Patil SB (2022) End-use applications of sugarcane trash: A comprehensive review. Sugar Tech 24(3):699–714. https://doi.org/10.1007/s12355-022-01107-5
36. Powar RV, Pawar PS, Powar TR, Sabale PR, and Patil SB (2023) Farmers' knowledge, perception, and practices toward the use of sugarcane trash. In Handbook of Energy Management in Agriculture, ed. Rahul Datta, Amitava Rakshit, Asim Biswas, Deepranjan Sarkar, and Vijay Singh Meena, 1–17. Springer. https://doi.org/10.1007/978-981-19-7736-7_32-1

Index

C. Trois, R. C. Ray (eds.), *Plantation Crop Wastes: Valorization for Economic Sustainability*, Waste as a Resource,
https://doi.org/10.1007/978-3-031-96805-1

V

W

MIX
Papier aus verantwortungsvollen Quellen
Paper from responsible sources
FSC® C105338

If you have any concerns about our products,
you can contact us on
ProductSafety@springernature.com

In case Publisher is established outside the EU,
the EU authorized representative is:
Springer Nature Customer Service Center GmbH
Europaplatz 3, 69115 Heidelberg, Germany

Printed by Libri Plureos GmbH
in Hamburg, Germany